Springer Series in Astrophysics and Cosmology

The series covers all areas of astrophysics and cosmology, including theory, observations, and instrumentation. It publishes monographs and edited volumes. All books are authored or edited by leading experts in the field and are primarily intended for researchers and graduate students.

Daniele Malafarina · Pankaj S. Joshi
Editors

New Frontiers in Gravitational Collapse and Spacetime Singularities

Springer

Editors
Daniele Malafarina
Department of Physics
Nazarbayev University
Astana, Kazakhstan

Pankaj S. Joshi
International Centre for Space
and Cosmology
Ahmedabad University
Ahmedabad, Gujarat, India

ISSN 2731-734X ISSN 2731-7358 (electronic)
Springer Series in Astrophysics and Cosmology
ISBN 978-981-97-1171-0 ISBN 978-981-97-1172-7 (eBook)
https://doi.org/10.1007/978-981-97-1172-7

This Springer imprint is published by the registered company Springer Nature Singapore Pte Ltd.
The registered company address is: 152 Beach Road, #21-01/04 Gateway East, Singapore 189721, Singapore

Paper in this product is recyclable.

Preface

Relativistic gravitational collapse provides a theoretical framework within the general theory of relativity (GR) to explain how space-time singularities and black holes originate from massive astrophysical objects. It may be argued that relativistic gravitational collapse lies at the foundation of black hole physics and our belief that such objects exist in the universe. In fact, the 2020 Nobel Prize in Physics was awarded, among others, to Roger Penrose for his work on space-time singularities that originate from gravitational collapse.

The first analytical model for relativistic collapse was proposed in 1939 by J. R. Oppenheimer and H. Snyder. They studied how pressureless, homogeneous matter collapsing under its own gravity produces what they called a 'frozen star'. Around the same time, B. Datt obtained the same results independently, and today the analytical solution for homogeneous dust is generally known as the Oppenheimer-Snyder-Datt (OSD) model. These works show how a matter cloud may collapse continually under its own gravity to form a black hole. However, at that time, space-time singularities and horizons were not yet fully understood and thus the final fate of such a collapse was not clear. In such a scenario, Oppenheimer and Snyder used the term 'frozen star' to describe the collapsing object as it reaches its final stages, as they did not make use of a coordinate system that allows us to see how particles cross the horizon and what happens afterwards.

In fact, the very nature of the event horizon in the Schwarzschild solution was not yet clear when the OSD model came to light. It was D. Finklestein in 1958 who showed that the horizon is a 'one way membrane' that can be crossed only in one direction. Around the same time, people also realised that the continued gravitational collapse of the OSD model produces a curvature singularity at the centre of symmetry of the system in a finite proper time. Similar singularities were studied and found in cosmology also, which identified with the origin of the universe. The question then arose whether such singularities were generic in gravitational collapse and cosmology within the framework of general relativity. The expectation was that singularities would disappear once restrictive constraints such as spherical symmetry were relaxed.

Instead, as it often happens, nature turned out to be more nuanced than expected. R. Penrose, S. Hawking and R. Geroch proved that space-time singularities in gravitational collapse and cosmology are inevitable, within GR, when certain natural physical conditions are met. When black holes form in gravitational collapse, they hide the curvature singularities at their centre. This led to a variety of new research directions aimed at answering questions such as

- Can the singularities forming at the end of collapse in GR be visible to far away observers?
- How is the singularity formation scenario affected by modifications of GR?
- Do quantum effects ensure that singularities do not form?
- What are the properties of the final endstate of collapse in these modified scenarios? Are they distinguishable from black holes?

Susequently, detailed analytical and numerical models for gravitational collapse were developed, which have been at the heart of modern black hole physics. Formation of event horizons as well as visibility or otherwise of the space-time singularities have been examined. For decades these have been fruitful areas of research on many fronts, ranging from astrophysics to the search for a theory of quantum gravity, from exact solutions in GR and alternative theories to numerical simulations.

Today more than ever analytical and numerical models of gravitational collapse provide an ideal tool to probe into the nature of Einstein's equations, explore the limits of the theory and the implications for astrophysical phenomena and for quantum-gravitational effects.

Many alternative ideas, improvements and modifications of the OSD model have been proposed. For example, from the point of view of astrophysics, the study of collapse of inhomogeneous fluids with astrophysically relevant equations of state provides insights into how black holes or singularities not covered within horizons may be born from massive dying stars. On the other hand, from the point of view of fundamental aspects of gravity, the study of collapse of scalar fields and attempts towards a quantization of collapse models provide insights into the features that a viable theory of quantum gravity should have.

Such new models help us towards a better understanding of gravity in extreme regimes and potentially may have consequences for astrophysical phenomena and experimental searches of quantum-gravity signatures.

In addition, recent experimental results, such as the detection of gravitational waves from binary black hole inspirals, and the first images of the shadow of supermassive ultracompact objects at galactic centres, prelude to the possibility of testing the validity of the various theoretical proposals that have been put forwards over the years. These are exciting times and there are still many unanswered questions that are driving research forward. The aim of this book is to take the pulse of the current state of research in gravitational collapse, the issue of formation and causal nature and structure of space-time singularities. Our hope is that it will be useful for new researchers starting to work in the field as well as experienced researchers interested in reviewing the current research trends.

Astana, Kazakhstan
Ahmedabad, India
November 2023

Daniele Malafarina
Pankaj S. Joshi

Contents

Chapter 1
After Collapse: On How a Physical Vacuum Can Change the Black Hole Paradigm

Julio Arrechea, Carlos Barceló, and Valentin Boyanov

1.1 Introduction

General Relativity (GR) is the best theory we currently have for describing gravitational phenomena. It is successful not only in terms of fitting observational data extremely well [118], but also due to the elegance of its formulation and the depth of its implications, which have given rise to mathematical and philosophical concepts that are still being developed to this day (see e.g. [32, 48]). However, it is generally accepted that GR is not the final theory of gravity. Much like Newtonian gravity before it, GR has limits in its range of applicability. This becomes apparent in problems which involve either strong gravitational fields on small scales (i.e. large densities and curvatures, or configurations involving black-hole horizons), or large-scale astrophysical and cosmological structures (i.e. the dark energy and, potentially, dark matter puzzles).

J. Arrechea (✉)
SISSA, Via Bonomea 265, 34136 Trieste, Italy
e-mail: julio.arrechea@sissa.it

INFN Sezione di Trieste, Trieste, Italy

IFPU – Institute for Fundamental Physics of the Universe, Via Beirut 2, 34014 Trieste, Italy

J. Arrechea · C. Barceló
Instituto de Astrofísica de Andalucía (CSIC), Glorieta de la Astronomía, 18008 Granada, Spain
e-mail: carlos@iaa.es

V. Boyanov
Departamento de Física Teórica and IPARCOS, Universidad Complutense de Madrid, 28040 Madrid, Spain
e-mail: vboyanov@ucm.es

CENTRA, Departamento de Física, Instituto Superior Técnico, Universidade de Lisboa, Avenida Rovisco Pais 1, 1049 Lisboa, Portugal

D. Malafarina and P. S. Joshi (eds.), *New Frontiers in Gravitational Collapse and Spacetime Singularities*, Springer Series in Astrophysics and Cosmology,
https://doi.org/10.1007/978-981-97-1172-7_1

For strong fields in compact regions—this being the topic of the present chapter—the problem comes from the seemingly inevitable appearance of singularities when matter is compressed beyond certain limits [91, 105], at least when this matter satisfies certain classical energy-positivity conditions [57]. However, neither singularities themselves, nor a purely classical description of matter at such densities, can be considered physically reasonable. Indeed, the quantum nature of matter is generally thought to be directly related to the potential resolution of the singularity issue, along with a possible quantum behaviour of spacetime itself.

With these general considerations in mind, it is tempting to search for a new theory of gravity which is valid in these strong-field regimes. One possibility is to attempt to directly work out a quantum gravitational theory from first principles; examples of this are the Loop Quantum Gravity and String Theory programs. However, thus far these approaches have had major difficulties in making a clear connection between theory and phenomenology. A more conservative approach involves the search for improved theories of gravity that retain the idea of a classical spacetime geometry as an appropriate effective notion. In other words, maintaining an effective metric as one of the dynamical variables to be solved for, and only changing the dynamical equations it satisfies.

Taking the latter approach, however, suggests keeping in mind its potential limitations, given that a field theory with no pathologies whatsoever is an unlikely outcome. For instance, the Einstein field equations are rather unique in that they do not by themselves generate shockwaves or, in more generic terms, weak solutions (these only appear when the source fields themselves produce them [99], or when a full-blown curvature singularity is approached [75]). Modifying the Einstein field equations might achieve improved behaviours in certain situations, such as a removal of the formation of singularities in simple models of gravitational collapse, but this can potentially come at the cost of pathological behaviours in other situations. Within this effective spacetime perspective, the philosophy is therefore that of finding progressive improvements which are also compatible with previous successful theories, but not that of finding the *final* dynamical theory for spacetime. Ultimately, the success of these improvements is then to be judged by how they accommodate the observed phenomenology.

The standard approach in modified gravity searches is to analyse sets of possibilities in theory space and classify them based on the phenomenology they present. The large variety of observed and measured gravitational phenomena then allows for the direct elimination of a substantial amount of these theories, while others have their free parameters constrained. Given the vastness of theory space, to even begin such an analysis calls for some formal or physical arguments which allow for a selection of a specific theory or set of theories. Different researchers have different tastes and criteria for such a selection, resulting in the dendritic exploration of theory space currently being carried out [28, 82, 96].

New theories can be broadly categorised according to two non-mutually-exclusive structural features: (a) those which modify how matter behaves in strong gravity situations while retaining the form of the Einstein equations, or (b) those which modify directly the gravitational equations, without necessarily modifying the matter sources. Within the latter set, proposals can be found that change the geometry significantly only in the surroundings of would-be classical singular regions, and others that can lead to alterations of the geometry well beyond such regimes. There are numerous interesting proposals out there, some of them discussed in the chapters of this book.

This chapter in particular is centred around one modification of GR which can be considered as part of category (b). The idea is the following: we assume that the zero-point fluctuations of the quantum fields permeating spacetime gravitate by some amount determined by the very deviation from flatness of the spacetime, generally in a non-local manner (we note that at this stage we will neglect the possible presence of a global contribution in the form of a cosmological constant); then, even in regions where no classical matter exists, there would be an average vacuum energy (as well as pressure, fluxes, etc.) which would be a source of gravity. As this new source ultimately depends on the geometry, we can interpret it as part of the geometric side of Einstein equations, maintaining that the only real source on the right-hand side is the classical stress-energy tensor. An alternative way of describing this approach is by saying that the standard vacuum used in GR is, in a sense, "too empty"; this approach instead aims to characterise a more physical vacuum as the stage on which all gravitational phenomena take place. Before we describe this theory in more detail, let us first make a brief digression and discuss the notion of a theory of quantum gravity.

It is clear that Nature has a way of melding together both the quantum behaviour that we observe for matter in the atomic and subatomic regimes and the gravitational behaviour that we observe in the macrocosm. A theory of quantum gravity would be a model that consistently incorporates both of these behaviours, furthering our understanding of each. Thus far, we have had no clear indication of how these two regimes come together; however, theoretical considerations do provide some clues. For instance, in the direction of *how gravity affects matter* there are strong indications that the causality provided by a geometric gravitational description affects the behaviour of quantum matter (see for example [81]). Under this hypothesis, analyses of quantum field theory over fixed curved background spacetimes have lead to some of the most important results in modern theoretical physics. On the one hand, the idea that a quantum particle is an observer-dependent notion leads to analysis of particle production in cosmology [55, 87], and subsequently to the idea of how an inflationary regime could lead to a primordial spectrum of fluctuations in the early universe [80, 108]. On the other hand, a calculation of particle production associated to the formation of a black hole (BH) lead Hawking to his famous result that BHs should evaporate [58, 59].

In the opposite direction of *how matter affects gravity*, GR tells us that the average effect of the stress-energy contained in a macroscopic lump of matter is to bend causality in particular, well-defined ways. However, we know nothing of how a single quantum lump of matter affects causality, or even whether a classical notion of causality would be valid at these scales. In fact, it would not be surprising if it turned out that gravity as we know it is not a relevant notion until matter starts behaving classically due to its aggregation [93, 94]. Indeed, such regimes are where

GR is less known and less clearly tested, to say nothing of when matter enters into full Planckian-density regimes. Putting aside the precise resolution of these issues, it is at least reasonable to believe that there is a regime in which an effective classical spacetime makes sense, with curvature being sourced by the average energy contained in quantum states. Then, aside from the intrinsic excitations of quantum fields, which we usually define as matter, one must also take into account the energy which can be generated by the very presence of curvature in the spacetime. Since on large scales matter can, for the most part, be described classically, in this regime the total stress-energy tensor in the Einstein equations should contain a classical contribution, where actual matter is localised, plus another source term taking into account the average energy of the vacuum fluctuations (as mentioned before, this last term can also be moved to the geometric side of the equations). This description would be self-consistent whenever the fluctuations of the energy around the average (or expectation value) are small [68], as one might expect to be the case for a vacuum state in standard situations, or in quasi-classical states.

When performing calculations within standard relativistic quantum field theory in Minkowski spacetime, there are no observables that couple to an absolute notion of energy—they only couple to differences of energy between different states. Thus, one can always subtract from all energy measurements an arbitrary reference value. In fact, a first naive calculation indicates that any quantum state has an infinite energy, as is shown in any introductory text on the subject [95]. However, there is the freedom to renormalise the energy of all states in such a way that the lowest energy state, i.e. the vacuum state, has zero energy. By subtracting the "same" infinite value from the energy of all states, we obtain finite definite values for finite-particle states (with respect to the chosen zero point). This can be done easily in flat spacetime because there is a unique natural notion of a vacuum state (the so-called Minkowski vacuum), and subtracting its expectation values from other states is a procedure which preserves the symmetry of the background.

However, the theory changes in an essential way in the presence of spacetime curvature, as there is no longer a natural notion of vacuum state. Indeed, extensions of the flat spacetime theory only lead to conclude that different observers can have a different perception of what the true vacuum state is. There is generally no longer a "vacuum" state which can be used for renormalisation by subtraction, as such a procedure would break the symmetries of GR. However, given that gravity is sensitive to the total energy of a state, rather than just energy differences, renormalisation is of particular importance. Extreme care is required in selecting how and what to subtract in search of sensible results.

With these observations in mind, the requirements for renormalisation indeed change quite a bit. Particularly, one preferably needs a local subtraction prescription which does not make use of knowledge of either the global structure of spacetime, or of the particular vacuum and particle states chosen for quantisation. Prescriptions of this sort do exist, but they generally leave a non-homogeneous, finite vacuum energy residue, which depends on the characteristics of the spacetime and of the chosen vacuum state [27, 114]. The subtraction can be made consistent with recovering asymptotic Minkowski spacetime when going arbitrarily far from the classical sources of curvature, or it can be made to leave an offset in the form of a cosmological constant. As in the present chapter we are not considering cosmological scenarios, we will neglect such terms, and set the cosmological constant to zero.

The central hypothesis of this chapter is that the theory of classical matter plus vacuum fluctuations sourcing gravity is applicable in the astrophysical scenarios we analyse below. Formally, this theory is constructed with the effectively classical stress-energy tensor (SET), and the residual vacuum energy, or zero-point fluctuations, of the quantum fields as sources in the Einstein field equations,

$$G_{\mu\nu} = 8\pi G(T^{\rm C}_{\mu\nu} + T^{\rm ZP}_{\mu\nu}), \tag{1.1}$$

were $T^{\rm C}_{\mu\nu}$ is the classical and $T^{\rm ZP}_{\mu\nu}$ the zero-point stress-energy source term. Equivalently, from a modified gravity perspective the equations can formulated as

$$G_{\mu\nu} - 8\pi G\, T^{\rm ZP}_{\mu\nu} = 8\pi G\, T^{\rm C}_{\mu\nu}. \tag{1.2}$$

These formal equations of motion constitute what is commonly referred to as *semi-classical gravity*. Indeed, the presence of a non-trivial zero-point SET is one of the most soundly motivated modifications of gravity we currently have. Though we have no direct observational evidence of this gravitational vacuum polarisation, we can straightforwardly make an analogy with the case of electromagnetism, where we know that a similar notion of charge polarisation does exist [70].

Up to this point we have talked about vacuum energy in a deliberately vague manner. If we knew the exact expression of $T^{\rm ZP}_{\mu\nu}$ in terms of simple analytic functions of the metric, we would just have to solve the new system of gravitational equations and analyse its phenomenology, testing against observations. However, things are not so simple, as we do not possess an indisputable prescription for calculating $T^{\rm ZP}_{\mu\nu}$ in generic scenarios. The best reasoned method we have for obtaining an appropriate $T^{\rm ZP}_{\mu\nu}$ is through the expectation value of a SET operator for different fields in a given vacuum state,

$$T^{\rm ZP}_{\mu\nu} = \langle \Psi_0 | \hat{T}^{\rm QF}_{\mu\nu} | \Psi_0 \rangle. \tag{1.3}$$

Ideally, $T^{\rm ZP}_{\mu\nu}$ would describe the SET operator associated with the complete standard model of particle physics, with all its interacting fields, together with any yet undiscovered fields, such as the possible constituents of dark matter. Additionally, $T^{\rm ZP}_{\mu\nu}$ would incorporate any fluctuating energy offset that might be contained within the gravitational field itself. Neither of these idealised requirements is realistically feasible as of yet. On the one hand, given the absence of a theory of quantum gravity, we are bound to hope that the gravitational contribution to these fluctuations is small enough to be negligible with respect to the ones associated to the standard model fields.[1] On the other hand, even the standard model zero-point SET is nearly

[1] It has been suggested that this assumption becomes more accurate when the number of quantum fields N is sufficiently large [1, 27].

impossible to calculate. Such difficulty arises from two fundamental aspects: (a) it is unknown how to treat interacting field theories beyond the S-matrix perturbative approach; (b) as the SET operator contains products of field operators at the same point, it is not a well defined operator in the quantum theory; therefore, to make sense of this object one has to resort to regularising and renormalising this expectation value, which poses its own difficulties in curved spacetimes. Faced with these problems, as a proxy to the qualitative form that $T_{\mu\nu}^{\mathrm{ZP}}$ might have, researchers have opted for calculating the renormalised SET (RSET) for free field theories (often as a one-loop approximation to interacting theories) in simple backgrounds. Among the test fields useful to understand semiclassical gravity, the free scalar field is the simplest, and indeed the most used in past and present literature. The term *semiclassical gravity* is typically used to refer to any theory that incorporates the effect of vacuum fluctuations, even in these simplified test field scenarios.

Even calculating the RSET of a free scalar field in simple, highly-symmetric geometries is not a trivial task; in fact, it can typically only be done numerically and with great difficulties. The complete problem of solving the semiclassical Einstein equations self-consistently and exactly is therefore not feasible at present. This problem has a large body of work addressing it, and we will briefly and non-exhaustively review it in the next section. For now let us just say that given the difficulties mentioned, there are essentially two strategies that can be adopted: (1) to develop incremental improvements in the method of calculating the RSET, or very close approximations thereof, and study its effects in physically relevant situations in an approximate perturbative manner; (2) to prescribe less precise but analytically simpler approximations to the RSET, such that one can readily investigate more complicated and realistic geometric situations, understanding meanwhile that the information extracted only provides a qualitative preview of what a full exact solution may look like. In the following sections of this chapter we present an application of the second strategy to gravitational collapse and compact object geometries.

Particularly, we will present work on two closely related physical problems. Firstly, we will address the question of whether semiclassical gravity allows for qualitatively different configurations of stellar equilibrium as compared to those present in GR. Then, we will look into the problem of gravitational collapse with a bit more scrutiny under a semiclassical lens. In other words, we will present results for both (meta-)stable configurations of semiclassical gravity which allow the existence of stellar objects of higher compactness than their classical counterparts [9, 10], as well as for a revised dynamical process of gravitational collapse which may lead to their formation [15, 17, 18].

Even considering the approximations and hypotheses involved, our philosophy with these analyses is to progressively build up and develop a modified gravitational model which is consistently treatable and comparable with GR on an equal footing. In this endeavour, both of the above-mentioned strategies can greatly contribute. While such a theory would by no means be a complete and exact treatment of matter at high densities (due to the approximations involved, as well as due to neglecting the fact that even classical matter at such densities would likely behave differently from what is known), it is our best attempt to push gravity onto the next stage of development.

The outline of this contribution is the following. In the next section we briefly review the status of research into calculating the RSET. Then, Sect. 1.3 will describe two qualitative approximation to the RSET which are analytically manageable: the Regularised Polyakov approximation and the Order-Reduced Anderson-Hiscock-Samuel tensor. Armed with these two RSETs, in Sect. 1.4 we will present an analysis of how the vacuum-induced changes in the equations of hydrostatic equilibrium lead to new families of stellar solutions absent in GR. Subsequently, in Sect. 1.5 we will analyse the dynamics of gravitational collapse, paying close attention to possible modifications to the standard picture due to horizon-related effects. Finally, we will summarise our findings and conclude with some final remarks.

1.2 Semiclassical Gravity: In Search of an Appropriate RSET

Within the semiclassical approach, the central and most important problem is the search for non-ambiguous and feasibly calculable RSETs. At the present stage, calculations are usually performed for simple test fields, as they provide an invaluable glimpse into the potential behaviours hidden within the semiclassical theory. Let us consider in particular the free scalar field. This is typically used in the literature [68] as the starting point for considering more complicated fields and interactions, which, while bringing about additional contributions, are not expected to lead to fundamental changes in the vacuum dynamics of semiclassical systems. For instance, particle creation processes in cosmology and gravitational collapse occur in a qualitatively robust way for a variety of different fields [27, 53, 85], showing that test-field semiclassical analyses suffice for a qualitative analysis, and even for some quantitative estimates [58]. Hereafter, the tool used for the entirety of the discussion will be the scalar field.

In general terms, we can define the RSET of a scalar field in some vacuum state as the result of applying a regularisation and renormalisation procedure $\mathscr{P}$ to the ill-defined object $\langle 0|\hat{T}_{\mu\nu}|0\rangle$. Symbolically we can write

$$\langle \hat{T}_{\mu\nu}\rangle_{\text{ren}} := \mathscr{P}\left[\langle 0|\hat{T}_{\mu\nu}|0\rangle\right] \tag{1.4}$$

Several regularisation techniques have been developed in the literature to find expressions for $\langle \hat{T}_{\mu\nu}\rangle_{\text{ren}}$. Examples of that are: covariant point separation (or point-splitting) [43, 44]; Hadamard regularisation [33]; dimensional regularisation [29]; Riemann ξ-function regularisation [60]; Pauli-Villars regularisation [90]; proper time [104] and adiabatic regularisation [89]. The choice of an appropriate method depends on the particularities of the system under consideration, but the underlying logic is common in all of them: the subtraction of local divergences is carried out in a way which preserves general covariance. Under some reasonable assumptions, it turns out that different regularisation procedures lead to RSETs that differ at most in locally constructed conserved quantities, as proven by Wald [115, 116]. After the divergent terms have been subtracted, one can find a well-defined RSET with the appropriate physical characteristics expected from a source term in the Einstein equations [114].

The task of identifying divergent terms in generic spacetimes so that they can be subtracted was completed by the end of the seventies [44]. Nonetheless, the problem of how to calculate the remaining finite terms in the most efficient and accurate manner has remained. The expectation values involved are constructed through a spectral decomposition of the field, as quantisation and the definition of a vacuum state themselves rely on choosing and obtaining a specific basis of modes which satisfy the field equation of motion. It is at this stage where we encounter the main difficulty of the problem: field modes cannot be calculated in closed form for most of the spacetimes of interest, let alone for generic spacetimes.

Given this situation, the approaches used for calculating the RSET are split into two categories: on the one hand, there are those which look for appropriate approximation schemes that are as accurate as possible, and on the other, those which implement progressively more efficient numerical schemes. For instance, the first methods used to calculate the RSET in Schwarzschild spacetime made use of Wentzel-Kramers-Brillouin (WKB) approximations to the modes. The WKB based method was originally devised in [35, 66, 67], and has been improved over the years [2]. The epitome of applying this method can perhaps be found in the work by Anderson-Hiscock-Samuel [3]. They were able to give expressions for the RSET in arbitrary static, spherically-symmetric spacetimes in the Hartle-Hawking vacuum state (there are results also for the Boulware vacuum state in [71]) for a scalar field with arbitrary mass and coupling to curvature. The RSET they found is split into two terms, each of which is conserved separately. The first has an analytic closed form, and the second is in general only obtainable numerically. Additionally, they showed that for massless fields the analytic part on its own constitutes a good approximation to the total RSET. In fact, for the conformally invariant field this approximate RSET was found before by Page [84] and later on by Frolov and Zelnikov [54].

However, this analytical approximation to the RSET, which we will refer to as the AHS-RSET, has a number of shortcomings for its implementation in the field equations of gravity. Chief among them is the fact that it depends on up to fourth order derivatives of the metric functions. Firstly, this makes the solutions of the semiclassical Einstein equations depend on too many boundary conditions, with no clear physical interpretation (such as initial momentum, reflective boundaries, etc.). Secondly, when analysing self-consistent solutions of the system of equations, spurious solutions connect to seemingly tame initial conditions, making it so that even Minkowski spacetime can be unstable under small perturbations, as was found even before this approximation to the RSET was obtained [65]. This is reminiscent to what occurs with the Lorentz-Abraham-Dirac self-force of classical electrodynamics, which contains self-accelerated solutions [101, 102]. In that case we can track the origin of these solutions and improve the system of equations introducing some integro-differential operator [101], or applying an order-reduction procedure [52, 106].

Another issue with the AHS-RSET is that the WKB expansion it relies on breaks down at the horizon of BH geometries. This results in the approximate RSET having logarithmic divergences at the horizon, present even in the supposedly regular Hartle-Hawking vacuum state. These divergences have been shown to disappear when one adequately adds the numerical part of the tensor, as outlined in [3] (alternatively, one can also deal with them by treating the first modes of the expansion separately [13]). However, these issues make it so that using the AHS-RSET directly to find self-consistent solutions is rather complicated and the results are somewhat untrustworthy. Nonetheless, certain physical scenarios do allow for the approximation as such to be useful, such as for the wormhole solutions found in [62].

In more recent times, two new methods for obtaining the RSET have been devised. One is the so-called *pragmatic mode-sum method* pioneer by Levi and Ori [73]. The other is the *extended-coordinates method* proposed by Taylor and Breen [109]. The first is a completion and generalisation of a method developed by Candelas [35]. It does not make use of WKB expansions in the corresponding Euclidean sector. In fact, it does not use WKB expansions at all, since in the Lorentzian sector high-order WKB approximations are very cumbersome to use, as they involve using asymptotic matching techniques to approximate the modes at turning radii at which the approximation breaks down. Instead, the idea is to construct generalised integrals in frequency which directly incorporate a subtraction of the divergences (based on high frequency information in Christensen counterterms [44]), in such a way that the convergence of the integral is efficient. The advantage of this method is that it can be applied to dynamical situations (e.g. Hawking evaporation) and also to axisymmetric configurations (e.g. in [72] it was used to calculate the RSET on a Kerr background).

The extended coordinate method [109, 110] is performed in the Euclidean sector, and so it is not suitable for dynamical configurations. It introduces some new useful coordinates to decompose the Hadamard parametrix into multipoles and Fourier components. This allows for a mode by mode subtraction in a manner that results in a numerically very efficient algorithm. Summing up a few tens of modes gives quite accurate results and the method can be applied equally well to higher dimensional spacetimes [31, 78, 79, 111].

These developments are part of an exciting progress trend and motivated on excellent grounds. Due to recent computational advances, it is expected that progress in the efficiency of calculating the RSET in scenarios of greater phenomenological interest will carry on. Nonetheless, it is difficult to foresee how these RSETs could be used to search for self-consistent solutions. The main obstruction here owes to the complexity of simultaneously finding the geometry of spacetime and the field modes propagating on (and being sources of) that very same spacetime.[2] For this reason, we believe it is worth considering approximations to the RSET which, despite being

[2] In this regard, we could aim for developing efficient grid-search algorithms that converge to a metric satisfying the semiclassical equations, but even this possibility seems to us computationally discouraging.

less accurate, prove better suited for finding self-consistent solutions in semiclassical gravity.

1.3 Approximate RSETs

Analytic approximations to renormalised stress-energy tensors are all based on a similar rationale of finding trade-offs between the functional complexity of these RSETs and the accuracy of the physics they encode. As mentioned above, we will use a free scalar field throughout the remainder of the chapter. This field obeys the equation of motion

$$\Box\phi - \left(m^2 + \xi R\right)\phi = 0, \tag{1.5}$$

where $\Box$ is the d'Alembertian operator, m is the field mass, and ξ the coupling to the Ricci scalar R. A commonly used method for obtaining analytic, approximate expressions for the RSET is based on fixing the field parameters (the mass and curvature coupling) and restricting analyses to particularly simple spacetimes. For instance, for conformally invariant fields ($m = 0,\ \xi = 1/6$) on conformally flat backgrounds, the RSET is determined by the local trace anomaly [33, 84]. This leads to the existence of explicit analytic expressions for the RSET in a variety of situations, such as Friedmann-Lemaître-Robertson-Walker cosmologies and stellar interiors of constant density [88] (for the particular quantisations which are formulated in accordance with the conformal symmetry), and even for fields of higher spin [54].

However, for the situations we will analyse below, such as the formation and subsequent evaporation trapped regions or the existence of ultracompact stellar configurations in equilibrium, spacetime is not conformally flat. In such cases the RSET includes contributions which are non-local in curvature, and that depend on the vacuum state under consideration. The simplest RSET that captures these state-dependent effects in spherical symmetry is the Polyakov approximation, which incorporates the essential features of the propagation of a massless minimally coupled scalar ($m = 0,\ \xi = 0$) in four spacetime dimensions via two-dimensional model, described by the line element

$$ds^2_{(2D)} = -\mathscr{C}(u, v)dudv, \tag{1.6}$$

where $u = t - r^*$, $v = t + r^*$ are radial null coordinates, with r^* the tortoise coordinate obtained by integrating $dr^*/dr = [h(r)/f(r)]^{1/2}$. The analogy between scalar field propagation in four and two spacetime dimensions becomes clear when one expands the field in spherical harmonics and restricts the analysis to the $l = 0$ (or s-wave) mode, which typically dominates long-distance effects. By considering the propagation of the s-wave over BH spacetimes and taking the near-horizon limit in the wave equation (1.5), the part of the equation that can be identified as a gravitational potential vanishes, and the (t, r) sector reduces to the two-dimensional free wave equation

$$\partial_u \partial_v \phi = 0. \tag{1.7}$$

This equation is manifestly conformally invariant, and admits an analytic basis of solutions in the form of plane waves. In fact, there are infinitely many such bases, one for each pair of possible null coordinates, and each can be used for performing quantisation [19]. Using any one of these quantisations, the two-dimensional RSET can be obtained in closed analytic form [47], in particular,

$$\begin{aligned}
\langle \hat{T}_{uu} \rangle^{(2D)} &= \frac{1}{24\pi}\left(\frac{\mathscr{C}_{uu}}{\mathscr{C}} - \frac{3\mathscr{C}_u^2}{2\mathscr{C}^2}\right), \\
\langle \hat{T}_{vv} \rangle^{(2D)} &= \frac{1}{24\pi}\left(\frac{\mathscr{C}_{vv}}{\mathscr{C}} - \frac{3\mathscr{C}_v^2}{2\mathscr{C}^2}\right), \\
\langle \hat{T}_{uv} \rangle^{(2D)} = \langle \hat{T}_{vu} \rangle^{(2D)} &= -\frac{R^{(2D)}}{96\pi}\mathscr{C},
\end{aligned} \tag{1.8}$$

where $R^{(2D)}$ is the two-dimensional Ricci scalar. The difference between the results obtained for the modes corresponding to one pair of null coordinates or another, i.e. between different choices of vacuum state, comes in the form of uu and vv flux terms,

$$\begin{aligned}
\langle \hat{T}_{uu} \rangle^{(2D)} &\xrightarrow[\text{vac. change}]{} \langle \hat{T}_{uu} \rangle^{(2D)} + \langle : \hat{T}_{uu} : \rangle, \\
\langle \hat{T}_{vv} \rangle^{(2D)} &\xrightarrow[\text{vac. change}]{} \langle \hat{T}_{vv} \rangle^{(2D)} + \langle : \hat{T}_{vv} : \rangle, \\
\langle \hat{T}_{uv} \rangle^{(2D)} &\xrightarrow[\text{vac. change}]{} \langle \hat{T}_{uv} \rangle^{(2D)}
\end{aligned} \tag{1.9}$$

The terms $\langle : \hat{T}_{\mu\nu} : \rangle$ incorporate all the dependence on the state in the RSET. For instance, in the case of BHs, when one switches between the Boulware and Unruh states, they contain the fluxes across horizons responsible for the phenomenon of Hawking evaporation [51]. They are obtained through the Schwarzian derivative between the null coordinates which encode the different quantisations [49, 51].

Having chosen a particular quantisation and obtained the two-dimensional RSET, the next step in the Polyakov approximation consists in defining a four-dimensional RSET from the components (1.8) through the relations

$$\langle \hat{T}^{\mu}_{\nu} \rangle^{\mathrm{P}} = \frac{F(r)}{4\pi} \delta^{\mu}_{a} \delta^{b}_{\nu} \langle \hat{T}^{a}_{b} \rangle^{(2D)} + (T_{\mathrm{AC}})^{\mu}_{\nu}, \tag{1.10}$$

where Greek and Latin indices take 4 and 2 values, respectively, P stands for Polyakov RSET, $F(r)$ is a radial function that up-scales the tensor to four dimensions, and $(T_{\mathrm{AC}})^{\mu}_{\nu}$ is a term which encodes angular pressures not contained in the two-dimensional theory. The standard Polyakov approximation is obtained with the choice $F(r) = 1/r^2$, which mimics the four dimensional behaviour of spherical modes, and $(T_{\text{AC}})^{\mu}_{\nu} = 0$. This already suffices to reproduce the behaviour of the RSET at BH horizons [46, 51], and is generally expected to work well far away from the origin.

The Polyakov RSET has been extensively used to study the semiclassical backreaction problem in a variety of situations, from dynamical BH formation and evaporation to static stars in equilibrium [37, 42, 50, 86]. The fact that this RSET is analytic, simple and has only up to second derivatives of the metric functions allows one to pose an evolution problem that is not too different from that of general relativity.

However, due to the divergence at the origin for this standard choice of $F(r)$, one can instead use a regularised version of it when dealing with systems which include $r = 0$. On its own, this comes at the price of breaking conservation, but this can be compensated by the introduction of an appropriate $(T_{\text{AC}})^{\mu}_{\nu}$ term. We will refer to this tensor as the *regularised* Polyakov RSET (RP-RSET), which we will use for stellar configurations in the next section. The $F(r)$ function can be fixed freely in static configurations (though the $1/r^2$ form should be retained far away from the origin in order to recover the s-wave behaviour), as the conservation equations

$$\nabla_{\mu}\langle \hat{T}^{\mu}_{r}\rangle^{\text{P}} = \partial_r \langle \hat{T}^{r}_{r}\rangle^{\text{P}} + \frac{2}{r}\left(\langle \hat{T}^{r}_{r}\rangle^{\text{P}} - \langle \hat{T}^{\theta}_{\theta}\rangle^{\text{P}}\right) + \frac{f'}{2f}\left(\langle \hat{T}^{r}_{r}\rangle^{\text{P}} - \langle \hat{T}^{t}_{t}\rangle^{\text{P}}\right) = 0, \quad (1.11)$$

can be satisfied with an appropriate choice of $(T_{\text{AC}})^{\mu}_{\nu}$. In dynamical situations, however, such a regularisation is not as straightforward, and one needs a more thorough deformation of the RSET components near the origin. In the following we only use the RP-RSET in static scenarios, while in dynamical ones we simply steer clear of the origin for now, focusing instead on the vicinity of horizons.

For completeness let us also mention that it is possible to incorporate into the Polyakov RSET the backscattering effects of the gravitational potential by considering a two-dimensional scalar coupled to a dilaton field (we refer the reader to [49, 51] for details on this approach). This method has a similar issue to the RP-RSET, in the sense that the resulting two-dimensional RSET is not conserved; this is again compensated with angular components in the four-dimensional tensor. We avoid the use of this approach in the following analyses, since the RSET it gives exhibits similar problems as the standard Polyakov RSET at $r = 0$, while at the same time containing terms with third order spatial derivatives of the metric functions.

The simplicity of the RP-RSET grants it a sort of malleability that makes it applicable to a variety of static scenarios. The static and spherically-symmetric geometries we will work with have a metric which can be written as

$$ds^2 = -f(r)dt^2 + h(r)dr^2 + r^2 d\Omega^2. \quad (1.12)$$

Leaving F unspecified and choosing the quantisation which respects staticity (corresponding to the Boulware vacuum), we arrive, through (1.11), at the components of the RP-RSET,

$$\langle \hat{T}_t^t \rangle^{\mathrm{P}} = \frac{F}{96\pi^2 h}\left[\frac{2f'h'}{fh} + 3\left(\frac{f'}{f}\right)^2 - \frac{4f''}{f}\right],$$
$$\langle \hat{T}_r^r \rangle^{\mathrm{P}} = -\frac{F}{96\pi^2 h}\left(\frac{f'}{f}\right)^2,$$
$$\langle \hat{T}_\theta^\theta \rangle^{\mathrm{P}} = -\frac{(2F + rF')}{192\pi^2 h}\left(\frac{f'}{f}\right)^2. \tag{1.13}$$

This will be the RSET we will use for the majority of the next section, where we analyse self-consistent semiclassical solutions. Before this, let us make some final remarks on other existing RSET approximations in static spacetimes.

Approximate RSETs based on dimensional reduction have been extensively used despite their problematic behaviour at $r = 0$. These problems are absent from RSET approximations that consider the four-dimensional field dynamics from the start, like the analytic approximation derived by Anderson, Hiscock and Samuel [3]. Although this approximation gives a well-behaved RSET at $r = 0$, the AHS-RSET has additional problems for massive fields. For example, in flat spacetime it contains non-zero contributions that cannot be renormalised away [11] nor be identified with a cosmological constant [76].

Some of the complications associated with the AHS-RSET can be circumvented by considering just the zero-mass case and applying to it a reduction-of-order procedure (see [11, 88]). After the reduction the resulting system of differential equations is again second order a thus adequate for backreaction analyses [6, 11]. In the next section we will use mostly the Regularised Polyakov approximation, but also this Order-Reduced AHS-RSET to find self-consistent solutions of the semiclassical equations both for classically empty spacetimes and for stars of constant density.

1.4 Stellar Equilibrium on a Physical Vacuum Soil

Equipped with a suitable RSET satisfying the desired properties of analyticity, regularity, conservation and low derivative order, the semiclassical equations (1.1) can be solved in a full, self-consistent manner. Since the RSET is a function of the spacetime metric and its derivatives, self-consistent solutions to the semiclassical equations will be those in which classical spacetime configurations are everywhere corrected by the backreaction of quantum vacuum polarisation.

Treating semiclassical gravity as a modified theory of gravity generates situations in which semiclassical corrections overcome their $\mathcal{O}(\hbar)$ suppression. When there is not classical matter this, for example, can lead to geometries that cannot be smoothly deformed into their general relativistic counterparts in the $\hbar \to 0$ limit. Notwithstanding the absence of a unique prescription to obtain analytic RSET approximations, one advantage of the semiclassical approach is that self-consistent solutions exhibit robust properties that appear to be independent of the RSET approximation adopted.

Before passing to the description of the solutions found, let us make two remarks regarding the selection of Boulware state and the surpassing of the Buchdahl limit.

1.4.1 *The Importance of the Boulware Vacuum*

Semiclassical gravity discloses its non-perturbative phenomenology at event horizons. In static situations, the natural vacuum state for the field is the Boulware vacuum [30], which reduces to the Minkowski vacuum at radial infinity, a characteristic that is consistent with the asymptotic flatness of spacetime. This modes defining the state are manifestly singular at the event horizon, such behaviour spreading to the RSET. In fact, the RSET has a physical divergence at $r = r_{\rm H}$ if the energy density measured by a freely-falling observer diverges there [74]. Let us illustrate this divergence in the Polyakov RSET. For a metric adopting the form

$$f(r) = h(r)^{-1} = \frac{r - r_{\rm H}}{r_{\rm H}} + \mathcal{O}\left(\frac{r - r_{\rm H}}{r_{\rm H}}^2\right) \tag{1.14}$$

near the event horizon, then the quantity

$$\mathcal{E} = \frac{\langle \hat{T}_r^r \rangle^{\rm P} - \langle \hat{T}_t^t \rangle^{\rm P}}{f} \propto -\frac{l_{\rm P}^2}{r_{\rm H}^2 (r - r_{\rm H})^2} + \mathcal{O}\left(\frac{r - r_{\rm H}}{r_{\rm H}}\right)^0, \tag{1.15}$$

is infinite (here $l_{\rm P} = 1/\sqrt{12\pi}$). A similar divergence is found for the AHS-RSET [11].

As a consequence of its divergent behaviour at the event horizon,[3] the Boulware vacuum is commonly dismissed as plainly non-physical, deemed as the natural state only for horizonless stellar configurations instead. This argument is based on the assumption that the background spacetime is unaffected by vacuum polarisation. At the level of the semiclassical equations, when the RSET is allowed to backreact on the spacetime, the Boulware vacuum is a perfectly self-consistent vacuum state, since its characteristic divergence gets absorbed by the background spacetime. As a consequence, the event horizon gets destroyed by the backreaction of vacuum polarisation. When there is no classical matter, this carries along additional pathologies that are absent in classical vacuum solutions, such as the presence of curvature singularities that are not concealed by event horizons. This is avoided in the presence of classical matter, which makes the geometry akin to stellar configurations.

At this stage, it is interesting to advance an intriguing difference between GR and its semiclassical counterpart [12]. In GR the eternal Schwarzschild (or Kerr) vacuum solution captures all the relevant aspects of the more realistic situation in which a BH is formed from the collapse of a previous stellar configuration. This can be taken

[3] This divergence is rooted to the choice of plane-wave mode solutions to Eq. (1.7). The null Eddington-Finkelstein coordinates diverge at even horizons making the modes to acquire infinite frequencies.

as suggesting that the behaviour of matter is not that relevant for analysing the end point of gravitational collapse (beyond imposing that matter does not violate energy conditions). On the contrary, in semiclassical GR vacuum eternal (or static) solutions exhibit more clearly their pathological nature. As we will see, in order to obtain physically reasonable configurations, some classical matter needs to be included in the spacetime. It is this classical material that seeds how vacuum polarisation is in turn excited.

1.4.2 Surpassing the Buchdahl Limit

We follow this line of thought to its ultimate consequences, assuming the existence of an additional material modelled as a classical SET describing an isotropic perfect fluid of constant density in equilibrium. Through this simple assumption, a window opens towards the possibility that the (on average repulsive) effects of vacuum polarisation generate new configurations in equilibrium that are more compact than those allowed by classical general relativity. We define the compactness function as

$$C(r) \equiv \frac{2m(r)}{r} = 1 - h(r)^{-1}, \tag{1.16}$$

where $m(r)$ is the Misner-Sharp mass [77]. The value $C(r) = 1$ denotes the compactness of the BH event horizon and $C(r) = 8/9$ denotes the largest surface compactness attainable by hydrostatic equilibrium configurations in general relativity, or Buchdahl limit [34, 112]. The Buchdahl compactness bound applies to stars satisfying the following: (i) the star has a Schwarzschild exterior, (ii) internal pressures in the angular directions do not surpass the pressure in the radial direction, and (iii) a density profile that is non-increasing outwards. Self-consistent semiclassical gravity has the potentiality to violate all three possibilities: the exterior spacetime is no longer Schwarzschild, RSETs are anisotropic by construction, and they have negative energy densities that can revert the tendency of the total density to be non-increasing outwards. In consequence, this theory stands out as a promising place in which to seek for new stages of stellar equilibrium that can solve the pathologies posed by vacuum solutions.

By seeking for RSET approximations that are adapted to describe stellar structures, it is possible to find families of RSETs whose backreaction effects support stars that overcome the Buchdahl limit. Once the Buchdahl limit is surpassed, these stars can have a surface lying extremely close to their gravitational radius (both surfaces being separated just by few Planck lengths). Their large interior redshifts makes them easily mistaken for BHs through electromagnetic observations [40]. Nonetheless, the presence of a surface inside their photon sphere could produce distinct gravitational-wave echoes [41]. This way we evidence that, by considering a more physical vacuum than the one from general relativity, together with the simplest material contents, semiclassical gravity allows for the existence of ultracompact alternatives to BHs.

These exotic compact objects, that we denoted as relativistic semiclassical stars, are realised in two ways: through exploring families of RP-RSETs where the regulator function $F(r)$ is distorted within some central stellar core, and through order-reduced versions of the AHS-RSET. Our philosophy in here is not to argue for a particular approximation scheme as the best one; it is more to put all the possibilities on the table to see what they can offer. Our point of view concerning semiclassical theories is more heuristic and closer to the phenomenological philosophy underneath modified theories of gravity: motivating a possible form for some modifications of general relativity and then analysing the new equations without caring how these equations might show up hierarchically from an even deeper description of spacetime. The existence of common features in the solutions to semiclassical equations sourced by unrelated RSETs evidences the robustness of semiclassical analyses.

1.4.3 Solutions with No Classical Matter

We now turn towards deriving the semiclassical counterpart to the Schwarzschild BH solution. By semiclassical counterpart, we refer to the solution that incorporates the effects of vacuum polarisation in a self-consistent way through the backreaction of the RSET. We will use two qualitative approximations to the RSET: the Regularised Polyakov approximation and an Order Reduced AHS approximation.

1.4.3.1 The Regularised Polyakov Approximation

For the metric (1.12), the tt and rr components of the semiclassical Einstein equations in vacuum are, respectively,

$$\begin{aligned} \frac{h(1-h)-rh'}{h^2r^2} &= 8\pi\hbar\langle\hat{T}_t^{\,t}\rangle^{\mathrm{P}}, \\ \frac{rf'f-fh}{fhr^2} &= 8\pi\hbar\langle\hat{T}_r^{\,r}\rangle^{\mathrm{P}}, \end{aligned} \tag{1.17}$$

where the RSET is described by the Regularised Polyakov approximation (1.13) with

$$F(r) = 1/\left(r^2+\alpha l_{\mathrm{P}}^2\right), \ \alpha > 1. \tag{1.18}$$

This simple choice of $F(r)$ acts as a cutoff to the magnitude of the Polyakov RSET, which becomes finite on regular spacetimes.

Equation (1.17) can be integrated as a boundary value problem from radial infinity (in practice, a distant referential radius) assuming the metric takes the asymptotic form

$$f(r) = h(r)^{-1} = 1 - \frac{2M}{r}, \quad M > 0, \tag{1.19}$$

which is consistent with the way the RSET components (1.13) decay at infinity in the Boulware vacuum [7]. Due to the presence of an additional source in the right-hand side of the semiclassical equations, the metric no longer obeys $f(r) = h(r)^{-1}$. As the semiclassical equations are integrated inwards, the spacetime geometry progressively deviates from the Schwarzschild solution. This deviation amounts to a redshift function $f(r)$ modified with respect to its Schwarzschild form (1.19) and a Misner-Sharp mass $m(r)$ that acquires a dependence on the radial coordinate, as if the whole spacetime was surrounded by an inhomogeneous cloud of negative mass. Notice however that at any macroscopic distance from the gravitational radius this modification is absolutely negligible (it is of order $\hbar$) leaving unchanged any gravitational test related to the Schwarzschild exterior metric. The magnitude of the RSET increases inwards until a special surface $r_{\rm B} > 2M$ where quantum corrections become non-perturbative is encountered. At $r_{\rm B}$, $h(r_{\rm B}) \to \infty$ and we find a coordinate singularity that corresponds to a minimal surface for r. This is demonstrated by adopting a change to the proper radial coordinate l,

$$\frac{dr}{dl} = \pm \frac{1}{\sqrt{h}}, \tag{1.20}$$

where the $\pm$ signs denote the two branches of the radial coordinate at each side of the minimal surface. Assuming the following behaviours for the metric functions

$$\begin{aligned} f(l) &= f_{\rm B} + f_1 (l - l_{\rm B}) + \mathcal{O}(l - l_{\rm B})^2, \\ r(l) &= r_{\rm B} + r_1 (l - l_{\rm B}) + r_2 (l - l_{\rm B})^2 + \mathcal{O}(l - l_{\rm B})^3, \end{aligned} \tag{1.21}$$

and replacing them in the semiclassical equations we find

$$f_1 = \frac{2 f_{\rm B} \sqrt{r_{\rm B}^2 + \alpha l_{\rm P}^2}}{l_{\rm P} r_{\rm B}}, \quad r_1 = 0, \quad r_2 = \frac{\left(r_{\rm B}^2 + \alpha l_{\rm P}^2\right)^2 + \alpha l_{\rm P}^4}{2 r_{\rm B} \left[r_{\rm B}^2 + (\alpha - 1) l_{\rm P}^2\right] \left(r_{\rm B}^2 + \alpha l_{\rm P}^2\right)}, \tag{1.22}$$

where $f_{\rm B}$ and $r_{\rm B}$ are positive constants baring a non-analytic relation to the ADM mass M and the parameter α. The functions (1.22) make the metric explicitly regular at $l = l_{\rm B}$. The radial function $r(l)$ is symmetric around the minimal surface $r_{\rm B}$, while the redshift function is positive and asymmetric, showing there is a wormhole neck that connects the asymptotically flat region of the spacetime with a new region of different characteristics. Figure 1.1 shows these metric functions in terms of l for an example integration of the semiclassical equations.

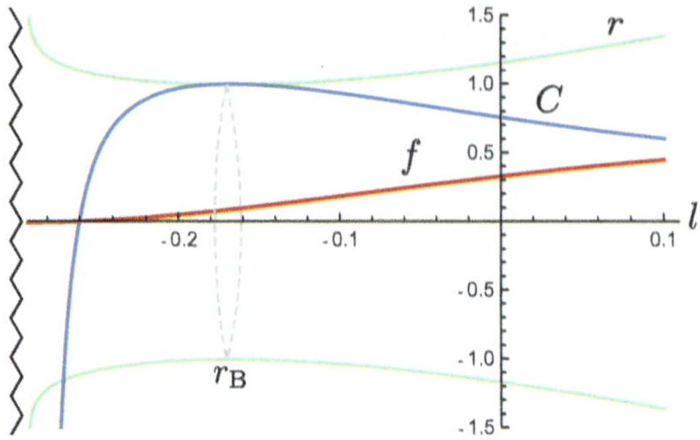

Fig. 1.1 Numerical plot of the semiclassical counterpart of the Schwarzschild vacuum geometry. The horizontal axis is the proper coordinate l while the above and below curves in green represent the radial coordinate r. The behavior of the redshift function, in red, and the compactness, in blue, are shown. The right side of the wormhole is asymptotically flat whereas the other is asymptotically singular. Both regions are joined by a minimal surface of radius $r = r_{\rm B}$. We have chosen $M = 0.1$ and $\alpha = 1.01$ for illustrative purposes. Geometrical characteristics are identical for larger ADM masses

Below the wormhole neck, vacuum polarisation enters into a runaway regime that makes the metric approach a null singularity at infinite r, but finite l. Through an asymptotic analysis of the semiclassical equations [7], the form of the metric nearing the singularity is found to be, in Schwarzschild coordinates,

$$\begin{aligned} ds^2 \simeq & \left(\frac{r}{l_{\rm P}}\right)^{1-4\alpha} e^{-\frac{2r^2}{l_{\rm P}^2}} \cdot \\ & \cdot \left\{-a_0\left(1-\frac{l_{\rm P}^2}{8r^2}\right)dt^2 + \frac{2\chi_0 r^2}{l_{\rm P}^2}\left[1-\frac{(9-32\alpha)l_{\rm P}^2}{8r^2}\right]dr^2\right\} + r^2 d\Omega^2, \end{aligned} \tag{1.23}$$

where a_0 and χ_0 are dimensionless positive constants. The vanishing of the conformal factor as $r \to \infty$ manifests the null character of this singularity. The Ricci scalar, defined as

$$\mathscr{R} = \frac{2}{r^2}\left(1-\frac{1}{h}\right) + \frac{2}{hr}\left(\frac{h'}{h} - \frac{f'}{f} + \frac{rf'h'}{4fh}\right) + \frac{1}{2h}\left[\left(\frac{f'}{f}\right)^2 - 2\frac{f''}{f}\right], \tag{1.24}$$

becomes negatively divergent at the singularity, i.e.,

$$\mathscr{R} \simeq -\frac{e^{2r^2/l_{\rm P}^2}(2\alpha-1)}{l_{\rm P}^2\chi_0}\left(\frac{r}{l_{\rm P}}\right)^{-5+4\alpha}. \tag{1.25}$$

Furthermore, this singular region is located at a finite proper distance $l_{\rm S} < l_{\rm B}$ from the throat, as shown by integrating the quantity

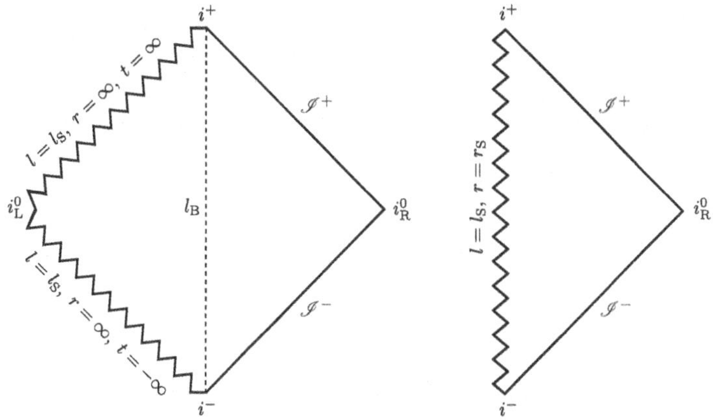

Fig. 1.2 Left panel: Penrose diagram corresponding to the singular wormhole solution for the RP-RSET. The dashed lines denote the location of the wormhole neck. To their right, the asymptotically flat portion of spacetime is depicted alongside its asymptotic regions. The left hand side of the diagram shows the internal past and future null singularities, which are located at finite proper distance from the neck $l_S - l_B$. The point i^0_L is singular as well, and is reached in finite proper time by spacelike geodesics. Right panel: Penrose diagram associated with the vacuum solution for the OR-RSET. In this case, the singularity is timelike and constitutes a naked singularity. While differences in the modelling of the semiclassical source result in singularities of different sorts, both models agree on the absence of event horizons

$$\left(\frac{dl}{dr}\right)^2 = 2\chi_0 \left(r/l_P\right)^{3-4\alpha} e^{-\frac{2r^2}{l_P^2}} \left[1 - \frac{(9-32\alpha)l_P^2}{8r^2}\right]. \tag{1.26}$$

Near asymptotically flat regions, far from any source of gravity, semiclassical corrections amount to extremely weak, thus perturbative, corrections. As the surface $r = 2M$ is approached, however, vacuum polarisation builds up and destroys the event horizon, generating instead a wormhole neck. While, as shown below, the specific features of the region of non-perturbative semiclassical corrections depend on the particular modelling of the RSET (see Fig. 1.2), the replacement of the event horizon by a singularity appears as a robust characteristic of the RSET not depending on the approximation, as it is a consequence of evaluating the RSET in the Boulware state. Similar characteristics are displayed by the semiclassical counterpart to the Reissner-Nordström sub-extremal BH [8].

The regularisation scheme we have adopted for the Polyakov RSET [see Eq. (1.18)] amounts to a cutoff to the magnitude of its components, whose strength is modulated by α. Increasing α brings the singularity at l_S closer to the wormhole neck l_B. Spacetime regions near $r = 0$, while unexplored in vacuum solutions, are present in stellar configurations. We will return to exploring more elaborate regularisation schemes for the Polyakov RSET later.

The results here presented are consistent with previous works [25, 50, 61]. A clear extension of this work is to consider the backreaction effects of an RSET approximation that does not rely on dimensional reduction. The Order Reduced version of the AHS-RSET stands on equal footing with the RP-RSET, in the sense that it is a quantity covariantly conserved, analytic, without higher-derivative terms, and well-defined at $r = 0$. We now briefly sketch the derivation of this RSET approximation and the characteristics of the corresponding vacuum solutions in the minimally coupled case.

1.4.3.2 The Order Reduced AHS-RSET

We start with the tt and rr components of the vacuum semiclassical equations (1.1), now sourced by the AHS-RSET,

$$\begin{aligned} \frac{h(1-h) - rh'}{h^2 r^2} &= 8\pi\hbar \langle \hat{T}_t^t \rangle^{\text{AHS}}, \\ \frac{rf'f - fh}{fhr^2} &= 8\pi\hbar \langle \hat{T}_r^r \rangle^{\text{AHS}}, \end{aligned} \tag{1.27}$$

where the right-hand side contains higher-derivative terms. The concrete and lengthy form of the AHS-RSET, which is not very illustrative, can be seen in [3]. To obtain a set of equations of the same derivative order as the classical ones, we subject the AHS-RSET to a perturbative reduction of order. The first step in this procedure consists in neglecting terms $\mathcal{O}(\hbar)$ in Eq. (1.27), leading to

$$\begin{aligned} \frac{h(1-h) - rh'}{h^2 r^2} &= \mathcal{O}(\hbar), \\ \frac{rf' + f - fh}{fhr^2} &= \mathcal{O}(\hbar). \end{aligned} \tag{1.28}$$

These expressions can be differentiated consecutively to derive recursion relations between f, h, and their higher-order derivatives $\{f^{(n)}\}_{n=1}^{\infty}$ and $\{h^{(n)}\}_{n=1}^{\infty}$. For h, said relations are obtained by solving the tt equation directly, which can then be used to derive the f relations from the rr equation. The resulting relations are

$$\begin{aligned} h^{(n)} &= (-1)^n \frac{n! h^n}{r^n}(h-1) + \mathcal{O}(\hbar), \\ f^{(n)} &= (-1)^{n+1} \frac{n! f}{r^n}(h-1) + \mathcal{O}(\hbar). \end{aligned} \tag{1.29}$$

Relations (1.29) are now inserted in the AHS-RSET components $\langle \hat{T}_t^t \rangle^{\text{AHS}}$ and $\langle \hat{T}_r^r \rangle^{\text{AHS}}$ until they only depend on f and h. After a lengthy but straightforward calculation using symbolic computation software, we arrive at

$$
\begin{aligned}
16\pi^2\langle\hat{T}_t^t\rangle^{\rm OR} &= \frac{(h-1)^2\left(h^2+6h+33\right)}{480h^2r^4} - \left(\xi-\frac{1}{6}\right)\frac{(h-1)^2\left(h^2+2h+5\right)}{8h^2r^4},\\
16\pi^2\langle\hat{T}_r^r\rangle^{\rm OR} &= -\frac{(h-1)^2\left(h^2+6h-15\right)}{1440h^2r^4} + \left(\xi-\frac{1}{6}\right)\frac{(h-1)^2(h+3)^2}{24h^2r^4},
\end{aligned}
\tag{1.30}
$$

where the suffix OR stands for Order-Reduced. Using the expressions for $\langle\hat{T}_t^t\rangle^{\rm OR}$ and $\langle\hat{T}_r^r\rangle^{\rm OR}$ we can deduce, through the conservation relation (1.11), the angular components necessary to force its divergence to vanish. Specifically, we find

$$
\begin{aligned}
16\pi^2\langle\hat{T}_\theta^\theta\rangle^{\rm OR} &= 16\pi^2\langle\hat{T}_\varphi^\varphi\rangle^{\rm OR} = \\
&-\frac{h-1}{1440fh^2r^4}\left\{f\left[h^3(r-1)+h^2(3r-5)+3h(r+7)-15(r+1)\right]\right.\\
&\qquad\left.+f'(h-1)\left(h^2+6h+21\right)\right\}\\
&+\left(\xi-\frac{1}{6}\right)\frac{h-1}{24fh^2r^4}\left\{f(h+3)\left[h^2(r-1)-2h+3(r+1)\right]\right.\\
&\qquad\left.+rf'(h-1)\left(h^2+3h+6\right)\right\}.
\end{aligned}
\tag{1.31}
$$

In this way, we construct a new approximate RSET (the OR-RSET) valid for back-reaction studies in vacuum spacetimes and which satisfies all Wald's axioms [115]. We can now assume the scalar field is minimally coupled to gravity ($\xi = 0$) to analyse a material content equivalent to the one described by the Regularised Polyakov approximation.

Regarding the semiclassical counterpart to the Schwarzschild spacetime in this prescription, it is possible to construct an analytical argument [11] proving that, when the order-reduced semiclassical equations are integrated from the asymptotically flat region inwards, the solution inevitably encounters a naked, timelike singularity at $r = r_{\rm D} > r_{\rm H}$. Roughly sketched, this argument is based on the fact that the tt equation enforces the h function to be monotonically increasing inwards. By ruling out any other possibility, we find it must diverge at $r = r_{\rm D}$. This implies, through the rr equation, that f has a global minimum at some radius $r > r_{\rm D}$, diverging positively at the singularity. Near the singular region, the metric behaves as

$$
f \simeq \left(\frac{f_{\rm D}}{r-r_{\rm D}}\right)^{1/9}, \quad h \simeq \left(\frac{h_{\rm D}}{r-r_{\rm D}}\right)^{1/3}, \tag{1.32}
$$

where $f_{\rm D}$, $h_{\rm D}$ are positive constants bearing a relation to the ADM mass M and $r_{\rm D}$, not relevant here. The behaviours (1.32) cause the curvature scalar (1.24) to diverge as

$$
\mathscr{R} \simeq -\chi_1 (r-r_{\rm D})^{-5/3}, \quad \chi_1 > 0. \tag{1.33}
$$

This divergence is slower than the one predicted by the Polyakov approximation, but it is reached at finite r and finite affine distance for all geodesic observers, giving the singularity a timelike character. The Penrose diagram associated to this solution is represented in Fig. 1.2.

Through two unrelated methods for obtaining proper RSET approximations in static situations for massless, minimally coupled fields, we have shown that the semiclassical counterparts to the Schwarzschild spacetime exhibit pathologies that make them inadequate to describe the late time outcome of gravitational collapse. Either they present null curvature singularities and a change in the topology of the spacetime, as for the RP-RSET, or they exhibit a naked curvature singularity, as in the OR-RSET. Note that this geometry is obtained for as long as $\xi < 11/60$.

There are two possibilities for making sense of semiclassical solutions in light of these results: Either we consider horizons are dynamic and undergo an evaporation process or, if we insist on exploring static possibilities within this theory, introducing classical matter becomes essential.

1.4.4 Stellar Equilibrium

After considering purely vacuum solutions in semiclassical gravity, i.e., geometries self-consistent with the cloud of vacuum polarisation that they generate, we turn to the analysis of stellar configurations in hydrostatic equilibrium. Adding classical matter makes the analysis of the semiclassical equations considerably more involved, since the classical and quantum SETs affect each other through the geometry of spacetime. In the aim to simplify this analysis, we model the stellar interior via the stress-energy tensor (SET) of an isotropic perfect fluid

$$T_{\nu}^{\mu} = (\rho + p)\, u^{\mu} u_{\nu} + p \delta_{\nu}^{\mu}, \tag{1.34}$$

with p and ρ denoting the pressure and energy density measured by an observer comoving with the static fluid with 4-velocity u^{μ}. Covariant conservation of the SET (1.34) translates into the relation

$$\nabla_{\mu} T_{r}^{\mu} = p' + \frac{f'}{2f} (\rho + p) = 0. \tag{1.35}$$

An equation of state for the classical fluid needs to be specified. We assume a simple uniform-energy-density fluid

$$\rho(r) \equiv \rho = \text{const}, \tag{1.36}$$

so that we are seeking for the semiclassical counterparts of Schwarzschild's stellar solutions [103]. In standard GR, the assumption of an interior with uniform classical density saturates the hypotheses behind the Buchdahl theorem: it produces the most compact regular stars, with $C(R) = 8/9$. In our analyses, any surpassing of the Buchdahl compactness bound will be caused by the semiclassical corrections we have incorporated through the RSET.

We start integrating the semiclassical equations (1.1) from the asymptotically flat region inwards in absence of any classical SET and with positive ADM mass. This vacuum geometry will correspond to either of the solutions found in the previous section, depending whether we consider the RP-RSET or the OR-RSET. Then, at a particular R we continue the integration but now in the presence of the classical SET, that is, we consider the equations

$$\begin{aligned} \frac{h(1-h)-rh'}{h^2r^2} &= 8\pi\left(-\rho+\hbar\langle\hat{T}_t^t\rangle\right), \\ \frac{rf'+f-fh}{fhr^2} &= 8\pi\left(p+\hbar\langle\hat{T}_t^t\rangle\right). \end{aligned} \tag{1.37}$$

The RSET will, in this interior region, be modelled in two ways that are natural generalisations of the RSETs we considered in the vacuum case.

Since the standard Polyakov approximation is not well-defined at $r = 0$, some regularisation of this approximation is mandatory to construct stellar equilibrium configurations, even in those situations where the semiclassical corrections received are small and perturbative. Therefore, the behaviour of the Polyakov approximation at $r = 0$ needs to be specified by hand through suitable choices of $F(r)$. To this aim, we consider modified $F(r)$ functions that distort the Polyakov approximation in some internal region of the star $r \leq r_{\text{core}} < R$ (the stellar core hereafter). Through minimal assumptions on the geometrical characteristics of this core, we find, among the space of modified $F(r)$ functions, whole families compatible with regular stars that surpass the Buchdahl limit (see next subsection).

Then, to add cumulative evidence to this result and erase all doubt of it being an ad hoc procedure, we derive the equivalent to the previous OR-RSET but now in the presence of a classical constant density fluid. Sourced by this RSET, we find regular stars that surpass the Buchdahl limit with remarkably similar properties to those obtained through the RP-RSET (Sect. 1.4.4.3).

Exploring the space of semiclassical stellar solutions is an exhaustive task than involves both regular and singular solutions. For didactic purposes, here we just outline the derivation of the corresponding RSETs supporting semiclassical stars that surpass the Buchdahl limit. In between, we discuss the main physical properties of these new objects and the implications stemming from them. We encourage the reader to consult the Refs. [6, 9, 10] for details.

1.4.4.1 Stellar Equilibrium in a Regularised Polyakov Approximation

We now consider the semiclassical equations (1.37) in presence of classical matter which occupies a finite portion of spacetime inside a sphere of radius R. The geometry exterior to this sphere is the Schwarzschild counterpart in the Regularised Polyakov approximation. For simplicity, let us place the star surface at the side of the wormhole that connects with the asymptotically flat region, although it is also possible to place it at the neck and even slightly inside.

The integration from the surface inwards takes $C(R)$ and ρ as parameters. The regulator function $F(r)$ is first taken, for simplicity, to be of the form (1.18). Sweeping through the parameter space we find for $C(R) < 8/9$, regular stars perturbatively close to their classical counterparts. This is to be expected from their small interior curvatures that generate a negligible polarisation of the vacuum. Increasing the surface compactness to the regime $1 > C(R) \geq 8/9$, we stop finding regular stars. Nonetheless, it is always possible to observe a finite set of ρ values for which stars contains a surface where $C(r_0) = 0$ at $0 < r_0 \ll R$. We denote these solutions as "quasi-regular" stars, given that a vanishing compactness at $r = 0$ is a requirement for spacetime regularity[4]. Since r_0 can be arbitrarily small, this is taken as an indication that semiclassical effects bring super Buchdahl stars closer to attaining regularity than what classical matter can do by itself (quasi-regular solutions are not found in the classical theory in the compactness range $1 > C(R) \geq 8/9$).

From a physical perspective, adding a small, positive ρ to the spacetime at $R > r_B$ effectively shrinks the size of the wormhole neck slightly, which now appears inside the region covered by matter. As ρ increases, this wormhole neck shrinks further, eventually disappearing for some critical density ρ_c. Quasi-regular stars are those displaying a wormhole neck extremely close to the centre of spherical symmetry. These configurations are realised due to a fine balance between classical and quantum contributions that eventually breaks down in the innermost regions of the star, precisely where the Polyakov approximation is less reliable.

The regularisation adopted in (1.18) is just the simplest among plenty of possibilities, some of which might even approximate the behaviour of an exact RSET in these regions. In what follows, we search for more elaborate regularisation schemes for the RP-RSET. These regularisations are characterised by the following: for any star belonging to or nearby the quasi-regular family, we select a spherical surface of radius $r_{\text{core}} \in (r_0, R)$ and, in the range $r \in [0, r_{\text{core}})$, we assume minimal deformations of the regulator function (1.18) that are compatible with the complete absence of curvature singularities.

To find the form of these deformations of the F function, we follow a reverse-engineering logic that consists in proposing an ansatz for a regular geometry in the range $r \in [0, r_{\text{core}})$ and then obtaining the regulator F that sources the geometry via the RP-RSET, in case there is any compatible with regularity. For this purpose, we derive an expression for h from the rr component of the semiclassical equations (1.37) and replace it in the tt component. Through the conservation relation of the classical SET (1.35), we arrive at the expression

[4] Strictly, regularity conditions impose that compactness must vanish at least as $C \simeq C_0 r^2 + \mathcal{O}r^3$ and redshift must go as $f \simeq f_0 + f_1 r^2 + \mathcal{O}r^3$. Conditions on the pressure are then derived from Eq. (1.35).

$$p'' = \mathscr{D}\left[\mathscr{A}_0 + \mathscr{A}_1 \left(p'\right) + \mathscr{A}_2 \left(p'\right)^2 + \mathscr{A}_3 \left(p'\right)^3\right], \tag{1.38}$$

where

$$\begin{aligned}
\mathscr{A}_0 &= -\,8\pi r \left(\rho + p\right)^3 \left(\rho + 3p\right), \\
\mathscr{A}_1 &= 4 \left(\rho + 3p\right)^2 \left[6\pi r^2 \left(\rho + p\right) + 16\pi^2 F l_{\mathrm{P}}^2 r^2 p - 1\right], \\
\mathscr{A}_2 &= -\,r \left(\rho + p\right) \left[16\pi r^2 \left(\rho - 2p\right) - 4\pi l_{\mathrm{P}}^2 \left(2F + rF'\right)\right. \\
&\quad \left. + 32\pi^2 F l_{\mathrm{P}}^2 r^2 \left(\rho + 5p\right) - 32\pi^2 F' l_{\mathrm{P}}^2 r^3 p - 6\right], \\
\mathscr{A}_3 &= F l_{\mathrm{P}}^2 r^2 \left[8\pi r^2 \left(\rho - p\right) - 4\pi l_{\mathrm{P}}^2 \left(2F - rF'\right) - 32\pi^2 F' l_{\mathrm{P}}^2 r^3 p - 2\right], \\
\mathscr{D} &= 2r \left(1 - 4\pi l_{\mathrm{P}}^2 F\right) \left(\rho + p\right)^2 \left(1 + 8\pi r^2 p\right).
\end{aligned} \tag{1.39}$$

Now we make use of this differential equation to find the geometry and the RP-RSET inside the core. By imposing an ansatz for the pressure and its derivatives in the region $r \in [0, r_{\mathrm{core}})$, Eq. (1.38) becomes a first order differential equation for F which, upon solving, determines the entire geometry inside the core. Naturally, if the resulting F and F' are everywhere finite within the core, $h(r)$ is regular from (1.37), and hence the complete spacetime metric.

We consider a pressure profile for the core [whose classical energy density is constant, recall Eq. (1.36)] that is everywhere finite and has a global maximum at $r = 0$. These are the minimal conditions necessary for regularity, given that we are just extending the behaviour of the pressure function in quasi-regular stars (which grows monotonically from the surface inwards) in the simplest way, that is, so that it reaches a global maximum at $r = 0$. At r_{core}, continuity of the metric enforces pressure to be continuous up to its second derivative. The simplest analytic function that satisfies these conditions is the fifth-order polynomial

$$p = p_0 + p_0'' r^2/2 + c_0 r^3 + c_1 r^4 + c_2 r^5, \tag{1.40}$$

where the pressure at the origin p_0 and its second derivative p_0'' are positive and negative constants, respectively. Determining the coefficients $\{c_i\}_{i=0}^{2}$ is straightforward given the aforementioned conditions of regularity and continuity.

The last step consists in taking a fixed numerical solution for the bulk region $r \in [r_{\mathrm{core}}, R)$. This amounts to choosing a particular pressure profile, which we typically consider to be the critical solution $\rho = \rho_{\mathrm{c}}$, but we can also select a nearby solution in the space of parameters. Given a core size r_{core}, the pressure function inside the core is determined upon fixing the two remaining free parameters $\{p_0, p_0''\}$ by hand.

We explore numerically a wide range of values of the parameters $\{p_0, p_0''\}$ that select a particular pressure profile in (1.40) and obtain their associated $F(r)$ through (1.38). We find that regular solutions exist for cores of any size r_{core}. Let us emphasise that the existence of regular semiclassical stars that surpass the Buchdahl limit is a remarkably non-trivial result. Of course, the Einstein equations guarantee that any exotic spacetime geometry, like that of a BH mimicker, is generated by a corresponding effective SET that will certainly violate energy conditions. This should not be mistaken with our approach here, as we are imposing that the pressure inside the core obeys a polynomial form and asking whether this is compatible with a stress-energy tensor that can be divided in two parts: a classical uniform density perfect fluid and a RP-RSET. As it is not guaranteed that a prescribed geometry will be compatible with the RP-RSET, it might have happened that no F existed for any regular pressure ansatz. Hence, that this compatibility is realised for the simple polynomial example in Eq. (1.40) is a strong indication that the Polyakov approximation is able to capture an important fraction of the relevant physics. Furthermore, the existence of these solutions for central cores of any size suggests that semiclassical physics is capable of arranging itself in the way necessary to attain equilibrium in each situation.

1.4.4.2 Physical Properties of Semiclassical Stars

Figure 1.3 shows the Misner-Sharp mass and classical pressure profiles for semiclassical relativistic stars with different core sizes. Stars with larger cores have less negative Misner-Sharp mass, although these values are distributed throughout a wider portion of the stellar interior. Increasing the size of the core also diminishes the magnitude of the central pressure p_0.

Semiclassical contributions are suppressed by $l_{\rm P}$ so, in order for vacuum polarisation to be an appreciable effect, there needs to be some additional scale compensating for this suppression. In the case of semiclassical stars, this scale is precisely the proximity (in compactness) to the Buchdahl limit. When the surface compactness of a classical constant density sphere reaches the value $C(R) = 8/9$, spacetime curvature diverges at $r = 0$, triggering a subsequent divergence in the RSET components [100]. The backreaction of this diverging RSET can then incorporate additional repulsive

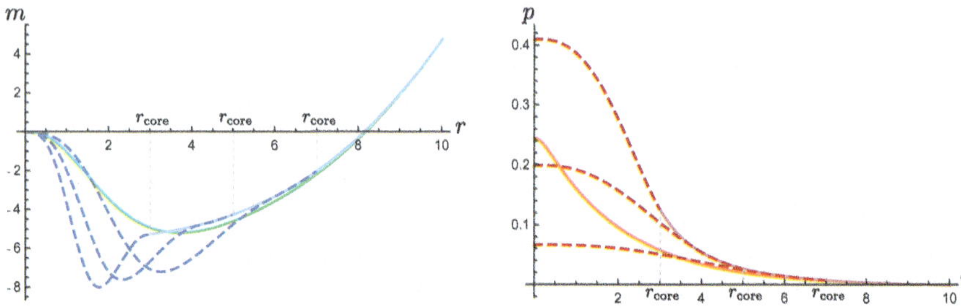

Fig. 1.3 Misner-Sharp mass (left plot) and classical pressure (right plot) of stars with $C(R) = 0.96$ and $R = 10$. The dashed curves represent solutions obtained through the Regularised Polyakov approximation. We have shown three solutions with $\rho \simeq 0.00246$ and different $r_{\rm core}$ values, below which the F function is distorted. We have also plotted a star obtained in the order-reduced prescription with $\rho \simeq 0.00248$ (turquoise and orange lines) that highlight the qualitative agreement obtained through both RSET approximations

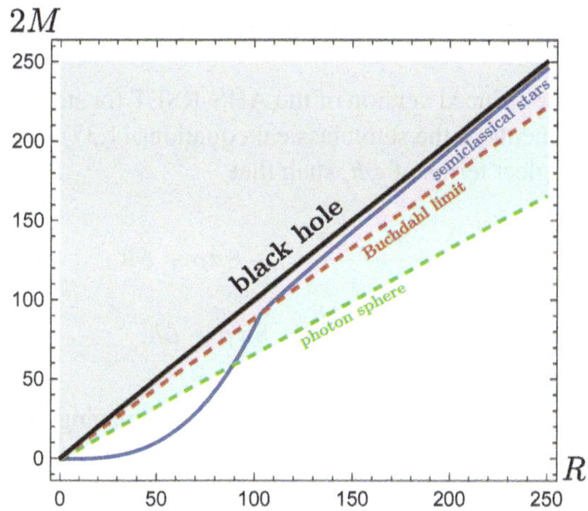

Fig. 1.4 Mass-to-Radius diagram of semiclassical stars in the Regularised Polyakov approximation with $\rho = 10^{-5}$ (in Planck units). The black line represents the compactness parameter of BHs, $C(R) = 1$, the dashed red line denotes the Buchdahl compactness bound $C(R) = 8/9$, and the dashed green line is the minimum compactness of objects that exhibit a photon sphere, $C(R) = 1/3$. The blue curve represents semiclassical stars. For those stars surpassing the Buchdahl limit, the total mass M grows approximately linearly with the radius R. Each point within the blue curve admits whole families of regulator functions F for which the entire geometry is regular. A qualitatively similar diagram is obtained making use of the MOR-RSET

effects that prevent the pressure from diverging, allowing the existence of regular stars at the Buchdahl limit and beyond. This marks a transition between regimes with negligible and significant quantum corrections, which can be observed in the mass-radius diagram from Fig. 1.4. In this diagram, we find, in addition to the perturbatively corrected stars with $C(R) < 8/9$, whole families of semiclassical stars whose surface compactness ranges from the Buchdahl limit $C(R) = 8/9$ to the BH limit $C(R) = 1$ itself. The phenomenological implications of this result are promising, as they allow to probe the observational properties of exotic compact objects at nearly unexplored compactness values [69].

Looking at the bigger picture, semiclassical stars are the result of a balance between classical and quantum effects. A promising extension of this work is to analyse whether similar solutions exist when using another RSET approximations such as the AHS-RSET, which is four dimensional [3] by construction. This RSET is a perfectly well-defined source in stellar spacetimes, with the shortcoming that it exhibits high-order terms in the derivatives of the metric. To conclude this section, we present below a method of reduction of order that transforms the high-derivative terms in the AHS-RSET into low-derivative ones, yielding a novel RSET approximation that is amenable to backreaction studies in stellar spacetimes. Using this approximation, we have found preliminary evidence for the existence of stellar spacetimes with akin characteristics (i.e. surpassing the Buchdahl limit by means of a negative mass interior).

1.4.4.3 Stellar Equilibrium in an Order-Reduced Prescription

To construct an order-reduced version of the AHS-RSET for stellar spacetimes, take the tt and rr components of the semiclassical equations (1.37) (now sourced by the AHS-RSET) and neglect terms of $\mathcal{O}\hbar$, such that

$$\frac{h(1-h)-rh'}{h^2r^2} = -\,8\pi\rho + \mathcal{O}\hbar,$$
$$\frac{rf'+f-fh}{fhr^2} = 8\pi p + \mathcal{O}\hbar. \tag{1.41}$$

Consecutively differentiating these expressions and replacing derivatives of p by derivatives of f through Eq. (1.35), we derive a set of expressions relating derivatives of any order of f, h, and p, with these functions themselves [these take a more involved form than the simple recursion relations from Eq. (1.29)]. Upon replacing said relations in the AHS-RSET, its $\langle \hat{T}_t^t \rangle$ and $\langle \hat{T}_r^r \rangle$ components are reduced to expressions without spatial derivatives. Lastly, we obtain low-derivative-order expressions for the $\langle \hat{T}_\theta^\theta \rangle$ component by imposing covariant conservation through (1.11), in a manner analogous to the OR-RSET. This leads to the Matter-Order-Reduced RSET (MOR-RSET), a semiclassical source that is analytic, conserved, low-order and well-defined at $r = 0$. These characteristics put the MOR-RSET on equal footing with regularised versions of the Polyakov approximation. More details on the procedure and full expressions are available in the PhD Thesis [6], and will appear in a forthcoming publication.

The AHS- and MOR-RSETs are valid to describe scalar fields with any curvature coupling. Here, we assume a massless, minimally coupled scalar field to allow for a faithful comparison with the RP-RSET. Semiclassical stars that surpass the Buchdahl limit are also found for non-minimally-coupled fields, but this is something that we leave for future studies on the subject.

Sourced by the MOR-RSET, the semiclassical equations are of the same derivative order as the classical Einstein equations and can be solved as a boundary value problem, assuming the exterior geometry is given by the OR-RSET. By specifying the integration parameters R, $C(R)$ and ρ, we perform an exploration of the space of solutions and find regular stars that surpass the Buchdahl limit. See Fig. 1.3 for a numerical integration of the semiclassical equations with the MOR-RSET. The principal qualities of these stars, such as a wide interior region of negative Misner-Sharp mass and classical pressures that grow inwards, are shared with the semiclassical stars found with the RP-RSET. Their associated mass-radius diagram is also qualitatively identical to Fig. 1.4.

1.4.4.4 Summary of the Section and Further Comments

We have followed two unrelated approaches to modelling the RSET of a massless, minimally coupled scalar field, whose backreaction allows, in both cases, for the existence of a new type of ultracompact object supported by quantum vacuum polarisation. The robustness of this result is manifested through its generality, as both approaches not only agree in the existence of semiclassical black stars, but also on their most basic properties. What makes their matter overcome gravitational contraction is the presence of a negative mass interior, generated by the negative energy densities characteristic of the vacuum. These negative masses exert an additional gravitational repulsion that supports the object, while relation (1.35) translates their inwards-increasing classical pressures into large interior redshifts. Semiclassical stars exhibit a photon sphere, so they could produce gravitational wave echoes. To give a rough estimate, for the star obtained through the MOR-RSET model from Fig. 1.3, the crossing time for a null ray emitted from the photon sphere to be reflected at $r = 0$ and return to the photon sphere is one order of magnitude larger (in units of the stellar mass M) than the time it takes to go from the photon sphere to the surface, i.e.,

$$\tau_{\text{echo}} = 2 \int_0^{r_{\text{ph}}} (h/f)^{1/2} \, dr' \simeq 306M \sim 52\tau_{\text{s}}, \tag{1.42}$$

where τ_{s} is the light-crossing time between the photon sphere and the star surface. Attending only to the geometrical features, this effect could result in gravitational-wave echoes clearly distinguishable from the characteristic signals associated to BH mergers.

These stars might be the end-state of gravitational collapse, if at some point after the initial collapsing phase the dynamics takes the configuration to situations allowing for the accumulation of significant vacuum polarisation. Such situations are in fact more plausible than one might think. Semiclassical effects may turn out to be capable of significantly altering the dynamics of trapping horizons on much shorter timescales than what was previously believed, altering the standard gravitational collapse paradigm by rendering generic trapped regions dynamically unstable. We explore this possibility, along with scenarios of collapse which tends to relax to staticity, in the following section.

1.5 Dynamical Situations Beyond the Hawking Evaporation Paradigm

There is by now a long-established picture of BH formation and evolution in semiclassical gravity. It is widely regarded that the classical collapse to a BH has negligible short-term semiclassical corrections, with the final configuration only deviating from its classical behaviour in the subsequent, very slow process of Hawking evaporation [59]. Within this picture, the static ultracompact solutions presented above would be of little physical interest, as they would not match with any part of such dynamical scenarios. Even if such objects could be formed when starting from some very particular initial conditions, if they were to be compact enough to mimic BHs, they would likely be dynamically unstable: perturbations around them would tend to form trapped surfaces which enclose them, independently of how the perturbation and object actually interact [56].

However, this can all change if the effective semiclassical evolution of generic trapped regions turns out to be different from what is commonly believed. Indeed, the only established fact is that once a trapped region is formed, the configuration quantum fields relax to leads to no significant short-term semiclassical corrections to the evolution of the geometry in the vicinity of the *outer horizon*. From there, making use of classical intuitions [57, 92], it is assumed that anything that happens inside the trapped region cannot affect the region external to the BH. The flaw in this logic is apparent: Hawking evaporation itself already breaks the causal principles which forbid contact between the interior and exterior of the BH. The RSET typically violates energy positivity conditions in the vicinity of horizons, rendering classical intuitions useless. Thus, in semiclassical gravity (and beyond), the external universe is not safe from what may transpire inside a BH, and a thorough analysis of the depths of these objects is called for.

The simplest BH model is the Schwarzschild solution, which is comprised of a spacelike singularity fully enclosed by a trapped region. Being the simplest BH, this solution has typically been used as a background in semiclassical toy models, due mainly to the notorious complexity of quantum field theory in less symmetric spacetimes (such as actual dynamical collapse scenarios). However, perhaps unsurprisingly, this type of toy model is not sufficient to get a full picture of semiclassical BH formation and evolution. On the one hand, the inner structure of generic classical BHs is quite different from the vacuum Schwarzschild geometry, generally having an inner horizon [117] (albeit a classically unstable one), as well as an approach toward a potentially oscillatory BKL singularity at their centre [24]. On the other hand, one finds that even the vicinity of the outer horizon can be subject to large semiclassical corrections, given the right dynamical circumstances (indeed, the very effects which sustain the ultracompact configurations seen previously).

In this final section of the present chapter, we will aim to convey two particular results which showcase the potential for deviations from the standard Hawking evaporation picture. The first result is the fact that a dynamically formed inner horizon suffers not only from a classical instability (mass inflation [97]), but also from a semiclassical one, which tends to have the opposite dynamical effect to its classical counterpart, reducing the size of the trapped region from the inside-out. The second result is the fact that, given an appropriate dynamical scenario, the "in" vacuum state of gravitational collapse can have a behaviour in therms of its energy content similar to the divergent Boulware state, which sustains horizonless ultracompact objects.

1.5.1 Inner Horizon Classical and Semiclassical Dynamics

Semiclassical analyses of BH evolution typically focus on the vicinity of the outer apparent horizon of the trapped region. This being the standard definition of the "size" of a BH, it would be a good enough criterion for the state of these objects if their interior were indeed causally separated from the external universe, as is the case classically. However, due to the fact that the RSET can violate energy positivity conditions [113], semiclassical analyses must include a description of the interior of these objects, particularly focusing on the dynamics of the trapped region as a whole. Since making sense of the high-curvature region at the centre of these objects is typically considered to go beyond the scope of the semiclassical approach, one should focus first on the region where this approach is potentially still applicable and, as we will see, necessary to understand the full dynamics of BHs: the *inner* apparent horizon.

The geometries of generic BHs, such as those with electric charge and/or angular momentum, posses an inner horizon. Classically, this horizon has been shown to be prone to dynamical instability. The so-called *mass inflation* instability [83, 97] is triggered when, after gravitational collapse, the newly formed BH with an inner horizon receives an ingoing flux of energy from the natural decay tails of fields in its vicinity. Semiclassically, a similar instability is triggered, sourced by flux terms present in the RSET of quantum fields [14, 64, 120]. While initially the flux terms which source the classical instability are larger, their decay in time implies that the semiclassical instability may eventually dominate the dynamics. Indeed, analyses on fixed backgrounds (in the absence of backreaction) do suggest such a late-time semiclassical dominance. This fact has previously been used to argue in favour of the cosmic censorship conjecture [64, 120] due to the elimination of potential extendability of the geometry past its Cauchy horizon. However, recent works suggest that this semiclassical dominance could have a much more drastic effect on the dynamics of the BH spacetime, potentially eliminating the trapped region through an outward displacement of the inner horizon on short timescales [17, 18].

In the following we will review the main features of both the classical and semiclassical inner horizon instabilities, with the aim of comparing them and building a clear picture of the possible outcomes of the full dynamical evolution of BHs.

1.5.1.1 Classical Mass Inflation Instability

A general characteristic of static and stationary BH geometries which posses an inner horizon is the presence of a Cauchy horizon, beyond which the evolution of spacetime can no longer be determined by initial conditions from a past (partial) Cauchy surface [57, 117]. The limit in which this latter horizon is formed is a useful tool to understand the origin of the instability of the inner horizon.

To get a general grasp of the features of such a geometry, let us look at a simple example. Consider the static spherically-symmetric metric, written in advanced Eddington-Finkelstein coordinates,

$$ds^2 = -f(r)dv^2 + 2dvdr + r^2 d\Omega^2. \tag{1.43}$$

This represents a BH with an inner and outer horizon if the redshift function $f(r)$ has two zeros, one with negative and one with positive slope. We can expand $f(r)$ around the inner horizon (the negative-slope root) r_i, as

$$f(r) = -2\kappa_\mathrm{i}(r - r_\mathrm{i}) + \mathcal{O}(r - r_\mathrm{i})^2, \tag{1.44}$$

with $\kappa_\mathrm{i} > 0$. The first important feature of the inner horizon is the accumulation of geodesics which approach it in time. Particularly, if we consider outgoing null geodesics, equating the line element (1.43) to zero and using (1.44) we find that the solutions for their radial motion in time is

$$r - r_\mathrm{i} \propto e^{-\kappa_\mathrm{i} v}. \tag{1.45}$$

The second feature of note is the evolution of the affine parameter σ along these geodesic trajectories. If we take one of the geodesic equations for radial motion $(v(\sigma), r(\sigma))$,

$$\ddot{v} = -\frac{\partial_r f}{2}\dot{v}^2, \tag{1.46}$$

where the dot indicates a derivative with respect to the affine parameter σ, with the expansion (1.44) we get that the relation between the advanced time v and the affine parameter σ of geodesics which remain in the vicinity of r_i is

$$\sigma \propto e^{-\kappa_\mathrm{i} v}. \tag{1.47}$$

For outgoing null geodesics which tend toward the inner horizon asymptotically, this relation becomes exact. The affine parameter of these geodesics therefore tends to a finite value when $v \to \infty$. In the absence of a singularity in this limit, a Cauchy horizon is formed, beyond which these geodesics, and the geometry itself, are extendable.

The relation (1.47) between proper time near the Cauchy horizon and far away from the BH can be related with a blueshift amplification of perturbations [107], as can be seen schematically in Fig. 1.5. If one were to send lightlike signals into the BH at regular time intervals form the outside, then the spacing between those signal as perceived by an observer approaching the Cauchy horizon would tend to decrease to zero. Said observer would then get hit with infinitely many signals compacted in a nearly infinitesimally thin region. Even if the strength of each successive signal were to decrease, say, as an inverse polynomial in time (such that the total energy thrown into the BH is finite) [98], the exponential blueshift would still lead to a divergent energy density accumulation at the Cauchy horizon.

Aside from the headache this would likely cause the observer crossing this horizon, one might also expect that spacetime itself would react violently to such an energy accumulation. Indeed, when one considers backreaction on the geometry, one finds that the so-called *mass inflation* instability is often triggered in such conditions.

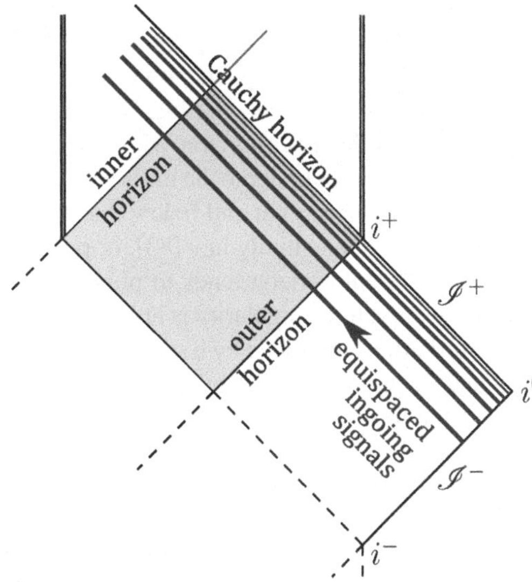

Fig. 1.5 Causal diagram of the (future extension of) a BH with an inner and Cauchy horizons (the past must be connected with a matter collapse in a physical scenario). A series of signals is sent into the BH on regular time intervals, as measured by v. Due to the discrepancy between the infinite external time v and the finite affine time it takes to cross the Cauchy horizon, the spacing between the signals, and hence their energy density, is infinitely blueshifted

While this argument gives a clear intuitive picture of the origin of mass inflation, before we describe the characteristics and consequences of this instability we must stress two important facts, which will be significant in later parts of this section. The first is that the blueshift described is a local effect produced in the vicinity of the inner horizon, and it is not dependant on the global causal structure of the spacetime being like the one depicted in Fig. 1.5. Mass inflation already has a substantial effect at finite values of the time coordinate v [75], implying that even spacetimes in which the trapped region has a finite lifetime are not exempt from it.

The second observation is the fact that not all inner horizons are prone to mass inflation. On the one hand, it is clear that the exponential blueshift argument presented above relies heavily on the surface gravity $\kappa_{\rm i}$ being non-zero. If the surface gravity were to be zero, i.e. if the expansion (1.44) started at second or higher order in $(r - r_{\rm i})$, then mass inflation would indeed be expected to not take place [39]. On the other hand, even when the surface gravity is non-zero, there are some requirements on the dynamical response of the horizon to a given perturbation which must be satisfied in order for the instability to be triggered. The shell-based model in [18] provides a clear example of this: for a generic metric of the type (1.43), the instability was only observed when the inner horizon position had an inverse-polynomial relation to the mass of the ingoing perturbations (or exponential with an exponent coefficient smaller than $\kappa_{\rm i}$). In other words, when the ingoing part of the perturbation alone leads to a displacement in time

$$\Delta r_i \gtrsim e^{-\kappa_{\rm i} v}. \qquad (1.48)$$

This is satisfied, for instance, by a Reissner-Nordström BH with an ingoing perturbation governed by the Price law [98]. Then, when one adds an (arbitrarily small) outgoing component to the perturbation inside the trapped region, the gravity-mediated exchange of mass between the ingoing and outgoing sectors, combined with the displacement of the inner horizon, triggers the non-linear amplification of curvature.

If a geometry does undergo the mass inflation process, then the Misner-Sharp mass (in spherical symmetry) at and below the initial position of the inner horizon tends to grow either exponentially in v [83], or polynomially in v [38]. As a consequence, the inner apparent horizon tends to plummet toward the origin, and the formation of a strong curvature singularity is approached. For a Reissner-Nordström geometry and perturbations modelled by a series of ingoing null spherical shells of decreasing mass and a single outgoing null shell inside the trapped region [18], the geometry takes on the form

$$ds^2 = A(v)\left[-A(v)F(v,r)dv^2 + 2dvdr\right] + r^2 d\Omega^2, \tag{1.49}$$

where $A(v) \propto e^{-\kappa_{\rm i} v}$, with $\kappa_{\rm i}$ being the initial (absolute) value of the inner horizon surface gravity, and $F(v,r)$ is the Reissner-Nordström redshift function, except with an exponentially growing mass term [18].

The inner apparent horizon of the geometry (1.49) approaches the origin as $e^{-\kappa_{\rm i} v}$, but not all outgoing null geodesics in the trapped region approach this horizon. Due to the presence of the function $A(v)$, the movement of geodesics inside the mass-inflated region is quickly frozen, and many reach the limit $v \to \infty$ while still far away from the inner horizon and the origin. There are thus two separate regions these geodesics can end up in: some do indeed approach the inner horizon and the origin, and end up falling into a strong curvature singularity, while others are frozen at a finite radius and reach a Cauchy horizon. This latter horizon is now of variable radius, extending from the initial radius of the inner horizon all the way down to the origin. Additionally, due to the exponentially growing mass in this region, this horizon is also a singular surface. However, it still constitutes a Cauchy horizon because it turns out to be only *weakly* singular, in the sense that tidal deformations across it are finite, and the metric is continuously extendable beyond it [45, 83]. Causal diagrams of two possible such outcomes are depicted in Fig. 1.6.

1.5.1.2 Semiclassical Backreaction on Static Inner Horizons

Even before the classical mass inflation process was fully understood, the possibility of semiclassical effects destabilising such configurations was also contemplated. Unsurprisingly, the blueshift effect discussed above also affects the modes which define a field quantisation. The result is that regular quantum states defined from appropriate initial conditions in the past end up developing a singular behaviour at the Cauchy horizon. This is most clearly encoded in the RSET, which tends to a divergence there.

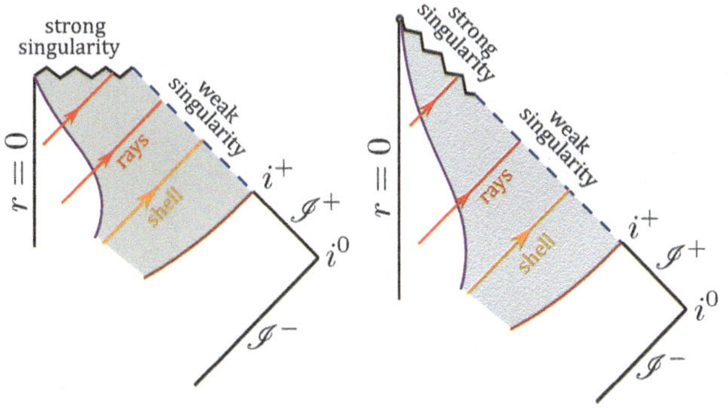

Fig. 1.6 Future part of the causal diagram of a mass inflation geometry [18]. The shaded part is the trapped region and the dashed line is the Cauchy horizon (and a weak singularity). The outgoing perturbation in the form of a shell is shown as well. Left: the inner apparent horizon reaches $r = 0$ at finite v and forms a Schwarzschild-type spacelike singularity. Right: the inner horizon only tends to $r = 0$ asymptotically in v, resulting in a strong null singularity at $v \to \infty$ and $r = 0$, at a finite affine distance for geodesics which fall into it

Following the spirit of classical arguments, the first semiclassical analysis of the problem [26] was formulated with the purpose of showing the impossibility of extending a universe through a Cauchy horizon, due to divergent values of the RSET backreacting on said horizon to create a singularity. Subsequent analyses [14, 63, 64, 119, 120] have continued in the same direction: calculating the RSET on a fixed static or stationary background (usually Reissner-Nordström or Kerr, with either flat or de Sitter asymptotics) in a physically reasonable vacuum state, and showing the divergent values of the RSET at the Cauchy horizon. The computational feat of performing such calculations in 3+1 dimensional backgrounds is tremendously impressive, and it is indeed clear by now that if a trapped region were to exist long enough for the formation of a Cauchy horizon to be approached, semiclassical backreaction would step in prevent it from happening.

However, it is worth remembering that semiclassically, trapped regions are not expected to last that long. So long as the outer horizon has a non-zero surface gravity, Hawking evaporation [59] is expected to reduce the size of the BH from the outside, making its lifetime long but finite. Additionally, much like how classical mass inflation changes the size of the trapped region at finite times, making the inner horizon move inwards, one may expect that whatever the energy content in the RSET is which tends to a divergence might also cause some significant backreaction long before a Cauchy horizon is formed. This does indeed appear to be the case, and it may be quite significant for our understanding of BH evolution.

The divergence in the RSET is contained in the $\langle T_{vv} \rangle$ flux term, where v is again the Eddington-Finkelstein ingoing null coordinate. The similarity with the classical case is apparent, since it is once again an ingoing flux which gets blueshifted infinitely. However, there are two crucial differences, as can be seen in e.g. [64, 120]. First, on a fixed background, the term $\langle T_{vv} \rangle$ becomes a constant when nearing

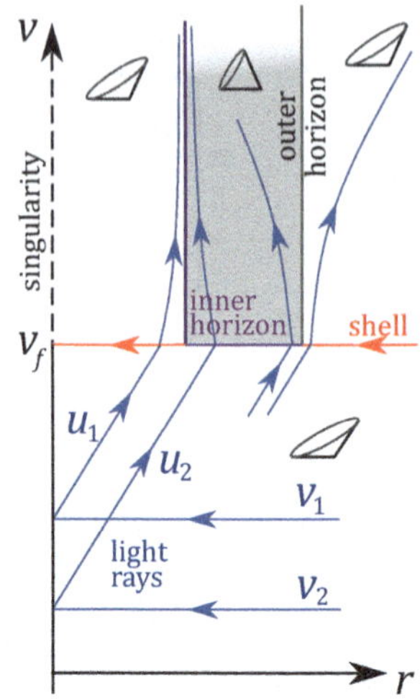

Fig. 1.7 Spacetime diagram of the formation of a BH with an outer and inner horizon from the collapse of a null shell. The coordinates are the Eddington-Finkelstein $\{v, r\}$, and the shaded part is the trapped region

the Cauchy horizon, unlike the classical perturbing flux which, tends to zero. Note that the v coordinate is singular at this horizon (1.47), making this a clear physical divergence. Second, the term seems to be *negative*. The RSET often violates energy conditions [113], so this is hardly a surprise. The important point is what this could imply for backreaction at finite times. Much like how a negative ingoing flux drives Hawking evaporation of the trapped region from the outside [46], it is quite likely that the flux at the inner horizon would lead to classically forbidden dynamics, and the instability of this horizon could exponentiate the timescale involved.

This is precisely what is found in [17]. The model considered there is quite simple: a BH with a geometry of the type (1.43) with an outer and inner horizon, formed from the collapse of a null spherical shell, as depicted in Fig. 1.7. The shell is used to simplify the initial conditions for the quantum modes in the BH region for the "in" vacuum state, and backreaction is analysed perturbatively from the point at which the inner horizon is formed. The RSET is calculated in the Polyakov approximation, and it has the expected negative ingoing flux at the inner horizon

$$\langle T_{vv} \rangle \left.\right|_{r=r_{\mathrm{i}}} = -\frac{l_{\mathrm{p}}^2 \kappa_{\mathrm{i}}^2}{192\pi^2 r_{\mathrm{i}}^2}, \tag{1.50}$$

where again κ_{i} is (the absolute value of) the surface gravity of this horizon, r_{i} is its radial position, and l_p is the Planck length.

A perturbation of the metric around the inner horizon (with a series expansion in r) sourced by the RSET leads to a modification of the geometry which tends to displace this horizon as

$$r_{\mathrm{i}} \simeq r_{\mathrm{i},0} + \frac{l_{\mathrm{p}}^2 \kappa_{\mathrm{i}}}{48\pi\lambda}(e^{\lambda v/r_{\mathrm{i},0}} - 1), \tag{1.51}$$

with $\lambda = 1 + 2\kappa_{\rm i} r_{\rm i,0}$, and assuming $v = 0$ is the moment the semiclassical evolution begins. The position of this horizon changes in an outward direction, and the dynamical tendency has an initially exponential behaviour in time. While the approximation used to obtain this estimate breaks down after a small time interval, is suffices to show that backreaction is indeed likely to be amplified by the unstable nature of this horizon. If we extrapolate from this initial tendency and assume that the exponential behaviour persists throughout the evolution of the horizon, then one sees that the lifetime of the trapped region would be much shorter than what can be expected from Hawking evaporation from the outside. Particularly, assuming that the inner horizon surface gravity is initially larger than that of the outer horizon (which is typically the case), then the time it would take the perturbation to overcome its Planck-scale suppression and for the trapped region to be eliminated by the exponential outwards motion of the inner horizon (1.51) would be

$$v_{\rm evap} \simeq \frac{1}{\kappa_{\rm i}\lambda} \log \frac{M}{l_{\rm p}} \lesssim \frac{M}{M_{\odot}} \times 10^{-5}\,{\rm s}, \tag{1.52}$$

where $M_{\odot}$ is the solar mass, and the logarithmic dependence has been omitted in the bound on the right-hand side due to the fact that for no astrophysically reasonable object would it increase the order of magnitude further.

This result, while being only an extrapolation of the initial tendency in a perturbative analysis, is already highly suggestive of the presence of an additional characteristic timescale in the semiclassical BH evolution problem which is much smaller than the notoriously large Hawking evaporation time [58]. It also suggests that if BHs do disappear in this way, they may do so in a rather explosive manner. Such a phenomenon could potentially be astrophysically observable, and should it occur, its characteristics should be studied in detail once full solutions of semiclassical evolution are available. Full solutions would of course also incorporate at least a transient period of classical mass inflation, which we will discuss in the following.

1.5.1.3 Full Semiclassical Evolution

As we have seen, there are two dynamical effects which control the evolution of a BH inner horizon. On the one hand, there is classical mass inflation [97], sourced by the inverse-polynomial decay of fields in a BH environment [98]. On the other hand, there is the possibility of a semiclassical outward inflation of the inner horizon [17], sourced by the negative ingoing flux present in the RSET [14]. While the semiclassical effect is initially suppressed by a Planck constant, unlike its classical counterpart the flux which sources it does not decay in time, making it likely that its effect would be quicker to accumulate. In fact, the flux is proportional to the surface gravity of the inner horizon, suggesting that its value may even increase as mass inflation goes on. On these grounds, it would appear that the RSET ends up becoming the dominant source of dynamics, as was already argued in the analyses of the vicinity of the Cauchy horizon [14, 64].

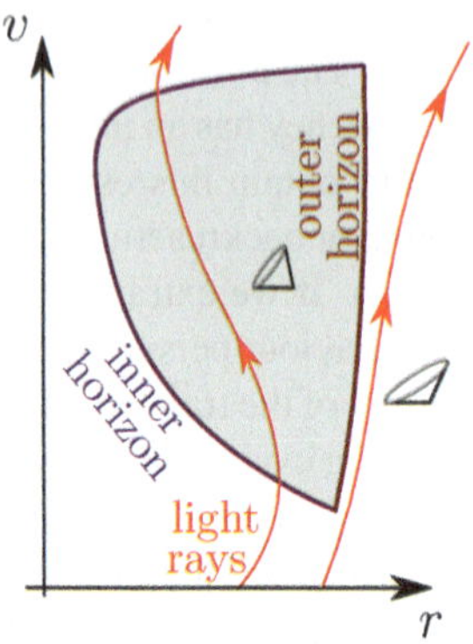

Fig. 1.8 Spacetime diagram of the extrapolated dynamics of the semiclassical BH model obtained in [18]. The inner horizon begins to move inwards due to mass inflation, but when the semiclassically sourced terms overcome the Planck scale suppression, it bounces back outwards and eventually meets the outer one

It is therefore crucial to understand what finite-time backreaction form the RSET looks like on a background undergoing mass inflation, and seeing whether and when an outwards inflation of the inner horizon takes place. A first attempt at discerning this was made in [18]. A family of particularly simple geometries of the type (1.49) which reproduce the causal properties of mass inflation was constructed (on purely geometric grounds), and backreaction from the RSET was solved for a small but finite time interval. What was found is that the motion of the inner horizon takes the form

$$r_{\mathrm{i}}(v) = \alpha e^{-\kappa_{\mathrm{i}} v} + l_{\mathrm{p}}^2 \beta e^{\kappa_{\mathrm{i}} v} + \cdots \tag{1.53}$$

with α and β positive coefficients. The first term on the right-hand side is the classical trajectory of an inner horizon in the initial stages of mass inflation, and the second term is the correction sourced by the RSET. The approximation does break down when the correction becomes of the order of the background, but the tendency for an outward bounce is quite clear. Extrapolating from this initial tendency, an enticing possibility for the full semiclassical evolution of a BH emerges. As depicted in Fig. 1.8, if the exponential tendency form the correction term is continued, then the trapped region may disappear in a rapid outward burst, on a timescale similar to the estimate (1.52).

In light of these results, the possibilities for the full semiclassical evolution of the trapped region can be divided into two categories. First, while the observed initial tendencies for an outward horizon inflation are highly suggestive, we must reiterate that as of yet there are no full semiclassical solutions describing the complete evolution of a trapped region, even in the Polyakov approximation. It is therefore still possible that classical mass inflation prevails in the end, and curvature continues to increase until the depths of the trapped region can only be consistently described in a full quantum gravity theory. In such a scenario, if the quantum gravity effects do not lead to a quick bounce themselves, then the dominant dynamics would simply be the Hawking evaporation of the outer horizon.

The second category is the one more in tune with the observed tendencies: the RSET becomes the dominant source of dynamics at the inner horizon, and this horizon begins to move outwards. In this scenario, the two apparent horizons end up meeting (near the initial position of the outer one), and either a semiclassically stable extremal BH is left, or the trapped region disappears completely. From an observational standpoint, this last case would be the most intriguing. Not only would the disappearance of the trapped region likely be accompanied by a large burst of energy, which could potentially be observed, but a general instability of trapped regions would imply that the astrophysically observed long-lived dark ultracompact objects must in fact be horizonless BH mimickers, much like the ones described earlier in this chapter.

1.5.2 Outer Horizon and the Vacuum State of Gravitational Collapse

Even with the above analysis, there are still missing pieces in the picture which connects gravitational collapse with ultracompact objects as a final configuration. First of all, it is of course still premature to claim with certainty that long-lived trapped regions are not a stable and physically realisable configuration. Then, even assuming they are not (be it due to semiclassical or full quantum gravity effects), it is not fully clear how the transition between a collapse/bounce geometry and a static horizonless object may occur. Particularly, while in the previous section it is shown that vacuum energy can sustain ultracompact configurations with a surface arbitrarily close to the would-be horizon when the configuration is already static, it is not obvious whether and when such large vacuum energies can be achieved when starting from dynamical collapse.

What is missing is an understanding of what, if any, dynamical scenarios in which classical matter collapses can produce large vacuum energies which, in turn, can potentially stop the collapse. In other words, unlike the above analysis of inner horizon dynamics in which a small value of the RSET leads to a large dynamical contribution due to the instability of the geometry itself, here we would be looking for cases in which the RSET directly acquires large enough values to overcome its Planckian suppression and affect short-term dynamics. This was the subject of the works [15, 16], and here we will briefly outline the results.

1.5.2.1 Vacuum State of Gravitational Collapse

Gravitational collapse geometries are typically constructed as an isolated system in an asymptotically flat spacetime. This not only simplifies the classical treatment of the system, but also the semiclassical one. Particularly, having an asymptotically flat region in the past (without horizons) gives an unambiguous choice for a quantum vacuum state: the Minkowski vacuum, constructed from the plane-wave mode solutions for the free field. Evolving this state from the past flat region into the future collapse geometry, one obtains the so-called "in" state, which is the standard choice in collapse scenarios.

When a trapped region forms, the "in" state quickly relaxes to the so-called Unruh state around and above the outer horizon [51]. This state contains the fluxes governing the Hawking effect, but its energy content is small enough to not produce any significant short-term backreaction in the vicinity of the outer horizon. Additionally, past analyses have established that in a standard collapse, in which matter generally moves inwards quite quickly, there are no significant vacuum energy contributions close to the formation of a trapped region even before the relaxation to the Unruh state [4, 23, 46, 86].

However, the "in" state can also have a different behaviour: if instead of forming a BH the matter content stabilises to a static configuration, then the "in" state relaxes to the so-called Boulware state [30]. This state is known for its divergence when defined on a BH spacetime, but it is perfectly well defined, and is in fact the physically reasonable choice, in stellar configurations. If the stellar object has a compactness close to that of a BH, then the vacuum energy does indeed approach a divergence, which turns out to work in favour of stabilising the configuration, as discussed in the previous chapter.

With this in mind, it becomes apparent that the the dynamical collapse scenarios most likely to produce large vacuum energies (in the "in" state) are ones in which the matter content satisfies appropriate conditions on both its position and velocity: it must be both close to forming a horizon, and close to staticity. In such regimes, one may expect that the "in" vacuum could have a behaviour somewhere in between the tendencies toward the Unruh and Boulware states, and this is indeed what is found.

1.5.2.2 Dynamical Approach to Horizon Formation: A Thin Shell Model

In [15] (followed up by [16]) we analysed a series of simplified collapse scenarios which approach the above conditions (staticity and horizon formation) in different ways. The first model we looked at was that of a thin spherical shell oscillating just above the Schwarzschild radius corresponding to its mass, paying particular attention to the part of its dynamics where it decelerates from collapse, halts and bounces back out. The second model was one in which a shell crosses its Schwarzschild radius, but does so at an arbitrarily low speed. The third was a case in which the formation of a single (marginally) trapped surface is only approached asymptotically in time. Let us briefly outline the results.

For the first two scenarios, the thin shell model simply consists of a Schwarzschild exterior and a Minkowski interior, matched together with appropriate junction conditions (which ensure the continuity of the metric) on a surface of variable radius $R(v)$ which we are free to fix. In double null coordinates the line element of this geometry is

$$ds^2 = \begin{cases} -f\,du_+ dv_+ + r^2 d\Omega^2 & \text{for } r > R(v), \\ -du_- dv_- + r^2 d\Omega^2 & \text{for } r < R(v), \end{cases} \tag{1.54}$$

where the redshift function is the standard $f = 1 - 2M/r$, with M the shell mass. The evolution of the "in" state is determined by the relations between the pairs of null coordinates at the shell surface, which we define as the functions

$$g = \frac{du_+}{du_-} = \frac{1}{\sqrt{|f|}} \sqrt{\frac{\alpha_-}{\alpha_+}}\Bigg|_{\text{shell}}, \qquad h = \frac{dv_+}{dv_-} = \frac{1}{\sqrt{|f|}} \sqrt{\frac{\alpha_+}{\alpha_-}}\Bigg|_{\text{shell}}, \tag{1.55}$$

where the parameters

$$\alpha_- \equiv \frac{dv_-}{du_-}\Bigg|_{\text{shell}}, \qquad \alpha_+ \equiv \frac{dv_+}{du_+}\Bigg|_{\text{shell}} \tag{1.56}$$

represent the velocity of the shell as seen from the inside and outside, in terms of the slope of its trajectory in the light cone ($\alpha_\pm = 0$ and $\alpha_\pm = \infty$ correspond to ingoing and outgoing light speed, respectively, while $\alpha_\pm = 1$ corresponds to staticity). The RSET is determined by the behaviour of these functions and their derivatives, as discussed in detail in [15].

The main conclusion from this analysis comes form the fact that the functions g and h both have a dependence on two distinct terms: a redshift-dependent term, and a velocity ($\alpha_\pm$) dependent term. The velocity dependence in particular is the novel result, as it turns out that it can in fact potentiate the large vacuum energy production when matter slows down just before horizon formation. As shown in Fig. 1.9, the slope of the quotient α_-/α_+, which is part of what determines the magnitude of the RSET, diverges at the point of zero velocity and zero redshift (the horizon position). In other words, if matter finds itself collapsing very slowly close to the formation of a horizon, then the vacuum energy produced is not only that given by a quasi-stationary approach toward the Boulware state, but is in fact larger due to the approach toward an extra divergence in the velocity terms.

The consequence of this for the oscillating shell is a production of large bursts of radiation whenever its trajectory passes through the region of large gradient of Fig. 1.9. If, for instance, this model were to represent a perturbed ultracompact object, then perhaps these bursts could serve to quickly dissipate the perturbation, and they may potentially be observable form a great distance. However, a complete model of dynamical BH mimickers would be necessary to establish the scale involved in a physical realisation of such a scenario.

The second model we analysed was the case in which the shell actually crosses its Schwarzschild radius, but at a low speed (with $\alpha_\pm \sim 1$). What we found is that at and around the outer horizon, just after the shell crosses it, the RSET can acquire vary large values, the magnitude of which is qualitatively given by

$$\langle T_{\mu\nu} \rangle \sim \frac{1}{(1 - \alpha_-)^4}. \tag{1.57}$$

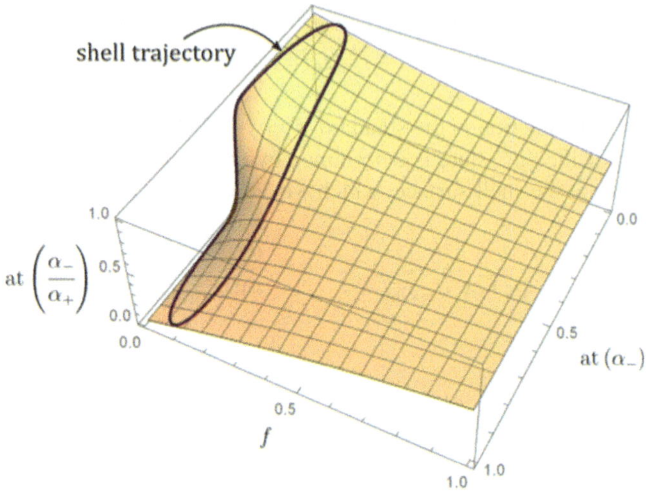

Fig. 1.9 Velocity-dependent term of the relations g and h between the null coordinates on each side of the shell. The horizontal axes are the redshift function f (as a compactification of the radial coordinate) and the velocity from the inside α_-. Both the vertical and α_- axes are rescaled by $\text{at}(x) = \frac{2}{\pi}\arctan(x)$. The closed curve shows an oscillatory shell trajectory

After this, however, the "in" state predictably relaxes to the Unruh state and the large RSET values decay to the small energy content of this state. As this decay happens exponentially, the time it takes for the BH to "thermalise" is given by

$$v_{\text{therm}} = 8M \log\left(\frac{1}{1-\alpha_-}\right). \tag{1.58}$$

The final dynamical model we looked at (and developed more thoroughly in [16]) is one in which the formation of a trapped surface is approached asymptotically in time. For this analysis we needed to go beyond the thin shell model. To summarise the results briefly, we found that in such cases the "in" state can have a hybrid behaviour between the Boulware and Unruh states: it can produce a thermal emission of radiation at infinity, but at a temperature which is lower than the surface gravity of the asymptotically formed horizon. This mismatch in temperature then leads to the approach toward a Boulware-like divergence in the vacuum energy at the asymptotic horizon. The implication this may have for horizonless ultracompact objects is the fact that they can potentially have phases of thermal Hawking-like emission, while also retaining the large vacuum polarisation which sustains them.

1.6 Summary and Conclusions

In this contribution we have discussed the potential effects that considering a physical vacuum, as opposed to the standard empty space of classical GR, could have on the dynamics of gravitational collapse and its end-point configurations. Considering the presence of quantum fields, as long as there is some spacetime curvature, even regions without classical matter should contain some average energy and pressures. This implies the presence of a new source term in the Einstein equations, which, being ultimately a function of the geometry itself, can alternatively be seen as a modification of the theory of gravity. One can picture the physical vacuum as a dielectric which can be polarised and, in extreme situations, even broken, giving rise to the creation of real particles.

Our analysis was restricted to spherically-symmetric situations, and we made use of qualitative approximations to vacuum effects based on simple analytic approximations to the Renormalised Stress-Energy Tensor (RSET) of a simple scalar field. Particularly, we used the (regularised) Polyakov approximation, and the Order-Reduced Anderson-Hiscock-Samuel approximation. We have recovered the standard Hawking result of BH evaporation of the outer horizon, and we have gone beyond it, identifying and analysing new situations which have previously not received much attention in the literature.

On the side of our analysis pertaining to static configurations, we have asked ourselves what effect a physical reactive vacuum can have on the fine balance of stellar equilibrium. After familiarising ourselves with the effects of the physical vacuum, we have searched for self-consistent regular static solutions of the stellar equilibrium equations for the simple case of constant-density stars. By self-consistent we mean that the vacuum polarisation effects generated by the geometry is the vacuum part of the source that generates this very geometry. In technical terms, in these analyses we have used the Boulware vacuum for the field as it is the only genuinely static vacuum compatible with asymptotic flatness. Our philosophy in these analyses is restricted to see what types of static configurations could exist, and not whether Nature could reach them or not and how.

We have found two families of solutions: (a) Stars with sub-Buchdahl compactness, which are just slight perturbations of the classical Schwarzschild stars; (b) Stars with super-Buchdahl compactness, which are absent in classical GR. The existence of this second group of stars is one of the main results of these analyses. Ultracompact stars (i.e. stars with compactness near that of BHs) are able to exist because vacuum polarisation generates a central core of negative energy able to support their structure against gravitational collapse (see Fig. 1.3). We have found similar ultracompact configurations through the use of our two rather different approximations; in both cases they exhibit the same qualitative features.

There is an extended belief that the dark and compact objects we know exist in the cosmos are GR BHs. This believe is based, on the one hand, on the fact that the BH model accommodates very well the observations, but also on the theoretical idea that no stellar equilibrium can exist beyond certain compactness. Here we have shown that this second idea is not as strong as typically assumed. Our analyses show that the "astrophysical BHs" that we observe might equally well be ultracompact stars than GR BHs.

On the side of our analysis pertaining to dynamical scenarios of gravitational collapse, we have analysed the effect of vacuum energy in scenarios which go beyond the standard Hawking evaporation paradigm. This paradigm rest on two essential pillars, one of which is indeed justified by quantitative estimates, while the other turns out to be on rather shaky ground, bringing the whole picture into question. The first pillar is the fact that in standard situations of stellar collapse, one finds no reason why matter should qualitatively change its behaviour when approaching horizon formation; from a local point of view matter does not even know that globally a horizon is approached and eventually formed. This is indeed the case physically, except if matter happens to linger near the point of horizon formation long enough for light to cross from one end of it to the other; then, non-local effects which relate to the global configuration could play a role, such as the potential approach of the dynamical vacuum state toward a behaviour more akin to the static Boulware state. However, as a typical collapse is assumed to be initiated from not very compact configurations (as compared to the BH limit), matter is expected to cross the horizon threshold at very high speeds. Under this high-speed condition one can show that semiclassical effects around the point of horizon formation are negligible and so are not necessary for analysing the dynamics of the collapse [15, 23, 36]. The second pillar is the assumption that after the outer horizon is formed, whatever it is that occurs in the depths of the BH is irrelevant for its short-term evolution as seen from the outside. This is an extension of the logic originating from the classical case, where energy positivity conditions are satisfied, but with the addition of the slow process of Hawking evaporation [58]; this process, however, only depends on the properties of the geometry at and above the outer horizon, and only brings about a negligible amount of contact between the interior and exterior of the trapped region due to the motion of this horizon, making it nearly compatible with the classical logic.

Our analysis brings strong evidence against the robustness of the second pillar, which in turn eventually makes the assumptions behind the first one crumble as well. First, we consider the more realistic situation in which the trapped region formed after gravitational collapse does not only have an outer horizon, but also an inner horizon. Much like how the inner horizon is unstable on a classical level (mass inflation), we find that semiclassical perturbations tend to destabilise it as well. We begin by analysing the effect of the RSET-sourced perturbation alone, finding that backreaction leads to an initially exponential tendency for this horizon to move outwards, shrinking the trapped region from the inside. If we extrapolate from this initial tendency, then instead of mass inflation we would have an inner horizon outwards inflation. This effect of course has to contend with its classical counterpart, and where the overall inner horizon instability leads to is still an open question. However, we find that due to the decreasing nature of the classical perturbation source leading to mass inflation, and the constant or increasing semiclassical source, it is likely that the semiclassical effect wins in the end. We observed this in one particular toy model which reproduces the causal properties of a classical mass inflation background, but a complete analysis of a full physical scenario is still pending. If the inner horizon does inflate outwards, then it would meet the outer one on a very short timescale (milliseconds for a Solar mass object), much shorter than the Hawking evaporation time. The analysis of what may come after such an evolution enters the realm of speculation, though given the limited number of possibilities, we can make some educated guesses. For instance, one could form an extremal BH, or end up eliminating the two horizons altogether. As extremal BHs are known to be unstable classically [5], it is tempting to consider this latter possibility as the most likely outcome, though the possibility of semiclassical effects recuperating the stability of extremal objects is not out of the question.

If the trapped region does disappear completely, if all this occurs before the interior has reached Planckian densities (ensuring the validity of the semiclassical approach), then the previously collapsing matter would be freed form its causal restriction and likely begin expanding outwards following the inner horizon.[5] Part of the energy of the initial configuration would be dissipated to infinity after such a process, partly through particle production from the vacuum, but also likely due to processes within the effectively classical matter. The leftover matter would then be left in a condition to undergo collapse for a second time. If the trapped region instability on short timescales is indeed generic, then this process would repeat until at some point, potentially after a number of cycles of bouncing, dissipation and recollapse, the situation will change in the sense that the hypothesis than matter approaches horizon formation while moving rapidly will not longer be justifiable. Thus, the first pillar supporting the Hawking paradigm will also come into question, opening the door to a myriad of different possible dynamical continuations of the process. Here, our analysis of low-speed approaches toward horizon formation can provide some insight. Particularly, we have shown that large vacuum energy is generated by such dynamics, larger in fact than what might be expected from a mere transition in behaviour of the "in" vacuum towards a Boulware-like behaviour. We have seen that the slow but present dynamics potentiate this growth of energy even further, making it more likely that semiclassical effects result in considerable deviations from the classical collapse dynamics. Additionally, we have also shown that cases in which the formation of a horizon is only approached, or even reversed periodically, can lead to the emission of Hawking-like radiation at infinity, leading to further dissipation until a stable endpoint is reached.

These results point towards the possibility that the final result of such dynamics would likely be an approach toward staticity, in which the "in" vacuum would end up with a full Boulware behaviour. In such a scenario, classical and semiclassical matter and energy sources would end up in one of the configurations of hydrostatic equilibrium analysed above. The analyses of near-horizon dynamics also point to the potential behaviour of the vacuum when such objects undergo perturbations: they can emit both a nearly thermal spectrum of Hawking-like radiation, and bursts

[5] Let us also mention that a bounce of the collapsing matter has also been proposed previously on different grounds, with a similar overall effect on the dynamics, though also including a transient anti-trapped region [20–22].

with much larger amplitudes, the latter being potentially observable if the object is sufficiently compact.

Overall, these results point towards the existence of an alternative theoretical paradigm for the ultimate fate of gravitational collapse. They also serve as a reminder of the fact that the standard paradigm is also to very large extent theoretical, relying on a series of not-so-obvious assumptions. It is therefore clearly necessary to continue working on the phenomenology of all these possibilities, preparing the grounds for their eventual verification or falsification through observation. From the perspective advocated in this contribution, GR BHs may just be idealised models of much more complicated stellar-like bodies.

Acknowledgements The authors would like to thank Raúl Carballo-Rubio and Luis J. Garay for enlightening discussions and for all their work in our multiple collaborations, without which this contribution would not have been possible. Financial support was provided by the Spanish Government through the projects PID2020-118159GB-C43, PID2020-118159GB-C44 (with FEDER contribution), and by the Junta de Andalucía through the project FQM219. J.A. and C.B. acknowledge financial support from the grant CEX2021-001131-S funded by MCIN/AEI/ 10.13039/501100011033. V.B. acknowledges financial support provided under the European Union's H2020 ERC Advanced Grant "Black holes: gravitational engines of discovery" grant agreement no. Gravitas-101052587.

References

1. P. Anderson, Effects of quantum fields on singularities and particle horizons in the early universe. Phys. Rev. D **28**, 271–285 (1983). https://doi.org/10.1103/PhysRevD.28.271
2. P.R. Anderson, $\langle\varphi^2\rangle$ for massive fields in schwarzschild spacetime. Phys. Rev. D **39**, 3785–3788 (1989). https://doi.org/10.1103/PhysRevD.39.3785
3. P.R. Anderson, W.A. Hiscock, D.A. Samuel, Stress-energy tensor of quantized scalar fields in static spherically symmetric spacetimes. Phys. Rev. D **51**, 4337–4358 (1995). https://doi.org/10.1103/PhysRevD.51.4337
4. B. Arderucio-Costa, W. Unruh, Model for quantum effects in stellar collapse. Phys. Rev. D **97**(2), 024,005 (2018). https://doi.org/10.1103/PhysRevD.97.024005
5. S. Aretakis, Horizon instability of extremal black holes. Adv. Theor. Math. Phys. **19**, 507–530 (2015). https://doi.org/10.4310/ATMP.2015.v19.n3.a1
6. J. Arrechea, Hydrostatic equilibrium in the semiclassical approximation. Ph.D. thesis, Granada (2023)
7. J. Arrechea, C. Barceló, R. Carballo-Rubio, L.J. Garay, Schwarzschild geometry counterpart in semiclassical gravity. Phys. Rev. D **101**, 064,059 (2020). https://doi.org/10.1103/PhysRevD.101.064059
8. J. Arrechea, C. Barceló, R. Carballo-Rubio, L.J. Garay, Reissner–Nordström geometry counterpart in semiclassical gravity. Class. Quant. Grav. **38**(11), 115,014 (2021). https://doi.org/10.1088/1361-6382/abf628
9. J. Arrechea, C. Barceló, R. Carballo-Rubio, L.J. Garay, Semiclassical constant-density spheres in a regularized Polyakov approximation. Phys. Rev. D **104**(8), 084,071 (2021). https://doi.org/10.1103/PhysRevD.104.084071
10. J. Arrechea, C. Barceló, R. Carballo-Rubio, L.J. Garay, Semiclassical relativistic stars. Sci. Rep. **12**(1), 15,958 (2022). https://doi.org/10.1038/s41598-022-19836-8
11. J. Arrechea, C. Barceló, R. Carballo-Rubio, L.J. Garay, Asymptotically flat vacuum solutions in order-reduced semiclassical gravity. Phys. Rev. D **107**, 085,005 (2023). https://doi.org/10.1103/PhysRevD.107.085005

12. J. Arrechea, C. Barceló, R. Carballo-Rubio, L.J. Garay, Vacuum semiclassical gravity does not leave space for safe singularities. Universe **7**(8) (2021). https://doi.org/10.3390/universe7080281. https://www.mdpi.com/2218-1997/7/8/281
13. R. Balbinot, A. Fabbri, V.P. Frolov, P. Nicolini, P. Sutton, A. Zelnikov, Vacuum polarization in the Schwarzschild space-time and dimensional reduction. Phys. Rev. D **63**, 084,029 (2001). https://doi.org/10.1103/PhysRevD.63.084029
14. R. Balbinot, E. Poisson, Mass inflation: the semiclassical regime. Phys. Rev. Lett. **70**, 13–16 (1993). https://doi.org/10.1103/PhysRevLett.70.13
15. C. Barceló, V. Boyanov, R. Carballo-Rubio, L.J. Garay, Semiclassical gravity effects near horizon formation. Class. Quant. Grav. **36**(16), 165,004 (2019). https://doi.org/10.1088/1361-6382/ab2e43
16. C. Barceló, V. Boyanov, R. Carballo-Rubio, L.J. Garay, Asymptotic horizon formation, space-time stretching and causality. Phys. Rev. D **102**(4), 045,001 (2020). https://doi.org/10.1103/PhysRevD.102.045001
17. C. Barceló, V. Boyanov, R. Carballo-Rubio, L.J. Garay, Black hole inner horizon evaporation in semiclassical gravity. Class. Quant. Grav. **38**(12), 125,003 (2021). https://doi.org/10.1088/1361-6382/abf89c
18. C. Barceló, V. Boyanov, R. Carballo-Rubio, L.J. Garay, Classical mass inflation versus semi-classical inner horizon inflation. Phys. Rev. D **106**(12), 124,006 (2022). https://doi.org/10.1103/PhysRevD.106.124006
19. C. Barceló, R. Carballo-Rubio, L.J. Garay, Two formalisms, one renormalized stress-energy tensor. Phys. Rev. D **85**, 084,001 (2012). https://doi.org/10.1103/PhysRevD.85.084001
20. C. Barceló, R. Carballo-Rubio, L.J. Garay, Where does the physics of extreme gravitational collapse reside? Universe **2**(2), 7 (2016). https://doi.org/10.3390/universe2020007
21. C. Barceló, R. Carballo-Rubio, L.J. Garay, G. Jannes, The lifetime problem of evaporating black holes: mutiny or resignation. Class. Quant. Grav. **32**(3), 035,012 (2015). https://doi.org/10.1088/0264-9381/32/3/035012
22. C. Barceló, L.J. Garay, G. Jannes, Quantum non-gravity and stellar collapse. Found. Phys. **41**, 1532–1541 (2011). https://doi.org/10.1007/s10701-011-9577-9
23. C. Barceló, S. Liberati, S. Sonego, M. Visser, Fate of gravitational collapse in semiclassical gravity. Phys. Rev. D **77**(4) (2008). https://doi.org/10.1103/physrevd.77.044032
24. V.A. Belinsky, I.M. Khalatnikov, E.M. Lifshitz, Oscillatory approach to a singular point in the relativistic cosmology. Adv. Phys. **19**, 525–573 (1970). https://doi.org/10.1080/00018737000101171
25. C. Berthiere, D. Sarkar, S.N. Solodukhin, The fate of black hole horizons in semiclassical gravity. Phys. Lett. B **786**, 21–27 (2018). https://doi.org/10.1016/j.physletb.2018.09.027
26. N.D. Birrell, P.C.W. Davies, On falling through a black hole into another universe. Nature **272**, 35 (1978). https://doi.org/10.1038/272035a0
27. N.D. Birrell, P.C.W. Davies, Quantum Fields in Curved Space. Cambridge Monographs on Mathematical Physics. Cambridge University Press, Cambridge, UK (1984). https://doi.org/10.1017/CBO9780511622632
28. C.G. Böhmer, E. Jensko, Modified gravity: A unified approach. Phys. Rev. D **104**, 024,010 (2021). https://doi.org/10.1103/PhysRevD.104.024010
29. C.G. Bollini, J.J. Giambiagi, Dimensional renormalization: the number of dimensions as a regularizing parameter. Nuovo Cim. B **12**, 20–26 (1972). https://doi.org/10.1007/BF02895558
30. D.G. Boulware, Quantum field theory in schwarzschild and rindler spaces. Phys. Rev. D **11**, 1404–1423 (1975). https://doi.org/10.1103/PhysRevD.11.1404
31. C. Breen, A.C. Ottewill, Hadamard renormalization of the stress energy tensor in a spherically symmetric black hole space-time with an application to lukewarm black holes. Phys. Rev. D **85**, 084,029 (2012). https://doi.org/10.1103/PhysRevD.85.084029
32. H. Brown, *Physical Relativity: Space-time structure from a dynamical perspective* (Clarendon Press, Oxford Scholarship Online. Philosophy module, 2005)
33. M.R. Brown, A.C. Ottewill, Effective actions and conformal transformations. Phys. Rev. D **31**, 2514–2520 (1985). https://doi.org/10.1103/PhysRevD.31.2514

34. H.A. Buchdahl, General relativistic fluid spheres. Phys. Rev. **116**, 1027–1034 (1959). https://doi.org/10.1103/PhysRev.116.1027
35. P. Candelas, Vacuum polarization in schwarzschild spacetime. Phys. Rev. D **21**, 2185–2202 (1980). https://doi.org/10.1103/PhysRevD.21.2185
36. P. Candelas, K.W. Howard, VACUUM (PHI**2) IN SCHWARZSCHILD SPACE-TIME. Phys. Rev. D **29**, 1618–1625 (1984). https://doi.org/10.1103/PhysRevD.29.1618
37. R. Carballo-Rubio, Stellar equilibrium in semiclassical gravity. Phys. Rev. Lett. **120**(6) (2018). https://doi.org/10.1103/physrevlett.120.061102
38. R. Carballo-Rubio, F. Di Filippo, S. Liberati, C. Pacilio, M. Visser, Inner horizon instability and the unstable cores of regular black holes. JHEP **05**, 132 (2021). https://doi.org/10.1007/JHEP05(2021)132
39. R. Carballo-Rubio, F. Di Filippo, S. Liberati, C. Pacilio, M. Visser, Regular black holes without mass inflation instability. JHEP **09**, 118 (2022). https://doi.org/10.1007/JHEP09(2022)118
40. R. Carballo-Rubio, F. Di Filippo, S. Liberati, M. Visser, Phenomenological aspects of black holes beyond general relativity. Phys. Rev. **D98**(12), 124,009 (2018). https://doi.org/10.1103/PhysRevD.98.124009
41. V. Cardoso, P. Pani, Testing the nature of dark compact objects: a status report. Living Rev. Rel. **22**(1), 4 (2019). https://doi.org/10.1007/s41114-019-0020-4
42. S. Chakraborty, S. Singh, T. Padmanabhan, A quantum peek inside the black hole event horizon. JHEP **06**, 192 (2015). https://doi.org/10.1007/JHEP06(2015)192
43. S.M. Christensen, Vacuum expectation value of the stress tensor in an arbitrary curved background: the covariant point-separation method. Phys. Rev. D **14**, 2490–2501 (1976). https://doi.org/10.1103/PhysRevD.14.2490
44. S.M. Christensen, Regularization, renormalization, and covariant geodesic point separation. Phys. Rev. D **17**, 946–963 (1978). https://doi.org/10.1103/PhysRevD.17.946
45. M. Dafermos, J. Luk, The interior of dynamical vacuum black holes I: The C^0-stability of the Kerr Cauchy horizon (2017)
46. P. Davies, S. Fulling, W. Unruh, Energy-momentum tensor near an evaporating black hole. Phys. Rev. D **13**(10), 2720 (1976). https://doi.org/10.1103/PhysRevD.13.2720
47. P.C.W. Davies, S.A. Fulling, Quantum vacuum energy in two dimensional space-times. Proc. Roy. Soc. Lond. Ser. A, Math. Phys. Sci. **354**(1676), 59–77 (1977)
48. J. Eisenstaedt, A. Sangalli, The Curious History of Relativity: How Einstein's Theory of Gravity Was Lost and Found Again. Princeton University Press (2006)
49. A. Fabbri, S. Farese, J. Navarro-Salas, Generalized Virasoro anomaly and stress tensor for dilaton coupled theories. Phys. Lett. B **574**, 309–318 (2003). https://doi.org/10.1016/j.physletb.2003.09.012
50. A. Fabbri, S. Farese, J. Navarro-Salas, G.J. Olmo, H. Sanchis-Alepuz, Semiclassical zero-temperature corrections to schwarzschild spacetime and holography. Phys. Rev. D **73**(10) (2006). https://doi.org/10.1103/physrevd.73.104023
51. A. Fabbri, J. Navarro-Salas, Modeling Black Hole Evaporation. Imperial College Press (2005)
52. E.E. Flanagan, R.M. Wald, Does back reaction enforce the averaged null energy condition in semiclassical gravity? Phys. Rev. D **54**, 6233–6283 (1996). https://doi.org/10.1103/PhysRevD.54.6233
53. L.H. Ford, Cosmological particle production: a review. Rept. Prog. Phys. **84**(11) (2021). DOI https://doi.org/10.1088/1361-6633/ac1b23
54. V.P. Frolov, A.I. Zel'nikov, Killing approximation for vacuum and thermal stress-energy tensor in static space-times. Phys. Rev. D **35**, 3031–3044 (1987). https://doi.org/10.1103/PhysRevD.35.3031
55. S.A. Fulling, Nonuniqueness of canonical field quantization in Riemannian space-time. Phys. Rev. D **7**, 2850–2862 (1973). https://doi.org/10.1103/PhysRevD.7.2850
56. B. Guo, S.D. Mathur, Are there echoes of gravitational waves? Int. J. Mod. Phys. D **31**(14), 2242,009 (2022). https://doi.org/10.1142/S0218271822420093
57. S. Hawking, G. Ellis, *The Large Scale Structure of Space-Time* (Cambridge University Press, Cambridge Monographs on Mathematical Physics, 1973)

58. S.W. Hawking, Black hole explosions? Nature **248**(5443), 30–31 (1974). https://doi.org/10.1038/248030a0
59. S.W. Hawking, Particle Creation by Black Holes. Commun. Math. Phys. **43**, 199–220 (1975). https://doi.org/10.1007/BF02345020. [Erratum: Commun. Math. Phys. 46, 206 (1976)]
60. S.W. Hawking, Zeta function regularization of path integrals in curved space-time. Commun. Math. Phys. **55**, 133 (1977). https://doi.org/10.1007/BF01626516
61. P.M. Ho, Y. Matsuo, Static black holes with back reaction from vacuum energy. Class. Quant. Grav. **35**(6), 065,012 (2018). https://doi.org/10.1088/1361-6382/aaac8f
62. D. Hochberg, A. Popov, S.V. Sushkov, Self-consistent wormhole solutions of semiclassical gravity. Phys. Rev. Lett. **78**(11), 2050–2053 (1997). https://doi.org/10.1103/physrevlett.78.2050
63. S. Hollands, C. Klein, J. Zahn, Quantum stress tensor at the Cauchy horizon of the Reissner–Nordström–de Sitter spacetime. Phys. Rev. D **102**(8), 085,004 (2020). https://doi.org/10.1103/PhysRevD.102.085004
64. S. Hollands, R.M. Wald, J. Zahn, Quantum Instability of the Cauchy Horizon in Reissner-Nordström-deSitter Spacetime. Class. Quant. Grav. **37**(11), 115,009 (2020). https://doi.org/10.1088/1361-6382/ab8052
65. G.T. Horowitz, Semiclassical relativity: The weak-field limit. Phys. Rev. D **21**, 1445–1461 (1980). https://doi.org/10.1103/PhysRevD.21.1445
66. K.W. Howard, Vacuum < T (MU) (NU) > in Schwarzschild space-time. Phys. Rev. D **30**, 2532–2547 (1984). https://doi.org/10.1103/PhysRevD.30.2532
67. K.W. Howard, P. Candelas, Quantum stress tensor in Schwarzschild space-time. Phys. Rev. Lett. **53**, 403–406 (1984). https://doi.org/10.1103/PhysRevLett.53.403
68. B.L.B. Hu, E. Verdaguer, Semiclassical and Stochastic Gravity: Quantum Field Effects on Curved Spacetime. Cambridge Monographs on Mathematical Physics. (Cambridge University Press, 2020). https://doi.org/10.1017/9780511667497
69. R.A., Konoplya, C. Posada, Z. Stuchlík, A. Zhidenko, Stable Schwarzschild stars as black-hole mimickers. Phys. Rev. D **100**(4), 044,027 (2019). https://doi.org/10.1103/PhysRevD.100.044027
70. W.E. Lamb, R.C. Retherford, Fine structure of the hydrogen atom by a microwave method. Phys. Rev. **72**, 241–243 (1947). https://doi.org/10.1103/PhysRev.72.241
71. A. Levi, Renormalized stress-energy tensor for stationary black holes. Phys. Rev. D **95**(2), 025,007 (2017). https://doi.org/10.1103/PhysRevD.95.025007
72. A. Levi, E. Eilon, A. Ori, M. van de Meent, Renormalized stress-energy tensor of an evaporating spinning black hole. Phys. Rev. Lett. **118**(14), 141,102 (2017). https://doi.org/10.1103/PhysRevLett.118.141102
73. A. Levi, A. Ori, Pragmatic mode-sum regularization method for semiclassical black-hole spacetimes. Phys. Rev. D **91**, 104,028 (2015). https://doi.org/10.1103/PhysRevD.91.104028
74. D.J. Loranz, W.A. Hiscock, P.R. Anderson, Thermal divergences on the event horizons of two-dimensional black holes. Phys. Rev. D **52**, 4554–4558 (1995). https://doi.org/10.1103/PhysRevD.52.4554
75. D. Marolf, A. Ori, Outgoing gravitational shock-wave at the inner horizon: The late-time limit of black hole interiors. Phys. Rev. D **86**, 124,026 (2012). https://doi.org/10.1103/PhysRevD.86.124026
76. J. Martin, Everything you always wanted to know about the cosmological constant problem (but were afraid to ask). C R Phys. **13**, 566–665 (2012). https://doi.org/10.1016/j.crhy.2012.04.008
77. C.W. Misner, D.H. Sharp, Relativistic equations for adiabatic, spherically symmetric gravitational collapse. Phys. Rev. **136**, B571–B576 (1964). https://doi.org/10.1103/PhysRev.136.B571
78. T. Morley, P. Taylor, E. Winstanley, Quantum field theory on global anti-de Sitter space-time with Robin boundary conditions. Class. Quant. Grav. **38**(3), 035,009 (2021). https://doi.org/10.1088/1361-6382/aba58a

79. T. Morley, P. Taylor, E. Winstanley, Vacuum polarization on topological black holes with Robin boundary conditions. Phys. Rev. D **103**(4), 045,007 (2021). https://doi.org/10.1103/PhysRevD.103.045007
80. V.F. Mukhanov, G.V. Chibisov, Quantum fluctuations and a nonsingular universe. JETP Lett. **33**, 532–535 (1981)
81. H. Muller, A. Peters, S. Chu, A precision measurement of the gravitational redshift by the interference of matter waves. Nature **463**, 926–929 (2010). https://doi.org/10.1038/nature08776
82. G.J. Olmo, *Introduction to Modified Gravity: From the Cosmic Speedup Problem to Quantum Gravity Phenomenology* (InTech Publishing, 2011). https://doi.org/10.5772/1838
83. A. Ori, Inner structure of a charged black hole: An exact mass-inflation solution. Phys. Rev. Lett. **67**, 789–792 (1991). https://doi.org/10.1103/PhysRevLett.67.789
84. D.N. Page, Thermal stress tensors in static einstein spaces. Phys. Rev. D **25**, 1499–1509 (1982). https://doi.org/10.1103/PhysRevD.25.1499
85. D.N. Page, Hawking radiation and black hole thermodynamics. New J. Phys. **7**, 203 (2005). https://doi.org/10.1088/1367-2630/7/1/203
86. R. Parentani, T. Piran, The Internal geometry of an evaporating black hole. Phys. Rev. Lett. **73**, 2805–2808 (1994). https://doi.org/10.1103/PhysRevLett.73.2805
87. L. Parker, Particle creation in expanding universes. Phys. Rev. Lett. **21**, 562–564 (1968). https://doi.org/10.1103/PhysRevLett.21.562
88. L. Parker, J.Z. Simon, Einstein equation with quantum corrections reduced to second order. Phys. Rev. D **47**(4), 1339–1355 (1993). https://doi.org/10.1103/physrevd.47.1339
89. L. Parker, D. Toms, *Quantum Field Theory in Curved Spacetime: Quantized Fields and Gravity* Cambridge Monographs on Mathematical Physics. (Cambridge University Press, 2009)
90. W. Pauli, F. Villars, On the invariant regularization in relativistic quantum theory. Rev. Mod. Phys. **21**, 434–444 (1949). https://doi.org/10.1103/RevModPhys.21.434
91. R. Penrose, Gravitational collapse and space-time singularities. Phys. Rev. Lett. **14**, 57–59 (1965). https://doi.org/10.1103/PhysRevLett.14.57
92. R. Penrose, Gravitational collapse and space-time singularities. Phys. Rev. Lett. **14**, 57–59 (1965). https://doi.org/10.1103/PhysRevLett.14.57
93. R. Penrose, On gravity's role in quantum state reduction. Gen. Rel. Grav. **28**, 581–600 (1996). https://doi.org/10.1007/BF02105068
94. R. Penrose, On the gravitization of quantum mechanics 1: quantum state reduction. Found. Phys. **44**, 557–575 (2014). https://doi.org/10.1007/s10701-013-9770-0
95. M.E. Peskin, D.V. Schroeder, *An Introduction to Quantum Field Theory* (Addison-Wesley, Reading, USA, 1995)
96. A.Y. Petrov, *Introduction to Modified Gravity* Springer Briefs in Physics. (Springer, 2020). https://doi.org/10.1007/978-3-030-52862-1
97. E. Poisson, W. Israel, Inner-horizon instability and mass inflation in black holes. Phys. Rev. Lett. **63**, 1663–1666 (1989). https://doi.org/10.1103/PhysRevLett.63.1663
98. R.H. Price, Nonspherical perturbations of relativistic gravitational collapse. 1. Scalar and gravitational perturbations. Phys. Rev. D **5**, 2419–2438 (1972). https://doi.org/10.1103/PhysRevD.5.2419
99. H.S. Reall, N. Tanahashi, B. Way, Shock formation in lovelock theories. Phys. Rev. D **91**(4), 044,013 (2015). https://doi.org/10.1103/PhysRevD.91.044013
100. I.A., Reyes, G.M. Tomaselli, Compact stars in Quantum Field Theory (2023)
101. F. Rohrlich, Classical selfforce. Phys. Rev. D **60**, 084,017 (1999). https://doi.org/10.1103/PhysRevD.60.084017
102. F. Rohrlich, The self-force and radiation reaction. Am. J. Phys. **68**(12), 1109–1112 (2000). https://doi.org/10.1119/1.1286430
103. K. Schwarzschild, On the gravitational field of a mass point according to Einstein's theory. Sitzungsber. Preuss. Akad. Wiss. Berlin (Math. Phys.) **1916**, 189–196 (1916)
104. J.S. Schwinger, On gauge invariance and vacuum polarization. Phys. Rev. **82**, 664–679 (1951). https://doi.org/10.1103/PhysRev.82.664

105. J.M.M. Senovilla, D. Garfinkle, The 1965 Penrose singularity theorem. Class. Quant. Grav. **32**(12), 124,008 (2015). https://doi.org/10.1088/0264-9381/32/12/124008
106. J.Z. Simon, The Stability of flat space, semiclassical gravity, and higher derivatives. Phys. Rev. D **43**, 3308–3316 (1991). https://doi.org/10.1103/PhysRevD.43.3308
107. M. Simpson, R. Penrose, Internal instability in a Reissner-Nordstrom black hole. Int. J. Theor. Phys. **7**, 183–197 (1973). https://doi.org/10.1007/BF00792069
108. A.A. Starobinsky, Spectrum of relict gravitational radiation and the early state of the universe. JETP Lett. **30**, 682–685 (1979)
109. P. Taylor, C. Breen, Mode-sum prescription for the vacuum polarization in odd dimensions. Phys. Rev. D **94**(12), 125,024 (2016). https://doi.org/10.1103/PhysRevD.94.125024
110. P. Taylor, C. Breen, Mode-sum prescription for vacuum polarization in black hole spacetimes in even dimensions. Phys. Rev. D **96**(10), 105,020 (2017). https://doi.org/10.1103/PhysRevD.96.105020
111. P. Taylor, C. Breen, A. Ottewill, Mode-sum prescription for the renormalized stress energy tensor on black hole spacetimes. Phys. Rev. D **106**(6), 065,023 (2022). https://doi.org/10.1103/PhysRevD.106.065023
112. A. Urbano, H. Veermäe, On gravitational echoes from ultracompact exotic stars. JCAP **04**, 011 (2019). https://doi.org/10.1088/1475-7516/2019/04/011
113. M. Visser, Gravitational vacuum polarization. ii. energy conditions in the boulware vacuum. Phys. Rev. D **54**(8), 5116–5122 (1996). https://doi.org/10.1103/physrevd.54.5116
114. R. Wald, *Quantum Field Theory in Curved Spacetime and Black Hole Thermodynamics*. Chicago Lectures in Physics. (University of Chicago Press, 1994)
115. R.M. Wald, The back reaction effect in particle creation in curved space-time. Commun. Math. Phys. **54**, 1–19 (1977). https://doi.org/10.1007/BF01609833
116. R.M. Wald, Trace anomaly of a conformally invariant quantum field in curved space-time. Phys. Rev. D **17**, 1477–1484 (1978). https://doi.org/10.1103/PhysRevD.17.1477
117. R.M. Wald, *General Relativity*. (Chicago University Press, 1984). https://doi.org/10.7208/chicago/9780226870373.001.0001 Chicago University Press, 1984). https://doi.org/10.7208/chicago/9780226870373.001.0001
118. C.M. Will, The confrontation between general relativity and experiment. Living Rev. Rel. **17**, 4 (2014). https://doi.org/10.12942/lrr-2014-4
119. N. Zilberman, M. Casals, A. Ori, A.C. Ottewill, Quantum fluxes at the inner horizon of a spinning black hole (2022)
120. N. Zilberman, A. Levi, A. Ori, Quantum fluxes at the inner horizon of a spherical charged black hole. Phys. Rev. Lett. **124**(17), 171,302 (2020). https://doi.org/10.1103/PhysRevLett.124.171302

Chapter 2
Quantum Matter Core of Black Holes (and Quantum Hair)

Roberto Casadio and Octavian Micu

2.1 Introduction

Quantum gravity is generically expected to eliminate the singularities predicted by general relativity [50] and several approaches to quantum gravity [55] have been proposed in this framework. It is suggestive to recall that the Planck scale corresponds to a black hole of mass M whose horizon radius $R_{\rm H} = 2\, G_{\rm N}\, M$ is of the order of its Compton length $\lambda_M = \hbar/M$.[1] This simple observation, for example, lies at the heart of generalised uncertainty principles [9] which modify the uncertainty relations of quantum mechanics in order to include gravity in scattering processes associated with measurements (like in the famous Heisenberg microscope [51]). The idea that gravity involves a minimum (fundamental) length $\ell_{\rm p}$ was also viewed as a possible cure for the ultraviolet divergences of quantum field theory [64], and regained notoriety with the increasing interest in quantum gravity and trans-Planckian effects [52]. Indeed, many candidates for quantum gravity exhibit a minimum length, from string theory to loop quantum gravity.

[1] We shall always use units with $c = 1$ and often write the Planck constant $\hbar = \ell_{\rm p}\, m_{\rm p}$ and the Newton constant $G_{\rm N} = \ell_{\rm p}/m_{\rm p}$, where $\ell_{\rm p}$ and $m_{\rm p}$ are the Planck length and mass, respectively.

Invited chapter for the edited book *New Frontiers in Gravitational Collapse and Spacetime Singularities* (Eds. P. Joshi and D. Malafarina, Springer Singapore, expected in 2023).

R. Casadio (✉)
Department of Physics and Astronomy "A. Righi", Bologna University, via Irnerio 46, 40126 Bologna, Italy
e-mail: casadio@bo.infn.it

O. Micu
Institute of Space Science, Subsidiary of INFLPR, Magurele, Romania
e-mail: octavian.micu@spacescience.ro

D. Malafarina and P. S. Joshi (eds.), *New Frontiers in Gravitational Collapse and Spacetime Singularities*, Springer Series in Astrophysics and Cosmology,
https://doi.org/10.1007/978-981-97-1172-7_2

It is consistent with the above picture that a Planck scale minimum length arises from the Feynman path integral that generates in-out amplitudes including gravity [61], hence in S-matrix elements via the Lehmann-Symanzik-Zimmermann formula. However, these amplitudes are acausal and complex since they are subjected to Feynman boundary conditions. A proper minimum spacetime length should instead be real to arbitrary loop orders as pertains to an expectation value. We will first clarify this point in the effective field theory on a fixed background geometry, with a general argument that support the existence only of a Planckian screening length in elementary scattering processes as a consequence of the gravitational dynamics described by the Einstein-Hilbert action [16, 24].

The next question is then whether such a dynamical scale remains of the Planck size in more complex systems, like very compact astrophysical objects eventually collapsing to form black holes. The nonlinearity of general relativity in fact does not guarantee that the above result carries on unaltered when one deals with a huge number of massive particles in a bound state. A hint of the role played by nonlinearity can be obtained from the bootstrapped Newtonian gravity [27, 28], which is defined by adding self-interaction terms to the Newtonian Lagrangian density for spherically symmetric compact objects. The main consequence of this approach can be characterised by modified relations between the mass M and radius R of stable configurations, and indeed the lack of a Buchdahl limit [10] (which requires $R > 9\,G_{\rm N}\,M/4$). A fully quantum version of it can also be considered [23] and first evidence for a maximum compactness for black holes is obtained [26].

We will next move on to full general relativity and study the Oppenheimer-Snyder model [60] of a collapsing ball of dust. Since the canonical quantisation of a large number of (dust) particles as (fundamental) field excitations is beyond our capability, we will quantise this model in analogy with the quantum mechanical treatment of the hydrogen atom. In particular, we shall retain the ball radius R as the only observable and show that it admits a (discrete) spectrum of bound states [14]. Furthermore, as a consequence of the nonlinearity of the governing (Einstein) equations, the ground state will be shown to have a surface area proportional to M^2, so that the mass is quantised according to Bekenstein's area law. This implies that no singularity ever appears but collapsing matter forms a quantum core of macroscopic size. Of course, the huge simplification in the model will reflect in a similarly large (configuration) entropy for such cores [32].

Having established that the collapse does not end in a singularity due to purely quantum effects, the geometry generated by the core in the ground state can be described by means of a coherent quantum state [15]. Regularity of the geometric quantum state is guaranteed by the finite size of the core, which removes all ultraviolet divergences and causes deviations from the Schwarzschild geometry. Since these deviations are regulated by the size of the core, one can clearly conceive that their presence modifies the thermodynamics and Hawking radiation depending on the matter state actually present inside the horizon [11]. Moreover, we will argue how a similar description can avoid the emergence of inner Cauchy horizons, both in spherical symmetry and for rotating systems [22], whose presence casts serious doubts about the physical relevance of classical models of regular black holes.

2.2 Minimum Length Scale in the Effective Field Theory of Gravity

In the effective field theory of gravity, the metric tensor defining the line element $\mathrm{d}s^2 = g_{\mu\nu}\,\mathrm{d}x^\mu\,\mathrm{d}x^\nu$ is promoted to an operator, but one assumes the existence of a classical configuration for matter sources and the corresponding spacetime metric $\bar{g}_{\mu\nu}$. Only "small" perturbations, below a certain (energy) threshold, around the chosen classical solution are allowed and described as excitations of quantum fields. We can therefore write

$$g_{\mu\nu} = \bar{g}_{\mu\rho}\left(e^{\sqrt{\frac{32\pi\,\ell_\mathrm{p}}{m_\mathrm{p}}}\,h}\right)^{\rho}_{\ \nu} \simeq \eta_{\mu\nu} + \sqrt{\frac{32\pi\,\ell_\mathrm{p}}{m_\mathrm{p}}}\,h_{\mu\nu} + \frac{16\pi\,\ell_\mathrm{p}}{m_\mathrm{p}}\,h_{\mu\rho}\,h^{\rho}_{\ \nu}\,, \tag{2.1}$$

where we assumed that the background metric $\bar{g}_{\mu\nu} = \eta_{\mu\nu}$ for simplicity (results can be generalised to any curved background by adopting normal coordinates). The Lagrangian governing the dynamics will include the usual Einstein-Hilbert term and possible quantum corrections, giving rise to a number of propagating modes $i = 1, \ldots$, in the linearised version for small perturbations.

The geometrical distance between two nearby points, say at coordinates x^μ and $y^\mu = x^\mu + \mathrm{d}x^\mu$, is computed from the expectation value of the above operator on a suitable quantum state. Hence, it can be technically obtained from the real in-in correlation function

$$\ell^2_{\text{in-in}}(x, y) = \left\langle 0_\mathrm{in}\right|\mathrm{d}s^2\left|0_\mathrm{in}\right\rangle\,, \tag{2.2}$$

where the "vacuum" $|0_\mathrm{in}\rangle$ is such that the expected classical background expression $\ell^2 = \eta_{\mu\nu}\,\mathrm{d}x^\mu\,\mathrm{d}x^\nu$ is recovered at the leading order $(\ell_\mathrm{p}/m_\mathrm{p})^0$. Note that ℓ^2 goes to zero when $\mathrm{d}x^\mu$ vanishes at coincident points, whereas terms of order $(\ell_\mathrm{p}/m_\mathrm{p})^{1/2}$ are linear in $h_{\mu\nu}$ and vanish everywhere by the usual Fock space construction.

One can now argue in a model-independent way for the absence of a minimum geometrical distance along physically allowed trajectories [16, 24]. For almost coincident points, Eq. (2.2) reads

$$\begin{aligned}
\ell^2_{\text{in-in}} &\simeq \frac{16\pi\,\ell_\mathrm{p}}{m_\mathrm{p}}\,\langle 0_\mathrm{in}|\,h_{\mu\rho}(x)\,h^{\rho}_{\ \nu}(y)\,|0_\mathrm{in}\rangle\,\mathrm{d}x^\mu\,\mathrm{d}x^\nu \\
&\simeq \frac{16\pi\,\ell_\mathrm{p}}{m_\mathrm{p}}\,G^{\text{ret}}_{\mu\nu\rho\sigma}(x, y)\,\mathrm{d}x^\mu\,\mathrm{d}x^\nu\,,
\end{aligned} \tag{2.3}$$

where we assumed that x^μ is in the future of y^μ. The retarded propagator is in general given by

$$G^{\text{ret}}_{\mu\nu\rho\sigma} = \sum_i\left[-\frac{\theta(x^0 - y^0)}{2\pi}\,\delta(\ell^2) + \theta(x^0 - y^0)\,\theta(\ell^2)\,\frac{m_i\,J_1(m_i\,\ell)}{4\pi\,\ell}\right]\hbar\,P^i_{\mu\nu\rho\sigma}\,, \tag{2.4}$$

where m_i is the (possibly vanishing) mass of the ith propagating mode and

$$P^i_{\mu\nu\rho\sigma} = \alpha_i\,\eta_{\mu\rho}\,\eta_{\nu\sigma} + \beta_i\,\eta_{\mu\sigma}\,\eta_{\nu\rho} + \gamma_i\,\eta_{\mu\nu}\,\eta_{\rho\sigma} \tag{2.5}$$

is the most general fourth rank tensor symmetric in $\{\mu\nu\}$ and $\{\rho\sigma\}$, with parameters α_i, β_i and γ_i depending on the particular gravitational Lagrangian for the ith mode. The contraction $P^{i\ \ \rho}_{\mu\rho\ \ \nu}\,\mathrm{d}x^\mu\,\mathrm{d}x^\nu \sim \ell^2$ and Eq. (2.3) in the coincident limit finally yields

$$\lim_{x\to y}\ \ell^2_{\text{in-in}} = 0\,. \tag{2.6}$$

This result relies solely on the analytic structure of the retarded propagator in position space. We can therefore conclude that the effective field theory (of gravity) shows no sign of a minimum geometrical length along the path of test particles, irrespectively of the specific (gravitational) Lagrangian.

By stressing that one of the main roles of the metric is to determine the geodesic motion of test particles, one can also conclude that free propagation of (quantum) excitations on the background state $|0_{\text{in}}\rangle$ is described by the geodesic equation in the (smooth) background $\bar{g}_{\mu\nu} = \eta_{\mu\nu}$. At the very least, this is a proof of consistency of the effective field theory which requires that the quantum state $|0_{\text{in}}\rangle$ entails the existence of a classical background spacetime.

2.2.1 *Length Scale in Scatterings*

We just used the fact that in-in amplitudes involve retarded propagators. For cross sections and decay rates, one instead assumes Feynman boundary conditions corresponding to the scattering of (particle) excitations of the initial state $|0_{\text{in}}\rangle$ into those of a final state $|0_{\text{out}}\rangle$. The Feynman amplitudes for the metric between two nearby points,

$$\ell^2_{\text{in-out}}(x, y) = \left\langle 0_{\text{out}} | \mathrm{d}s^2 | 0_{\text{in}} \right\rangle\,, \tag{2.7}$$

can then be used to determine a length scale for interaction processes involving gravity. In fact, for almost coincident points, one finds

$$\begin{aligned} \ell^2_{\text{in-out}} &\simeq \frac{16\pi\,\ell_{\text{p}}}{m_{\text{p}}}\,\langle 0_{\text{out}}|\,h_{\mu\rho}(x)\,h^\rho_{\ \nu}(y)\,|0_{\text{in}}\rangle\,\mathrm{d}x^\mu\,\mathrm{d}x^\nu \\ &\simeq \frac{16\pi\,\ell_{\text{p}}}{m_{\text{p}}}\,G^{\text{F}}_{\mu\nu\rho\sigma}(x, y)\,\mathrm{d}x^\mu\,\mathrm{d}x^\nu\,, \end{aligned} \tag{2.8}$$

where the Feynman propagator has the general short distance behaviour

$$G^{\mathrm{F}}_{\mu\nu\rho\sigma} \simeq \sum_i \frac{i\,\hbar\,P^i_{\mu\nu\rho\sigma}}{4\,\pi^2\,(x-y)^2}\,. \tag{2.9}$$

Eq. (2.8) then yields

$$\lim_{x\to y}\,\ell^2_{\text{in-out}} \sim \lim_{\ell\to 0}\,e^{-\frac{\ell_{\mathrm{p}}^2}{\ell^2}\,|\Sigma|}\,\ell^2 = 0 \tag{2.10}$$

if $\Sigma \equiv \sum_i(\alpha_i + 4\beta_i + \gamma_i) \le 0$ and

$$\lim_{x\to y}\,\ell^2_{\text{in-out}} \simeq i\,\frac{2}{\pi}\,\Sigma\,\ell_{\mathrm{p}}^2 \sim \ell_{\mathrm{p}}^2\,, \tag{2.11}$$

if $\Sigma > 0$.

For positive Σ, one therefore expects the emergence of a finite length scale proportional to ℓ_{p} in the modulus of transition amplitudes that will dynamically screen effects in the ultraviolet.

The above result again follows from general properties of the Feynman propagator for metric perturbations and the information about the details of the gravitational dynamics is contained solely in the parameters α_i, β_i and γ_i. In particular, the massless spin-2 field (the graviton) is the only degree of freedom in general relativity and $\Sigma = 4$. The corresponding length scale is therefore given by

$$\ell_{\text{in-out}}(x,x) \simeq \sqrt{\frac{8}{\pi}}\,\ell_{\mathrm{p}}\,. \tag{2.12}$$

Another interesting example is Stelle's theory [69], whose spectrum contains additional degrees of freedom needed to prove the renormalisability. In this case, there is a surprising accidental cancelation of the parameters so that $\Sigma = 0$ and a minimum length scale does not appear in scattering processes.[2]

Renormalisable theories possess no natural scale, whereas non-renormalisable theories contain an intrinsic scale which is then used as a cut-off to define the effective field theory and its range of validity. For lengths below this scale, the effective field theory breaks down and ultraviolet phenomena cannot be described without a ultraviolet completion. General relativity is not renormalisable and we indeed find a minimum scale of the order of ℓ_{p}, which is precisely the scale used to perform the effective field theory expansion. On the other hand, Stelle's theory is renormalisable and does not need any intrinsic scale. This suggests an interesting correspondence between the renormalisability of a theory (of gravity) and the non-existence of a minimum length scale. However, in a perspective like the one of the asymptotic safety scenario [7] and classicalization [38, 42], one might instead argue that the

[2] For a discussion of cases with $\Sigma < 0$, see Ref. [16].

dynamical emergence of the length scale ℓ_p does not imply the need of a ultraviolet completion but that the (effective) theory is self-complete and simply rearranges its degrees of freedom so that no new physics appears below ℓ_p. This realises the old idea that ℓ_p acts as a natural (albeit dynamical rather than geometrical) regulator that removes all ultraviolet divergences [62, 64] and allows to treat general relativity as a fundamental theory of gravity. From this viewpoint, it is not general relativity that fails at the Planck scale, but rather physics beyond ℓ_p becomes (operationally) meaningless.

2.2.2 *Minimum Length Scale and Generalised Uncertainty Principle*

From the operational point of view, scattering processes are generically involved in quantum mechanical measurements, as Heisenberg argued in his famous description of the quantum microscope [51]. The position x_e and momentum p_e of a particle, say an electron, can be determined by scattering a photon off the electron and measuring properties of the photon after the scattering. If λ is the photon wavelength, the uncertainty in the position of the electron is (at least) $\Delta x_e \simeq \lambda$. Moreover, the photon carries a momentum $p = h/\lambda$, which is partially transferred to the electron in an unknown magnitude and direction. This implies that, just after the scattering, the uncertainty in the electron momentum amounts to (at most) $\Delta p_e \simeq p = h/\lambda$. Heisenberg thus concluded that

$$\Delta x_e \, \Delta p_e \simeq \lambda \cdot \frac{h}{\lambda} \simeq h \, . \tag{2.13}$$

Later on, Schrödiger and Robinson formulated the uncertainty principle for canonically conjugated variables, like the x and p of a particle, as

$$\Delta x \, \Delta p \geq \frac{\hbar}{2} \, , \tag{2.14}$$

which is the most common form used at present.

Heisenberg's heuristic approach paved the way to the formulation of various generalised uncertainty principles [34, 52] from taking into account the gravitational effects in the photon-particle interaction. Indeed, if the centre-of-mass energy is sufficiently large to result in the formation of a black hole, the corresponding horizon radius acts a screening length below which position measurements become meaningless [66]. Generalised uncertainty principles can be mathematically derived from modified quantum mechanical commutators, and one is tempted to extend such modifications to quantum field theory commutators. However, we have seen that a minimum length scale can be naturally obtained without modifying the quantum field theory dynamics, hence without altering the field commutators. It would look more

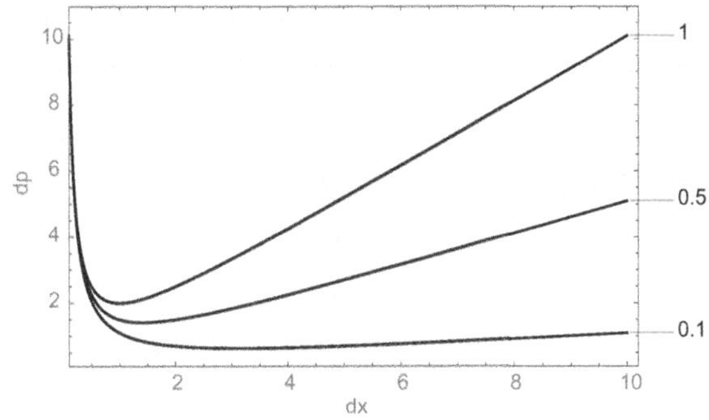

Fig. 2.1 Generalised uncertainty principle (2.15) and minimum length scale for different values of δ_0 (lengths and momenta in Planck units)

physically sound that generalised uncertainty principles therefore emerge effectively in quantum mechanics as the non-relativistic sector of quantum field theory of gravity without modifying the field propagators *ad hoc*.

The simplest form of generalised uncertainty principle is given by (see Fig. 2.1)

$$\Delta x\, \Delta p \geq \frac{\hbar}{2}\left(1+\frac{\delta_0}{m_{\mathrm{p}}^2}\,\Delta p^2\right), \tag{2.15}$$

where $\Delta x^2 \equiv \langle\, \hat{x}^2\, \rangle - \langle\, \hat{x}\, \rangle^2$, and similarly $\Delta p^2 \equiv \langle\, \hat{p}^2\, \rangle - \langle\, \hat{p}\, \rangle^2$, and δ_0 is a dimensionless deforming parameter expected to emerge from candidate theories of quantum gravity. Uncertainty relations can be associated with commutators of conjugated variables by means of the general inequality

$$\Delta A\, \Delta B \geq \frac{1}{2}\left|\langle\, [\hat{A}, \hat{B}]\, \rangle\right| . \tag{2.16}$$

For instance, one can derive Eq. (2.15) from the commutator

$$\left[\hat{x}, \hat{p}\right] = i\,\hbar\left(1+\frac{\delta_0}{m_{\mathrm{p}}^2}\,\hat{p}^2\right), \tag{2.17}$$

for which Eq. (2.16) yields

$$\Delta x\, \Delta p \geq \frac{\hbar}{2}\left[1+\frac{\delta_0}{m_{\mathrm{p}}^2}\left(\Delta p^2+\langle\, \hat{p}\, \rangle^2\right)\right]. \tag{2.18}$$

We stress that the derivation of Eq. (2.15) only makes use of the algebraic structure of the commutator (2.17) through the general inequality (2.16) and no specific representation of the physical operators $\hat{x}$ and $\hat{p}$ (in whatsoever form) is needed. This immediately implies that the generalised uncertainty principle (2.15) holds for any quantum state, since $\langle\, \hat{p}\, \rangle^2 \geq 0$. In particular, in the centre-of-mass frame of a scattering process, one can just consider the so-called mirror-symmetric states satisfying $\langle \hat{p} \rangle = 0$, and the inequality (2.18) reduces to the generalised uncertainty principle (2.15).

The uncertainty relation (2.15) readily implies the existence of a minimum (effective) length

$$\ell = \ell_{\mathrm{p}} \sqrt{\delta_0} \,. \tag{2.19}$$

By comparing with Eq. (2.11), we obtain

$$\delta_0 = \frac{2}{\pi} \Sigma \,, \tag{2.20}$$

which represents an exact expression for the parameter of the generalised uncertainty principle emerging from a general class of gravity theories.

For general relativity, one finds $\delta_0 = 8/\pi$ and $\ell \simeq 1.6\, \ell_{\mathrm{p}}$. Values of the deformation parameter δ_0 can be obtained also from other approaches [57, 68] but all available results agree in order of magnitude. Furthermore, experimental upper bounds on δ_0 exist (see Ref. [2, 67] and references therein), but they are typically too weak ($\delta_0 \lesssim 10^{36}$) to provide any useful information about the gravitational propagator. On the other hand, the theoretical value given by Eq. (2.20) can be viewed as a (general) lower bound. In any case, it is hard to conceive ways of detecting direct evidence for the existence of such a vanishingly short length.

The dynamical origin of the gravitational scale we discussed in this Section opens up the possibility that much larger lengths appear in dynamical processes involving macroscopically large amounts of matter. For those cases, one must however face the hurdle of dealing with the nonlinearity of the gravitational interaction beyond the weak-field approximation and perturbative methods.

2.3 Bootstrapped Newtonian Stars and Black Holes

General relativity is a nonlinear theory already at the classical level, which is what makes it so hard to provide a quantum description of macroscopic systems. Starting from the Fierz-Pauli action for a spin-2 field in Minkowski spacetime, one can indeed reconstruct the full Einstein-Hilbert action [37], but this bootstrap procedure is not free of ambiguities [63]. Moreover, assuming a Minkowski background makes this approach impractical for analysing compact sources of astrophysical size for which the nonlinear character of gravity should be more prominent. However, one can try and take advantage of the fact that Newtonian gravity is a well-tested approximation as a different starting point for the bootstrap process [18].

The bootstrapped Newtonian gravity [27, 28] is defined by considering the well-known Newtonian Lagrangian for a spherically symmetric and static source

$$L_{\rm N}[V] = -4\pi \int_0^\infty r^2\, dr \left[\frac{(V')^2}{8\pi\, G_{\rm N}} + \rho\, V \right] , \tag{2.21}$$

where $\rho = \rho(r)$ is the matter energy density. This Lagrangian yields the Poisson equation for the Newtonian potential $V = V_{\rm N}$, that is

$$r^{-2} \left(r^2\, V'\right)' \equiv \Delta V = 4\pi\, G_{\rm N}\, \rho\ . \tag{2.22}$$

The action for bootstrapped Newtonian gravity is obtained by first adding a gravitational self-coupling term proportional to the gravitational energy $U_{\rm N}$ per unit volume

$$\mathcal{J}_V \simeq \frac{d U_{\rm N}}{d\mathcal{V}} = -\frac{\left[V'(r)\right]^2}{2\pi\, G_{\rm N}}\ . \tag{2.23}$$

One next observes that the system must be kept in equilibrium by static (and for simplicity isotropic) pressure $p = p(r)$, which becomes increasingly large as the compactness $X \equiv G_{\rm N}\, M/R$ increases. This contribution is accounted for via a potential energy U_p such that

$$\mathcal{J}_p \simeq -\frac{dU_p}{d\mathcal{V}} = 3\, p\ . \tag{2.24}$$

In the above expressions, M represents an ADM-like mass and is what one would measure when studying orbits [20], whereas the source of density ρ and pressure p will be assumed to have finite radius R.

The pressure component adds to the density by effectively shifting $\rho \to \rho + 3\, q_p\, p$, where the non-relativistic limit is obtained for the positive coupling $q_p \to 0$. A higher-order term $\mathcal{J}_\rho = -2\, V^2$ is also added [18], which couples to the matter source, and finally leads to the total Lagrangian

$$\begin{aligned} L[V] &= L_{\rm N}[V] - 4\pi \int_0^\infty r^2\, dr \left[q_V\, \mathcal{J}_V\, V + q_p\, \mathcal{J}_p\, V + q_\rho\, \mathcal{J}_\rho \left(\rho + q_p\, \mathcal{J}_p\right)\right] \\ &= -4\pi \int_0^\infty r^2\, dr \left[\frac{(V')^2}{8\pi\, G_{\rm N}} (1 - 4 q_V\, V) + \left(\rho + 3\, q_p\, p\right) V \left(1 - 2\, q_\rho\, V\right) \right], \end{aligned} \tag{2.25}$$

where q_V, q_p and q_ρ are positive constants which track the effects of each contribution. For instance, the Newtonian limit is recovered when all couplings vanish.

The complete Euler-Lagrange equation for the potential V reads

$$\Delta V = 4\pi\, G_{\rm N} \left(\rho + 3\, q_p\, p\right) \frac{1 - 4\, q_\rho\, V}{1 - 4\, q_V\, V} + \frac{2\, q_V \left(V'\right)^2}{1 - 4\, q_V\, V}\ , \tag{2.26}$$

which can now be considered for sources of interest. In particular, one must first find solutions for the outer vacuum and then boundary conditions at $r = R$ will be used to constrain the inner solutions.

2.3.1 Outer Solution

Outside the source $\rho = p = 0$ and the general solution of Eq. (2.26) is given by

$$V_c = \frac{1}{4\,q_V}\left[1 - c_1\left(1 + \frac{c_2}{r}\right)^{2/3}\right], \tag{2.27}$$

where c_1 and c_2 are integration constants, and it must be noted that the potential V_c cannot be shifted by an arbitrary constant because of the nonlinearity.

The Newtonian behaviour in terms of the ADM-like mass M must be recovered asymptotically for $r \to \infty$, which fixes the two integration constants and one obtains

$$V_{out} = \frac{1}{4\,q_V}\left[1 - \left(1 + \frac{6\,q_V\,G_N\,M}{r}\right)^{2/3}\right]. \tag{2.28}$$

It is easy to check that the large r expansion of V_{out} contains the general relativity post-Newtonian term of order G_N^2 of the Schwarzschild metric for $q_V = 1$ [18]. The analytic solution from Eq. (2.28) also tracks the Newtonian potential

$$V_N = -\frac{G_N\,M}{r}, \tag{2.29}$$

for all $r > 0$. The most significant difference is that, for $r \to 0$,

$$\frac{V_{out}}{V_N} \sim \left(\frac{r}{R_H}\right)^{1/3} \tag{2.30}$$

where R_H is the event horizon, and V_{out} diverges more slowly than V_N towards the centre (see Fig. 2.2).

2.3.2 Boundary Conditions

The interior solutions of Eq. (2.26), $V_{in} = V(0 \le r < R)$, must match smoothly the outer solution $V_{out} = V(R \le r)$ from Eq. (2.28) across the boundary $r = R$ of the source. We therefore require that

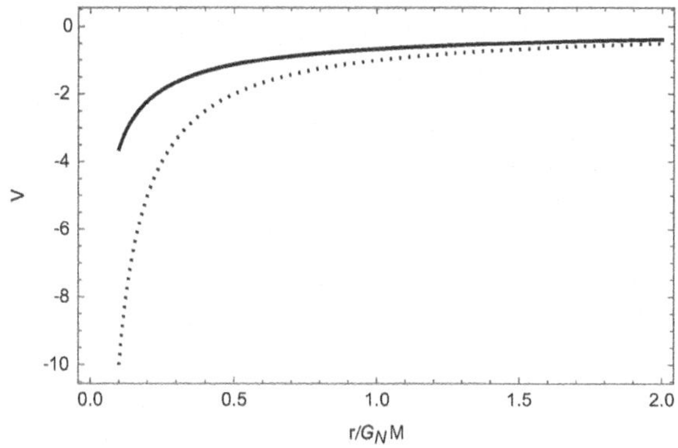

Fig. 2.2 Bootstrapped Newtonian potential (2.28) (solid line) compared to the Newtonian potential (dotted line)

$$V_{\rm in}(R) = V_{\rm out}(R) \equiv V_R = \frac{1}{4\,q_V}\left[1 - (1 + 6\,q_V\,X)^{2/3}\right] , \tag{2.31}$$

and

$$V_{\rm in}'(R) = V_{\rm out}'(R) \equiv V_R' = \frac{X}{R\,(1 + 6\,q_V\,X)^{1/3}} , \tag{2.32}$$

where the expressions on the right hand side are obtained by evaluating the outer potential at the boundary and $X = G_{\rm N}\,M/R$ is the outer compactness defined earlier. Moreover, to avoid a singular centre, the classical density profiles must be finite at $r = 0$, which translates into the regularity condition

$$V_{\rm in}'(0) = 0 . \tag{2.33}$$

Since Eq. (2.26) is a second order (ordinary) differential equation, the three constraints discussed above will uniquely fix the potential $V_{\rm in}$ and also determine the proper mass M_0 defined below.

2.3.3 *Uniform Sphere*

It is clear from Eq. (2.26) that finding interior solutions is going to be very complicated. The simplest case is that of a compact homogeneous ball of matter whose density vanishes outside the sphere of radius $r = R$,

$$\rho = \rho_0 \equiv \frac{3\,M_0}{4\,\pi\,R^3}\,\Theta(R - r) , \tag{2.34}$$

with Θ representing the Heaviside step function. The proper mass of the object can be defined according to the Newtonian expression

$$M_0 = 4\pi \int_0^R r^2 \rho(r)\, dr \ , \tag{2.35}$$

and it must be emphasised that, in bootstrapped Newtonian gravity, M_0 is not the same as the proper mass in general relativity (nor as the ADM mass, as we shall see below).

We must also assume a conservation equation which will further constrain the problem by relating the pressure to the density. We write this constraint as

$$p' = -V' \left(\rho + q_p\, p\right) \ , \tag{2.36}$$

so that it can be seen as an approximation of the Tollman-Oppenheimer-Volkoff equation of general relativity which reduces to the Newtonian formula for $q_p = 0$.

For the sake of simplicity, the three couplings shall be given numerical values $q_V = q_\rho = 1$, and for convenience $3\, q_p = 1$. These assumptions reduce Eq. (2.26) to

$$\triangle V = 4\pi\, G_N \left(\rho + p\right) + \frac{2\left(V'\right)^2}{1 - 4\, V} \ . \tag{2.37}$$

In the range $0 \leq r \leq R$, the conservation Eq. (2.36) then has the solution

$$p = \rho_0 \left[e^{V_R - V} - 1\right] \ , \tag{2.38}$$

where the boundary condition for the pressure $p(R) = 0$ was imposed, V_R is given in Eq. (2.31) and ρ_0 is defined in Eq. (2.34). Complete analytic solutions for a general value of the compactness remain too difficult to obtain and it is convenient to separate two compactness regimes, namely $X \lesssim 1$ and $X \gg 1$.

For low or intermediate compactness, $X \ll 1$ or $X \simeq 1$, reliable approximate analytic solutions can be found by perturbative methods. In particular, the potential $V_{\rm in}$ can be expanded near $r = 0$ as

$$V_{\rm s} \simeq V_0 + \frac{G_N\, M_0}{2\, R^3}\, e^{V_R - V_0}\, r^2 \ . \tag{2.39}$$

where $V_0 \equiv V_{\rm in}(0) < 0$ and all odd powers of r in the Taylor expansion must vanish due to the regularity condition (2.33). The approximate solution which matches the unique exterior (2.28) can then be written in terms of the compactness X as

$$V_{\rm s} \simeq \frac{\left[(1 + 6\, X)^{1/3} - 1\right] + 2\, X \left[(r/R)^2 - 4\right]}{4\,(1 + 6\, X)^{1/3}} \ , \tag{2.40}$$

along with the proper mass

$$M_0 \simeq \frac{M\, e^{-\frac{X}{2(1+6\,X)^{1/3}}}}{(1+6\,X)^{1/3}} \,. \tag{2.41}$$

This approximation of the potential can be shown to be in very good agreement with the numerical solutions for both small and intermediate compactness, the smaller the compactness X the less V_s differs from the numerical solution.

For high compactness, $X \gg 1$, perturbative methods fail, and even the numerical analysis becomes more difficult, because any slight error in the estimation of the proper mass $M_0 = M_0(M, R)$ induces large errors in the potential profile. In this regime, one can use comparison methods [36, 56], according to which upper and lower approximate bounding functions $V_\pm = V_\pm(r)$ must be somehow found such that $E_+(r) < 0$ and $E_-(r) > 0$ for the entire interval $0 \le r \le R$, where

$$E_\pm \equiv \triangle V_\pm - \frac{3\,G_\mathrm{N}\,M_0^\pm(M)}{R^3}\,e^{V_R - V_\pm} - \frac{2\,(V_\pm')^2}{1-4\,V_\pm} \,. \tag{2.42}$$

Mathematical theorems then ensure that the actual solution must lie in between the two bounding functions,

$$V_- < V_\mathrm{in} < V_+ \,. \tag{2.43}$$

As a further clarification, the approximate solutions $V_\pm$ are not required to have the same functional form of the exact solution V_in. For our case, a good starting point is the simpler equation

$$\psi'' = \frac{3\,G_\mathrm{N}\,M_0}{R^3}\,e^{V_R - \psi} \,, \tag{2.44}$$

whose relevant solutions are given by

$$\psi(r; M, R) \simeq \frac{X^{2/3}}{6^{1/3}} \left(\frac{r}{R} - \frac{5}{2}\right) , \tag{2.45}$$

with

$$\frac{M_0}{M} \simeq \left(\frac{2\,X}{3^5}\right)^{1/3} e^{-\frac{X^{2/3}}{6^{1/3}}} \,. \tag{2.46}$$

The bounding functions for the full Eq. (2.26) can now be expressed as

$$V_\pm = C_\pm\,\psi(r; A_\pm, B_\pm) \,, \tag{2.47}$$

where the paramaters $A_{\pm}$, $B_{\pm}$ and $C_{\pm}$ must be calculated using the boundary conditions (2.33), (2.31) and (2.32) with the additional constraint of Eq. (2.42).

For highly compact objects with homogeneous densities, the results show a potential that is practically linear, except in the vicinity of $r = 0$ where it becomes quadratic according to Eq. (2.33). In the linear approximation

$$V_{\text{lin}} \simeq V_R + V_R' (r - R) , \tag{2.48}$$

where V_R and V_R' are given in (2.31) and (2.32), one can also find

$$\frac{M_0}{M} \simeq \frac{2(1+5X)}{3(1+6X)^{4/3}} \sim \frac{1}{X^{1/3}} . \tag{2.49}$$

The bootstrapped Newtonian gravity does not account for the geometrical aspects of gravitation and differs from general relativity in other aspects. For example, in general relativity, black hole geometries form when matter collapses beyond the Schwarzschild radius $R_{\text{H}} = 2\, G_{\text{N}}\, M$ and the separation between isotropic stars (with $R \gg R_{\text{H}}$) and black holes (with $R \lesssim R_{\text{H}}$) is provided by the Buchdahl limit [10] $R > (9/8)\, R_{\text{H}}$, beyond which the necessary pressure would diverge.

Bootstrapped Newtonian objects always have a finite pressure, even if their surface lies behind the horizon, so there is no Buchdahl limit analogue. The horizon radius R_{H} can still be defined using Newtonian arguments as the radius at which the escape velocity of test particles is the same as the speed of light, or $2\, V(r_{\text{H}}) = -1$. The lowest value of X for which one has an horizon is then found by imposing

$$2\, V_{\text{in}}(r_{\text{H}} = 0) = -1 , \tag{2.50}$$

which gives $X \simeq 0.46$ if we use $V(0) = V_0$ from Eq. (2.40). With increasing compactness, the radius r_{H} grows equal to the radius R of the source for

$$2\, V_{\text{in}}(r_{\text{H}} = R) = 2\, V_{\text{out}}(R) = -1 , \tag{2.51}$$

where $V_{\text{out}}(R) = V_R$ is given in Eq. (2.31). The compactness for which this happens is $X \simeq 0.69$ and the horizon radius $r_{\text{H}} \simeq R \simeq 1.43\, G_{\text{N}}\, M \simeq 3\, G_{\text{N}}\, M_0$. For larger X, the horizon is located in the outer potential (2.28), and its value only depends on M.

In summary, both the pressure and density contribute to the gravitational potential for bootstrapped Newtonian stars. Because no equivalent to the Buchdahl limit exists, the source core can remain in equilibrium due to a large (and finite) pressure, regardless of the compactness, with the following regimes

$$\begin{cases} \text{no horizon} & \text{for } X \lesssim 0.46 \\ 0 < r_{\text{H}} \lesssim R \simeq 1.4\, G_{\text{N}}\, M & \text{for } 0.46 \lesssim X \lesssim 0.69 \\ r_{\text{H}} \simeq 1.4\, G_{\text{N}}\, M & \text{for } X \gtrsim 0.69 . \end{cases} \tag{2.52}$$

We remark that $r_H \neq R_H$, which is a clear prediction of the bootstrapped Newtonian gravity.

2.3.4 Polytropic Stars and Other Results

The simple model of homogenous balls was generalised to more realistic cases of stars governed by a polytropic equation of state [29]

$$p = \gamma\, \rho^n \ . \tag{2.53}$$

One can show numerically that the density profiles are well approximated by Gaussian functions. Such matter distributions can then be employed in order to investigate the interplay between the compactness X of the source, the width of the density profile $\rho = \rho(r)$ and the polytropic index n. Results showed that, in the high compactness regime, bootstrapped Newtonian stars tend to be more compact and massive than in general relativity for identical polytropic equations of state and central densities. No Buchdahl limit exists in this case either and finite pressure can support a star of arbitrarily high compactness. For objects with low compactness instead, the density profiles become practically identical in the two frameworks and agree with the Newtonian approximation.

One-loop corrections to the bootstrapped Newtonian potential were computed for small sources [23]. On the other hand, it was shown that the potential for large homogeneous sources is not sensitive to one-loop corrections for the matter coupling to gravity [30]. However, the ratio between the proper mass M_0 and ADM mass M depends on this coupling and can be either smaller or larger than one, depending on a critical value of the coupling which, in turn, is a function of the compactness.

Another case worth mentioning is the investigation of binary mergers [25], where several constraints resulting from the difference between the ADM mass M and the proper mass M_0 can be imposed. These constraints change depending on the types of stars that merge and resulting final object. The more interesting finding regards the GW150914 signal measured by LIGO: contrary to some other claims found in the literature, LIGO's findings do not violate the mass gap in bootstrapped Newtonian gravity, but typical stellar black hole masses easily fit the experimental data [1].

This classical picture can change significantly if one considers the quantum nature of matter and, consequently, of gravity. A first indication that the existence of a dynamical length similar to the one discussed in Sect. 2.2 leads to an upper bound on the compactness was discussed in Ref. [26] within the bootstrapped Newtonian picture. However, it is indeed simpler to reach that conclusion by starting from full general relativity, as we are going to see next.

2.4 Quantum Matter Core

Semiclassical models of gravitational collapse generically predict a bounce at a minimum radius [13, 44, 47, 48], but the nonlinearity of the Einstein equations makes it impossible to study realistic models analytically, which renders the problem of a quantum description intractable. For this reason, one can only study very simplified models obtained by forcing a strong symmetry and unphysical equation of state for the collapsing matter. A paramount example is given by the gravitational collapse of a ball of dust (matter with no other interaction but gravity) originally investigated by Oppenheimer and collaborators [60]. A similarly simple case is given by a shell of matter collapsing under its weight or towards a central source.

In fact, these systems are simple enough to allow for a canonical analysis [55] of the effective action obtained by restricting the Einstein-Hilbert action to metrics satisfying the assumed symmetry properties. Some key features are more readily obtained with the approach of Refs. [14, 32, 35], in which the geodesic equation for the areal radius of the ball provides an effective quantum mechanical description similar to the usual quantum model of the hydrogen atom obtained by quantising the electron's position. This approach straightforwardly leads to the existence of a discrete spectrum of bound states, with the ground state characterised by a macroscopically large surface area quantised according to Bekenstein's area law [5]. It is important to remark from the very beginning that the areal radius of the ball is not a fundamental degree of freedom for the matter in the collapsing core, which should instead be described by quantum excitations of fields in the Standard Model of particle physics. Although these fundamental degrees of freedom are neglected for the purpose of defining a tractable mathematical problem, their existence must be encoded in a suitable entropy that can be computed from the states of the effective quantum mechanical theory.

2.4.1 General Relativistic Discrete Spectrum

Let us consider a perfectly isotropic ball of dust with total ADM mass M and areal radius $R = R(\tau)$, where τ is the proper time measured by a clock comoving with the dust. Dust particles inside this collapsing ball will follow radial geodesics $r = r(\tau)$ in the Schwarzschild space-time metric

$$\mathrm{d}s^2 = -\left(1 - \frac{2\,G_\mathrm{N}\,M_\mu}{r}\right)\mathrm{d}t^2 + \left(1 - \frac{2\,G_\mathrm{N}\,M_\mu}{r}\right)^{-1}\mathrm{d}r^2 + r^2\,\mathrm{d}\Omega^2\,, \quad (2.54)$$

where $M_\mu = M_\mu(r)$ is the (constant) fraction of ADM mass inside the sphere of radius $r = r(\tau)$. In particular, we can consider the outermost (thin) layer of (average) radius $r = R(\tau)$ and mass $\mu = \varepsilon\, M$, with $0 < \varepsilon < 1$. The evolution of R is then determined by the radial equation[3]

$$\frac{E_\mu^2}{\mu^2} - \dot{R}^2 + \frac{2\, G_N\, M_\mu}{R} = 1\,, \tag{2.55}$$

where $M_\mu = (1-\varepsilon)\, M$ and E_μ is the conserved momentum conjugated to $t = t(\tau)$.[4] Eq. (2.55) can be written as

$$H \equiv \frac{P^2}{2\,\varepsilon\, M} - \frac{\varepsilon\,(1-\varepsilon)\, G_N\, M^2}{R} = \frac{\varepsilon\, M}{2}\left(\frac{E_\mu^2}{\varepsilon^2\, M^2} - 1\right) \equiv \mathscr{E}\,, \tag{2.56}$$

where $P = \mu\, \dot{R}$ is the momentum conjugated to $R = R(\tau)$.

We can next apply the canonical quantisation prescription $\hat{P} = -i\,\hbar\,\partial_R$, which allows us to write Eq. (2.56) as the time-independent Schrödinger equation

$$\hat{H}\,\Psi_{\bar{n}} = \left[-\frac{\hbar^2}{2\,\varepsilon\, M}\left(\frac{\mathrm{d}^2}{\mathrm{d}R^2} + \frac{2}{R}\frac{\mathrm{d}}{\mathrm{d}R}\right) - \frac{\varepsilon\,(1-\varepsilon)\, G_N\, M^2}{R}\right]\Psi_{\bar{n}} = \mathscr{E}_{\bar{n}}\,\Psi_n\,. \tag{2.57}$$

Since this is formally the same as the equation for the s-states of the hydrogen atom, solutions are given by the eigenfunctions

$$\Psi_{\bar{n}} = \sqrt{\frac{\varepsilon^6\,(1-\varepsilon)^3\, M^9}{\pi\,\ell_p^3\, m_p^9\, \bar{n}^5}}\,\exp\!\left(-\frac{\varepsilon^2\,(1-\varepsilon)\, M^3\, R}{\bar{n}\, m_p^3\, \ell_p}\right) L_{\bar{n}-1}^1\!\left(\frac{2\,\varepsilon^2\,(1-\varepsilon)\, M^3\, R}{\bar{n}\, m_p^3\, \ell_p}\right), \tag{2.58}$$

where $L_{\bar{n}-1}^1$ are Laguerre polynomials with $\bar{n} = 1, 2\,\ldots$, corresponding to the eigenvalues

$$\mathscr{E}_{\bar{n}} = -\frac{\varepsilon^3\,(1-\varepsilon)^2\, M^5}{2\, m_p^4\, \bar{n}^2}\,. \tag{2.59}$$

The normalisation is defined in the scalar product which makes $\hat{H}$ Hermitian on the above spectrum, that is

$$\langle\, \bar{n} \mid \bar{n}'\, \rangle = 4\,\pi \int_0^\infty R^2\, \Psi_{\bar{n}}^*(R)\, \Psi_{\bar{n}'}(R)\, \mathrm{d}R = \delta_{\bar{n}\bar{n}'}\,, \tag{2.60}$$

and the expectation value of the areal radius is thus given by

---

[3] One should notice that Eq. (2.55) holds both in a black hole and in a white hole background. Classically, geodesics falling into the black hole singularity cannot be continued into geodesics emerging from the white hole singularity. However, quantum physics bypasses this limitation and one finds bouncing solutions which do precisely that in the semiclassical analysis of the model.

[4] Of course, the conserved angular momentum vanishes for purely radial motion. For a (slowly and rigidly) rotating ball see Ref. [35].

$$\bar{R}_{\bar{n}} \equiv \langle \bar{n} | \hat{R} | \bar{n} \rangle = \frac{3\, m_{\mathrm{p}}^3\, \ell_{\mathrm{p}}\, \bar{n}^2}{2\, \varepsilon^2\, (1-\varepsilon)\, M^3} . \tag{2.61}$$

So far the quantum picture is the same that one would have in Newtonian physics, with the ground state $\bar{n} = 1$ having a width $\bar{R}_1 \sim \ell_{\mathrm{p}}\, (m_{\mathrm{p}}/M)^3$ and energy $\mathscr{E}_1 \sim -M\, (M/m_{\mathrm{p}})^4$. This state is practically indistinguishable from a point-like singularity for any macroscopic black hole of mass $M \gg m_{\mathrm{p}}$.

In fact, the only general relativistic feature that the model retains is given by the relation for $\mathscr{E} = \mathscr{E}(E_\mu)$ in Eq. (2.56). On assuming that E_μ is real for the allowed quantum states, we obtain $E_\mu^2 \geq 0$, which yields the bound

$$\bar{n} \geq N_M \equiv \varepsilon\, (1-\varepsilon) \left(\frac{M}{m_{\mathrm{p}}} \right)^2 . \tag{2.62}$$

The fundamental state of the outer layer saturates the inequality (2.62), which is again reminiscent of Bekenstein's area law [5].[5] This result implies that the probability to find the ball of dust near $r = 0$ is practically zero for $M \gg m_{\mathrm{p}}$ and the singularity is avoided. Classically, there are no reasons for a ball of dust evolving solely under the gravitational force to stop contracting, but quantum mechanics instead shows that, in order to have a well-defined energy spectrum, the proper ground state compatible with general relativity has a width

$$R_M \equiv \bar{R}_{N_M} = \frac{3}{4}\, (1-\varepsilon)\, R_{\mathrm{H}} , \tag{2.63}$$

where $R_{\mathrm{H}} = 2\, G_{\mathrm{N}}\, M$ is again the classical Schwarzschild radius of the ball.

From the wavefunction $\Psi = \Psi(R)$, we can determine the probability that the ball be inside its own gravitational radius,

$$P(R \leq R_{\mathrm{H}}) \equiv \int_0^{R_{\mathrm{H}}} \mathscr{P}(R)\, \mathrm{d}R = 4\, \pi \int_0^{R_{\mathrm{H}}} |\Psi(R)|^2\, R^2\, \mathrm{d}R , \tag{2.64}$$

which can be viewed as the probability that the dust ball is a black hole (when the mass M is treated as a fixed parameter[6]). For the ground state and values of $M \gg m_{\mathrm{p}}$, the probability density $\mathscr{P} = \mathscr{P}_{N_M}(R)$ narrowly peaks around a value of $R = R_M$ below R_{H} and one thus finds

$$P(R \leq R_{\mathrm{H}}) \simeq 1 , \tag{2.65}$$

which confirms that the ground state is (very likely) a black hole.

[5] It is interesting to recall that a similar bound was also found for stable configurations of boson stars [54, 65].

[6] An alternative viewpoint is considered in Refs. [31, 33], where the mass M is quantised.

We further notice that the uncertainty in the areal radius is given by

$$\frac{\Delta R_{\bar{n}}}{\bar{R}_{\bar{n}}} \equiv \frac{\sqrt{\langle \bar{n}|\,\hat{R}^2\,|\bar{n}\rangle - \bar{R}_n^2}}{\bar{R}_n} = \frac{\sqrt{\bar{n}^2+2}}{3\,\bar{n}} \,, \tag{2.66}$$

which approaches a minimum of $1/3$ for $\bar{n} \to \infty$. This supports the conclusion that the matter core of a black hole is indeed in a relatively "fuzzy" quantum state (whereas excited states are more and more classical). Since $0 < \varepsilon < 1$, one can conclude that $R_M < R_{\mathrm{H}}$ and a finite number of states (2.58) with $\bar{n} = N_M + n$ and $n = 0, 1, \ldots$ will exist inside the horizon.

Although we (formally) know the analytical expression for the spectrum of the quantum dust ball, it is technically very difficult to perform explicit calculations for $M \gg m_{\mathrm{p}}$. For instance, for $M = M_{\odot} \simeq 10^{30}\,$kg, one finds that the ground state has $N_M \sim 10^{76}$ nodes, which makes it impossible to handle the wavefunction (2.58), either analytically or numerically.

2.4.2 *Entropy and Thermodynamics*

The bound on the compactness following from Eq. (2.62) is caused by the non-linearity of general relativity and qualitatively agrees with the results obtained by adding a gravitational self-interaction term to the Newtonian theory described in the previous Sect. 2.3. It also agrees with those results following from the quantum description of the gravitational radius and black hole horizon [31, 33]. In particular, the latter approach leads to very similar conclusions when the self-gravitating object is described by an extended many-body system with a very large occupation number of order $N_M \sim M^2/m_{\mathrm{p}}^2$ in its ground state [19]. The first excited modes could then be populated thermally and reproduce the Hawking radiation [49].

In fact, we notice that, for $\bar{n} \gtrsim N_M$, the quantum of the Hamiltonian H in Eq. (2.57) is given by

$$\delta H \equiv |\mathscr{E}_{n+1} - \mathscr{E}_n| \simeq m_{\mathrm{p}}\,\frac{m_{\mathrm{p}}}{M} \,, \tag{2.67}$$

so that $\delta H \ll m_{\mathrm{p}}$ for a macroscopic object of mass $M \gg m_{\mathrm{p}}$. This is the quantum of horizon area $\delta M \simeq \delta H$ predicted by Bekenstein and the typical energy of Hawking quanta. Furthermore, it appears that the proper source "energy" $\delta E \equiv |E_{n+1} - E_n| \simeq m_{\mathrm{p}}$ is naturally quantised in units of the fundamental Planck mass m_{p}, but this quantum is redshifted down to the much smaller δH measured by outer observers. It is intriguing that Eq. (2.62) allows for recovering the fundamental scaling relations of the corpuscular description of black holes [39, 40] in which the Hawking evaporation process is described as the depletion of the quantum state of gravity [39, 40].

In this perspective, the wavefunction Ψ_{N_M} appears as the non-perturbative ground state for self-gravitating macroscopic objects of mass M and should thus be the closest possible to a classical black hole. This result is consistent with the fact that a static gravitational field must be fully determined by the state of the source [31, 33]. It can further be interpreted as the fact that the quantum state of a macroscopic self-gravitating system of mass M is very far from the vacuum Ψ_0 of quantum gravity, for which $N_M = 0$. The number $N_M \sim M^2/m_{\rm p}^2$ hence provides a quantitative measure for this "distance" from the vacuum in the Hilbert space of quantum gravity states. The appearance of this ground state, in turn, can be interpreted as a form of *classicalization* ensuring the ultraviolet self-completeness of gravity [38, 42], because states with spatial momenta much larger than $\hbar/R_M \sim \hbar/R_{\rm H}$ have $\bar{n} < N_M$ and cannot be populated at energy scales of the order of the mass M.

Given that the collapsing objects should contain a very large number of matter field excitations, whose states are completely neglected, two different types of information entropy were employed to probe this scenario in Ref. [32], namely the configurational entropy and the continuous limit of Shannon's entropy. Both can only be computed numerically and for relatively small principal quantum numbers $\bar{n} = N_M + n$, for which they increase with M, in agreement with the intuition that, the larger the mass the more constituents in the dust core, with consequently larger uncertainty in its microstates. Excited states (with $\bar{n} > N_M$) display (much) higher information entropy than the ground state for the same mass M, which also agrees with the intuition that excited states are spatially less localised and can be realised by larger numbers of microstates. These results are in particular compatible with the classical dynamics of a ball of dust, which is necessarily going to collapse under its own weight without loss of energy encoded by the ADM mass M. One can then reproduce the time evolution of the areal radius $R(\tau) = \bar{R}_{\bar{n}}$ with decreasing values of $\bar{n} = N_M + n$ in correspondence with increasing proper time τ. The existence of a ground state halts this process at a finite macroscopic size $\bar{R} = R_M$, after which the ball can only shrink further by losing energy M, so that N_M decreases. This behaviour is qualitatively very different from the hydrogen atom, in which an electron bound to the nucleus will jump into smaller quantum states necessarily by emitting energy in the form of radiation. The dust ball instead collapses without emitting (gravitational) energy, because of the strict spherical symmetry, until it reaches a minimum (quantum) size (although one expects that a realistic collapse is instead always associated with the emission of energy). Finally, both entropies are smaller for larger fractions $\varepsilon = \mu/M$ of dust in the outermost layer. Since larger values of the entropy should correspond to more unstable configurations, this result seems to favour the accumulation of matter in the outermost layer, due to quantum pressure.

2.5 Coherent State for the "Hairy" Geometry

The main conclusion of the previous Section was that matter at the end of the gravitational collapse forms an extended quantum core of width $R_M \lesssim R_{\rm H} = 2\,G_{\rm N}\,M$. It is conceivable that this feature reflects in the quantum state of the outer geometry, by showing some sort of deviation (or quantum "hair") from the classical general relativity solutions, albeit without changing the fundamental dynamics of general relativity. In fact, while the existence of particular quantum states can radically change the description of a system, at the same time it can leave the underlying fundamental dynamics unaffected. One must also bear in mind that, for an astrophysical black hole, the resulting effectively classical description should be consistent with the phenomenology of spacetime that is already confirmed experimentally [1, 46].

In the vast landscape of quantum models of black holes, the corpuscular picture [39, 40] is among those in which geometry emerges at some macroscopic scale from the microscopic quantum theory of gravitons [37, 43]. In this model, soft (virtual) gravitons are marginally bound in the potential well that they create and are characterised by a Compton-de Broglie wavelength λ_{G} equal to the horizon radius R_{H} of the black hole. One can easily infer that the energy scale of the gravitons is on the order of $\varepsilon_{\mathrm{G}} \sim \hbar/\lambda_{\mathrm{G}}$. For a total number N_{G} of gravitons, the total mass is $M \simeq N_{\mathrm{G}}\,\varepsilon_{\mathrm{G}}$, and one can infer the scaling relations $N_{\mathrm{G}} \sim M^2/m_{\mathrm{p}}^2 \sim R_{\mathrm{H}}^2/\ell_{\mathrm{p}}^2$. The nonlinearity of the gravitational interaction is apparent from the (negative) gravitational energy for an object of mass M contained within a sphere of radius R, that is $U_{\mathrm{N}} \sim M\, V_{\mathrm{N}}(R)$, with V_{N} given in Eq. (2.29). Most gravitons in this black hole model should have roughly the same wavelength λ_{G} and individual binding energy

$$\varepsilon_{\mathrm{G}} \sim \frac{U_{\mathrm{N}}}{N_{\mathrm{G}}} \sim -\frac{\ell_{\mathrm{p}}\, m_{\mathrm{p}}}{R}\,, \tag{2.68}$$

from which the Compton-de Broglie length $\lambda_{\mathrm{G}} \sim R$. The graviton self-interaction energy then reproduces the positive post-Newtonian energy

$$U_{\mathrm{GG}} \sim N_{\mathrm{G}}\,\varepsilon_{\mathrm{G}}\, V_{\mathrm{N}}(R) \sim \frac{G_{\mathrm{N}}^2\, M^3}{R^2}\,. \tag{2.69}$$

Assuming that the gravitons trapped inside the black hole are marginally bound, that is $U_{\mathrm{N}} + U_{\mathrm{GG}} \simeq 0$, one finally finds $\lambda_{\mathrm{G}} \sim R \simeq R_{\mathrm{H}}$ [17, 18]. Like for all bound states in quantum physics, modes of arbitrarily small wavelengths do not appear and, as a result, the classical central singularity is not realised. This is another form of *classicalization* of gravity [38, 42] and, as we recalled in previous Sections, this feature can also be explained from the quantum nature of the source.

Modelling black holes so simply as quantum states of gravitons with similar wavelengths λ_{G} is quite restrictive, since the inhomogeneous gravitational field in the space outside the object cannot be reproduced, not even in the Newtonian approximation. General relativity is a metric theory and without going as far as a full quantum theory of gravity, this problem can be approached in a simpler way for static and spherically symmetric sources [17, 21, 45]. More precisely, the quantum description can be limited to the metric function (2.29) that appears in the Schwarzschild metric

$$\mathrm{d}s^2 = -\,(1 + 2\, V_{\mathrm{N}})\,\mathrm{d}t^2 + (1 + 2\, V_{\mathrm{N}})^{-1}\,\mathrm{d}r^2 + r^2\,\mathrm{d}\Omega^2\,. \tag{2.70}$$

We remind that V_N represents the potential in the radial geodesic equation (2.56) from which we derived the discrete spectrum in Sect. 2.4. However, while r in Eq. (2.70) is the areal radius, the (post-)Newtonian expressions above and in Sect. 2.3 should be given in terms of the harmonic radius [14, 20, 70].

We next assume the quantum vacuum $|0\rangle$ to be a state of the universe with no excited modes (of matter or gravity). The Minkowski metric $\eta_{\mu\nu}$ can be associated to this absolute vacuum, because it is used to describe linearised gravity and for the definition of matter and gravitational excitations under these circumstances. Small matter sources with energy $M \ll m_p$ should be described accurately in this regime. Also, the Newtonian potential should be recovered from tree-level graviton exchanges in the non-relativistic limit. The potential V_N therefore emerges from the (non-propagating) longitudinal polarisation of virtual gravitons in this limit [43]. For sources with masses much larger than the Planck mass $M \gg m_p$ the complete classical dynamics can be in principle reconstructed [37, 63]. However, calculating the proper quantum state starting from the excitations of the linearised theory looks impossible in a highly nonlinear regime. This problem is usually bypassed (although not solved) in the effective field theory approach by assuming the existence of a classical background geometry which replaces $\eta_{\mu\nu}$ with $g_{\mu\nu}$ and which should be a solution to the classical Einstein equations.

This assumption is necessary if one wants to compare the theory with experimental data available in the weak-field regime, where the Einstein equations well describe the dynamics at macroscopic scales. We shall therefore assume that the relevant quantum states of gravity for static and spherically symmetric black holes effectively reproduce the Schwarzschild geometry (2.70). With this in mind, what is proposed next can be seen as quantising the longitudinal mode of the gravitational field using coherent states (which have the property of minimising the quantum uncertainty) in a suitable Fock space that is built starting from the Minkowski background vacuum [6, 8, 15, 59].

2.5.1 Quantum Coherent States for Classical Static Configurations

Given the previous discussion, a generic static metric function $V = V(r)$ can be conveniently described as the mean field of the coherent state of a free massless scalar field [18]. The dimensionless potential V must first be rescaled to arrive at a canonically normalised real scalar field $\Phi = G_N V = \sqrt{m_p/\ell_p}\, V$. The next step is to quantise Φ as a massless field which must satisfy the wave equation

$$\left[-\frac{\partial^2}{\partial t^2} + \frac{1}{r^2}\frac{\partial}{\partial r}\left(r^2\frac{\partial}{\partial r}\right)\right]\Phi(t,r) \equiv \left(-\partial_t^2 + \triangle\right)\Phi = 0\,, \qquad (2.71)$$

whose solutions can be expressed as

$$u_k(t,r) = e^{-i\,k\,t}\, j_0(k\,r) \,, \tag{2.72}$$

where $k > 0$ and $j_0 = \sin(k\,r)/k\,r$ represents the spherical Bessel function satisfying the orthogonality condition

$$4\,\pi \int_0^\infty r^2\,\mathrm{d}r\; j_0(k\,r)\; j_0(p\,r) = \frac{2\,\pi^2}{k^2}\,\delta(k-p)\,. \tag{2.73}$$

The quantum field operator, respectively its conjugate momentum are of course given by

$$\hat{\Phi}(t,r) = \int_0^\infty \frac{k^2\,\mathrm{d}k}{2\,\pi^2}\sqrt{\frac{\hbar}{2\,k}}\left[\hat{a}_k\, u_k(t,r) + \hat{a}_k^\dagger\, u_k^*(t,r)\right] , \tag{2.74}$$

$$\hat{\Pi}(t,r) = i\int_0^\infty \frac{k^2\,\mathrm{d}k}{2\,\pi^2}\sqrt{\frac{\hbar\,k}{2}}\left[\hat{a}_k\, u_k(t,r) - \hat{a}_k^\dagger\, u_k^*(t,r)\right] . \tag{2.75}$$

The two operators must satisfy the equal time commutation relations,

$$\left[\hat{\Phi}(t,r), \hat{\Pi}(t,s)\right] = \frac{i\,\hbar}{4\,\pi\,r^2}\,\delta(r-s)\,, \tag{2.76}$$

so that the creation and annihilation operators satisfy the commutation rules

$$\left[\hat{a}_k, \hat{a}_p^\dagger\right] = \frac{2\,\pi^2}{k^2}\,\delta(k-p)\,. \tag{2.77}$$

As usual, the Fock space is constructed from the vacuum defined as $\hat{a}_k\,|0\rangle = 0$ for all $k > 0$. The background we have chosen to correspond to the state $|0\rangle$ is the flat Minkowski metric which defines the d'Alembert operator in Eq. (2.71), precisely because this vacuum $|0\rangle$ is supposed to describe a spacetime that is completely empty. One could in fact have started from a space of any dimension, since without events, the notions of spatial dimension and time are empty (and the entire spacetime manifold cannot be distinguished from a point). Knowing that the system we eventually want to describe is static and three-dimensional, we explicitly introduced three spatial dimensions and one time from the onset.

The classical configurations of Φ which can be obtained in the quantum theory must be associated with certain states in this Fock space. In particular, coherent states

$$\hat{a}_k\,|g\rangle = g_k\, e^{i\,\gamma_k(t)}\,|g\rangle \tag{2.78}$$

represent a good option provided the expectation values of $\hat{\Phi}$ reproduce the classical potential, that is

$$G_{\rm N}\, \langle g|\, \hat{\Phi}(t,r)\, |g\rangle = V(r)\ . \tag{2.79}$$

From Eq. (2.74) one finds

$$\langle g|\, \hat{\Phi}(t,r)\, |g\rangle = \int_0^\infty \frac{k^2\, {\rm d}k}{2\,\pi^2} \sqrt{\frac{2\,\ell_{\rm p}\, m_{\rm p}}{k}}\, g_k\, \cos[\gamma_k(t) - k\,t]\, j_0(k\,r)\ . \tag{2.80}$$

By writing the potential as

$$V = \int_0^\infty \frac{k^2\, {\rm d}k}{2\,\pi^2}\, \tilde{V}(k)\, j_0(k\,r)\ , \tag{2.81}$$

the condition of perfect staticity is obtained for $\gamma_k = k\,t$ and

$$g_k = \sqrt{\frac{k}{2}}\, \frac{\tilde{V}(k)}{\ell_{\rm p}}\ . \tag{2.82}$$

The above choice of phases is a rather utilitarian way to eliminate the time dependence from the normal modes (2.72) and recover static configurations. From a physical perspective, no real system is perfectly static and this choice is a good approximation as long as the actual time evolution occurs on time scales much longer than $\Delta t \sim k^{-1}$.

Putting together these results, the coherent state becomes

$$|g\rangle = e^{-N_{\rm G}/2}\, \exp\left\{\int_0^\infty \frac{k^2\, {\rm d}k}{2\,\pi^2}\, g_k\, \hat{a}_k^\dagger\right\} |0\rangle\ , \tag{2.83}$$

with

$$N_{\rm G} = \int_0^\infty \frac{k^2\, {\rm d}k}{2\,\pi^2}\, g_k^2 \tag{2.84}$$

formally representing the total occupation number. A more explicative meaning is perhaps that the value of $N_{\rm G}$ measures the "distance" in the Fock space between $|g\rangle$ and the vacuum $|0\rangle$, the latter corresponding to $N_{\rm G} = 0$. Another quantity of interest is given by

$$\langle\, k\,\rangle = \int_0^\infty \frac{k^2\, {\rm d}k}{2\,\pi^2}\, k\, g_k^2\ , \tag{2.85}$$

from which the "average" wavelength $\lambda_{\rm G} = N_{\rm G}/\langle\, k\,\rangle$ is obtained.

2.5.2 Quantum Schwarzschild Black Holes

The Schwarzschild metric (2.70) is completely defined by $V_{\rm N} = \sqrt{G_{\rm N}}\,\Phi$ in Eq. (2.29), and all the necessary quantities can be calculated from the occupation numbers $|g_k|^2$ of the modes u_k. Eq. (2.81) can be easily inverted to obtain

$$\tilde{V}_{\rm N} = -4\,\pi\,G_{\rm N}\,\frac{M}{k^2} \tag{2.86}$$

and the coefficients

$$g_k = -\frac{4\,\pi\,M}{\sqrt{2\,k^3}\,m_{\rm p}}\,. \tag{2.87}$$

The total occupation number for such a coherent state is then formally given by

$$N_{\rm G} = 4\,\frac{M^2}{m_{\rm p}^2}\int_0^\infty \frac{{\rm d}k}{k}\,, \tag{2.88}$$

which suffers of logarithmic divergences both in the infrared and in the ultraviolet. Likewise, we can compute the average wavenumber

$$\langle\,k\,\rangle = 4\,\frac{M^2}{m_{\rm p}^2}\int_0^\infty {\rm d}k\,, \tag{2.89}$$

which only diverges (linearly) in the ultraviolet.

The ultraviolet divergences appear in the above expression because the Schwarzschild geometry (2.70) was assumed to hold for all values of $r > 0$, corresponding to a vanishing size of the central matter source. On the other hand, a diverging $N_{\rm G}$ implies that the coherent state $|g\rangle$ is not normalisable and one therefore arrives at the strong conclusion that the classical vacuum Schwarzschild geometry cannot be realised in quantum gravity [15]. Equivalently, the very existence of a proper quantum state $|g\rangle$ requires that the coefficients g_k depart from their classical expression (2.87) for $k \to 0$ and $k \to \infty$.

An obvious solution are sources with a finite size R, such as the quantum dust cores investigated in Sect. 2.4, or even with regular density profiles like the ones in Sect. 2.3. Either case would lead to occupation numbers $|g_k|^2$ which are sufficiently suppressed for $k \gg R^{-1}$ to remove the ultraviolet divergences. (In the case of black holes, R must be smaller than $R_{\rm H}$.) A detailed knowledge of the (effective) energy-momentum tensor of these sources would be necessary to compute explicitly the coefficients g_k, but this information is not yet available for the quantum dust core. For the sole purpose of removing the ultraviolet divergences, it is however enough to impose a cut-off $k_{\rm UV} \sim 1/R$, where we can eventually set $R \simeq R_M$ in Eq. (2.63) for a formed black hole.

In the infrared regime, we can impose a cut-off $k_{\text{IR}} = 1/R_\infty$ which accounts for the finite life-time $\tau \sim R_\infty$ of a realistic source [18]. For R_∞ one can also consider as an upper bound the size of the observable Universe [45]. However, this does not affect the results very much because modes with wavelength k^{-1} that are orders of magnitude larger than the horizon radius R_{H} do not impact the expectation value in Eq. (2.79) significantly [26, 41]. The occupation number now becomes

$$N_{\text{G}} = 4\,\frac{M^2}{m_{\text{p}}^2}\int_{k_{\text{IR}}}^{k_{\text{UV}}}\frac{\mathrm{d}k}{k} = 4\,\frac{M^2}{m_{\text{p}}^2}\ln\left(\frac{R_\infty}{R}\right) \tag{2.90}$$

and the average wavenumber

$$\langle\, k\,\rangle = 4\,\frac{M^2}{m_{\text{p}}^2}\int_{k_{\text{IR}}}^{k_{\text{UV}}}\mathrm{d}k = 4\,\frac{M^2}{m_{\text{p}}^2}\left(\frac{1}{R}-\frac{1}{R_\infty}\right). \tag{2.91}$$

The expression for N_{G} shows the same leading dependence on M^2 as the corpuscular scaling relation and Bekenstein's area law. The other essential result is that

$$\lambda_{\text{G}} = \frac{N_{\text{G}}}{\langle\, k\,\rangle} \sim \ell_{\text{p}}\,\frac{M}{m_{\text{p}}}\,, \tag{2.92}$$

for $R \simeq R_M \sim R_{\text{H}}$. One can further speculate that, because of the logarithm in Eq. (2.90), the number N_{G} will grow during the collapse, when the radius R of the matter source decreases, until the maximum is reached for $R \simeq R_M$ corresponding to the core ground state.

For the model to be realistic and comparable with experimental bounds, the mean field must still accurately reproduce the classical V_{N}, at least in the region outside the black hole horizon. Therefore, the coherent state $|g_{\text{BH}}\rangle$ which represents astrophysical black holes should give

$$G_{\text{N}}\,\langle g_{\text{BH}}|\,\hat{\Phi}(t,r)\,|g_{\text{BH}}\rangle \simeq V_{\text{N}}(r) \qquad \text{for } r \gtrsim R_{\text{H}}\,, \tag{2.93}$$

where the approximate equality was used to suggest that this is up to the experimental precision. Because of the previous discussion, the mean field in the left hand side of Eq. (2.93) will necessarily involve deviations from the classical function in the right hand side. In fact, one can easily compute the effective quantum potential

$$V_{\text{QN}} \simeq \int_{k_{\text{IR}}}^{k_{\text{UV}}}\frac{k^2\,\mathrm{d}k}{2\,\pi^2}\,\tilde{V}_{\text{N}}(k)\,j_0(k\,r) \simeq -\frac{2\,\ell_{\text{p}}\,M}{\pi\,m_{\text{p}}\,r}\int_0^{r/R}\mathrm{d}z\,\frac{\sin z}{z}\,, \tag{2.94}$$

where, in the second line, the substitution $z = k\,r$ was made and for the infrared cut-off the limit $k_{\text{IR}} = 1/R_\infty \to 0$ was taken. Finally, one obtains

$$V_{\rm QN} \simeq V_{\rm N} \left\{ 1 - \left[1 - \frac{2}{\pi} \,{\rm Si} \left(\frac{r}{R} \right) \right] \right\} , \tag{2.95}$$

where Si represents the sine integral function.

2.5.3 Geometry

The outer spacetime metric can now be written using the effective quantum potential from Eq. (2.95) as [15]

$$\mathrm{d}s^2 = -\left(1 + 2\, V_{\rm QN}\right) \mathrm{d}t^2 + \frac{\mathrm{d}r^2}{1 + 2\, V_{\rm QN}} + r^2 \,\mathrm{d}\Omega^2 \,. \tag{2.96}$$

Since $V_{\rm QN}$ is a function of R, this result can be directly interpreted as a quantum violation of the no-hair theorem [11]. The metric component $g_{tt} = 1 + 2\, V_{\rm QN}$ is plotted in Fig. 2.3, where one can see that the quantum corrected metric remains finite for $r \to 0$ and oscillates around the classical expression for sufficiently large r. These oscillations are in fact a mere consequence of the use of a hard ultraviolet cut-off, but the amplitude of such oscillations is still indicative of the size of the deviations from the classical geometry. Near the centre $r = 0$, the metric function is approximately given by

$$V_{\rm QN} \simeq -\frac{2\, G_{\rm N}\, M}{\pi\, R} \left[1 - \frac{r^2}{18\, R^2} \right] , \tag{2.97}$$

which is a bounded function with a derivative that vanishes in $r = 0$. One can conclude that this quantum effective gravitational potential leads to forces that remain finite everywhere.

The above result is confirmed by the form of the Kretschmann scalar for the quantum corrected metric in the limit $r \to 0$, which reads

$$R_{\alpha\beta\mu\nu}\, R^{\alpha\beta\mu\nu} \simeq \frac{64\, G_{\rm N}^2\, M^2}{\pi\, R^2\, r^4} \,. \tag{2.98}$$

This expression differs significantly from that for the classical Schwarzschild metric (2.70), namely $R_{\alpha\beta\mu\nu}\, R^{\alpha\beta\mu\nu} \sim r^{-6}$ in the same limit $r \to 0$. As a result the radial tidal forces are finite all the way into the centre. Another way of reaching the same conclusion is by looking at the relative acceleration between radial geodesics separated by δr, to wit

$$\frac{\ddot{\delta r}}{\delta r} = -R^1{}_{010} \simeq \frac{8\, G_{\rm N}^2\, M^2}{9\, \pi^2\, R^4} \left(1 - \frac{\pi\, R}{4\, G_{\rm N}\, M} \right) . \tag{2.99}$$

In the Schwarzschild spacetime, where the "spaghettification" of matter approaching the central singularity is supposed to happen, the corresponding relative acceleration is $\ddot{\delta r}/\delta r \sim r^{-4}$ and the difference between the two is obvious.

In the quantum corrected geometry, $r = 0$ is an integrable singularity [58]. Some geometric invariants still diverge in this case, but with no harmful effects to matter, as we have just shown. Moreover, the Einstein tensor $G_{\mu\nu}$ obtained from the metric (2.96) can be used to calculate the effective energy-momentum tensor $T_{\mu\nu}$ which, in turn, leads to the effective energy density

$$\rho = -\frac{G^0{}_0}{8\,\pi\,G_\mathrm{N}} = \frac{M}{2\,\pi^2\,r^3}\,\sin\!\left(\frac{r}{R}\right)\,, \tag{2.100}$$

as well as the effective radial pressure

$$p_r = \frac{G^1{}_1}{8\,\pi\,G_\mathrm{N}} = -\rho \tag{2.101}$$

and the effective tension

$$p_t = \frac{G^2{}_2}{8\,\pi\,G_\mathrm{N}} = \frac{M}{4\,\pi^2\,r^3}\left[\sin\!\left(\frac{r}{R}\right) - \frac{r}{R}\,\cos\!\left(\frac{r}{R}\right)\right]\,. \tag{2.102}$$

The integrals of both, the density and the pressure, over all space are finite with the following results

$$4\,\pi\int_0^\infty r^2\,\rho(r)\,\mathrm{d}r = -4\,\pi\int_0^\infty r^2\,p_r(r)\,\mathrm{d}r = M \tag{2.103}$$

and

$$4\,\pi\int_0^\infty r^2\,p_t(r)\,\mathrm{d}r = \frac{M}{2}\,, \tag{2.104}$$

which explains the integrable quantum nature of the metric.

The quantum corrected metric still contains a horizon when the cut-off scale $R \lesssim R_\mathrm{H}$. The horizon radius $r = r_\mathrm{H}$ is given by the solution of $2\,V_\mathrm{QN} = -1$, which can be computed numerically for different values of R. A remarkable result is that the horizon radius r_H shrinks for R approaching R_H. In fact, $r_\mathrm{H} = R$ for $R \simeq 0.6\,R_\mathrm{H}$ and r_H would further vanish for $R \simeq 0.7\,R_\mathrm{H}$. This means that the material core cannot be too close in size to the classical gravitational radius R_H in order for it to lie inside the actual horizon. Moreover, the quantum corrected geometry does not contain an (inner) Cauchy horizon (when an outer event horizon exists). Therefore, some potentially serious casual issues that appear for regular black holes [4, 12] are avoided in this scenario. Remarkably, this result extends to electrically charged black holes [21] and can be further generalised for rotating systems [22].

Fig. 2.3 Quantum metric component $g_{tt} = 1 + 2\,V_{QN}$ in Eq. (2.96) compared to $g_{tt} = 1 + 2\,V_N$ (dotted line) for $R = R_H/10$ (solid line) and $R = R_H/20$ (dashed line)

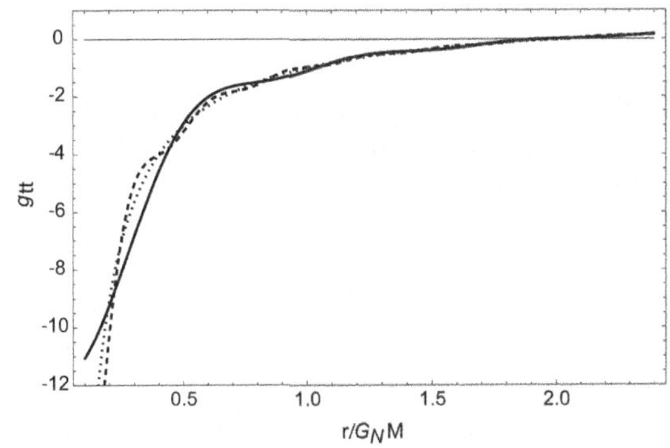

With enough experimental precision, the deviations of the quantum metric from the classical geometry could be observed on test bodies located outside the horizon at $r > r_H$. The scale of these deviations depends on the (quantum) size R of the matter source. Outside the horizon (for $r > r_H \sim R_H$) their amplitude decreases with the decrease of the ratio R/R_H (see Fig. 2.3). Therefore, these departures of V_{QN} from V_N become too small to be measured by a distant observer for small enough (but finite) values of the ratio R/R_H. (They should also converge to the expected amplitudes obtained from the effective field theory approach in the weak-field regime.) This effect can also be interpreted as a damping of transients, which happens as the source radius R shrinks and the collapse continues behind the horizon until $R \simeq R_M$.

The departure of r_H from the standard Schwarzschild radius $R_H = 2\,G_N\,M$ will also result in modifications of the horizon area $\mathscr{A}_H$, which ultimately means changes of the Bekenstein-Hawking entropy [5]

$$S_{QBH} = \frac{\mathscr{A}_H}{4\,\ell_p^2} = \frac{\pi\,r_H^2}{\ell_p^2} \tag{2.105}$$

and black hole temperature [49]

$$T_Q = \frac{\hbar\,\kappa}{2\,\pi} = \frac{\hbar}{2\,\pi}\,\frac{\partial V_{QN}}{\partial r}\bigg|_{r=r_H}\,, \tag{2.106}$$

where κ is the horizon surface gravity (see Ref. [15] for more details). Such changes will affect the Hawking emission in a way that depends on the state of the inner quantum core [11].

2.6 Outlook

In the previous sections we provided a comprehensive scenario supporting the conclusion that black holes are indeed quantum objects that cannot be described accurately within the effective field theory of gravity. In particular, the collapsed matter core displays a quantised surface areal radius in qualitative agreement with Bekenstein's area law and the same result is obtained by requiring that the outer geometry be derived as the mean-field approximation of a coherent quantum state.

One of the features stemming from the latter result is that the effective energy density inside the core behaves like $\rho \sim r^{-2}$ for $r \to 0$, which would be usually ruled out as a classical regular distribution. However, if one considers that one usually has $\rho \sim |\Psi|^2$ in quantum physics, where $\Psi = \Psi(r)$ is a wavefunction in position space, the only fundamental requirement is the integrability of the quantum state, that is

$$\int_0^r \rho(x)\, x^2\, \mathrm{d}x \sim \int_0^r |\Psi(x)|^2 x^2\, \mathrm{d}x < \infty\,, \tag{2.1}$$

for all $r > 0$. This condition leads to a mass function $m \sim r$ near the centre, which shows that $r = 0$ does not contain any singular source.

A classical energy density is supposed to vanish in $r = 0$, which instead implies that $m \sim r^3$. There is of course no singular contribution at $r = 0$ in this case either, but it becomes generically impossible to avoid the presence of a Cauchy horizon when the event horizon exists. Remarkably, the quantum nature of black holes with $m \sim r$ could solve this issue as well as removing the central singularity, at least in spherically symmetric configurations [15, 21].

Not surprisingly, for axially symmetric (rotating) black holes, the case is more involved. Assuming a constant specific angular momentum $a = J/M$ inevitably brings back the inner horizon, even if the mass function $m \sim r$ and there is no central singularity. However, there is no reason to not allow for $a = a(r)$ in such a way that it vanishes sufficiently fast for $r \to 0$ to again avoid the inner horizon [22].

From all of the above preliminary considerations, it looks promising to develop a more refined description of black holes as quantum objects which are safe from pathologies like the central singularity and Cauchy horizons, and still reproduce the well-tested geometries in the exterior. Of course, to validate this picture, one must find testable predictions that deviate from the classical general relativity phenomenology. The outer hair of these black holes is the natural candidate for further investigations.

Acknowledgements R.C. is partially supported by the INFN grant FLAG and his work has also been carried out in the framework of activities of the National Group of Mathematical Physics (GNFM, INdAM). O.M. was supported by Romanian Ministry of Research, Innovation and Digitalisation under Romanian National Core Program LAPLAS VII—contract no. 30N/2023.

References

1. B.P. Abbott et al., LIGO scientific and Virgo. Phys. Rev. X **9**, 031040 (2019). ([arXiv:1811.12907 [astro-ph.HE]])
2. S. Aghababaei, H. Moradpour, E.C. Vagenas, Eur. Phys. J. Plus **136**, 997 (2021). ([arXiv:2109.14826 [gr-qc]])
3. R.L. Arnowitt, S. Deser, C.W. Misner, Phys. Rev. **116**, 1322 (1959)
4. J. Bardeen, in *Proceedings of International Conference GR5* (Tbilisi, USSR, 1968), p. 174
5. J.D. Bekenstein, Phys. Rev. D **7**, 2333 (1973)
6. L. Berezhiani, G. Dvali, O. Sakhelashvili, Phys. Rev. D **105**, 025022 (2022). ([arXiv:2111.12022 [hep-th]])
7. A. Bonanno, A. Eichhorn, H. Gies, J.M. Pawlowski, R. Percacci, M. Reuter, F. Saueressig, G.P. Vacca, Front. Phys. **8**, 269 (2020). ([arXiv:2004.06810 [gr-qc]])
8. S. Bose, A. Mazumdar, M. Toroš, Nucl. Phys. B **977**, 115730 (2022). ([arXiv:2110.04536 [gr-qc]])
9. M.P. Bronstein, Phys. Zeitschr. der Sowjetunion **9**, 140 (1936)
10. H.A. Buchdahl, Phys. Rev. **116**, 1027 (1959)
11. X. Calmet, R. Casadio, S.D.H. Hsu, F. Kuipers, Phys. Rev. Lett. **128**, 111301 (2022). ([arXiv:2110.09386 [hep-th]])
12. R. Carballo-Rubio, F. Di Filippo, S. Liberati, C. Pacilio, M. Visser, JHEP **05**, 132 (2021). ([arXiv:2101.05006 [gr-qc]])
13. R. Casadio, Int. J. Mod. Phys. D **9**, 511 (2000). ([arXiv:gr-qc/9810073 [gr-qc]])
14. R. Casadio, Eur. Phys. J. C **82**, 10 (2022). ([arXiv:2103.14582 [gr-qc]])
15. R. Casadio, Int. J. Mod. Phys. D **31**, 2250128 (2022). ([arXiv:2103.00183 [gr-qc]])
16. R. Casadio, W. Feng, I. Kuntz, F. Scardigli, Phys. Lett. B **838**, 137722 (2023). ([arXiv:2210.12801 [hep-th]])
17. R. Casadio, A. Giugno, A. Giusti, Phys. Lett. B **763**, 337 (2016). ([arXiv:1606.04744 [gr-qc]])
18. R. Casadio, A. Giugno, A. Giusti, M. Lenzi, Phys. Rev. D **96**, 044010 (2017). ([arXiv:1702.05918 [gr-qc]])
19. R. Casadio, A. Giugno, A. Orlandi, Phys. Rev. D **91**, 124069 (2015). ([arXiv:1504.05356 [gr-qc]])
20. R. Casadio, A. Giusti, I. Kuntz, G. Neri, Phys. Rev. D **103**, 064001 (2021). ([arXiv:2101.12471 [gr-qc]])
21. R. Casadio, A. Giusti, J. Ovalle, Phys. Rev. D **105**, 124026 (2022). ([arXiv:2203.03252 [gr-qc]])
22. R. Casadio, A. Giusti, J. Ovalle, Quantum rotating black holes. [arXiv:2303.02713 [gr-qc]]
23. R. Casadio, I. Kuntz, Eur. Phys. J. C **80**, 581 (2020). ([arXiv:2003.03579 [gr-qc]])
24. R. Casadio, I. Kuntz, Eur. Phys. J. C **80**, 958 (2020). ([arXiv:2006.08450 [hep-th]])
25. R. Casadio, I. Kuntz, O. Micu, Phys. Lett. B **834**, 137455 (2022). ([arXiv:2206.13588 [gr-qc]])
26. R. Casadio, M. Lenzi, A. Ciarfella, Phys. Rev. D **101**, 124032 (2020). ([arXiv:2002.00221 [gr-qc]])
27. R. Casadio, M. Lenzi, O. Micu, Phys. Rev. D **98**, 104016 (2018). ([arXiv:1806.07639 [gr-qc]])
28. R. Casadio, M. Lenzi, O. Micu, Eur. Phys. J. C **79**, 894 (2019). ([arXiv:1904.06752 [gr-qc]])
29. R. Casadio, O. Micu, Phys. Rev. D **102**, 104058 (2020). ([arXiv:2005.09378 [gr-qc]])
30. R. Casadio, O. Micu, J. Mureika, Mod. Phys. Lett. A **35**, 2050172 (2020). ([arXiv:1910.03243 [gr-qc]])
31. R. Casadio, O. Micu, F. Scardigli, Phys. Lett. B **732**, 105 (2014). ([arXiv:1311.5698 [hep-th]])
32. R. Casadio, R. da Rocha, P. Meert, L. Tabarroni, W. Barreto, Class. Quant. Grav. **40**, 075014 (2023). ([arXiv:2206.10398 [gr-qc]])
33. R. Casadio, F. Scardigli, Eur. Phys. J. C **74**, 2685 (2014). ([arXiv:1306.5298 [gr-qc]])
34. R. Casadio, F. Scardigli, Phys. Lett. B **807**, 135558 (2020). ([arXiv:2004.04076 [gr-qc]])
35. R. Casadio, L. Tabarroni, Eur. Phys. J. Plus **138**, 104 (2023). ([arXiv:2212.05514 [gr-qc]])
36. C. De Coster, P. Habets, *Two-Point Boundary Value Problems: Lower and Upper Solutions* (Elsevier, Oxford, 2006)

37. S. Deser, Gen. Rel. Grav. **1**, 9 (1970). ([gr-qc/0411023])
38. G. Dvali, G.F. Giudice, C. Gomez, A. Kehagias, JHEP **08**, 108 (2011). ([arXiv:1010.1415 [hep-ph]])
39. G. Dvali, C. Gomez, Fortsch. Phys. **61**, 742 (2013). ([arXiv:1112.3359 [hep-th]])
40. G. Dvali, C. Gomez, Phys. Lett. B **719**, 419 (2013). ([arXiv:1203.6575 [hep-th]])
41. G. Dvali, C. Gomez, L. Gruending, T. Rug, Nucl. Phys. B **901**, 338 (2015). ([arXiv:1508.03074 [hep-th]])
42. G. Dvali, D. Pirtskhalava, Phys. Lett. B **699**, 78 (2011). ([arXiv:1011.0114 [hep-ph]])
43. R.P. Feynman, F.B. Morinigo, W.G. Wagner, B. Hatfield, *Feynman Lectures on Gravitation* (Addison-Wesley Publishing Co., 1995)
44. V.P. Frolov, G.A. Vilkovisky, Phys. Lett. B **106**, 307 (1981)
45. A. Giusti, S. Buffa, L. Heisenberg, R. Casadio, Phys. Lett. B **826**, 136900 (2022). ([arXiv:2108.05111 [gr-qc]])
46. C. Goddi et al., EHT. The Messenger **177**, 25 (2019). ([arXiv:1910.10193 [astro-ph.HE]])
47. P. Hajicek, B.S. Kay, K.V. Kuchar, Phys. Rev. D **46**, 5439 (1992)
48. P. Hajicek, C. Kiefer, Int. J. Mod. Phys. D **10**, 775 (2001). ([arXiv:gr-qc/0107102 [gr-qc]])
49. S.W. Hawking, Commun. Math. Phys. **43**, 199 (1975). [Erratum: Commun. Math. Phys. **46**, 206 (1976)]
50. S.W. Hawking, G.F.R. Ellis, *The Large Scale Structure of Space-Time* (Cambridge University Press, Cambridge, 1973)
51. W. Heisenberg, *The Physical Principles of the Quantum Theory* (University of Chicago Press, Chicago, 1930)
52. S. Hossenfelder, Living Rev. Rel. **16**, 2 (2013). ([arXiv:1203.6191 [gr-qc]])
53. P. Jizba, H. Kleinert, F. Scardigli, Phys. Rev. D **81**, 084030 (2010). ([arXiv:0912.2253 [hep-th]])
54. D.J. Kaup, Phys. Rev. **172**, 1331 (1968)
55. C. Kiefer, *Quantum Gravity* (Oxford University Press, 2007)
56. A.C. King, J. Billingham, S.R. Otto, *Differential Equations: Linear, Nonlinear, Ordinary, Partial* (Cambridge University Press, Cambridge, 2003)
57. G.G. Luciano, L. Petruzziello, Eur. Phys. J. C **79**, 283 (2019). ([arXiv:1902.07059 [hep-th]])
58. V.N. Lukash, V.N. Strokov, Int. J. Mod. Phys. A **28**, 1350007 (2013). ([arXiv:1301.5544 [gr-qc]])
59. W. Mück, Can. J. Phys. **92**, 973 (2014). ([arXiv:1306.6245 [hep-th]])
60. J.R. Oppenheimer, H. Snyder, Phys. Rev. **56**, 455 (1939)
61. T. Padmanabhan, Gen. Rel. Grav. **17**, 215 (1985)
62. T. Padmanabhan, Ann. Phys. **165**, 38 (1985)
63. T. Padmanabhan, Int. J. Mod. Phys. D **17**, 367 (2008). ([arXiv:gr-qc/0409089 [gr-qc]])
64. W. Pauli, Letter of Heisenberg to Peierls (1930), in *Scientific Correspondence*, Vol. II, ed. by K. von Meyenn (Springer, 1985) p. 15
65. R. Ruffini, S. Bonazzola, Phys. Rev. **187**, 1767 (1969)
66. F. Scardigli, Phys. Lett. B **452**, 39 (1999). ([arXiv:hep-th/9904025 [hep-th]])
67. F. Scardigli, R. Casadio, Eur. Phys. J. C **75**, 425 (2015). ([arXiv:1407.0113 [hep-th]])
68. F. Scardigli, G. Lambiase, E. Vagenas, Phys. Lett. B **767**, 242 (2017). ([arXiv:1611.01469 [hep-th]])
69. K.S. Stelle, Phys. Rev. D **16**, 953 (1977)
70. S. Weinberg, *Gravitation and Cosmology: Principles and Applications of the General Theory of Relativity* (Wiley, 1972)

Chapter 3
Covariant Collapse

Martin Bojowald and Erick I. Duque

3.1 Introduction

Gravitational collapse leads to regimes of large curvature and is therefore a popular testing ground for approaches to quantum gravity. Since some of them are based on canonical quantization, a description of collapse with canonical methods is of interest. As we demonstrate here, canonical gravity, compared with the more common formulation of gravity based on space-time tensors, is not only an alternative formulation, it also presents a previously unrecognized opening toward new theories of modified gravity in which novel effects of gravitational collapse may be possible.

An important ingredient of any canonical formulation of gravitational phenomena in modified form is an analysis of space-time structure. General covariance is not manifest in the canonical approach and therefore cannot be presupposed in modified gravitational theories of canonical form. Mathematically possible modifications are not guaranteed to preserve this important symmetry, and modifications that may respect other consistency conditions, such as maintaining gauge transformations, cannot always be expressed as properties of space-time. The notion of quantum space-time may take various forms that differ from the classical continuum picture, as proposed in several approaches to background-independent quantum gravity. Nevertheless, there should be at least a semiclassical region in which a continuum description of space-time in the standard form is possible, based on 4-dimensional line elements. Such line elements are not necessarily fundamental if they are used to

M. Bojowald (✉) · E. I. Duque
Institute for Gravitation and the Cosmos, The Pennsylvania State University, 104 Davey Lab, University Park, PA 16802, USA
e-mail: bojowald@psu.edu

E. I. Duque
e-mail: eqd5272@psu.edu

D. Malafarina and P. S. Joshi (eds.), *New Frontiers in Gravitational Collapse and Spacetime Singularities*, Springer Series in Astrophysics and Cosmology,
https://doi.org/10.1007/978-981-97-1172-7_3

describe solutions of quantum gravity, but in practice they often play an important role in analyzing or just defining models of quantum gravitational collapse.

If the dynamics of the underlying geometry is modified, as expected in quantum gravity, the preservation of general covariance, making it possible to use line elements, turns out to be a highly non-trivial question. In practice, collapse models often go even further and presuppose the existence of well-defined line elements not only in semiclassical regimes but even at large curvature, implicitly assuming that solutions of some theory of quantum gravity are compatible with a continuum space-time picture (although with non-classical dynamics). In the canonical approach, it is possible to analyze both ingredients—modified dynamics and modified space-time structure—and to determine how consistency requirements on the latter restrict possible versions of the former. The results presented here, worked out in explicit examples and based on our more detailed [1, 2], apply to all approaches to gravitational collapse based on modifications suggested by canonical quantum gravity, including loop quantum gravity. They demonstrate that potential effects of modified canonical gravity are highly but not completely restricted by the condition that the concept of space-time remain meaningful.

After a brief review of canonical gravity and its possible consistent modifications, we apply a class of candidate theories to the standard descriptions of non-rotating black holes, including a static slicing (usually used for the region outside of the horizon), a spatially homogeneous but time dependent slicing (usually used for the interior), and a suitably generalized Painlevé–Gullstrand slicing that is able to connect the two former solutions. We will explicitly demonstrate that these slicings are related by coordinate transformations even in modified canonical gravity, and highlight crucial features of a covariance analysis that guarantees this important property. We will also discuss the possibility of signature change in these models, which differs from a similar phenomenon found in models of loop quantum gravity [3, 4]. Our conclusions will briefly revisit the question of covariance in models of loop quantum gravity.

3.2 Canonical Gravity

Canonical gravity describes space-time as a family of evolving spatial geometries on equal-time hypersurfaces, together with their extrinsic curvature. This approach has the advantage that the structure of space-time need not be presupposed, as done implicitly by using an action principle that requires a well-defined 4-volume element for space-time integrations. Instead, space-time structure is encoded non-trivially in algebraic properties of the generators of evolution, subject to consistency conditions to ensure that the evolving spatial geometries described by solutions of the canonical theory can consistently be viewed as spacelike hypersurfaces in a single space-time solution.

The underlying evolution equations are of Hamiltonian form, generated by linear combinations of the Hamiltonian constraint $H[N]$ and the diffeomorphism constraint

$D[\mathbf{M}]$, where N is the spatial lapse function and $\mathbf{M}$ the spatial shift vector field [5–7]. These coefficients determine the space-time direction in which a time derivative is taken for the evolution equations. Classically, the resulting solutions are generally covariant and can be interpreted as spatial slices in an unambiguous space-time geometry if the generators $H[N]$ and $D[\mathbf{M}]$, expressed as functions on the phase space of the theory, are constraints, providing additional equations that the spatial metric and extrinsic curvature must obey. There is a description of canonical solutions through space-time line elements because the classical constraints are such that their gauge transformations, applied to solutions to the constraints and evolution equations, are equivalent to coordinate transformations in a space-time line element constructed from the spatial metric, as well as lapse and shift.

In this picture, the constraints serve multiple, partially independent roles: They (i) generate Hamilton's equations, (ii) impose constraints on the phase-space variables, (iii) directly imply gauge transformations on phase space, (iv) indirectly imply gauge transformations on lapse and shift by making sure that gauge transformations are compatible with evolution equations, and (v) have a combined set of gauge transformations on the spatial metric and lapse and shift as components of a space-time line element that is compatible with coordinate transformations on space-time. The first three points are common to all completely constrained gauge theories, while the fourth and fifth are special to gravitational (or, more generally, geometrical) theories. (Point (iv), the gauge transformation of multipliers in the constraints, may also be applied in non-gravitational theories, but in such cases rarely has physical relevance. Gravity is special because the lapse function and shift vector not only serve as multipliers of constraints but also appear as time components of the space-time line element.)

Consistent modifications of gauge theories require that the number of independent gauge transformations generated by modified constraints be the same as the original number, making sure that the modification maintains the same number of physical degrees of freedom. The constraints are then first class. If this condition is met, the modified constraints can consistently be imposed as constraints and generate evolution at the same time. Moreover, it is possible to remove non-physical degrees of freedom as gauge. The first three roles played by the constraints are therefore interlinked. The fourth and in particular the fifth one do not appear for non-gravitational gauge theories, but they are of particular importance for modified gravity in canonical form. Even if roles (i), (ii) and (iii) are consistently preserved by modifications, using modified or quantized constraints that are first-class, role (iv) can still be implemented, but role (v) is in general not realized. Modifications of gravitational theories therefore have to impose two independent and often tricky conditions, given by the first-class requirement together with the condition that gauge transformations not only remain valid but also be of a specific form that is compatible with coordinate transformations of some space-time line element. Formally, the distinguishing feature of gravitational theories is that they have constraints with multipliers and structure functions related to components of the space-time metric. Gauge properties therefore determine and restrict possible structures of space-time geometries given by solutions of the theory.

Explicit conditions for the covariance requirement can be determined by looking at the classical constraints of canonical gravity, in particular at their Poisson brackets

$$\{D[\mathbf{M}_1], D[\mathbf{M}_2]\} = D[\mathcal{L}_{\mathbf{M}_1}\mathbf{M}_2] \quad (3.1)$$

$$\{H[N], D[\mathbf{M}]\} = -H[\mathcal{L}_{\mathbf{M}}N] \quad (3.2)$$

$$\{H[N_1], H[N_2]\} = D[N_1\nabla N_2 - N_2\nabla N_1] \quad (3.3)$$

with Lie derivatives in the first two lines, and the gradient $\nabla^a = q^{ab}\partial_b$ in the last line, using the spatial metric q^{ab}. The lapse functions N, N_1, N_2 and the shift vector fields $\mathbf{M}$, $\mathbf{M}_1$, $\mathbf{M}_2$ in these brackets are, at this point, assumed to depend on the position x^a on a spatial slice as well as on the slice within a foliation labeled by some "time" coordinate t, but not independently on the spatial metric or its momentum.

On-shell, when the constraints and the equations of motion they generate are satisfied, this specific form of the brackets describes deformations of spatial hypersurfaces in space-time, and it guarantees that q^{ab} can be interpreted as the inverse of a metric induced on spatial slices in a space-time line element

$$\mathrm{d}s^2 = -N^2\mathrm{d}t^2 + q_{ab}(\mathrm{d}x^a + M^a\mathrm{d}t)(\mathrm{d}x^b + M^b\mathrm{d}t) \quad (3.4)$$

if spatial slices are evolved by Hamilton's equations generated by $H[N] + D[\mathbf{M}]$. More specifically than before, the conditions on modified canonical gravity are therefore that (i) the modified constraints remain first class, (ii) in such a form that they have brackets resembling (3.1)–(3.3) with a suitable candidate for an inverse spatial metric (in a sense to be specified below).

These conditions are highly non-trivial and do not allow much room for modifications. In fact, there are uniqueness theorems that demonstrate how the brackets determine compatible dynamical theories for q_{ab} as a spatial metric to be strictly of the classical form [8]. Nevertheless, there is a small opening because, unlike in the classical theory and still assumed in [8], the spatial tensor that appears in the bracket (3.3) through the gradient, raising the index of a partial derivative, need not be identical with the fundamental field q_{ab} (or a triad version) that defines the canonical phase space and appears in the constraint functionals. Consistency with coordinate transformations then requires that the spatial tensor in a modified (3.3), but not the fundamental field, supplies the spatial part of a space-time line element (3.4). Similarly, the remaining components of the space-time line element need not be identical with (though still related to) the multipliers of the evolution generator $H[N] + D[\mathbf{M}]$. If there is a difference in any of these objects, the space-time line element is not based directly on fundamental fields of the canonical formulation, but rather emergent because its coefficients must be derived by using a subset of the field equations. We will demonstrate this new possibility by a specific example, using spherical symmetry.

3.3 Spherical Symmetry

To simplify calculations, we work with spherically symmetric line elements in (densitized) triad variables e_I, $I = 1, 2$, such that a generic space-time line element takes the form

$$ds^2 = -N(t,x)^2 dt^2 + \frac{e_2(t,x)^2}{e_1(t,x)}(dx + M(t,x)dt)^2 + e_1(t,x)d\Omega^2 \tag{3.5}$$

with coefficients depending only on t and x. In this reduction, the constraints have phase-space expressions [9]

$$D[M] = \int dx\, M\left(k_2' e_2 - k_1 e_1'\right) \tag{3.6}$$

and

$$H[N] = \int dx N\left(\frac{(e_1')^2}{8\sqrt{e_1}e_2} - \frac{\sqrt{e_1}}{2e_2^2}e_1'e_2' + \frac{\sqrt{e_1}}{2e_2}e_1'' - \frac{e_2}{2\sqrt{e_1}} - \frac{e_2 k_2^2}{2\sqrt{e_1}} - 2\sqrt{e_1}k_1k_2\right) \tag{3.7}$$

where k_1 is canonically conjugate to e_1 and k_2 to e_2. (The classical equations of motion then demonstrate that k_1 and k_2 are components of extrinsic curvature, but this relationship is usually not maintained by modifications.)

Because all brackets in (3.1)–(3.3) are non-zero, they may be modified if the constraints are rearranged by linear combinations, such as replacing $H[N]$ with $H[\alpha N] + D[\beta N]$ with phase-space functions α and β (while leaving $D[M]$ unchanged). Usually, one does not expect linear combinations of constraints or gauge generators to change the underlying physics, but in gravitational theories, as it turns out, there is a possibility for this to happen. While $(H[N], D[M])$ and $(H[\alpha N] + D[\beta N], D[M])$ have the same solutions as constraints, they generate different gauge transformations. More importantly, since their brackets have additional terms from derivatives of α and β by the phase-space degrees of freedom, they do not necessarily imply the same emergent metric q_{ab}, derived from structure functions (3.3), as the original constraints. Space-time solutions can then be different from the classical ones, even if phase-space solutions of the constraints, derived on a fixed spatial slice, remain the same. If gauge transformations are modified, the spatial slices evolve non-classically, and they are lined up in an emergent space-time that differs from the classical composition. This outcome is indeed what happens if α and β are phase-space dependent, that is, dependent on e_1 and e_2 as well as their momenta, k_1 and k_2. (Mathematically, linear combinations with phase-space dependent coefficients may change space-time results because the brackets (3.1)–(3.3) with structure functions do not describe a Lie bracket but rather an L_∞-bracket [10]. Phase-space dependent lapse and shift have a home in the same algebraic structure, but compared with the classical description they determine an emergent line element on a non-constant section.)

A detailed calculation [2] shows that covariance is preserved if

$$\alpha(e_1, k_2) = \mu(e_1)\sqrt{1 - s\lambda(e_1)^2 k_2^2} \tag{3.8}$$

and

$$\beta(e_1, k_2) = \mu(e_1)\frac{\sqrt{e_1}}{2e_2^2}\frac{\partial e_1}{\partial x}\frac{s\lambda(e_1)^2 k_2}{\sqrt{1 - s\lambda(e_1)^2 k_2^2}} \tag{3.9}$$

with two free functions, $\lambda(e_1)$ and $\mu(e_1)$, and a parameter $s = \pm 1$, provided the emergent spatial metric equals

$$q_{xx}^{\rm em} = \frac{1}{\mu(e_1)^2}\left|1 + \frac{1}{4e_2^2}\frac{s\lambda(e_1)^2}{1 - s\lambda(e_1)^2 k_2^2}\left(\frac{\partial e_1}{\partial x}\right)^2\right|^{-1}\frac{e_2^2}{e_1} \tag{3.10}$$

and a signature factor

$$\epsilon = \mathrm{sgn}\left(1 + \frac{1}{4e_2^2}\frac{s\lambda(e_1)^2}{1 - s\lambda(e_1)^2 k_2^2}\left(\frac{\partial e_1}{\partial x}\right)^2\right) \tag{3.11}$$

is used in the emergent line element

$$ds_{\rm em}^2 = -\epsilon N^2 dt^2 + q_{xx}^{\rm em}(dx + M dt)^2 + x^2 d\Omega^2\,. \tag{3.12}$$

(The role of the signature factor has been discussed in [11]). The resulting class of canonical space-time theories, obtained from linear combinations of the classical constraints, is unique up to canonical transformations. For instance, the covariant models derived in [12–15] are related by a canonical transformation of (e_2, k_2) to our case of $s = +1$.

If one imposes first-class brackets of the constraints, one obtains the relationship

$$\beta(e_1, k_2) = -\frac{\sqrt{e_1}e_1'}{2e_2^2}\frac{\partial\alpha}{\partial k_2} - 2\sqrt{e_1}k_2\frac{\partial\alpha}{\partial e_1'} \tag{3.13}$$

between α and β, and a corresponding modified structure function given by $\epsilon/q_{xx}^{\rm em}$. The dependence on k_2 of the remaining function α is undetermined by this condition, unless the inverse of the structure function is required to transform like the inverse of a spatial metric under gauge transformations generated by the new constraints. Imposing this additional covariance condition restricts α to be of the form given above. This result shows that the first-class nature of modified constraints allows theories that are not covariant, given by linear combinations of the constraints where α is related to β by (3.13) but not of the form (3.8).

The existence of some remaining freedom after imposing covariance, parameterized by λ, μ and s, shows that classical spherically symmetric gravity is not uniquely determined by this condition, unless one imposes the additional condition that the radial metric of the emergent line element strictly equal the fundamental phase-space degree of freedom q_{xx}: Setting (3.10) equal to e_2^2/e_2 requires $\mu = 1$ and $s\lambda^2 = 0$. Uniqueness theorems for the classical field equations in a canonical formulation, given in [8], are circumvented by allowing for the existence of a new emergent metric distinct from the fundamental fields.

In this construction, the phase-space functions are subject to the constraints and canonical equations of motion, generated by $H[\alpha N] + D[\beta N + M]$. This combination can be interpreted as the old

$$D[M] = \int \mathrm{d}x\, M\left(k_2' e_2 - k_1 e_1'\right) \tag{3.14}$$

with a new Hamiltonian constraint

$$\begin{aligned} H^{(\mathrm{new})}[N] = \int \mathrm{d}x N\mu\sqrt{1 - s\lambda^2 k_2^2}\Bigg(&\left(\frac{1}{8\sqrt{e_1}e_2} - s\lambda^2 \frac{\sqrt{e_1}}{2e_2^2}\frac{k_1 k_2}{1 - s\lambda^2 k_2^2}\right)(e_1')^2 \\ &- \frac{\sqrt{e_1}}{2e_2^2} e_1' e_2' + \frac{\sqrt{e_1}}{2e_2} e_1'' + s\lambda^2 \frac{\sqrt{e_1}}{2e_2^2}\frac{e_2 k_2}{1 - s\lambda^2 k_2^2} e_1' k_2' \\ &- \frac{e_2}{2\sqrt{e_1}} - \frac{e_2 k_2^2}{2\sqrt{e_1}} - 2\sqrt{e_1} k_1 k_2 \Bigg). \end{aligned} \tag{3.15}$$

After performing a linear combination with phase-space dependent coefficients, the constraint equations imply the same relationships as the classical constraints, but equations of motion may have additional contributions from α and β. They can be solved for in suitable gauge choices. For the sake of simplicity, in what follows we present solutions only for constant μ and λ.

3.3.1 *Schwarzschild-Like Exterior*

If we look for static solutions we have $\dot{e}_1 = 0 = \dot{e}_2$ as well as $M = 0$, and we can fix the spatial gauge by choosing $e_1(x) = x^2$. The Hamiltonian constraint then implies that $e_2(x) = x/\sqrt{1 - 2m/x}$ takes the classical form. Equations of motion lead to additional conditions by requiring that all phase-space functions are independent of time. In particular, they determine the x-dependence of $N = \sqrt{1 - 2m/x}$, which also takes the classical form. (The factor α is constant in this case because $k_2 = 0$, and therefore does not contribute to the equations of motion.)

It may therefore seem that the classical Schwarzschild form is completely unmodified. However, there is an additional modification in the space-time line element because covariance requires that it be based on the emergent spatial metric

$$q_{xx}^{\text{em}} = \frac{e_2^2}{\mu^2 e_1|1 + s\lambda^2 e_1/e_2^2|} \tag{3.16}$$

(using $k_2 = 0$) and possibly a signature factor if $1 + s\lambda^2 e_1/e_2^2 < 0$. The corresponding emergent line element is given by

$$\mathrm{d}s_{\text{em}}^2 = -\epsilon\left(1 - \frac{2m}{x}\right)\mathrm{d}t^2 + \frac{\mathrm{d}x^2/\mu^2}{(1 - 2m/x)|1 + s\lambda^2 - 2ms\lambda^2/x|} + x^2\mathrm{d}\Omega^2\,. \tag{3.17}$$

Unless μ and λ are related by $\mu^2(1 + s\lambda^2) = 1$, this line element is not asymptotically flat in a strict sense because q_{xx}^{em} approaches $\mu^{-2}|1 + s\lambda^2|^{-1}$ for $x \gg 2m$. It is quasi-asymptotically flat [16], as also observed for other collapse models in [17].

The signature factor always equals $\epsilon = +1$ if $s = +1$. If $s = -1$ and $\lambda > 1$, we have $\epsilon = +1$ in a range of x determined by $1 - \lambda^2 + 2m\lambda^2 x^{-1} > 0$, or $x < 2m/(1 - \lambda^{-2}) = x_{\max}^- > 2m$. In this range, the emergent line element with explicitly positive spatial coefficients reads

$$\mathrm{d}s_{\text{em}}^2 = -\left(1 - \frac{2m}{x}\right)\mathrm{d}t^2 + \frac{\mathrm{d}x^2/(\mu^2(\lambda^2 - 1))}{(1 - 2m/x)(x_{\max}^-/x - 1)} + x^2\mathrm{d}\Omega^2\,. \tag{3.18}$$

For $x > x_{\max}^-$, $\epsilon = -1$ and we have the emergent line element

$$\mathrm{d}s_{\text{em}}^2 = \left(1 - \frac{2m}{x}\right)\mathrm{d}t^2 + \frac{\mathrm{d}x^2/(\mu^2(\lambda^2 - 1))}{(1 - 2m/x)(1 - x_{\max}^-/x)} + x^2\mathrm{d}\Omega^2\,. \tag{3.19}$$

These line elements, obtained for $s = -1$ and $\lambda > 1$, describe curved 4-dimensional Euclidean-signature space surrounding a Lorentzian black-hole geometry. The value $x_{\max}^-$ approaches the Schwarzschild radius in the limit of $\lambda \to \infty$. Maintaining asymptotic flatness, $\mu = 1/\sqrt{\lambda^2 - 1} \to 0$ in this case.

3.3.2 *Homogeneous Interior*

The emergent line element changes only the spatial part of the metric but provides no additional factor for the lapse function, other than the signature. The lapse function is modified at most through consistency relations that follow from modified equations of motion in a given gauge. It is therefore not clear that the usual procedure still applies in which one extends an exterior solution obtained for $x > 2m$, such as (3.17), to an interior solution by using the same metric components for $x < 2m$ and flipping the time and radial coordinates. We will independently derive a modified interior solution by imposing spatial homogeneity instead of staticity, and then compare with (3.17).

We assume that e_1, e_2, k_1, k_2 and N depend only on a time coordinate T, while $M = 0$. Instead of fixing the time gauge by specifying $N(T)$, we postulate that

$e_1(T) = T^2$ to make sure that the solution is written in a gauge that may be obtained from (3.17) by flipping space and time. The equation of motion for e_1,

$$\dot{e}_1 = -2\mu N\sqrt{e_1}k_2\sqrt{1 - s\lambda^2 k_2^2} \tag{3.20}$$

then provides a relationship between N and k_2:

$$N = -\frac{1}{\mu k_2\sqrt{1 - s\lambda^2 k_2^2}}\,. \tag{3.21}$$

With this relationship, the equation of motion for k_2,

$$\dot{k}_2 = \frac{\mu N}{2\sqrt{e_1}}\sqrt{1 - s\lambda^2 k_2^2}(1 + k_2^2) = -\frac{1}{2T}\frac{1 + k_2^2}{k_2} \tag{3.22}$$

decouples from the other equations and can be solved by

$$k_2(T) = \sqrt{\frac{2m}{T} - 1} \tag{3.23}$$

where we already identified the integration constant with twice the mass. The Hamiltonian constraint then determines the ratio

$$\frac{k_1}{k_2} = -\frac{me_2}{2T^2(2m - T)} \tag{3.24}$$

as a function of e_2, which simplifies the equation of motion

$$\dot{e}_2 = -\mu N\sqrt{1 - s\lambda^2 k_2^2}\left(\frac{e_2 k_2}{\sqrt{e_1}} + 2\sqrt{e_1}k_1\right) = \frac{1}{2}e_2\left(\frac{1}{T} - \frac{1}{2m - T}\right) \tag{3.25}$$

solved by

$$e_2(T) = B\sqrt{T(2m - T)} \tag{3.26}$$

with an integration constant B. This integration constant could be absorbed in the spatial coordinate, but we will set it equal to $B = \mu$ in order to eliminate a constant factor of $1/\mu^2$ from the emergent spatial metric derived below.

With these solutions, the emergent line element has lapse function

$$N = -\frac{1}{\mu\sqrt{(2m/T - 1)(1 - s\lambda^2(2m/T - 1))}} \tag{3.27}$$

and radial metric component

$$q_{XX}^{\mathrm{em}} = \frac{e_2^2}{\mu^2 e_1} = \left(\frac{2m}{T} - 1\right) \tag{3.28}$$

and therefore equals

$$\mathrm{d}s_{\mathrm{em}}^2 = -\frac{\mathrm{d}T^2/(\mu^2|1+s\lambda^2|)}{(2m/T-1)|1-2m(1+s\lambda^{-2})^{-1}/T|} + \left(\frac{2m}{T}-1\right)\mathrm{d}X^2 + T^2\mathrm{d}\Omega^2\,. \tag{3.29}$$

The new time coordinate takes its maximal value $T_{\max} = 2m$ at the horizon, and ends at the minimal value $T_{\min}^- = 0$ for $s = -1$, and

$$T_{\min}^+ = \frac{2m\lambda^2}{1+\lambda^2} \tag{3.30}$$

for $s = +1$, in which case k_2 reaches its upper bound of λ^{-1} required for the Hamiltonian constraint to be real. In these ranges, the line element always has Lorentzian signature.

The line element (3.29) is indeed obtained from (3.17) by flipping space and time, $T = x$ and $X = t$, as long as the latter is not subject to signature change. If there is signature change in the exterior, as discussed for $s = -1$ and $\lambda > 1$, the interior solution does not receive an additional sign factor because the emergent metric q_{XX}^{em} is always positive in the given range of T, and N in (3.27) is always real. Signature change, therefore, does not occur in this homogeneous model.

Looking back at the solution (3.29), some of the coefficients of the emergent interior line element could have been derived directly from a homogeneous model representing the interior geometry in this slicing. Since $e_1' = 0$ thanks to homogeneity, the emergent spatial metric does not differ much from the classical field e_2^2/e_1, except for an additional factor of μ^{-2} that can be absorbed in other functions as long as it is constant, as assumed here. The interior geometry could therefore have been derived without knowledge of the emergent spatial metric, but a complete interpretation of the remaining constants, such as m or a suitable relationship between μ and λ, requires a comparison with a compatible exterior. The non-trivial emergent spatial metric in the exterior then correctly matches the modified lapse dynamics in the interior upon flipping space and time coordinates, confirming that the constant m used in the interior solution is indeed the mass of a black hole. This matching can be done only with a full covariance analysis, and it would be impossible if modifications had been used that do not allow a covariant emergent space-time line element.

Even though covariance conditions trivialize in a homogeneous model, this property refers only to a single slicing in the Schwarzschild interior. Making sure that a valid description of space-time is obtained from such a treatment requires a full discussion of covariance, allowing for various slicings of the interior as well as extending the geometry to an exterior region. The latter can then be used, for instance, to relate μ to λ by imposing asymptotic flatness, or to identify integration constants with the mass. In our next example, we will show that the exterior (3.17) and the interior (3.29) are indeed two parts of a single space-time.

3.3.3 Painlevé–Gullstrand Line Element

A modified version of space-time in the Painlevé–Gullstrand gauge can be derived from the condition $e_1 = x^2$ together with uniform lapse, $N = 1$, replacing staticity. The component e_2 as well as the radial component of the emergent line element are then determined by the equations of motion, and it turns out that, unlike in the classical solution, the emergent spatial metric is not flat for non-zero λ. As before, we will present solutions for the case of constant λ independent of e_1 or x.

Since the gauge is not static, k_1 and k_2 are non-zero, but related to each other by the diffeomorphism constraint:

$$k_1 = \frac{e_2}{2x} k_2' . \tag{3.31}$$

The Hamiltonian constraint implies

$$\frac{e_2}{2x} k_2^2 + e_2 k_2 k_2' + \frac{e_2}{2x} - \frac{3x}{2e_2} + \frac{x^2}{e_2^2} e_2' = 0 . \tag{3.32}$$

Since $\dot{e}_1 = 0$ in this gauge, the corresponding equation of motion with $N = 1$ implies

$$\frac{M}{\mu} x + x k_2 \sqrt{1 - s\lambda^2 k_2^2} + s\lambda^2 \frac{x^3}{e_2^2} \frac{k_2}{\sqrt{1 - s\lambda^2 k_2^2}} = 0 \tag{3.33}$$

from which we obtain the shift vector M as a function of e_2 and k_2, which is related to e_2 by the Hamiltonian constraint.

In order to solve for e_2, we assume that it is time independent but not necessarily equal to x. The equation of motion $\dot{e}_2 = 0$ then implies

$$s\lambda^2 \left(\frac{x^2}{e_2} \frac{k_2}{\sqrt{1 - s\lambda^2 k_2}} \right)' + \frac{e_2 k_2}{x} \sqrt{1 - s\lambda^2 k_2} + 2x k_2 \sqrt{1 - s\lambda^2 k_2} = 0 \tag{3.34}$$

which can be combined with (3.33) and simplified to

$$x e_2' - e_2 + s\lambda^2 k_2 (e_2 k_2 - x e_2' k_2 - x e_2 k_2') = 0 . \tag{3.35}$$

The ansatz $e_2(x) = x f(x)$ then requires that $f(x)$ obeys the equation

$$(1 - s\lambda^2 k_2^2) f' - s\lambda^2 k_2 k_2' f = 0 \tag{3.36}$$

solved by $f\sqrt{1-s\lambda^2k_2^2} = $ constant. For our purposes, it is sufficient to assume that this constant equals one, and therefore

$$e_2(x) = \frac{x}{\sqrt{1-s\lambda^2k_2(x)^2}}. \tag{3.37}$$

(There may be other applications of this slicing that require a different choice of the constant, for instance if one is interested in an extension to the Euclidean region of the exterior.) Inserting this expression into the Hamiltonian constraint then implies

$$2xk_2k_2'(1+s\lambda^2) + k_2^2(1+s\lambda^2) = 0 \tag{3.38}$$

solved for $1+s\lambda^2 \neq 0$ by

$$k_2(x) = \sqrt{\frac{2C}{x}}, \tag{3.39}$$

retaining its classical form for all λ, up to an integration constant $C > 0$ that will be determined from asymptotic properties of the resulting emergent line element. Therefore

$$e_2(x) = \frac{x}{\sqrt{1-2Cs\lambda^2/x}}. \tag{3.40}$$

This slicing is available only in a range of x such that $1-2Cs\lambda^2/x > 0$.

With these solutions, the emergent line element has the radial component

$$q_{xx}^{\mathrm{em}} = \frac{1}{\mu^2|1+s\lambda^2|}\frac{1}{1-2Cs\lambda^2/x} \tag{3.41}$$

with the signature factor

$$\epsilon_{\mathrm{PG}} = \mathrm{sgn}(1+s\lambda^2) \tag{3.42}$$

from (3.11). (The second fraction in (3.41) is derived from e_2, which may also change sign, but this is not included in the definition of the signature factor because the effective spatial metric depends on e_2^2.) The shift vector equals

$$\begin{aligned} M &= -\mu k_2\left(\sqrt{1-s\lambda^2k_2^2} + s\lambda^2\frac{x^2}{e_2^2}\frac{1}{\sqrt{1-s\lambda^2k_2^2}}\right) \\ &= -k_2\sqrt{1-s\lambda^2k_2^2}\,\mu(1+s\lambda^2) = -\sqrt{\frac{2C}{x}}\sqrt{1-\frac{2Cs\lambda^2}{x}}\,\mu(1+s\lambda^2) \end{aligned} \tag{3.43}$$

as a consequence of (3.33).

We arrive at the emergent line element

$$\begin{aligned} ds_{\rm em}^2 &= -\epsilon_{\rm PG}{\rm d}\bar{t}^2 + q_{xx}^{\rm em}({\rm d}x + M{\rm d}\bar{t})^2 + x^2{\rm d}\Omega^2 \\ &= -\left(\epsilon_{\rm PG} - \frac{2C}{x}|1+s\lambda^2|\right){\rm d}\bar{t}^2 - \frac{2\epsilon_{\rm PG}}{\sqrt{x/(2C) - s\lambda^2}}\frac{{\rm d}x{\rm d}\bar{t}}{\mu} \\ &\quad + \frac{1}{1-2Cs\lambda^2/x}\frac{{\rm d}x^2}{\mu^2|1+s\lambda^2|} + x^2{\rm d}\Omega^2\,. \end{aligned} \tag{3.44}$$

For $s = -1$ and $\lambda > 1$, the slicing describes only a Euclidean region but not the Lorentzian part expected for larger x from the Schwarzschild exterior. The line element (3.44) then shows that x should be limited to a finite range to exclude the coordinate singularity where $1 - 2Cs\lambda^2/x = 0$. If we set $C = m/(\lambda^2 - 1)$, we obtain the same condition $x > x_{\rm max}^-$ as before where the Schwarzschild-type solution turns Euclidean.

In the Lorentzian case where $\epsilon_{\rm PG} = 1$, or $s = +1$ or $s = -1$ and $\lambda < 1$, the Newtonian limit at large x can be applied to show that $C(1 + s\lambda^2)$ equals the mass m. The Lorentzian line element therefore reads

$$\begin{aligned} ds_{\rm em}^2 &= -\left(1 - \frac{2m}{x}\right){\rm d}\bar{t}^2 - 2\frac{\sqrt{2m/x}}{\sqrt{1+s\lambda^2 - 2ms\lambda^2/x}}\frac{{\rm d}x{\rm d}\bar{t}}{\mu} \\ &\quad + \frac{1}{1+s\lambda^2 - 2ms\lambda^2/x}\frac{{\rm d}x^2}{\mu^2} + x^2{\rm d}\Omega^2 \end{aligned} \tag{3.45}$$

using $1 + s\lambda^2 > 0$.

We can now confirm that this line element is related to (3.17) for $x > 2m$ and $x < x_{\rm max}^-$. The angular term implies that the same radial coordinate x should be used in both cases. We then look for a transformation of the time coordinates such that

$${\rm d}\bar{t} = {\rm d}t + g(x){\rm d}x \tag{3.46}$$

with a function $g(x)$ to be determined. Inserting this transformation in (3.45) shows that the cross-term is eliminated provided that

$$g(x) = -\frac{\sqrt{2m/x}}{\mu(1 - 2m/x)\sqrt{1 + s\lambda^2(1 - 2m/x)}}\,. \tag{3.47}$$

(If $s = -1$, this function is real only if $x < x_{\rm max}^-$, where the Schwarzschild-type solution is Lorentzian.) The resulting coefficient of ${\rm d}x^2$ then equals

$$-(1-2m/x)g(x)^2-2\frac{\sqrt{2m/x}}{\sqrt{1+s\lambda^2(1-2m/x)}}\frac{g(x)}{\mu}+\frac{1}{\mu^2}\frac{1}{1+s\lambda^2(1-2m/x)}$$
$$=\frac{1}{\mu^2(1-2m/x)(1+s\lambda^2(1-2m/x))} \tag{3.48}$$

in agreement with (3.17).

Similarly, the transformation to new coordinates (T, X) such that $x = T$ and $X(\bar{t}, x)$ fulfills the equation

$$\mathrm{d}\bar{t} = \mathrm{d}X + \frac{\sqrt{2m/T}}{(2m/T-1)\sqrt{1-s\lambda^2(2m/T-1)}}\mathrm{d}x \tag{3.49}$$

maps (3.45) to the interior line element (3.29). This transformation confirms that the homogeneous interior and the static exterior are indeed two parts of a single space-time solution.

3.4 Conclusions

We have explicitly demonstrated general covariance and slicing independence in a model of modified canonical gravity. The presence of regions of signature change for some of the possible parameter choices demonstrates that these theories cannot be mimicked by higher-curvature effective actions. Since they are based on a canonical formulation, a likely origin of the modifications can be found in canonical or loop quantum gravity, where similar (but not identical) versions have indeed appeared, for instance related to the curvature bound seen here for $s = +1$.

More specifically, a rather large number of "quantum black holes" have been proposed based on loop modifications of the homogeneous Schwarzschild interior, starting with [18]. In this context, one treats the homogeneous model in isolation, implementing modifications of the Hamiltonian constraint that are largely unrestricted by covariance conditions thanks to exact spatial homogeneity. However, the resulting dynamics can be viewed as part of a black-hole space-time only if the homogeneous model can be consistently embedded in an exterior space-time. In an effective description, this condition requires the construction of a consistent modified line element that implements slicing independence. Our examples show how this feature can be achieved, but the underlying constructions also demonstrate that slicing independence is very restrictive. Most canonical modifications that could be imagined for a homogeneous model are ruled out by the requirement that it be embedded in a complete space-time.

Explicit examples of loop-modified black-hole models had already been subject to no-go theorems about covariance, for instance by [19] addressing claims in [20]. In fact, the direct conclusion of [19] could be evaded by the existence of new theories of modified canonical gravity, which had not been taken into account in the no-go result shown there. (The theorem used the class of generalized dilaton gravity [21, 22] as possible modifications of spherically symmetric models, which are of higher-curvature form.) However, since the required emergent metric had not been taken into account in [20] (or similar papers), the no-go result still applies. Using the specific form of modified canonical gravity with the covariance condition imposed, the no-go theorem of [19] can be generalized because the specific forms of modifications in black-hole models motivated by loop quantum gravity are incompatible not only with generalized dilaton gravity, but also with the new versions of modified gravity.

Developing consistent models of quantum black holes based on canonical quantum gravity is challenging. The encouraging message of our analysis, as already suggested by [14], is that consistent space-time models exist that are not classical and are not of higher-curvature form. Canonical quantum gravity can therefore lead to unique effects that may be testable by observations or at least have important conceptual implications.

Acknowledgements This work was supported in part by NSF grant PHY-2206591. An M-dependent term is missing in Eq. (3.34). The full equation and a detailed treatment can be found in arXiv:2312.09217 by the authors and Dennis Hartmann.

References

1. M. Bojowald, E.I. Duque, Emergent modified gravity, in preparation
2. M. Bojowald, E.I. Duque, Emergent modified gravity: covariance regained. Phys. Rev. D **108**, 084066 (2023). arXiv:2310.06798
3. M. Bojowald, G.M. Paily, Deformed general relativity and effective actions from loop quantum gravity. Phys. Rev. D **86**, 104018 (2012). (arXiv:1112.1899)
4. J. Mielczarek, Signature change in loop quantum cosmology. Springer Proc. Phys. **157**, 555 (2014). (arXiv:1207.4657)
5. P.A.M. Dirac, The theory of gravitation in Hamiltonian form. Proc. Roy. Soc. A **246**, 333–343 (1958)
6. J. Katz, Les crochets de Poisson des contraintes du champ gravitationne. Comptes Rendus Acad. Sci. Paris **254**, 1386–1387 (1962)
7. R. Arnowitt, S. Deser, C.W. Misner, *The Dynamics of General Relativity* (Wiley, New York, 1962), Reprinted in [23]
8. S.A. Hojman, K. Kuchař, C. Teitelboim, Geometrodynamics regained. Ann. Phys. (New York) **96**, 88–135 (1976)
9. M. Bojowald, Spherically symmetric quantum geometry: states and basic operators. Class. Quantum Grav. **21**, 3733–3753 (2004). (gr-qc/0407017)
10. C. Blohmann, M. Schiavina, A. Weinstein, A Lie-Rinehart algebra in general relativity, arXiv:2201.02883
11. M. Bojowald, S. Brahma, D.-H. Yeom, Effective line elements and black-hole models in canonical (loop) quantum gravity. Phys. Rev. D **98**, 046015 (2018). (arXiv:1803.01119)
12. A. Alonso-Bardají, D. Brizuela, Holonomy and inverse-triad corrections in spherical models coupled to matter. Eur. Phys. J. C **81**, 283 (2021). (arXiv:2010.14437)
13. A. Alonso-Bardají, D. Brizuela, Anomaly-free deformations of spherical general relativity coupled to matter. Phys. Rev. D **104**, 084064 (2021). (arXiv:2106.07595)
14. A. Alonso-Bardají, D. Brizuela, R. Vera, An effective model for the quantum Schwarzschild black hole. Phys. Lett. B **829**, 137075 (2022). (arXiv:2112.12110)

15. A. Alonso-Bardají, D. Brizuela, R. Vera, Nonsingular spherically symmetric black-hole model with holonomy corrections. Phys. Rev. D **106**, 024035 (2022). (arXiv:2205.02098)
16. U. Nucamendi, D. Sudarsky, Quasi-Asimptotically Flat Spacetimes and Their ADM Mass. Class. Quantum Grav. **14**, 1309–1327 (1997). (gr-qc/9611043)
17. M. Bouhmadi-López, S. Brahma, C.-Y. Chen, P. Chen, D.-H. Yeom, Asymptotic non-flatness of an effective black hole model based on loop quantum gravity. Phys. Dark Univ. **30**, 100701 (2020). (arXiv:1902.07874)
18. A. Ashtekar, M. Bojowald, Quantum geometry and the Schwarzschild singularity. Class. Quantum Grav. **23**, 391–411 (2006). (gr-qc/0509075)
19. M. Bojowald, No-go result for covariance in models of loop quantum gravity. Phys. Rev. D **102**, 046006 (2020). (arXiv:2007.16066)
20. A. Ashtekar, J. Olmedo, P. Singh, Quantum transfiguration of Kruskal black holes. Phys. Rev. Lett. **121**, 241301 (2018). (arXiv:1806.00648)
21. G. Kunstatter, H. Maeda, T. Taves, New 2D dilaton gravity for nonsingular black holes. Class. Quant. Grav. **33**, 105005 (2016). (arXiv:1509.06746)
22. K. Takahashi, T. Kobayashi, Generalized 2D dilaton gravity and KGB. Class. Quant. Grav. **36**, 095003 (2019). (arXiv:1812.08847)
23. R. Arnowitt, S. Deser, C.W. Misner, The dynamics of general relativity. Gen. Rel. Grav. **40**, 1997–2027 (2008)

Chapter 4
Quantum Black Holes: A Survey

Viqar Husain

4.1 Introduction

Black holes are stable solutions of general relativity [36, 44, 49], but are unstable "semiclassically" [23, 24] in the setting of quantum fields in curved spacetime [9], where they glow like a black body and eventually lose all their mass. The equation describing this mass loss follows from three observations: the energy flux from a black body follows the Stefan-Boltzmann law; the mass M of black hole is proportional to its horizon radius R, and inversely proportional to its temperature T. This leads to

$$\frac{dM}{dt} \sim M^{-2}, \tag{4.1}$$

which predicts a black hole lifetime of order M^3. This heuristic mass loss formula may be visualized, up to a point, by a continuous sequence of Schwarzschild metrics with $M = M(t)$; it includes a type of "backreaction" on the geometry in that the mass is time dependent; the singularity of the Schwarzschild metric at $r = 0$ remains.

It is widely believed that quantum gravitational effects become important in the late stages of Hawking radiation, with singularity avoidance being the most important one. This has led to two separate lines of investigation, "derivations" of non-singular static black holes from various approaches to quantum gravity, and the recalculation of Hawking radiation from either derived or postulated non-singular static black holes (see e.g. [15]). Another approach involves postulating time dependent metrics that describe matter inflow, singularity avoiding bounce, followed by matter outflow as phenomenological models for dynamical collapse with singularity avoidance [25, 26].

V. Husain (✉)
Department of Mathematics and Statistics, University of New Brunswick, Fredericton, Canada
e-mail: vhusain@unb.ca

D. Malafarina and P. S. Joshi (eds.), *New Frontiers in Gravitational Collapse and Spacetime Singularities*, Springer Series in Astrophysics and Cosmology,
https://doi.org/10.1007/978-981-97-1172-7_4

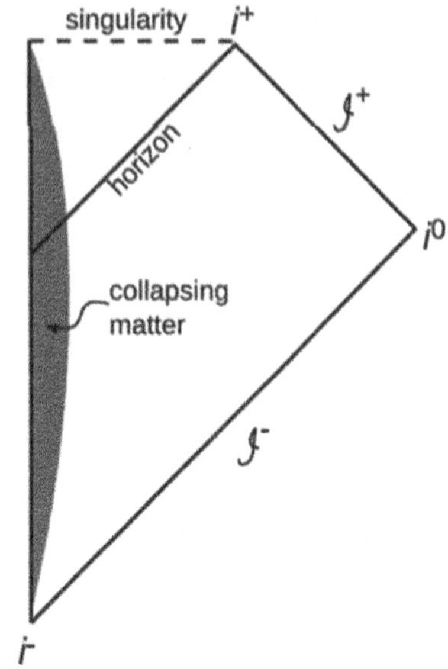

Fig. 4.1 Conformal diagram for classical gravitational collapse of matter in

There are numerous works on classical gravitational collapse in spherical symmetry beginning with the Oppenheimer-Snyder (OS) model [42], collapse of thin shells, and various generalizations that attempt to include a singularity avoiding core. Perhaps the most physically revealing classical work is the numerical study of scalar field in spherical symmetry [11], where it was shown that black hole formation exhibits critical behaviour in the sense that initial data must be above a critical threshold for black holes to form; below the threshold a black hole does not form; and at the threshold is a naked singularity. This work and its generalizations to perfect fluids establish the difference between the global structure of spacetimes with collapsing matter on the one hand and the global structure of vacuum static and stationary solutions on the other. Figure 4.1 is the Penrose diagram for matter that collapses to form a black hole.

In the semiclassical proposal due to Hawking shown in Fig. 4.2, matter collapse is followed by black hole evaporation—an important feature of this proposal is that gravitational collapse produces a spacelike singularity at the origin which then becomes regular after the black hole has evaporated; this is the celebrated "information loss" Penrose diagram, so named because in such a diagram an initial pure quantum state of matter on a Cauchy surface before the singularity forms would evolve to an impure thermal state on a Cauchy surface after the singularity has formed; this would mean that there is a loss of unitarity for quantum fields.

The essence of singularity avoidance via a quantum gravity mechanism is that a singularity does not form during gravitational collapse. This requires an upper bound on curvature. One way to see how such a scenario might arise in a theory of quantum gravity is to quantize a classical model system that contains collapse solutions where the long time limit of collapse is a static or stationary singular geometry. There are

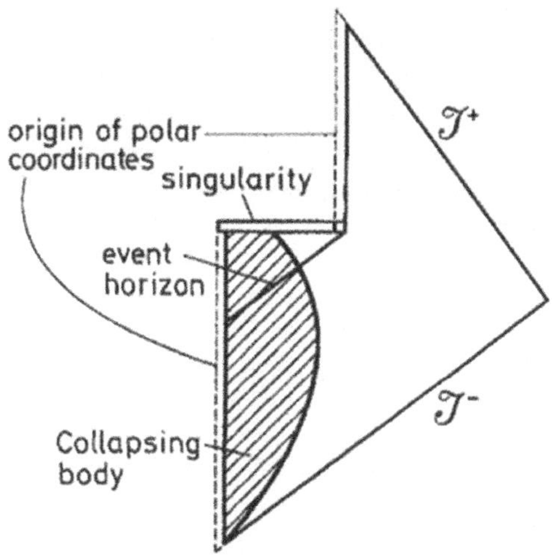

Fig. 4.2 Hawking's proposed Penrose diagram for black hole formation and evaporation (from [24])

two two types of models that can be used to study "quantum dynamical collapse," particle or field theoretic; the former are typically OS models where the radius of the star or shell becomes a quantum degree of freedom; the simplest of the latter type is spherically symmetric gravity with a perfect fluid or scalar field matter. (An alternative to quantization of a gravitational theory with matter are proposals for classical theories that have a built in limiting curvature–I will not discuss such theories here—see e.g. [16].)

The quantum OS models may be viewed as the Bohr atoms for gravitational collapse; both are spherically symmetric one particle systems that incorporate singularity avoidance. On the other hand spherically symmetric models with matter are two-dimensional field theories; their quantization would constitute at least a small milestone in understanding quantum gravitational collapse. Furthermore, apriori there is no reason to think that a complete quantization of a matter-gravity system would not be unitary by definition of quantization. This point of view suggests that semiclassical considerations are insufficient to settle the question of the end point of black hole evaporation, and that the semiclassical proposal without singularity avoidance as depicted in Fig. 4.2 is not what quantum gravity would give. Spacetime conformal diagrams that take into account singularity avoidance have been proposed; these include those depicted in Figs. 4.3 and 4.4—the first is a depiction of Hawking radiation from a non-singular black hole with dynamical horizons; the second is a proposal that includes a quantum gravity region in place of the curvature singularity. Neither is based on actual calculations from a quantum dynamical collapse model, but nevertheless each contains plausible physics that might emerge from a quantum theory of gravity. Lastly Fig. 4.5 is the conformal diagram of the non-singular metric [26]

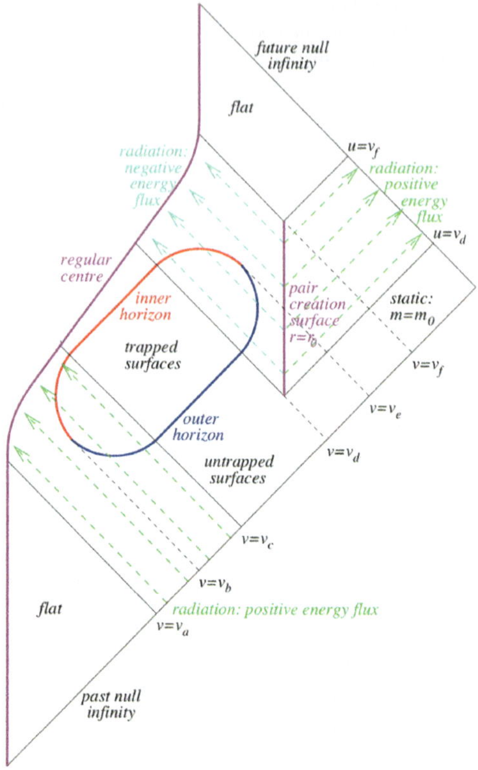

Fig. 4.3 A proposed conformal diagram [25] for formation and evaporation of non-singular black holes: infalling matter produces a pair of dynamical horizons which then annihilate as pair creation leads to evaporation

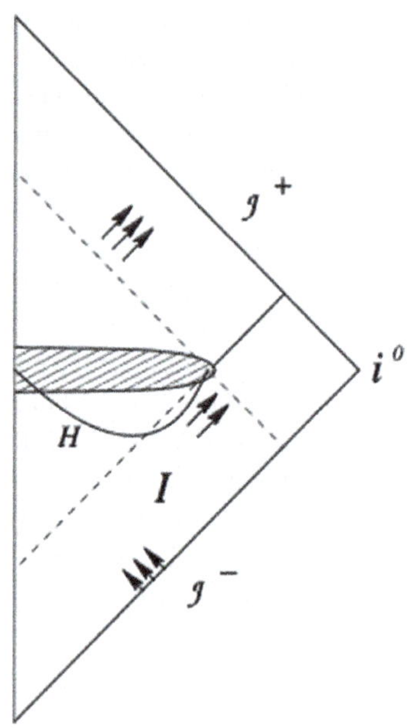

Fig. 4.4 Another proposed conformal diagram [4] for non-singular gravitational collapse: the shaded region depicts the quantum gravity region that replaces the spacelike singularity; *H* denotes a dynamical horizon

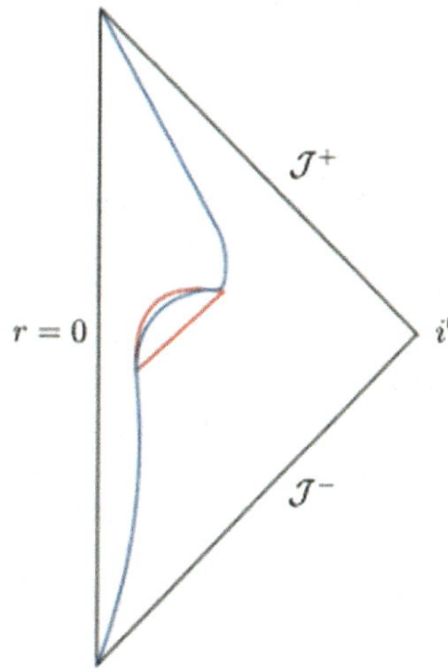

Fig. 4.5 Conformal diagram for matter collapse and bounce from [26]. The red lines show the horizons. The outer horizon segment is null. The inner horizon is timelike before the bounce and becomes spacelike after the bounce. The blue line is the peak of the density profile. It is timelike outside the bubble region (between the horizons) and spacelike inside it

$$ds^2 = -dt^2 + \left(dr + \sqrt{\frac{2m(r,t)}{r}}dt\right)^2 + r^2 d\Omega^2 \tag{4.2}$$

with mass function

$$m(r,t) = M_0 \left[1 + \tanh\left(\frac{r - r_0 - v(r,t)t}{\alpha\, l_0}\right)\right]^a \tanh\left(\frac{r^b}{l_0^b}\right). \tag{4.3}$$

and velocity profile

$$v(r,t) = \frac{A}{(1+r)^n} \tanh\left(\frac{t - t_0}{l_0}\right); \tag{4.4}$$

$M_0, r_0, l_0, \alpha, a, b, n$ and A are constants. The velocity function is designed to produce an inflow, bounce and outflow. The parameters may be adjusted to produce black hole lifetime that includes the M^3 suggested by (4.1).

With this outline and introduction, the next section reviews non-singular black hole proposals that are either postulated, or derived from a "quantization" of Schwarzschild spacetime; Section 4.3 is on quantum mechanical collapse models including OS and shell collapse; Sect. 4.4 reviews field theoretic models, gravity with a scalar field, and with dust; it also summarizes an effective theory of dust collapse that suggests black holes disappear in a shock wave [28, 29]; lastly Section 4.5 discusses open questions concerning quantum dynamical collapse and quantum gravity.

4.2 Quantizing a Black Hole

4.2.1 Schwarzschild Minisuperspace

Among the approaches to understanding quantum black holes is that of quantizing the Schwarzschild black hole starting from the canonical formulation of Einstein gravity. Such an approach follows the Wheeler-DeWitt approach with the constraints of gravity imposed as Dirac quantization conditions. In vacuum spherical symmetry the canonical theory has no local degrees of freedom; the only degrees of freedom are the mass and its conjugate momentum, both of which may be viewed as the action-angle variables of the solved canonical theory. Kuchar performed such an analysis [39] by using a time dependent canonical transformation on an ADM parametrization of canonical variables. As Kuchar points out, such an approach for the vacuum case is dynamically trivial, but it provides a potentially useful formalism for adding matter fields. The staring point of this analysis is the ADM canonical action for the spherically symmetric sector of the Einstein equations with the spatial metric in the form

$$ds^2 = \Lambda(r)dr^2 + R^2(r)d\Omega^2. \tag{4.5}$$

This defines the two metric functions; their conjugate momenta are P_Λ and P_R. The ADM canonical action is

$$S = \int dt \int_0^\infty dr \left[P_\Lambda \dot{\Lambda} + P_r \dot{R} - N^r C_r - NH\right], \tag{4.6}$$

where N, N^r are the lapse and radial shift functions, and C_r and H are the radial diffeomorphism and Hamiltonian constraints which are function of the canonical variables and their derivatives; their exact form does not matter for this discussion. Kuchar's analysis involves imposing boundary conditions appropriate to asymptotic flatness, and then making a canonical transformation to write the Schwarzschild mass and time coordinate as a function of these canonical variables. The end result of this detailed analysis is that, on the reduced phase space, the physical Hamiltonian vanishes leaving only the action angle variables—the mass and its conjugate momentum. Upon quantization, the quantum state of the Schwarzschild black hole is any normalizable function of its mass: $\Psi(m)$. This rather simple result is not surprising : if a classical system can be completely solved to its action angle variables, then quantization of the resulting action angle variables, with their zero Hamiltonian, does not contain any dynamics. A lesson for quantum gravity is that it is useful to have a time gauge choice with respect to which dynamics can be observed.

4.2.2 *Effective Equations*

Another approach to quantizing the Schwarzscild spacetime comes from the observation that the region interior to the horizon is isometric to the Kantowski-Sachs cosmology. Application of a quantization procedure to this interior cosmology [40, 41] and then matching to the exterior Schwarzschild results in a non-singular black hole. There have been several investigations of such "interior" quantizations—for a sample see [6, 12, 13] and references therein. The essence of this approach in Loop Quantum Gravity (LQG) is to view the underlying geometry as discrete. This leads to "effective equations" where the gravitational momentum term in the Hamiltonian constraint is bounded. This in turn bounds the matter density through the effective constraint. There are many choices of such discretizations, with each one leading to its corresponding non-singular black hole.

The basic idea underlying effective equations is illustrated by considering the parametrized particle with canonical action

$$S = \int ds \left[p_\tau \dot{\tau} + p\dot{x} - N(p_\tau + H(x, p))\right]. \tag{4.7}$$

This action is invariant under the reparametrization $s \to f(s)$. It has features in common with the canonical action for a homogeneous cosmological model, with the difference that the constraint obtained by varying with respect to the lapse function N is linear in the momentum p_τ; $H(x, p)$ can be any function, but is usually taken to have the standard form $H(x, p) = p^2/2m + V(x)$. The gauge fixed action obtained by making the time gauge choice $s = \tau$ and solving the constraint gives the usual particle action. Quantization of the parametrized action using Dirac's method entails solving the corresponding Wheeler-DeWitt equation

$$\left(-i\frac{\partial}{\partial\tau} + \hat{H}(\hat{x}, \hat{p})\right)\Psi(x, \tau) = 0; \tag{4.8}$$

it is the linearity of the constraint in p_τ that gives the time dependent Schrodinger equation.

Let us now assume that the configuration space is a lattice and consider quantization on this lattice. The natural Poisson algebra for quantization that replaces $\{x, p\} = 1$ is

$$\{x, e^{i\lambda p}\} = i\lambda e^{i\lambda p}, \tag{4.9}$$

where λ is a parameter. Such an algebra becomes necessary on a lattice since it is not possible to define the momentum as a derivative operator; it must be replaced by a translation operator $U_\lambda = \exp[i\lambda p]$; the algebra (4.9) corresponds to uniform lattice with spacing λ. There is a more general elementary algebra that can be used for quantization. This is

$$\{f(x), \exp[i\lambda p/f'(x)]\} = i\lambda \exp[i\lambda p/f'(x)], \tag{4.10}$$

where $f(x)$ is any once differentiable function. It corresponds to a non-uniform "adaptive mesh" lattice with spacing that varies according to the function $f'(x)$.

Let's consider a lattice quantization of the parametrized particle that uses the first algebra (4.9) since that is sufficient to illustrate the path to effective equations. A quantization of this algebra is defined by

$$\hat{x}|n\rangle = n\lambda|n\rangle, \quad \hat{U}_\lambda|n\rangle = |n+1\rangle, \tag{4.11}$$

where $\hat{U}_\lambda \equiv \widehat{e^{i\lambda p}}$ is like a ladder operator. It is readily verified that this satisfies the commutator relation $[\hat{x}, \hat{U}] = \lambda\hat{U}$ and gives the correspondence $\{\ ,\ \}/i \leftrightarrow [\ ,\ ]$. This type of quantization is referred to as "polymer quantization;" while we have considered a fixed lattice, this is a special case of the full non-separable Hilbert space consisting of all possible lattices [5, 21]; in practice for computation a specific lattice is deployed.

With this in hand it is possible to define an effective classical dynamics. This is the type of dynamics that underlies the non-singular black hole in effective LQG. The first point to note is that kinetic energy operator must be defined using the operator $\hat{U}_\lambda$. One way to do this is by noting that $(U - U^*)/2i\lambda \approx p$ for $\lambda \ll 1.$, and therefore a lattice dependent "momentum" operator may be defined by

$$\hat{p}_\lambda = \frac{\hat{U}_\lambda - \hat{U}_\lambda^\dagger}{2i\lambda} \equiv \frac{\widehat{\sin\lambda p}}{\lambda}; \tag{4.12}$$

there are other choices, but this is the simplest. The main point is that this is a bounded operator so the kinetic energy constructed from it is bounded above. *The prescription for obtaining effective equations is to replace all momenta in the classical constraint by the corresponding sine functions.* This rule of thumb is made precise by taking the expectation value of the constraint operator in semiclassical states [33]. Following this procedure, the simplest effective dynamics for parametrized particle is defined by the effective Hamiltonian constraint

$$H_{\text{eff}} = \frac{\sin(\lambda p_\tau)}{\lambda} + \frac{\sin^2(\lambda p)}{2\lambda^2} + V(x) = 0. \tag{4.13}$$

The λ dependent evolution equations follow from this using the Poisson brackets $\{\tau, p_\tau\} = 1$ and $\{x, p\} = 1$, e.g.

$$\begin{aligned} \dot{\tau} &= \{\tau, H_{\text{eff}}\} = \cos(\lambda p_\tau) \approx 1 - \frac{p_\tau^2\lambda^2}{2}; \\ \dot{x} &= \{x, H_{\text{eff}}\} = \frac{1}{\lambda}\sin(\lambda p)\cos(\lambda p) \approx p\left(1 - \frac{2}{3}\lambda^2 p^2\right). \end{aligned} \tag{4.14}$$

It is immediate that the $\lambda \ll 1$ approximation of the constraint gives the continuum result without the discreteness corrections that come from this type of quantization (here with lapse $N = 1$). However if the discreteness is considered fundamental, the limit is not taken and one seeks effective solutions that may exhibit continuum behavior in some regime.

A similar procedure is applied to the Hamiltonian constraint for the Kantowski-Sachs cosmology which describes the interior of the Schwarzschild metric; the difference is that the same λ and algebra need not underlie the quantization for different degrees of freedom, i.e. that both the algebras (4.9) and (4.10) with different λ parameters and choice of f may be used for different canonical pairs; another difference is that the connection-triad variables are used as the canonical variables—details of such applications, along with solutions of the corresponding effective equations appear in the recent review [7]. It is the solutions of such effective equations that give rise to non-singular black holes in LQG. This example of the parametrized particle serves only to describe the basic idea in a simple setting.

Effective equations of this type which contain a fundamental discreteness scale (λ in the above example) are also curvature singularity avoiding. This may be seen by noting that an effective Hamiltonian constraint in which the gravitational momentum term p^2 is replaced by $\sin^2(\lambda p)/\lambda^2$ leads to bounded energy density. For instance the ADM Hamiltonian constraint for FRW cosmology and scalar field with phase space variables $(v = a^3, p_v)$ and (ϕ, p_ϕ) is

$$H = -\frac{3}{8} v p_v^2 + \frac{p_\phi^2}{2v} + \frac{v}{2} m^2 \phi^2 = 0 \tag{4.15}$$

gives the effective constraint

$$-\frac{3v}{8\lambda^2} \sin^2(\lambda p_v) + \frac{1}{2v} p_\phi^2 + +\frac{v}{2} m^2 \phi^2 = 0; \tag{4.16}$$

from this it is immediate that the scalar field energy density

$$\rho_\phi = \frac{p_\phi^2}{2v^2} + \frac{1}{2} m^2 \phi^2 \tag{4.17}$$

is bounded above.

There are other examples of curvature limiting theories that lead to regular black holes. However the question remains whether such black holes form as the end point of gravitational collapse. One way to approach this problem is to write effective equations for models such as the gravity-scalar field system in spherical symmetry [27, 37, 51, 52], and to determine whether and how such models emerge from a theory of quantum gravity.

4.3 Quantizing Gravitational Collapse: Particle Models

As mentioned in the introduction there are two types of dynamical collapse models, the quantum mechanical OS models and their variants, and field theoretic models such as the 1+1 field theory for scalar field collapse in spherical symmetry. There is a vast literature on the former, and relatively little on the latter due to its inherent difficulty.

4.3.1 Quantum Oppenheimer-Snyder Models

Let us briefly summarize the simplest OS model [1, 43] to set the stage for its variants and their quantization. The star is modelled as an FRW cosmology in coordinates (t, r, θ, ϕ) with metric

$$ds^2 = -dt^2 + a^2(t)\left(dr^2 + r^2 d\Omega\right), \tag{4.18}$$

and its exterior in coordinates (T, R, θ, ϕ) with metric

$$ds^2 = -f(R)dT^2 + f(R)^{-1}dR^2 + R^2 d\Omega^2. \tag{4.19}$$

Let the motion of the star's surface be defined by the equations $R = R(t)$ and $T = T(t)$ as observed from the exterior, where t is the proper time of comoving observers on the star's surface, and let this surface be at $r = r_0$ as seen from the interior. The equations of motion of the OS model arises from matching the interior and exterior metrics induced on the star's surface. These are

$$R(t) = r_0 a(t), \quad f(R)\dot{T}^2 - f^{-1}(R)\dot{R}^2 = 1. \tag{4.20}$$

From these equations it is evident that the trajectory of the star's surface is determined by a given $a(t)$. With $R(t)$ determined from the first equation, the second gives $T(t)$, and hence the trajectory of the star's surface as seen in the exterior (T, R) coordinates.

The quantum version of this model now follows by simply using a form of $a(t)$ determined by one's favorite model of a quantum FRW universe. Typically such a quantum universe is regular with the curvature singularity replaced by a quantum gravity induced bounce. As seen from the exterior this would appear initially as a collapsing star which then starts to expand outward after reaching a minimum radius of the order of the Planck length. This is the basic intuition.

One such model comes from an effective Hamiltonian constraint that arises from considerations similar to those discussed above and first derived from loop quantum cosmology (see e.g. [2] for a brief review). Here the effective constraint can be put into the form

$$\left(\frac{\dot{a}}{a}\right)^2 = \frac{4\pi G}{3}\rho\left(1 - \frac{\rho}{\rho_c}\right), \tag{4.21}$$

where ρ_c is of order the Planck density. It is evident from this equation that the universe undergoes a bounce at the maximum density $\rho = \rho_c$.

A specific case where this arises is in the spherically symmetric Lemaitre-Tolman-Bondi (LTB) class of models which describe gravitational collapse of dust. The quantization of this model has been studied in the ADM canonical variables [47]. One approach to studying this model uses the dust time gauge [30] which yields a physical Hamiltonian for the system starting from the Painleve-Gulstrand (PG) coordinates for the canonical theory [32]. These ingredients were combined in a recent OS model [34] where the interior of the star is described by an LTB metric with spatially constant density, and an effective physical Hamiltonian, constructed along the lines described above, is used to solve the effective classical equations for the star's surface. The result is that the star collapses, forms a black hole, and then it's surface bounces outward and exits the event horizon; the time scale for disappearance of the event horizon is order mass square of the black hole that formed. Although at the effective equation level, this result indicates what might happen in field theory models. Such an extension is described in some detail in the next section [29].

Another approach that does not use effective equations, but utilizes a canonical quantization based on Kuchar's geometrodynamics (summarized above [39]). This also results in a quantum bounce of the star [35]. There are many other papers on the quantization of this model, too numerous to review here. All are concerned ultimately with the effect of an interior quantization on the motion of a star's surface, or on the quantization of the surface dynamics.

4.3.2 *Collapsing Shells*

A related class of models in spherical symmetry describe collapsing thin shells which may be timelike or null. The interior star of the OS model is replaced by flat spacetime, and the exterior remains Schwarzschild. The interior, surface and exterior metrics are respectively

$$\begin{aligned} ds^2_{int} &= -dt^2 + dr^2 + r^2 d\Omega^2 , \\ d\sigma^2_{\Sigma} &= -d\tau^2 + X^2(\tau) d\Omega^2 , \\ ds^2_{ext} &= -f(R)dT^2 + f^{-1}(R)dR^2 + R^2 d\Omega^2 . \end{aligned} \tag{4.22}$$

The spherical coordinates (θ, ϕ) are the same in the three regions, τ is the proper time of the shell, $X(\tau)$ is the shell's radius (the configuration variable of the model), $f(R) = 1 - 2E/R$ and the exterior Schwarzschild parameter E is the total gravitational mass of the shell as seen at infinity. The junction condition on the metrics at Σ are

$$g_{int}|_\Sigma = \sigma_\Sigma = g_{ext}|_\Sigma. \tag{4.23}$$

These give $X(\tau) = r(\tau) = R(\tau)$, and

$$\dot{t}^2 - \dot{r}^2 = 1 = f\dot{T}^2 - f^{-1}\dot{R}^2. \tag{4.24}$$

where a dot denotes the derivative with respect to τ.

The shell stress energy tensor is taken to be that of pressureless dust $T_{ab} = \sigma u_a u_b$, which is confined to the boundary between the interior Minkowski metric and the exterior Schwarzschild metric. To derive T_{ab}, the jump in the extrinsic curvature from the interior Minkowski metric to the exterior Schwarzschild metric, is computed using the outward pointing normals $n_a^{int}dx^a = -\dot{r}dt + \dot{t}dr$ and $n_a^{ext}dx^a = -\dot{R}dT + \dot{T}dR$. This leads to the result [43]

$$4\pi X^2(\tau)\sigma = \text{constant} \equiv m \tag{4.25}$$

and

$$E = m\sqrt{1+\dot{X}^2} - V, \qquad V \equiv \frac{m^2}{2X}. \tag{4.26}$$

This is the final dynamical equation for the thin shell. It also provides an expression for the Hamiltonian $H(X, \dot{X}; m) \equiv E$ as a function of the shell's mass, m, its radius $X(\tau)$, and velocity $\dot{X}$ with respect to shell proper time τ. The Hamiltonian is just the ADM mass of the exterior Schwarzschild solution.

An alternative form of the Hamiltonian uses X as a function of interior Minkowski time t rather than τ. This is obtained from (4.26) using the first equation in (4.24), which gives

$$H \equiv E = \frac{m}{\sqrt{1-(dX/dt)^2}} - V. \tag{4.27}$$

Taking the shell radius $X(t)$ as a phase space variable, its conjugate momentum P is obtained using the Hamilton equation

$$P = \int dH \left(\frac{dX}{dt}\right)^{-1} = \int dH \left(1 - \frac{m^2}{(H+V)^2}\right)^{-\frac{1}{2}}. \tag{4.28}$$

This gives a quadratic expression for the Hamiltonian

$$(H+V)^2 = P^2 + m^2. \tag{4.29}$$

4.3.3 Fermionic Quantization of a Collapsing Shell

Like the OS model, there is large literature on the quantization of shell collapse; a sampling of such works is in Refs. [19, 20, 50]. A feature of these papers, and indeed most works we are aware of, use bosonic quantization. It is curious that a fermionic quantization is also possible [22]. We summarize here one such method applicable to the quantum collapse of a thin shell using the Hamiltonian (4.29).

With the canonical commutation relation $[X, P] = i$, a Hamiltonian operator may be defined by

$$HI = P\sigma_3 + m\sigma_2 - V(X)I = \begin{bmatrix} P - V & -im \\ im & -P - V \end{bmatrix}. \tag{4.30}$$

where I is the 2×2 identity and σ_i are the Pauli matrices. Squaring the operator gives $(H + V)^2 = P^2 + m^2$. With this quantization the thin shell dynamics turns out to be the Dirac equation on the half line for a Coulomb potential with charge given by the rest mass m. This Hamiltonian can be realized as a self-adjoint operator on a Hilbert space of 2-spinor functions

$$\Phi(X) \equiv \begin{bmatrix} \phi_1 \\ \phi_2 \end{bmatrix} \tag{4.31}$$

with inner product

$$(\Psi, \Phi) := \int_0^\infty dX\ \Psi^\dagger \Phi = \int_0^\infty dX\ \left(\psi_1^* \phi_1 + \psi_2^* \phi_2\right), \tag{4.32}$$

where $\Psi^\dagger \equiv \Psi^{T*}$, and the standard representation

$$\hat{X}\phi = X\phi, \qquad \hat{P}\phi = -i\partial_X \phi. \tag{4.33}$$

The expression for the Hamiltonian operator on this Hilbert space is then

$$\hat{H} = \begin{bmatrix} -i\partial_X - V & -im \\ im & i\partial_X - V \end{bmatrix}. \tag{4.34}$$

Let us recall that the energy E (4.27) is the ADM mass of the exterior Schwarzschild solution. The spectrum of the Hamiltonian operator then gives the quantized ADM mass values. Although a bit involved, the spectrum can be determined [22]. It turns out that the Hamiltonian operator has a one parameter family of self-adjoint extensions with a discrete spectrum for $|E| < m$ and continuum scattering states for $|E| > m$; the former correspond to bound states of the shell that are black holes, and the latter to states where the collapsing shell bounces.

An important feature of the bound state spectrum is that for sufficiently large m (the rest mass of the shell given (4.25)) the ground state has negative ADM mass.

This shows that the positive mass theorem does not survive in quantization, at least with this quantization. It shows furthermore that the singularity is avoided just as in atomic physics.

The scattering states for $E > m$ may thus be viewed as exemplifying the form of singularity avoidance usually discussed in the quantum gravity literature: the physical picture is one of an ingoing shell that bounces near the origin. There is no classical analog of such scattering: for an initial ingoing velocity with $E > m$, the classical shell crosses the Schwarzschild radius and forms a black hole.

The last three subsections summarized techniques and results that have come from effective classical theories for OS models with fundamental discreteness and from quantizing gravitational collapse models that are essentially mechanical systems. We now turn to a discussion of field theoretic models.

4.4 Quantizing Gravitational Collapse: Field Theory Models

The simplest field theory model for gravitational collapse is the scalar field in spherical symmetry. This is a two dimensional field theory on the half line and it is well studied classically. Numerical results show that there are two types of solutions for asymptotically flat initial data: black hole formation or scalar field scattering with no black hole formation [11]; the dividing line between the two types of solution is, surprisingly, a finely tuned naked singularity.

There is no known quantum theory of this system—a situation that may be viewed as an important unsolved problem where the classical physics is fairly well understood. The only partial understanding of this system is through proposed "effective" theories which provide a modified set of field equations [27, 37, 52], but it is not understood how such a description might arise from a complete quantization of the field theory.

In the following we review two approaches aimed at understanding the classical field theory from a Hamiltonian perspective. The first is the ADM canonical formulation of the scalar field and gravity system in the Painleve-Gullstrand (PG) coordinates. This is useful because the coordinates are horizon penetrating and provides an alternative to the formulation in Schwarzschild coordinates [8, 46] and the issues involved with quantization. The second is the canonical formulation of LTB spacetimes in PG coordinates in the dust time gauge. This provides a way to derive effective equations from a physical Hamiltonian; solutions of these equations provide a surprising new perspective on what becomes of black holes in quantum theory [28, 29].

4.4.1 Scalar Field in Spherical Symmetry

The black hole metric in PG coordinates is

$$ds^2 = -dt^2 + \left(dr + \sqrt{\frac{2M}{r}}\,dt\right)^2 + r^2 d\Omega^2. \tag{4.35}$$

The spatial metric e_{ab} given by the constant t slices is flat, and the extrinsic curvature of the slices is

$$K_{ab} = -\sqrt{\frac{2M}{r^3}}\left(e_{ab} - \frac{3}{2}\,s_a s_b\right), \tag{4.36}$$

where $s^a = x^a/r$ for Cartesian coordinates x^a. The mass parameter is contained only in the extrinsic curvature. These coordinates motivate the following prescription for the falloff conditions for asymptotic flatness:

$$q_{ab} = e_{ab} + \frac{f_{ab}(\theta,\phi)}{r^{3/2+\varepsilon}} + \mathcal{O}\left(r^{-2}\right), \tag{4.37}$$

$$\pi^{ab} = \frac{g^{ab}(\theta,\phi)}{r^{3/2}} + \mathcal{O}\left(r^{-3/2-\varepsilon}\right), \tag{4.38}$$

where $\varepsilon > 0, f^{ab}, g^{ab}, h^{ab}$ are symmetric tensors, $\pi^{ab} = \tilde{\pi}^{ab}/\sqrt{q}$, and $q = \det(q_{ab})$. The leading terms correspond to the black hole solution in PG coordinates; the next terms containing ε come from the requirement of a well defined canonical variational principle. These falloff conditions provide the starting point for an analysis of the Hamiltonian dynamics for spherically symmetric gravity coupled to a scalar field with a metric of the form

$$ds^2 = -f(r,t)^2 dt^2 + (dr + g(r,t)dt)^2 + r^2 d\Omega^2. \tag{4.39}$$

A starting point is the canonical ADM action with scalar field. The phase space of the model is defined by prescribing a form of the gravitational phase space variables q_{ab} and $\tilde{\pi}^{ab}$, together with falloff conditions in r for these variables, and for the lapse and shift functions N and N^a, such that the ADM 3 + 1 action for general relativity minimally coupled to a massless scalar field (with $8\pi G = 1$) is

$$S = \int d^3x dt \left[\tilde{\pi}^{ab}\dot{q}_{ab} + \tilde{P}_\phi \dot{\phi} - NH - N^a C_a\right]; \tag{4.40}$$

$$\mathscr{H} = \frac{1}{\sqrt{q}}\left(\tilde{\pi}^{ab}\tilde{\pi}_{ab} - \frac{1}{2}\tilde{\pi}^2\right) - \sqrt{q}R(q) + \left(\frac{1}{\sqrt{q}}\tilde{P}_\phi^2 + \sqrt{q}q^{ab}\partial_a\phi\partial_b\phi\right) = 0; \tag{4.41}$$

$$\mathscr{C}_a = D_c\tilde{\pi}_a^c - \tilde{P}_\phi\partial_a\phi = 0; \tag{4.42}$$

and $\tilde{\pi} = \tilde{\pi}^{ab}q_{ab}$. Reduction to spherical symmetry comes from the parametrization

$$q_{ab} = \Lambda(r,t)^2\, s_a s_b + \frac{R(r,t)^2}{r^2}\,(e_{ab} - s_a s_b), \tag{4.43}$$

$$\tilde{\pi}^{ab} = \frac{P_\Lambda(r,t)}{2\Lambda(r,t)}\, s^a s^b + \frac{r^2 P_R(r,t)}{4R(r,t)}\,(e^{ab} - s^a s^b). \tag{4.44}$$

Substituting these into the 3+1 ADM action shows that the pairs (R, P_R) and (Λ, P_Λ) are canonically conjugate variables; the reduced ADM action is

$$S_R = \int dtdr\left(P_R\dot{R} + P_\Lambda\dot{\Lambda} + P_\phi\dot{\phi}\right) - \int dtdr\left(NH + N^r C_r\right) + \text{ surface term} \tag{4.45}$$

after angular integration. The constraints reduce to the first class pair

$$H = \frac{1}{R^2\Lambda}\left[\frac{1}{8}(P_\Lambda\Lambda)^2 - \frac{1}{4}(P_\Lambda\Lambda)(P_R R)\right] + \frac{2}{\Lambda^2}\left[2RR''\Lambda - 2RR'\Lambda' - \Lambda^3 + \Lambda R'^2\right] + \left[\frac{P_\phi^2}{2\Lambda R^2} + \frac{R^2}{2\Lambda}\phi'^2\right]. \tag{4.46}$$

$$C_r = P_R R' - \Lambda P_\Lambda' + P_\phi\phi' = 0\,; \tag{4.47}$$

the surface term is derived from the requirement that the canonical action yield a well-defined variational principle [32] by deducing the fall-off conditions on the reduced variables from (4.37–4.38). This system can be reduced by fixing the gauge condition $\Lambda = 1$ and $R = r$, solving the diffeomorphism and Hamiltonian constraints. But the resulting physical Hamiltonian that describes the dynamics of the scalar field is non-local. This situation also arises for the Schwarzschild gauge fixing [46]. Thus this path to quantization is effectively blocked: it appears impossible to write an operator corresponding to a non-local Hamiltonian. The other approach, via Dirac quantization is also challenging given the form of the constraints. This is the reason there has been little progress on quantization this system, although it has been shown that evolution from an initial product state of geometry and matter produces an entangled state [31].

(There is another gauge fixing that gives a local Hamiltonian [17], but it too remains largely unexplored for quantization.)

4.4.2 Scalar Field and Dust in Spherical Symmetry

There is a variant of the gravity-scalar field system with an additional field—a pressureless dust [30]. The corresponding canonical theory has the interesting feature that in the dust time—a special type of matter reference system [10, 18]—the physical Hamiltonian is exactly the same expression as the Hamiltonian constraint, but is no longer constrained to vanish. This immediately yields a local Hamiltonian density, together with the spatial diffeomorphism constraint. The local degree of freedom count is now different since the dust field adds a local degree of freedom. Thus in the dust time gauge, in addition to the scalar field, another degree of freedom remains in the metric; the canonical action after fixing the dust time gauge is

$$S = \int dtdr \left(P_R\dot{R} + P_\Lambda\dot{\Lambda} + P_\phi\dot{\phi}\right) - \int dtdr \left(H + N^r C_r\right) + \text{ surface term.} \tag{4.48}$$

The dust time gauge fixes the lapse to unity and the expression for H and C_r are as in (4.46–4.47). Choosing now the radial gauge $R = r$ and solving the diffeomorphism constraint gives the fully gauge fixed system

$$S = \int dtdr \left(P_\Lambda\dot{\Lambda} + P_\phi\dot{\phi} - H_{\text{phys}}\right) + \text{ surface term,} \tag{4.49}$$

where

$$H_{\text{phys}} = \frac{1}{r^2\Lambda}\left[\frac{1}{8}\left(P_\Lambda\Lambda\right)^2 - \frac{r}{4}P_R(P_\Lambda\Lambda)\right] + \frac{2}{\Lambda^2}\left(-2r\Lambda' - \Lambda^3 + \Lambda\right) + \left(\frac{P_\phi^2}{2\Lambda r^2} + \frac{r^2}{2\Lambda}\phi'^2\right). \tag{4.50}$$

with $P_R = \Lambda P_\Lambda - P_\phi\phi'$. Thus, while simpler in form than without the dust field and dust time gauge, the physical Hamiltonian is still unwieldly for quantization. The situation is similar in the connection-triad canonical variables.

This system, perhaps the simplest field theory with gravity, serves to illustrate the difficulty of the full quantum gravity problem from the canonical perspective, whether one uses ADM variables or the LQG approach [45].

4.4.3 Shock Waves from Quantum Black Holes

Although there is no complete quantization known for the field theoretic models reviewed in the last section, it is possible to formulate an effective Hamiltonian theory for the dust+gravity system, but with no scalar field. The resulting non-linear partial differential equations take a particularly simple form that is amenable to efficient numerical integration. The physical picture that emerges is one where a collapsing dust cloud first crosses its Schwarzschild radius, bounces near the origin due to quantum gravity effects, and then emerges from the event horizon as a shock wave—this is a discontinuity in the metric analogous to the density discontinuity seen in the more familiar fluid shock wave. The outer horizon exists for a period proportional to the square of the black hole mass, unlike that predicted by the Hawking evaporation formula (4.1).

The class of metrics for which the effective equations are derived is the LTB spacetimes

$$\mathrm{d}s^2 = -\mathrm{d}t^2 + \frac{\left(\mathrm{d}x + N^r(r,t)\mathrm{d}t\right)^2}{1+\omega(r,t)} + r^2\mathrm{d}\Omega^2. \tag{4.51}$$

The Hamilton equations for this class of metrics in the connection-triad variables in the dust time gauge follow from the action

$$S_{GF} = \int \mathrm{d}t \int \mathrm{d}r\,\left(\dot{b}E^b - \mathcal{H}_{\text{phys}}\right), \tag{4.52}$$

$$\mathcal{H}_{\text{phys}} \equiv -\frac{1}{2}\left[\frac{E^b}{r}\partial_r(rb^2) + \frac{E^b}{r} + \frac{r}{E^b}\right]; \tag{4.53}$$

$\mathcal{H}_{\text{phys}}$ is the true physical Hamiltonian (with G and the Immirzi parameter γ set to unity); the phase space variables (b, E^b) are the reduced connection-triad variables [28, 29].

The next step is a derivation of the effective Hamiltonian. In outline this is like the procedure described above for the parametrized particle: a polymer (or lattice) quantization of the system is followed by a replacement of an operator by a sinusoidal function. The details require care in that a physical length that depends on the connection component b is set to a multiple of the Planck length (as elaborated upon in [29]); this gives a specific choice for the function f in (4.10). The result of these considerations is the effective physical Hamiltonian

$$\mathcal{H}^{\text{eff}}_{\text{phys}} = -\frac{1}{2}\left[\frac{E^b}{\ell_{\text{Pl}}^2 r}\partial_r\left(r^3\sin^2\frac{\ell_{\text{Pl}}b}{r}\right) + \frac{E^b}{r} + \frac{r}{E^b}\right]; \tag{4.54}$$

this comes from the replacement

$$b \longrightarrow \frac{r}{\ell_{\mathrm{Pl}}} \sin\left(\frac{\ell_{\mathrm{Pl}} b}{r}\right) \tag{4.55}$$

in $\mathscr{H}_{\mathrm{phys}}$. The evolution equations for b and E^b are now derived from this effective Hamiltonian. Setting $E^b = r$ in the resulting equations gives a metric of the form (4.51) with $\omega(r, t) = 1$, and leaves a single equation to solve numerically:

$$\dot{b} + \frac{1}{2r}\partial_r\left(r^3 \sin^2 \frac{b}{r}\right) = 0. \tag{4.56}$$

In this sector of the solution space of the effective equations, the dust energy density and effective Hamiltonian are related by

$$\rho = -\frac{\mathscr{H}^{\mathrm{eff}}_{\mathrm{phys}}}{4\pi r^2} = \frac{1}{8\pi r^2}\,\partial_r\left(r^3 \sin^2 \frac{b}{r}\right). \tag{4.57}$$

Equation (4.56) may be written as a 1+1 conservation law in a manner similar to the well-known Burger's equation $\dot{u} + (u^2)' = 0$ for the function $u(x, t)$; one of it's features is the crossing of characteristics, indicating shock wave formation. Equation (4.56) exhibits the same feature.

A typical numerical solution of (4.56) for initial data with a Gaussian density profile is shown in Fig. 4.6; the left column shows the evolution of the dust profile and the right column shows the apparent horizon function

$$\Theta_+ = |\nabla r|^2, \tag{4.58}$$

The roots of this function give the radial locations of apparent horizons; black hole lifetime is inferred as the time duration of the outermost horizon. Figure 4.7 gives this lifetime as a function of the black hole mass, which is the integral of the density ρ over the numerical grid—this quantity coincides with twice the radius of the outermost horizon since all the mass falls through the outer horizon in the simulation; the mass is varied by changing the width of Gaussian initial data. It is evident from the fit given in this figure that black hole lifetime

$$T_{\mathrm{BH}} \sim M^2. \tag{4.59}$$

Lastly Fig. 4.8 shows the conformal diagram inferred from these numerical simulations; this shows a horizon bubble in red where the outer horizon is null; it annihilates with the inner horizon after the matter bounce; the post bounce indicates two matter lines in blue with a shaded region excised; this is drawn to indicate that the metric across the matter shock wave is discontinuous.

It is worth emphasizing that these results are a direct computational consequence of the effective dynamics of dust collapse with fundamental discreteness corrections. A general feature that might be inferred from this result is that if there is a matter bounce described by any pde, then it is reasonable to assume that characteristic

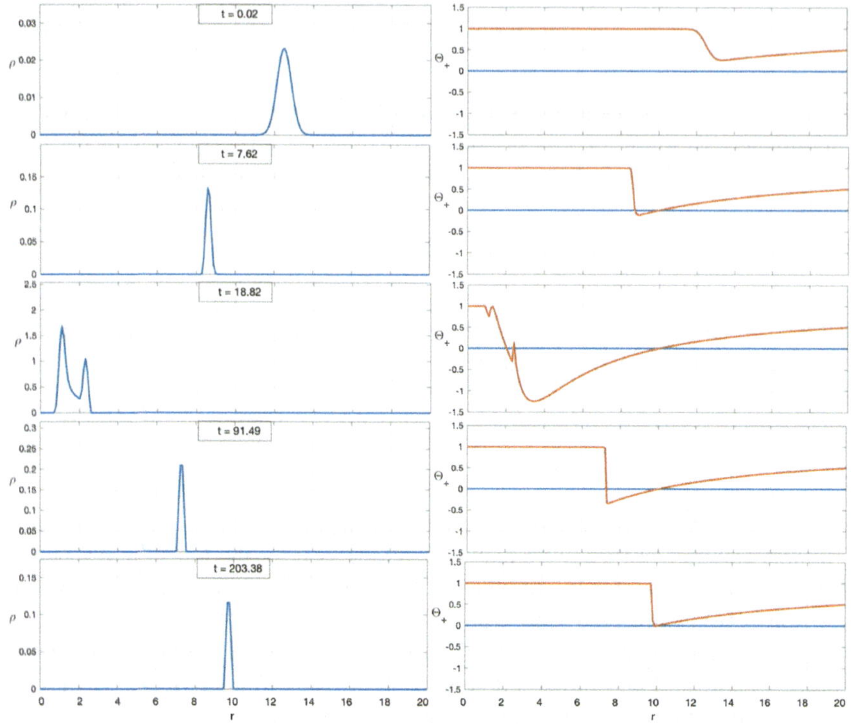

Fig. 4.6 Effective evolution of a Gaussian dust profile showing an ingoing pulse (top) to bounce (middle) and outgoing shock wave. The right column shows the apparent horizon function; it's roots indicate horizon formation; the second frame shows a pair of horizons; the bottom frame shows disappearance of the outer horizon as the shock moves outward [28]

crossing occurs because outgoing matter flow lines that have bounced will intersect ingoing matter flow lines that have not yet bounced. It is precisely such crossing of characteristics that result in shock wave formation in fluid flow. Thus shock waves might be a generic feature of curvature singularity avoidance via a matter bounce.

4.5 Open Problems

We have provided a survey of a few research directions aimed at understanding what is a quantum black hole. The underlying idea is that quantization of gravity with matter should provide a mechanism for *dynamical* singularity avoidance. This means that "quantizing the Schwarzschild black hole," or any other static classical solution is no more appropriate than quantizing a static solution of any theory—after all we normally quantize theories rather than solutions of theories characterized by

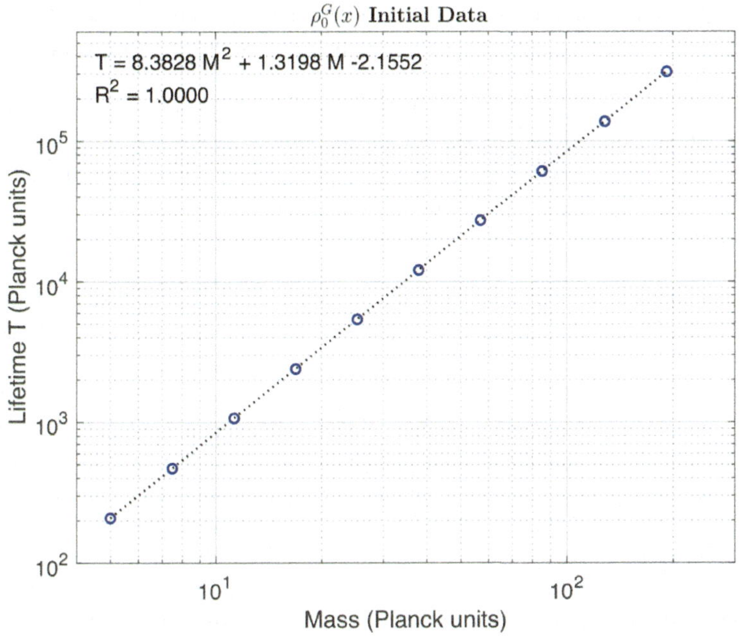

Fig. 4.7 Black hole lifetime as a function of mass [29] for Gaussian initial density profiles; the fit indicates a lifetime or order M^2

Fig. 4.8 Conformal diagram deduced from effective dust collapse simulations [29]: the blue lines shows the peak of the dust profile, the red lines show the horizon bubble; the outer horizon is null; the shaded region is excised and the blue lines are excised indicating that the metric is discontinuous across the outgoing shock wave

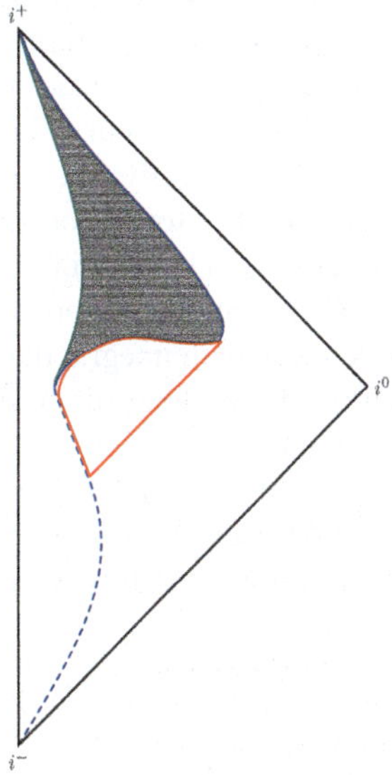

a few integration constants. Furthermore, non-singular black hole metrics may not be stable unlike their singular counterparts, therefore their use for quantum gravity appears limited.

The quantum mechanical OS and shell collapse models appear to have yielded all they can. Their usefulness for understanding quantum gravitational collapse therefore appears to be limited given the existing literature. While it may turn out that a new technical detail of collapse and bounce might emerge, the question is really whether they can provide any new insights. One of the lessons learned is that OS models are limited. For instance, by assumption of metric continuity they exclude potentially interesting phenomena such as gravitational shock waves, a phenomena known to arise in solutions of non-linear PDEs describing fluids. Another lesson learned is to exercise caution with conjectured conformal diagrams such as the one depicting Hawking's information loss picture.

What would be much more appealing is a complete quantization of a field theory model such as the gravity-scalar field system in spherical symmetry. By definition of quantization such a theory would be unitary. And if this could be accomplished, there would be no need to "add a scalar field on an effective semiclassical background" to study Hawking radiation, for the full gravity-scalar field quantum system would in principle contain all physical effects; the problem would be to extract them. This scenario would include the s-wave sector of Hawking radiation. We have outlined a few approaches to this problem and indicated the difficulties, which are many.

These include all the well-established problems of quantum gravity [38]: are there useful gauge invariant observables? Does a particular gauge fixing for time and/or spatial diffeomorphisms excise potential physical effects? Equivalently, what extracted physical features if any are gauge independent? There are examples even in classical gravity where a choice of time gauge can miss interesting physics, or suggest a result that turns out to be a gauge artifact [48]. The same considerations apply to the type of effective theories we have reviewed. Although some of the results, such as shock wave formation, are interesting and suggest further study, these overarching issues should be kept in mind.

There are other approaches to quantum aspects of black holes coming from the Euclidean path integral that concern Hawking radiation and a derivation of the Page curve [3]; the focus in these works is not on singularity avoidance and quantization of gravitational collapse, and so falls beyond the scope of this review; the same applies to work on black holes related to the AdS/CFT conjecture, where apparently all properties of quantum gravitational collapse are contained in the Hilbert space of a conformal field theory (see e.g. [14]), waiting to be extracted.

Acknowledgements The work reported here was supported in part by the Natural Science and Engineering Research Council of Canada. I thank Moeez Hassan, Jarod Kelly, Samantha Hergott, Saeed Rastgoo, Robert Santacruz, Danny Terno, Edward Wilson-Ewing and Jonathan Ziprick for discussions and collaboration.

References

1. R.J. Adler, J.D. Bjorken, P. Chen, J.S. Liu, Simple analytic models of gravitational collapse. Am. J. Phys. **73**, 1148–1159 (2005)
2. I. Agullo, P. Singh. *Loop Quantum Cosmology*. (WSP, 2017), pp. 183–240
3. A. Almheiri, N. Engelhardt, D. Marolf, H. Maxfield, The entropy of bulk quantum fields and the entanglement wedge of an evaporating black hole. JHEP **12**, 063 (2019)
4. A. Ashtekar, M. Bojowald, Black hole evaporation: a paradigm. Class. Quant. Grav. **22**, 3349–3362 (2005)
5. A. Ashtekar, S. Fairhurst, J.L. Willis, Quantum gravity, shadow states, and quantum mechanics. Class. Quant. Grav. **20**, 1031–1062 (2003)
6. A. Ashtekar, J. Olmedo, P. Singh, Quantum transfiguration of Kruskal black holes. Phys. Rev. Lett. **121**(24), 241301 (2018)
7. A. Ashtekar, J. Olmedo, P. Singh, *Regular Black Holes from Loop Quantum Gravity*, vol. 1 (2023)
8. B.K. Berger, D.M. Chitre, V.E. Moncrief, Y. Nutku, Hamiltonian formulation of spherically symmetric gravitational fields. Phys. Rev. D **5**, 2467–2470 (1972)
9. N.D. Birell, P.C.W. Davies, *Quantum Fields in Curved Space* (Cambridge University Press, 1982)
10. J. David Brown, K.V. Kuchar, Dust as a standard of space and time in canonical quantum gravity. Phys. Rev. D **51**, 5600–5629 (1995)
11. M.W. Choptuik, Universality and scaling in gravitational collapse of a massless scalar field. Phys. Rev. Lett. **70**, 9–12 (1993)
12. A. Corichi, P. Singh, Loop quantization of the Schwarzschild interior revisited. Class. Quant. Grav. **33**(5), 055006 (2016)
13. J. Cortez, W. Cuervo, H.A. Morales-Técotl, J.C. Ruelas, Effective loop quantum geometry of Schwarzschild interior. Phys. Rev. D **95**(6), 064041 (2017)
14. L. Fidkowski, V. Hubeny, M. Kleban, S. Shenker, The Black hole singularity in AdS / CFT. JHEP **02**, 014 (2004)
15. V.P. Frolov, A. Zelnikov, Quantum radiation from an evaporating nonsingular black hole. Phys. Rev. D **95**(12), 124028 (2017)
16. V.P. Frolov, A. Zelnikov, Spherically symmetric black holes in the limiting curvature theory of gravity. Phys. Rev. D **105**(2), 024041 (2022)
17. R. Gambini, J. Pullin, Spherically symmetric gravity coupled to a scalar field with a local Hamiltonian: the complete initial-boundary value problem using metric variables. Class. Quant. Grav. **30**, 025012 (2013)
18. K. Giesel, T. Thiemann, Scalar material reference systems and loop quantum gravity. Class. Quant. Grav. **32**, 135015 (2015)
19. P. Hajicek, Quantum mechanics of gravitational collapse. Commun. Math. Phys. **150**, 545–559 (1992)
20. P. Hájíček, B. Kay, K.V. Kuchař, Quantum collapse of a self-gravitating shell: equivalence to coulomb scattering. Phys. Rev. D **46**, 5439 (1992)
21. H. Halvorson, Complementarity of representations in quantum mechanics. Studies in History and Philosophy of Science Part B: Studies in History and Philosophy of Modern Physics **35**(1), 45–56 (2004). (Mar)
22. S.M. Hassan, V. Husain, J. Ziprick, Quantum gravitational collapse as a Dirac particle on the half line. Phys. Rev. D **97**(10), 104032 (2018)
23. S.W. Hawking, Black hole explosions. Nature **248**, 30–31 (1974)
24. S.W. Hawking, Particle creation by black holes. Commun. Math. Phys. **43**, 199–220 (1975). [Erratum: Commun. Math. Phys. 46, 206 (1976)]
25. S.A. Hayward, Formation and evaporation of regular black holes. Phys. Rev. Lett. **96**, 031103 (2006)

26. S. Hergott, V. Husain, S. Rastgoo, Model metrics for quantum black hole evolution: gravitational collapse, singularity resolution, and transient horizons. Phys. Rev. D **106**(4), 046012 (2022)
27. V. Husain, *Critical Behaviour in Quantum Gravitational Collapse*, vol. 8 (2008)
28. V. Husain, J.G. Kelly, R. Santacruz, E. Wilson-Ewing. Fate of quantum black holes. Phys. Rev. D **106**(2), 024014 (2022)
29. V. Husain, J.G. Kelly, R. Santacruz, E. Wilson-Ewing, Quantum gravity of dust collapse: shock waves from black holes. Phys. Rev. Lett. **128**(12), 121301 (2022)
30. V. Husain, T. Pawlowski, Time and a physical Hamiltonian for quantum gravity. Phys. Rev. Lett. **108**, 141301 (2012)
31. V. Husain, D.R. Terno, Dynamics and entanglement in spherically symmetric quantum gravity. Phys. Rev. D **81**, 044039 (2010)
32. V. Husain, O. Winkler, Flat slice Hamiltonian formalism for dynamical black holes. Phys. Rev. D **71**, 104001 (2005)
33. V. Husain, O. Winkler, Semiclassical states for quantum cosmology. Phys. Rev. D **75**, 024014 (2007)
34. J.G. Kelly, R. Santacruz, E. Wilson-Ewing, Black hole collapse and bounce in effective loop quantum gravity. Class. Quant. Grav. **38**(4), 04LT01 (2021)
35. C. Kiefer, H. Mohaddes, *From Classical to Quantum Oppenheimer-Snyder Model: Non-marginal Case*, vol. 3 (2023)
36. S. Klainerman, J. Szeftel, *Kerr Stability for Small Angular Momentum*
37. A. Kreienbuehl, V. Husain, S.S. Seahra, Modified general relativity as a model for quantum gravitational collapse. Class. Quant. Grav. **29**, 095008 (2012)
38. K.V. Kuchar, *Canonical Quantum Gravity*, vol. 4 (1993)
39. K.V. Kuchar, Geometrodynamics of Schwarzschild black holes. Phys. Rev. D **50**, 3961–3981 (1994)
40. L. Modesto, Disappearance of black hole singularity in quantum gravity. Phys. Rev. D **70**, 124009 (2004)
41. L. Modesto, Loop quantum black hole. Class. Quant. Grav. **23**, 5587–5602 (2006)
42. J.R. Oppenheimer, H. Snyder, On Continued gravitational contraction. Phys. Rev. **56**, 455–459 (1939)
43. E. Poisson, *A Relativist's Toolkit: The Mathematics of Black-Hole Mechanics* (Cambridge University Press, 2004)
44. T. Regge, J.A. Wheeler, Stability of a schwarzschild singularity. Phys. Rev. **108**, 1063–1069 (1957). (Nov)
45. T. Thiemann, K. Giesel, *Hamiltonian Theory: Dynamics*, vol. 3 (2023)
46. W.G. Unruh, Notes on black hole evaporation. Phys. Rev. D **14**, 870 (1976)
47. C. Vaz, L. Witten, T.P. Singh, Toward a midisuperspace quantization of LeMaî tre-tolman-bondi collapse models. Phys. Rev. D **63**(10) (2001)
48. R.M. Wald, V. Iyer, Trapped surfaces in the Schwarzschild geometry and cosmic censorship. Phys. Rev. D **44**, R3719–R3722 (1991)
49. F.J. Zerilli, Effective potential for even parity Regge-Wheeler gravitational perturbation equations. Phys. Rev. Lett. **24**, 737–738 (1970)
50. J. Ziprick, J. Gegenberg, G. Kunstatter, Polymer Quantization of a Self-Gravitating Thin Shell. Phys. Rev. D **94**(10), 104076 (2016)
51. J. Ziprick, G. Kunstatter, Dynamical singularity resolution in spherically symmetric black hole formation. Phys. Rev. D **80**, 024032 (2009)
52. J. Ziprick, G. Kunstatter, Quantum corrected spherical collapse: a phenomenological framework. Phys. Rev. D **82**, 044031 (2010)

Chapter 5
Numerical Simulations of Spacetime Singularities

David Garfinkle

5.1 Spacetime Singularities

Spacetime singularities occur under very general circumstances, as shown by the theorems of Penrose and Hawking [19, 26]. However, these theorems give very little information about the nature of such singularities, noting only that some observer or light ray ends in finite affine parameter. In contrast, the conjectures of Belinskii, Khalatnikov, and Lifschitz [1] (collectively known as BKL) give detailed behavior about the asymptotic behavior as the singularity is approached, but no methods to determine whether the conjecture is true.

To address this issue, Berger and Moncrief performed numerical simulations of the approach to spacetime singularities, starting with the Gowdy spacetimes, [4] and then moving on to spacetimes with $U(1)$ symmetry [5]. The goal of the research program they started [3] was to simulate the development of generic initial data whose evolution results in a singularity and to see whether the approach to the singularity is well described by the BKL conjecture.

A numerical simulation involves a foliation dividing spacetime up into space and time. The simulation must end when a slice of the foliation reaches the singularity. Thus we must choose some foliation that never reaches the singularity, but comes arbitrarily close to it. A good choice turns out to be the constant mean curvature (CMC) foliation. In CMC foliation, each slice is a surface of constant mean curvature. We expect that in this foliation the mean curvature beomes arbitrarily large as the slices come arbitrarily close to the singularity.

As the singularity is approached various physical quantities blow up. But we expect them to blow up like particular values of the mean curvature. This suggests that certain scale invariant variables, physical quantities divided by appropriate powers

D. Garfinkle (✉)
Department of Physics, Oakland University, Rochester, MI 48306, USA
e-mail: garfinkl@oakland.edu

D. Malafarina and P. S. Joshi (eds.), *New Frontiers in Gravitational Collapse and Spacetime Singularities*, Springer Series in Astrophysics and Cosmology,
https://doi.org/10.1007/978-981-97-1172-7_5

of the mean curvature, remain bounded as the singularity is approached. It is helpful to have a simulation that is done in terms of such scale invariant variables. Such a system, using a tetrad (rather than coordinate) version of the Einstein field equations was devised by Uggla et al. [31] The system we use is a variation of the system of [31].

The reason for using a tetrad system is that we know from the example of Kasner spacetimes that tetrad quantities all blow up in the same way as the singularity is approached, whereas coordinate components of the metric all behave as different powers of the mean curvature in the approach to the singularity.

In Sect. 5.2 we will introduce the scale invariant tetrad variables and their equations of motion. The behavior of vacuum solutions and free scalar field solutions in the approach to the singularity (including the behavior of spikes) will be covered in Sect. 5.3. Section 5.4 will cover the ekpyrotic cosmological scenario, which consists of a scalar field with a negative exponential potential. Conclusions are given in Sect. 5.5.

5.2 Scale Invariant Variables and Equations of Motion

The spacetime is described in terms of a coordinate system (t, x^i) and a tetrad $(\mathbf{e}_0, \mathbf{e}_\alpha)$ where both the spatial coordinate index i and the spatial tetrad index α go from 1 to 3. Choose $\mathbf{e}_0$ to be hypersurface orthogonal with the relation between tetrad and coordinates of the form $\mathbf{e}_0 = N^{-1}\partial_t$, and $\mathbf{e}_\alpha = e_\alpha{}^i\partial_i$, where N is the lapse and the shift is chosen to be zero. Choose the spatial frame $\{\mathbf{e}_\alpha\}$ to be Fermi propagated along the integral curves of $\mathbf{e}_0$. The commutators of the tetrad components are decomposed as follows:

$$[\mathbf{e}_0, \mathbf{e}_\alpha] = \dot{u}_\alpha \mathbf{e}_0 - (H\delta_\alpha{}^\beta + \sigma_\alpha{}^\beta)\mathbf{e}_\beta, \tag{5.1}$$

$$\left[\mathbf{e}_\alpha, \mathbf{e}_\beta\right] = (2a_{[\alpha}\delta_{\beta]}{}^\gamma + \varepsilon_{\alpha\beta\delta}n^{\delta\gamma})\mathbf{e}_\gamma, \tag{5.2}$$

where $n^{\alpha\beta}$ is symmetric, and $\sigma^{\alpha\beta}$ is symmetric and trace free.

For those cases where we have matter, it will be a scalar field ϕ and its potential (when there is one) will be denoted V.

The scale invariant tetrad variables are defined by $\boldsymbol{\partial}_0 \equiv \mathbf{e}_0/H$ and $\boldsymbol{\partial}_\alpha \equiv \mathbf{e}_\alpha/H$ while scale invariant versions of the other gravitational variables are given by

$$\{E_\alpha{}^i, \Sigma_{\alpha\beta}, A^\alpha, N_{\alpha\beta}\} \equiv \{e_\alpha{}^i, \sigma_{\alpha\beta}, a^\alpha, n_{\alpha\beta}\}/H. \tag{5.3}$$

Note that the relation between the scale invariant tetrad variables and the coordinate derivatives is

$$\boldsymbol{\partial}_0 = \mathcal{N}^{-1}\partial_t, \tag{5.4}$$

$$\boldsymbol{\partial}_\alpha = E_\alpha{}^i \partial_i, \tag{5.5}$$

where $\mathcal{N} = NH$ is the scale invariant lapse.

The scale invariant matter variables are given by

$$W = \boldsymbol{\partial}_0 \phi, \tag{5.6}$$

$$S_\alpha = \boldsymbol{\partial}_\alpha \phi, \tag{5.7}$$

$$\bar{V} = V/H^2. \tag{5.8}$$

The time coordinate t is chosen so that

$$e^{-t} = 3H. \tag{5.9}$$

Here we have used the scale invariance of the physical system to make both t and H dimensionless quantities. Note that Eq. (5.9) means that the surfaces of constant time are constant mean curvature surfaces. Note also that the singularity is approached as $t \to -\infty$.

Due to Eq. (5.9) the scale invariant lapse satisfies an elliptic equation

$$-\boldsymbol{\partial}^\alpha \boldsymbol{\partial}_\alpha \mathcal{N} + 2A^\alpha \boldsymbol{\partial}_\alpha \mathcal{N} + \mathcal{N}(3 + \Sigma_{\alpha\beta}\Sigma^{\alpha\beta} + W^2 - \bar{V}) = 3. \tag{5.10}$$

The gravitational quantities $E_\alpha{}^i$, A_α, $N^{\alpha\beta}$ and $\Sigma_{\alpha\beta}$ satisfy the following evolution equations.

$$\partial_t E_\alpha{}^i = E_\alpha{}^i - \mathcal{N}(E_\alpha{}^i + \Sigma_\alpha{}^\beta E_\beta{}^i), \tag{5.11}$$

$$\partial_t A_\alpha = A_\alpha + \tfrac{1}{2}\Sigma_\alpha{}^\beta \boldsymbol{\partial}_\beta \mathcal{N} - \boldsymbol{\partial}_\alpha \mathcal{N} + \mathcal{N}\left(\tfrac{1}{2}\boldsymbol{\partial}_\beta \Sigma_\alpha{}^\beta - A_\alpha - \Sigma_\alpha{}^\beta A_\beta\right), \tag{5.12}$$

$$\begin{aligned}\partial_t N^{\alpha\beta} &= N^{\alpha\beta} - \varepsilon^{\gamma\delta(\alpha}\Sigma_\delta{}^{\beta)}\boldsymbol{\partial}_\gamma \mathcal{N} \\ &\quad + \mathcal{N}\left(-N^{\alpha\beta} + 2N^{(\alpha}{}_\gamma \Sigma^{\beta)\gamma} - \varepsilon^{\gamma\delta(\alpha}\boldsymbol{\partial}_\gamma \Sigma_\delta{}^{\beta)}\right),\end{aligned} \tag{5.13}$$

$$\begin{aligned}\partial_t \Sigma_{\alpha\beta} &= \Sigma_{\alpha\beta} + \boldsymbol{\partial}_{<\alpha}\boldsymbol{\partial}_{\beta>}\mathcal{N} + A_{<\alpha}\boldsymbol{\partial}_{\beta>}\mathcal{N} + \varepsilon_{\gamma\delta(\alpha}N_{\beta)}{}^\delta \boldsymbol{\partial}^\gamma \mathcal{N} \\ &\quad + \mathcal{N}\Big[-3\Sigma_{\alpha\beta} - \boldsymbol{\partial}_{<\alpha}A_{\beta>} - 2N_{<\alpha}{}^\gamma N_{\beta>\gamma} + N^\gamma{}_\gamma N_{<\alpha\beta>} \\ &\quad + \varepsilon_{\gamma\delta(\alpha}(\boldsymbol{\partial}^\gamma N_{\beta)}{}^\delta - 2A^\gamma N_{\beta)}{}^\delta) + S_{<\alpha}S_{\beta>}\Big].\end{aligned} \tag{5.14}$$

Here parentheses around a pair of indices denote the symmetric part, while angle brackets denote the symmetric trace-free part.

The equations of motion for the matter variables are as follows:

$$\partial_t \phi = \mathcal{N} W, \tag{5.15}$$

$$\partial_t S_\alpha = S_\alpha + W \partial_\alpha \mathcal{N} + \mathcal{N}\left[\partial_\alpha W - (S_\alpha + \Sigma_\alpha{}^\beta S_\beta)\right], \tag{5.16}$$

$$\partial_t W = W + S^\alpha \partial_\alpha \mathcal{N} + \mathcal{N}\left(\partial^\alpha S_\alpha - 3\,W - 2A^\alpha S_\alpha - \frac{\partial \bar{V}}{\partial \phi}\right). \tag{5.17}$$

In addition, the variables are subject to the vanishing of the following constraint quantities

$$(\mathcal{C}_{\text{com}})^{\lambda i} = \varepsilon^{\alpha\beta\lambda}[\partial_\alpha E_\beta{}^i - A_\alpha E_\beta{}^i] - N^{\lambda\gamma} E_\gamma{}^i, \tag{5.18}$$

$$(\mathcal{C}_J)^\gamma = \partial_\alpha N^{\alpha\gamma} + \varepsilon^{\alpha\beta\gamma}\partial_\alpha A_\beta - 2A_\alpha N^{\alpha\gamma}, \tag{5.19}$$

$$(\mathcal{C}_C)_\alpha = \partial_\beta \Sigma_\alpha{}^\beta - 3\Sigma_\alpha{}^\beta A_\beta - \varepsilon_{\alpha\beta\gamma} N^{\beta\delta}\Sigma_\delta{}^\gamma - W S_\alpha, \tag{5.20}$$

$$\begin{aligned}\mathcal{C}_G &= 1 + \tfrac{2}{3}\partial_\alpha A^\alpha - A^\alpha A_\alpha - \tfrac{1}{6}N^{\alpha\beta}N_{\alpha\beta} + \tfrac{1}{12}(N^\gamma{}_\gamma)^2 \\ &\quad - \tfrac{1}{6}\Sigma^{\alpha\beta}\Sigma_{\alpha\beta} - \tfrac{1}{6}W^2 - \tfrac{1}{6}S^\alpha S_\alpha - \tfrac{1}{3}\bar{V},\end{aligned} \tag{5.21}$$

$$(\mathcal{C}_S)_\alpha = S_\alpha - \partial_\alpha \phi. \tag{5.22}$$

An explanation of these constraint quantities, including their labels, can be found in [31]. For example the "com" constraint in Eq. (5.18) comes from applying the commutators of the tetrad vector fields to the coordinates, while the "J" constraint of Eq. (5.19) comes from the Jacobi identity applied to the tetrad vector fields.

Initial data is chosen to satisfy the constraints of Eqs. (5.18)–(5.22), which are then preserved (to within numerical truncation error) under evolution. The data are evolved using the evolution equations, Eqs. (5.11)–(5.17), where to obtain a hyperbolic system a multiple of Eq. (5.20) is added to the right hand side of Eq. (5.12) [16].

To find initial data we the York method [33]. We use Cartesian coordinates (x, y, z) with each coordinate periodically identified with period 2π. We assume a spatial metric of the form $\psi^4\delta_{ij}$ where δ_{ij} is the Kronecker delta. That is, we are assuming a conformally flat spatial metric. Note that this choice leads to the following expression for the tetrad variables $E_\alpha{}^i$, $N_{\alpha\beta}$, and A_α.

$$E_\alpha{}^i = H^{-1}\psi^{-2}\delta_\alpha{}^i, \tag{5.23}$$

$$N_{\alpha\beta} = 0, \tag{5.24}$$

$$A_\alpha = -2\psi^{-1}E_\alpha{}^i\partial_i\psi. \tag{5.25}$$

We define the quantities $Z_{\alpha\beta}$ and Q by

$$\Sigma_{\alpha\beta} = \psi^{-6}Z_{\alpha\beta}, \tag{5.26}$$

$$W = \psi^{-6}Q. \tag{5.27}$$

Then the momentum constraint (Eq. 5.20) becomes

$$\partial^i Z_{ik} = Q \partial_k \phi. \tag{5.28}$$

while the Hamiltonian constraint (Eq. 5.21) becomes

$$\partial^i \partial_i \psi = (\tfrac{3}{4} H^2 - \tfrac{1}{4} V)\psi^5 - \tfrac{1}{8}(\partial^i \phi \partial_i \phi)\psi - \tfrac{1}{8}(Q^2 + Z^{ik} Z_{ik}) H^2 \psi^{-7}. \tag{5.29}$$

We solve Eq. (5.28) algebraically as follows: we specify Q and ϕ in terms of Fourier modes and then algebraically compute a particular solution of Eq. (5.28) to which we add a divergence-free Z_{ij} of the following form.

$$\begin{aligned}
Z_{11} &= b_2 + a_2 \cos(y + \phi_y) + a_3 \cos(z + \phi_z), \\
Z_{22} &= b_1 + a_1 \cos(x + \phi_x) - a_3 \cos(z + \phi_z), \\
Z_{33} &= -b_1 - b_2 - a_1 \cos(x + \phi_x) - a_2 \cos(y + \phi_y), \\
Z_{12} &= Z_{21} = \kappa_1, \\
Z_{13} &= Z_{31} = \kappa_2, \\
Z_{23} &= Z_{32} = 0.
\end{aligned} \tag{5.30}$$

Here, a_1, a_2, a_3, b_1, b_2, ϕ_x, ϕ_y, ϕ_z, κ_1 and κ_2 are constants. In many of the simulations presented here we will assume symmetry in the z direction, in which case $a_3 = 0$, or symmetry in both the y and z directions, in which case $a_2 = a_3 = 0$.

Equation (5.29) is solved using relaxation methods or multigrid methods [6], or (in the case of symmetry in two directions) tridiagonal methods. The evolution equations are implemented using the iterated Crank-Nicholson method with spatial derivatives computed using second order finite differences.

5.3 Vacuum and Free Scalar Field

The BKL conjecture is that as the singularity is approached spatial derivatives become negligible compared to time derivatives. Clearly a statement of this form only makes sense in a particular class of foliations. The original BKL formalism used synchronous coordinates with $t = 0$ at the singularity. However, such a foliation is not suitable for numerical simulations since we wouldn't know where to set $t = 0$ until the singularity forms. Nonetheless, any foliation that is suitable for numerical simulations is also a candidate for testing the BKL conjecture. The reason for this is that a numerically suitable foliation must have the property that it never hits the singularity but comes arbitrarily close to it. In this sense all numerically suitable foliations asymptotically behave the same (up to rescaling of the time coordinate) as the BKL synchronous coordinates. Foliations that have been shown to be numerically suitable in this sense include harmonic coordinates, [11] inverse mean curvature flow, [12] and CMC slicing [14].

The scale invariant tetrad system also provides a nice way of being specific about the notion that space derivatives become negligible. In the scale invariant tetrad system all spatial coordinate derivatives appear in the field equations only multiplied by the scale invariant triad components ${E_\alpha}^i$. Thus the statement "space derivatives become negligible in the field equations" becomes the statement that ${E_\alpha}^i$ become small as the singularity is approached. (The BKL conjecture also requires that the tetrad variable A_α becomes negligible as the singularity is approached). Note that this does *not* mean that spatial derivatives are becoming small or that space is becoming homogeneous (quite the opposite as we will see in Sect. 5.3). It just means that the space derivatives have negligible effect in the field equations because they are multiplied by the small quantity ${E_\alpha}^i$.

Thus the first task of a numerical simulation of the approach to the singularity is to see whether ${E_\alpha}^i$ and A_α are becoming small as the singularity is approached. Such simulations were performed for the case of no symmetry in [12] with inverse mean curvature slicing and in [14] with CMC slicing. In both sets of simulations ${E_\alpha}^i$ and A_α become negligibly small as the singularity is approached, and thus the BKL conjecture is satisfied.

In physical terms what is going on is that the light cones become increasingly narrow as the singularity is approached, and thus the dynamics at each spatial point becomes decoupled from the dynamics at neighboring points.

To get a feel for what this dynamics is like it is helpful to pick a spatial point and look at the evolution of the scale invariant quantities $\Sigma_{\alpha\beta}$ and $N_{\alpha\beta}$. This evolution is shown in Figs. 5.1 and 5.2. Note that the time coordinate goes to $-\infty$ as the singularity is approached. This means that these figures should be read from right to left. That is, the initial configuration is represented at the right side of the graph and as we go leftward on the graph we get ever closer to the singularity.

Note that in Fig. 5.1 we start with a somewhat complicated dynamics. In this period ${E_\alpha}^i$ and A_α are not yet negligible and therefore the simplified dynamics proposed by BKL has not yet taken hold. However, things eventually settle down to a simple dynamics of long stretches in which the components of $\Sigma_{\alpha\beta}$ are approximately constant, with short transitions between each of these long stretches.

The long stretches are called "Kasner epochs" for reasons that we will now explain: the Kasner solution [21] is a solution of the vacuum Einstein equation of the form

$$ds^2 = -dt^2 + t^{2p_1}dx^2 + t^{2p_2}dy^2 + t^{2p_3}dz^2, \tag{5.31}$$

where the constants p_i satisfy the relations

$$p_1 + p_2 + p_3 = p_1^2 + p_2^2 + p_3^2 = 1. \tag{5.32}$$

In terms of our variables, the Kasner spacetime corresponds to $\Sigma_{\alpha\beta}$ being a constant matrix and $N_{\alpha\beta} = 0$, in other words exactly the behavior that occurs in the Kasner

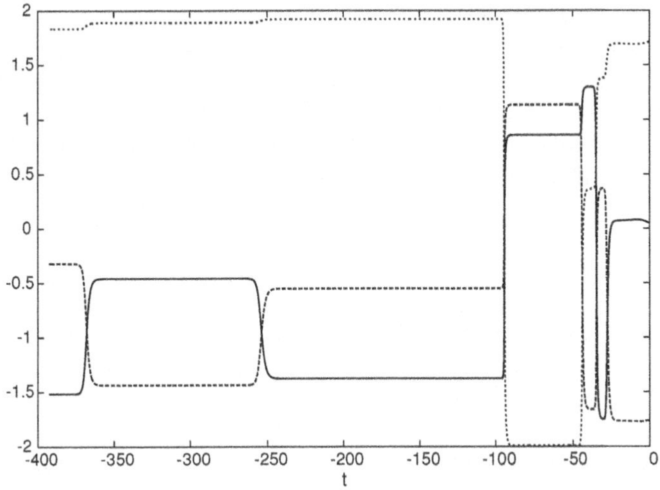

Fig. 5.1 Plot of the components of $\Sigma_{\alpha\beta}$ *vs* t

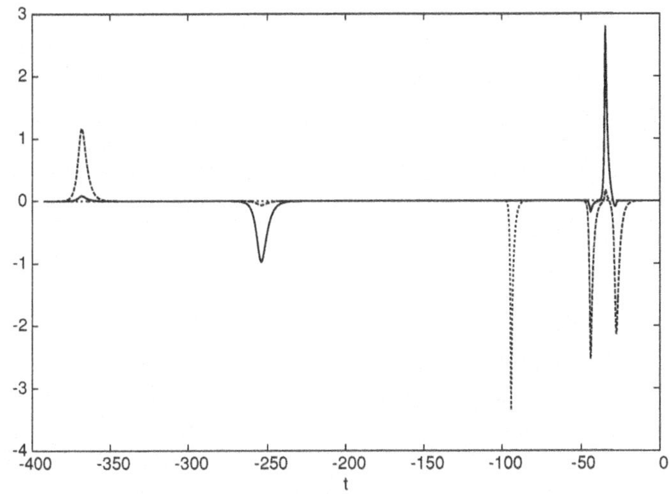

Fig. 5.2 Plot of the components of $N_{\alpha\beta}$ *vs* t

epochs. The short transitions between Kasner epochs are usually called "bounces" though that terminology can lead to confusion so we will use the term "transitions" instead.

Note that the during the Kasner epochs $N_{\alpha\beta}$ is negligible, while during each transition one eigenvalue of $N_{\alpha\beta}$ grows large and then small again. This behavior can be understood as follows: due to the condition in Eq. (5.32) one of the p_i must be negative. Thus in the approach to the singularity only two spatial directions are shrinking while the third direction is growing. Shrinking directions are associated with shrinking eigenvalues of $N_{\alpha\beta}$ while the growing direction is associated with a growing eigenvalue of $N_{\alpha\beta}$. Thus during any Kasner epoch two eigenvalues of $N_{\alpha\beta}$ are shrinking and one is growing. Eventually the growing eigenvalue of $N_{\alpha\beta}$ becomes large enough to trigger a transition to the next Kasner epoch, at which point it becomes a shrinking eigenvalue and some other eigenvalue grows. Eventually that growing eigenvalue will trigger another transition, and so on ...

Thus the approach to the singularity consists of an infinite sequence of Kasner epochs punctuated by short transitions. If this infinite sequence sounds overly complicated, it is helpful to consider the following: classical (*i.e.* non-quantum) general relativity cannot be trusted past the Planck scale. Due to the fact that our time coordinate is essentially a logarithmic time, the Planck scale is reached in a fairly small number of transitions. Thus in physical terms what our simulations say about the approach to spacetime singularities is that there are a small number of Kasner epochs, leading to the occurrence of Planck scale curvature, after which one would need a quantum theory of gravity (and a way of doing reliable calculations in that theory) to say what happens next.

Similarly, it is often said that BKL predict a chaotic approach to the singularity. It is true that the dynamical system of Kanser epochs and transitions is formally chaotic [9], but that chaos occurs after a large number of transitions while the Planck scale is reached in a small number of transitions. Thus in physical terms the "chaos" of spacetime singularities is moot.

It is helpful to take a similar attitude to the infinite curvature that occurs at spacetime singularities, which is sometimes regarded as a disturbing property (or even a fatal flaw) of general relativity. Once again, classical GR is not to be trusted past the Planck scale, so the prediction of GR for the approach to the singularity is not that infinite curvature actually arises but rather that Planck scale curvature arises, after which one must consult quantum gravity.

The behavior of singularities in the case of a free scalar field [15] is remarkably similar to the vacuum case and is illustrated in Figs. 5.3 and 5.4. (Here all quantities are plotted vs. $-t$ so the graph can be read from left to right). Once again there is a sequence of Kasner epochs punctuated by short transitions. However, in this case there is a last transition after which there is a Kasner epoch that lasts all the way to the singularity. This behavior has the following explanation: the Kasner solution for the Einstein-scalar system still takes the form of Eq. (5.31) but now only the first of the two conditions in Eq. (5.32) needs to be satisfied. This leaves open the possibility that all the p_i are positive and thus all eigenvalues of $N_{\alpha\beta}$ are shrinking. Indeed after a finite number of transitions this possibility is realized leading to a last Kasner epoch. Thus as far as the behavior of singularities is concerned the Einstein-scalar system is simpler than the Einstein vacuum system.

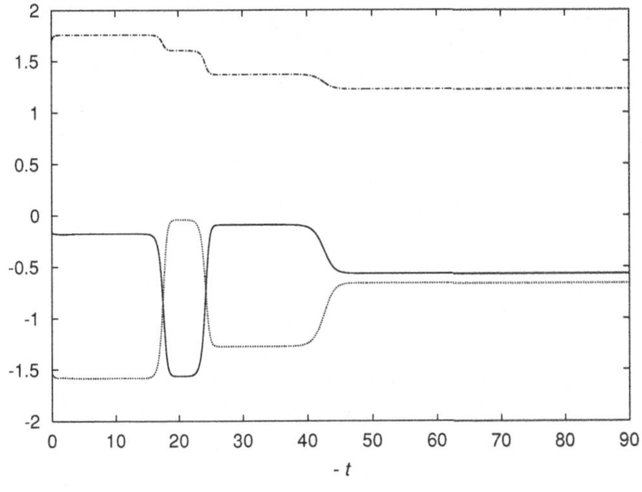

Fig. 5.3 Components of $\Sigma_{\alpha\beta}$ in the asymptotic frame *vs* $-t$ Σ_1 (solid line), Σ_2 (dotted line) and Σ_3 (dot-dashed line)

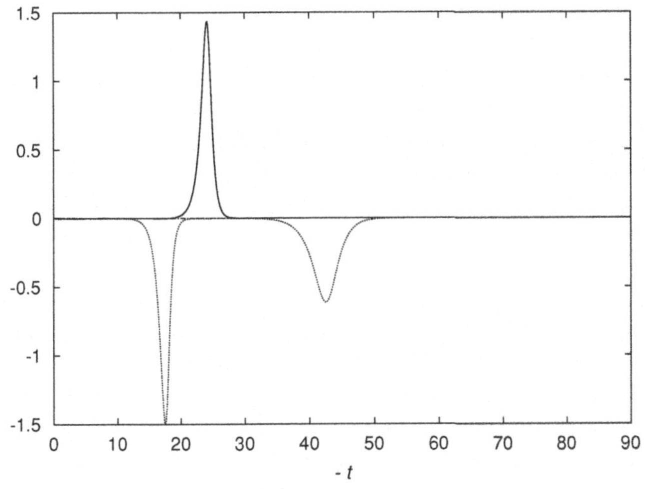

Fig. 5.4 Components of $N_{\alpha\beta}$ in the asymptotic frame *vs* $-t$ N_1 (solid line), N_2 (dotted line) and N_3 (dot-dashed line)

Spikes

One striking aspect of the simulations of [4] was the appearance of narrow features that became ever narrower as the singularity was approached. These features are called "spikes" and have been studied by several authors [2, 13, 23, 24, 28]. Though their appearance was a surprise, in hindsight it turns out that spikes are a straightforward consequence of the BKL conjecture. To see why, we will consider the behavior of the components of $N_{\alpha\beta}$ during a Kasner epoch. Equation (5.13) in the BKL approximation (*i.e.* neglecting $E_\alpha{}^i$ and A_α) simplifies to become

$$\partial_t N^{\alpha\beta} = N^{\alpha\beta} + \mathcal{N}\Big(-N^{\alpha\beta} + 2N^{(\alpha}{}_\gamma \Sigma^{\beta)\gamma}\Big), \tag{5.33}$$

while Eq. (5.20) in the BKL approximation ensures that $\Sigma_{\alpha\beta}$ and $N_{\alpha\beta}$ commute and thus have the same eigenvectors. It then follows from Eq. (5.33) that the eigenvalue of $N_{\alpha\beta}$ associated with the smallest eigenvalue of $\Sigma_{\alpha\beta}$ grows while the other two eigenvalues of $N_{\alpha\beta}$ decay. Note however, that in the special case where the relevant eigenvalue of $N_{\alpha\beta}$ starts out zero, it follows from Eq. (5.33) that it remains zero.

Now consider a configuration where the relevant eigenvalue of $N_{\alpha\beta}$ is positive in one region of space and negative in another. Then by continuity there will be points (in fact a surface of co-dimension 1) at which this eigenvalue starts out at zero and therefore remains zero. We will call these special points spike points. Since the transition to the new Kasner phase is triggered by the growth of $N_{\alpha\beta}$, spike points will remain stuck in the old Kasner epoch, while all around them other spatial points are making the transition to the new Kasner epoch. Furthermore, by continuity $N_{\alpha\beta}$ starts out small near spike points and thus these points take longer to make their transition to the new Kasner epoch. Thus at any given time we have the following: the spike point stuck in the old Kasner epoch, points far from the spike point that have made the transition to the new Kasner epoch, points near the spike point that are in the process of making the transition to the new Kasner epoch. This gives rise to a characteristic shape of the spike, and that shape becomes ever narrower as more and more points complete the transition to the new Kasner epoch.

In order to test this picture of spikes, it is necessary to perform numerical simulations [18]. Such simulations involve stringent numerical challenges. Since the spikes are ever narrowing, as the simulation proceeds it takes an increasingly fine grid spacing to resolve the spikes. Nor is the technique of adaptive mesh refinement helpful in this case: since the spike points are on surfaces of co-dimension one, too many regions would have to be refined. In the case of symmetry in two directions the simulations have effectively only one spatial dimension and one can simply choose a large number of grid points with a resulting fine spatial resolution. In the case of symmetry in only one direction one can use parallel code to run the simulation on many processors and thus provide adequate spatial resolution. Both methods are used in [18].

Some results from this study are presented in Figs. 5.5, 5.6 and 5.7. Here Figs. 5.5 and 5.6 are from a simulation where there is symmetry in two directions, while Fig. 5.7 is for a simulation in which there is symmetry in only one direction. In all cases the quantity that is being graphed is $N^\alpha{}_\alpha$.

Though the range of the coordinates for the simulations is $(0, 2\pi)$, the graph in Fig. 5.5 has a narrow range of the coordinate x offset in such a way that the center of the spike is at $x = 0$. The graph shows $N^\alpha{}_\alpha$ at several different times. Note that

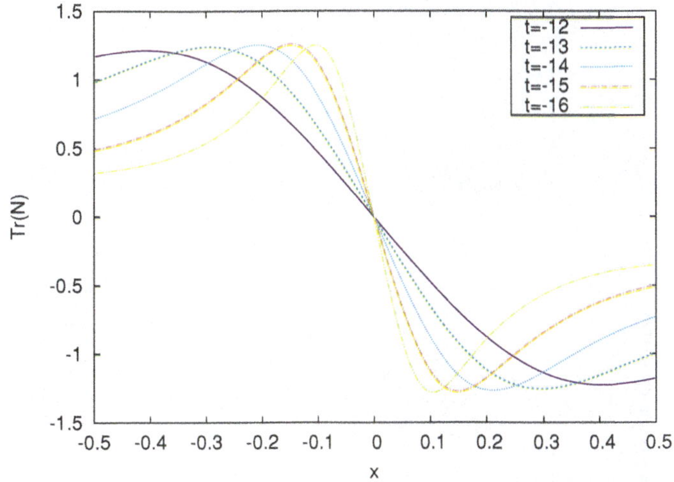

Fig. 5.5 $N^{\alpha}{}_{\alpha}$ versus x for $t = -12, -13, -14, -15$ and -16 for a spike. Here we have translated x so that zero is the center of the spike

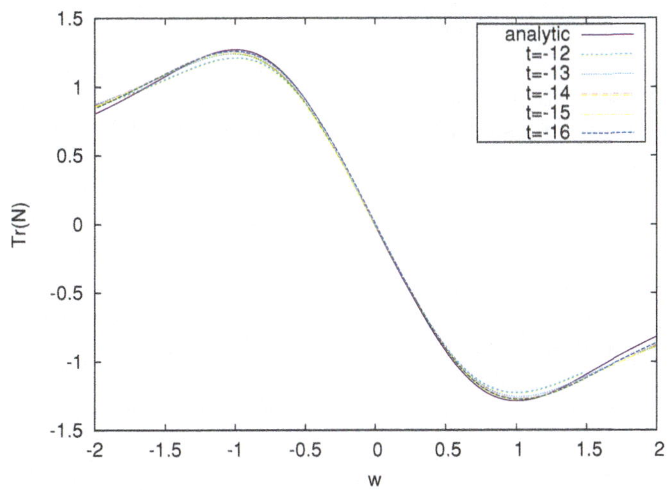

Fig. 5.6 $N^{\alpha}{}_{\alpha}$ versus the rescaled coordinate w for $t = -12, -13, -14, -15$ and -16 for the same data depicted in Fig. 5.5, along with the spike formula

the singularity is approached as $t \to -\infty$ and that as we approach the singularity the spike becomes ever narrower while retaining its shape. This point is illustrated in Fig. 5.6 which graphs the same data as in Fig. 5.5 but now *vs.* a rescaled coordinate w. Here w is just x centered on the spike and rescaled by the amount that the BKL approximation says the spike should shrink in the amount of time elapsed. Note that all the curves in Fig. 5.6 lie on top of each other. In addition, Fig. 5.6 also has a

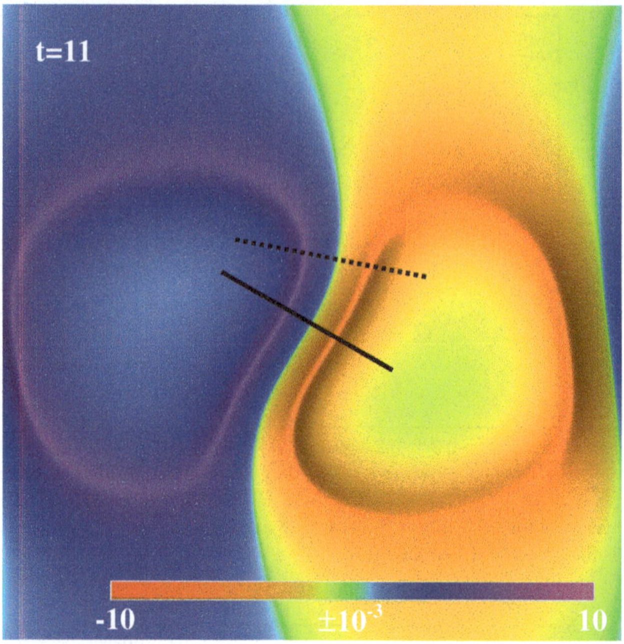

Fig. 5.7 Snapshot of $N^{\alpha}{}_{\alpha}$ at $|t| = 11$ of the 2D simulation. The solid [dashed] line illustrates the direction perpendicular to the spike surface that can be used to take a slice to compare to the one-dimensional spikes of Fig. 5.5 The range of the coordinates is $x = [0, 2\pi]$ and $y = [0, 2\pi]$

curve for the prediction of the BKL approximation for the shape (as a function of the rescaled coordinate) of the spike. Thus the spikes behave exactly as the BKL approximation says they should.

Though $E_{\alpha}{}^{i}$ is small, eventually the spike becomes so narrow that the product of $E_{\alpha}{}^{i}$ and spatial derivatives is no longer negligible. At this point these spatial derivative terms act to smooth out the spike and thus cause it to disappear. Thus each spike is a striking, but ultimately ephemeral, feature of the approach to the singularity.

Figure 5.7 shows $N^{\alpha}{}_{\alpha}$ as a function of x and y for the case of symmetry in only one direction. Here the spike surface is a curve, and there is slow variation along the curve and rapid variation in the direction perpendicular to the curve. This makes clear that spikes are a co-dimension one phenomenon. The spike behaves the same as in the case covered in the one-dimensional simulation, as a function of the perpendicular direction.

5.4 Ekpyrotic Scenario

One striking feature of the early universe (say at the time from which we see the cosmic microwave background) is that it is nearly homogeneous and isotropic. In order to account for this fact, cosmologists often invoke a mechanism to smooth out a much more inhomogeneous and anisotropic previous phase of the universe.

The most commonly invoked smoothing mechanism is inflation: approximately exponential expansion driven by a scalar field slowly rolling down a potential. However an alternative to inflation is the so-called ekpyrotic scenario [7, 22, 29] which involves a contracting phase of the universe.

Since the universe is currently expanding, the ekpyrotic scenario must also invoke a "bounce mechanism" to turn a contracting universe into an expanding one. While numerical simulations involving a bounce mechanism have been done [8, 32] they are beyond the scope of this chapter and we will only present results involving numerical simulations of the contracting phase of the ekpyrotic scenario.

From our point of view then, the contracting phase of the ekpyrotic scenario is just another case of the approach to the singularity: in this case with matter that is a scalar field with the potential $V(\phi) = -V_0 e^{-k\phi}$ where V_0 and k are positive constants. We will present the results of [17] which imposes symmetry in two directions in order to obtain very fine spatial resolution. However, we note that there are also simulations that impose symmetry in only one direction [20].

The simulations begin much as those of [14, 15]: after an initial period of somewhat complicated dynamics the quantities ${E_\alpha}^i$ and A_α become small indicating the onset of BKL dynamics. That period is followed by the formation of two different types of regions. One of those regions is much as in [14, 15]: highly inhomogeneous and anisotropic. However, in the other type of region the spacetime becomes to an excellent approximation homogeneous and isotropic.

In order to understand the behavior of these two different types of regions, it is helpful to introduce analogs of the usual cosmological density parameters Ω_i that give the ratio of the density of various types of matter to the critical density. Specifically we define an Ω for matter (Ω_m), shear (Ω_s) and curvature (Ω_k) by

$$\Omega_m \equiv \tfrac{1}{6} W^2 + \tfrac{1}{6} S^\alpha S_\alpha + \tfrac{1}{3} \bar{V}, \tag{5.34}$$

$$\Omega_s \equiv \tfrac{1}{6} \Sigma^{\alpha\beta} \Sigma_{\alpha\beta}, \tag{5.35}$$

$$\Omega_k \equiv -\tfrac{2}{3} \partial_\alpha A^\alpha + A^\alpha A_\alpha + \tfrac{1}{6} N^{\alpha\beta} N_{\alpha\beta} - \tfrac{1}{12} ({N^\gamma}_\gamma)^2, \tag{5.36}$$

where $\Omega_m + \Omega_s + \Omega_k = 1$ by (5.21). Figure 5.8 shows these quantities plotted at select times during the evolution (which was stopped at $t = -150$). Note that all features in the solution are locally smooth—apparent step functions in some of the plots are simply due to the large size of the domain relative to the size of the feature. As an example, Fig. 5.9 shows a zoom-in of the last panel of Fig. 5.8 about one of the late-time spike structures that formed in the anisotropic regime.

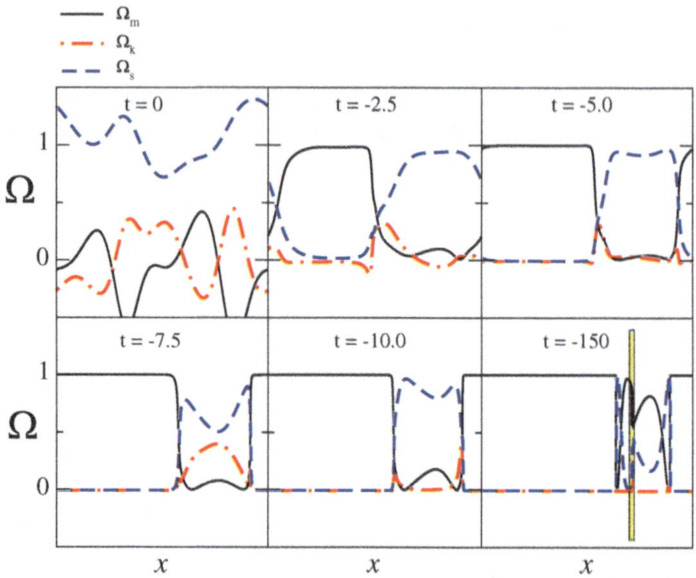

Fig. 5.8 t = const. snapshots of the normalized energy density in matter Ω_m (solid line), curvature Ω_k (dot-dash line) and shear Ω_s (dashed line) for $0 \leq x \leq 2\pi$ at several times during the evolution of the initial data described in Sect. 5.2. Time runs from left to right along the top row and continues along the bottom row. The shaded slit (dotted outline) in the last panel ($t = -150$) indicates the range of x shown in the blow-up in Fig. 5.9

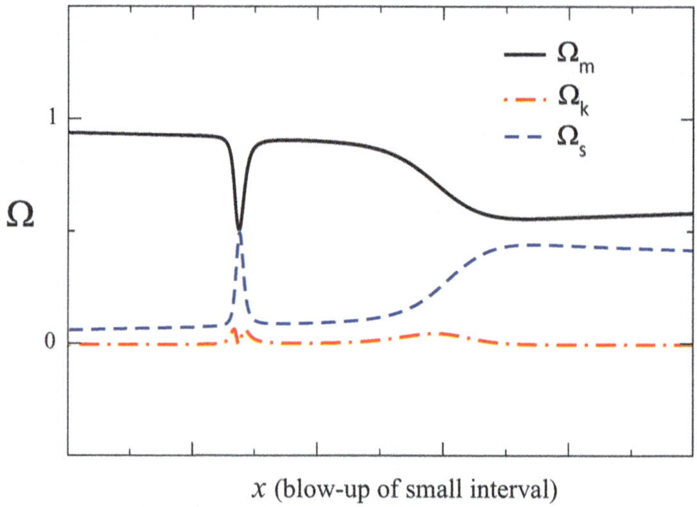

Fig. 5.9 Zooming in on one of the spike structures that formed in the anisotropic region at $t = -150$, as shown in Fig. 5.8

The inhomogeneous and anisotropic regions have comparable amounts of Ω_m and Ω_s, while the homogeneous, isotropic regions have $\Omega_m = 1$ and the other two Ω quantities essentially zero. Thus homogeneity and isotropy is brought about by the domination of the scalar field energy over all other types of energy.

Furthermore, an examination of the proper volume of the two types of region shows that the homogeneous, isotropic region has vastly more volume. Thus, despite the presence of the two types of region, the outcome is that (as measured by volume) the universe is homogeneous and isotropic.

The ekpyrotic mechanism is thus a very effective smoothing mechanism and could provide a compelling explanation for why our universe is (at early times) homogeneous and isotropic.

5.5 Conclusions

The numerical simulations reported in this chapter provide compelling evidence for the BKL picture of the approach to spacetime singularities. The use of the scale invariant tetrad method is very effective for numerical simulations because even though all physical quantities are blowing up as the singularity is approached, the variables used in the simulation remain bounded.

The use of CMC slicing allows the simulations to come arbitrarily close to the singularity without hitting it and thus provide a foliation in which one can make sense of the BKL notion that time derivatives are more important than space derivatives.

In the case of the ekpyrotic scenario, our methods also provide compelling evidence for the ekpyrotic mechanism as an effective smoothing mechanism giving rise to homogeneity and isotropy.

Our use of periodic boundary conditions for the coordinates means that we are considering the case of compact Cauchy surfaces. In particular, that means that our spacetimes are not asymptotically flat and therefore cannot capture the properties of the null singularities that form on black hole inner horizons in the asymptotically flat case [10, 25, 27]. Nonetheless, since BKL dynamics are local, it is likely that even in the asymptotically flat case part of the singularity is well described by the BKL picture.

Acknowledgements It is a pleasure to thank Beverly Berger, Vince Moncrief, Jim Isenberg, Frans Pretorius, Paul Steinhardt, and Anna Ijjas for helpful discussions. DG was supported by NSF grant PHY-2102914.

References

1. V. Belinskii, E. Lifschitz, E. Khalatnikov, Adv. Phys. **19**, 523 (1970)
2. B. Berger, D. Garfinkle, Phys. Rev. D **57**, 4767 (1997)
3. B.K. Berger, D. Garfinkle, J. Isenberg, V. Moncrief, M. Weaver, Mod. Phys. Lett. A **13**, 1565 (1998)
4. B. Berger, V. Moncrief, Phys. Rev. D **48**, 4676 (1993)
5. B. Berger, V. Moncrief, Phys. Rev. D **58**, 064023 (1998)
6. A. Brandt, Math. Comput. **31**, 333 (1977)
7. E.I. Buchbinder, J. Khoury, B.A. Ovrut, Phys. Rev. D **76**, 123503 (2007)
8. M. Corman, W. East, and. J. Ripley JCAP **09**, 063 (2002)
9. N. Cornish, J. Levin, Phys. Rev. Lett. **78**, 998 (1997)
10. M. Dafermos, Commun. Math. Phys. **332**, 729 (2014)
11. D. Garfinkle, Phys. Rev. **D65**, 044029 (2002)
12. D. Garfinkle, Phys. Rev. Lett. **93**, 161101 (2004)
13. D. Garfinkle, Class. Quant. Grav. **21**, S219 (2004)
14. D. Garfinkle, Class. Quant. Grav. **24**, S295 (2007)
15. D. Garfinkle, J. Curtis, Phys. Rev. **D72**, 064003 (2005)
16. D. Garfinkle, C. Gundlach, Class. Quant. Grav. **22**, 2679 (2005)
17. D. Garfinkle, W.C. Lim, F. Pretorius, P.J. Steinhardt, Phys. Rev. D **78**, 083537 (2008)
18. D. Garfinke, F. Pretorius, Phys. Rev. **D102**, 124067 (2020)
19. S.W. Hawking, R. Penrose, Proc. Roy. Soc. Lond. **A314**, 529 (1970)
20. A. Ijjas, W.G. Cook, F. Pretorius, P.J. Steinhardt, E.Y. Davies, JCAP **08**, 030 (2020)
21. E. Kasner, Amer. J. Math. **43**, 217 (1921)
22. J. Khoury, B.A. Ovrut, P.J. Steinhardt, N. Turok, Phys. Rev. D **64**, 123522 (2001)
23. W.C. Lim, Class. Quantum Grav. **25**, 045014 (2008)
24. W.C. Lim, L. Andersson, D. Garfinkle, F. Pretorius, Phys. Rev. D **79**, 123526 (2009)
25. A. Ori, Phys. Rev. Lett. **67**, 789 (1991)
26. R. Penrose, Phys. Rev. Lett. **14**, 57 (1965)
27. E. Poisson, W. Israel, Phys. Rev. Lett. **63**, 1663 (1989)
28. A. Rendall, M. Weaver, Class. Quant. Grav. **18**, 2959 (2001)
29. P.J. Steinhardt, N. Turok, Science **296**, 1436 (2002)
30. S. Hern, J. Stewart, Class. Quantum Grav. **15**, 1581 (1998)
31. C. Uggla, H. van Elst, J. Wainwright, G.F.R. Ellis, Phys. Rev. D **68**, 103502 (2003)
32. B. Xue, D. Garfinkle, F. Pretorius, P. Steinhardt, Phys. Rev. D **88**, 083509 (2013)
33. J.W. York, Phys. Rev. Lett. **26**, 1656 (1971)

Chapter 6
Gravitational Collapse of a Spherical Scalar Field

Roberto Giambò

6.1 Introduction and Historical Context

From the second half of last century's 60 s, the so called "singularity theorems" of Stephen Hawking and Roger Penrose, together with the first observational evidence of black holes candidates, shifted attention to the formation of singularities as an outcome of a *dynamical* process. Actually, despite the official reasons that were given half a century later for awarding the Nobel Prize to Roger Penrose, it is always worth remembering that the theorems of Hawking and Penrose are rather results that assert the geodesic *incompleteness* of a Lorentzian manifold. It was Hawking himself that supported (see e.g. [59, pag. 258]) the notion of causal geodesic incompleteness as a mathematically viable definition of spacetime singularity that, however, is not always completely satisfactory under the physical point of view [38]. In fact, one can conceive solutions where initial data on a would–be Cauchy hypersurface evolve into an incomplete spacetime that terminates at a Cauchy horizon, i.e. a boundary of the maximal future development—here, solution can be possibly extended but not in an unique way. This surprising feature is observed even in seemingly simple examples as the Kerr spacetime, indicating that the formation of high curvature regions evolving into a trapped surface from regular data is far from being solved by simply resorting

R. Giambò (✉)
Scuola di Scienze e Tecnologie, Mathematics Division, University of Camerino (Italy), Camerino, Italy
e-mail: roberto.giambo@unicam.it

INAF, Sezione di Roma, Rome, Italy

INdAM – GNAMPA, Rome, Italy

INFN, Sezione di Perugia, Perugia, Italy

D. Malafarina and P. S. Joshi (eds.), *New Frontiers in Gravitational Collapse and Spacetime Singularities*, Springer Series in Astrophysics and Cosmology,
https://doi.org/10.1007/978-981-97-1172-7_6

to geodesic incompleteness. For a comprehensive account of the history, stemming from the Penrose and Hawking theorems, let me recommend the beautiful review [69].

As the interest in the dynamical features of spacetimes grew, it became clear that Schwarzschild (or Kerr) spacetime could only be seen as an *asymptotic* state of more realistic configurations that could potentially develop a singularity. Therefore, it was necessary to consider solutions with richer features in order to gain insight into the problem of relativistic collapse. However, the nonlinear complexity of the Einstein Field Equations suggested the need for a trade-off between mathematical and physical reasons. As a result, researchers turned their attention to the collapse of *spherically symmetric* models, with a particular focus on the study of collapse in dust clouds, radiation, and scalar fields. This review will specifically examine the latter model.

Studies on scalar fields also date back to the late 1960s, with the introduction of the concept of spin-zero particles described by a scalar wave, which led to the development of the so-called *boson star*—for a comprehensive review, see [74]. From a mathematical standpoint, as we will see, the problem has been modelled in many different ways, using various gauges and assumptions. However, almost every approach shares some common features:

1. a metric g on the 4–dimensional manifold $M = \mathscr{Q} \times S^2$, that is given by the product metric of a Lorentzian metric on the 2-dimensional spacetime $\mathscr{Q}$ and a conformal metric to the induced Riemannian metric on $S^2 \hookrightarrow \mathbb{R}^3$, with conformal factor depending on $\mathscr{Q}$;
2. an energy momentum tensor arising from a Lagrangian density

$$L = \frac{1}{2} g(\nabla\phi, \nabla\phi) + \frac{1}{2} m^2 \phi^2 + \widetilde{V}(\phi), \tag{6.1}$$

where $\phi : \mathscr{Q} \to \mathbb{R}$ is the scalar field, ∇ is the gradient induced by g, m is the constant mass of the scalar field ($m = 0$ is the so–called *massless* case) and a *potential* function $\widetilde{V} : \mathbb{R} \to \mathbb{R}$ that can be possibly set to zero (*free* case) is also introduced. The Lagrangian may contain other terms when the scalar field is coupled with other energy sources, for example an electromagnetic field (see Sect. 6.2.4).

Therefore, in its more general extent, in the present paper (M, g) will be such that

$$R_{\mu\nu} - \frac{1}{2} g_{\mu\nu} R = 8\pi T_{\mu\nu}, \tag{6.2}$$

holds and, using an effective potential $V(\phi)$ possibly embodying the mass contribution, (6.1) gives

$$T_{\mu\nu} = \phi_{,\mu}\, \phi_{,\nu} - \left(\frac{1}{2} g^{\alpha\beta} \phi_{,\alpha}\, \phi_{,\beta} + V(\phi) \right) g_{\mu\nu}. \tag{6.3}$$

It is straightforward to calculate the conservation law

$$T^{\mu}_{\nu;\mu} = \phi,_{\nu}\left(\Box\phi - V'(\phi)\right) = 0, \tag{6.4}$$

where $\Box\phi = g^{\mu\nu}\phi_{,\mu;\nu}$ is the curved wave operator induced by g. The equation $\Box\phi = V'(\phi)$ is often referred to as *Klein–Gordon equation*.

To trace the beginning of research on scalar field collapse, we must start with the earlier works of Demetrios Christodoulou. As he recounts in his monograph [20], the problem of the evolutionary character of trapped surfaces was posed to him by his Ph.D. mentor at Princeton, John A. Wheeler. After working on stars filled with dust matter, where gravitational attraction is counterbalanced by outward pushing forces due to internal nuclear reactions, he turned his attention to the study of the spherical scalar field. As explained in [19], one reason why he chose to study this model was because the dynamics governing its evolution are precisely described by the wave equation. Therefore, one could view this model as a perturbation of some background spacetime.

Christodoulou's first work on the subject, published in [14], focused on the massless free case ($V = 0$ in (6.3)), as did his subsequent works. The main result of this paper is a theorem that shows that if one considers C^1 initial data sufficiently small with respect to a properly defined norm (see [14, Theorem 3]), they evolve into a globally unique solution that is null geodesically complete and asymptotically equivalent to flat spacetime. This result had a significant impact on the field of research, as it showed that despite the absence of forces counterbalancing gravitational attraction, as in the case of a dust star, the collapse of a scalar field—at least in the massless free case—does not necessarily lead to the formation of spacetime singularities.

At the time, Christodoulou was not the only one working on scalar field collapse. Matthew W. Choptuik had already addressed the problem using numerical methods in his Ph.D. thesis at British Columbia, under William Unruh supervision, and had published a couple of papers in which the presumed inconsistency of some numerical codes useful for numerically solving the Cauchy problem for some relativistic models was discussed and fixed. Choptuik got in touch with Christodoulou at the end of the 80 s, when the latter was at Syracuse. On one hand, a picture was emerging that suggested the existence of a sort of phase transition that governed initial data of solutions, with some evolving into a singularity and others regularly dispersing in the infinite future. However, it was still unclear at the time whether one could fine–tune the data to develop a horizon and obtain black holes of arbitrarily small mass, and whether one could fully capture the causal behaviour of this hypothetical phase transition between black hole formation and dispersion.

Indeed, Choptuik's numerical analysis published in [10, 11] confirmed the existence of what would have been subsequently known as the *black hole threshold*, dividing the two classes of initial data. In other words, considering various 1–parameters family of initial data, there exists a critical value of the parameter separating the two possible evolutions—i.e. trapped surface formation versus dispersion. This behaviour was found later to be common to many collapsing models, and for a complete account on the subject the main reference is the Living Review by Carsten Gundlach and José M. Martín–García [55].

The solution corresponding to the critical value of the parameter can be shown to exhibit surprising features [85]. To begin, it is a discretely self–similar solution. Moreover, it contains a point singularity that can communicate with the infinite future boundary—it is, therefore, an example of a *naked singularity*, thereby violating the cosmic censorship conjecture made by Roger Penrose [93]. By the way, it was Choptuik's discovery that in 1997 put an end to the first bet on naked singularity between Hawking, who was unfavourable to their existence, and John Preskill and Kip Thorne, who supported the possibility of naked singularities. Numerical analysis were suggesting that the initial data leading to naked singularities might not be generic, and this led to a new version of the bet. Furthermore, it boosted research towards an analytical approach to the analysis of naked singularities. And this bring us back to Christodoulou's work.

After some follow-ups to [14], in the attempt to prove Penrose conjecture, he had published in 1991 the first of four papers that are still nowadays considered cornerstones of the subject. The result [15, Theorem 5.1] expresses sufficient condition on the initial data of the problem leading to trapped surface formation, hence to a geodesically incomplete spacetime. That is a key result for the topic under exam here and for all following research (see discussion after Theorem 6.1 below).

The solutions considered so far were sufficiently regular, i.e. at least differentiable. The second paper [16] of the series improves previous results leading to dispersing behaviour but mostly takes into account bounded variation solutions, that prove to be extremely important for the other two papers of the series. Indeed, in [17] an analytical example of a scalar field solution collapsing to a singularity without horizon formation, hence naked, is found out. In the same work it is announced, for "a subsequent paper", that the space of initial data leading to solutions with this feature is unstable in the larger space of bounded variation solutions considered in [16]. Curiously enough this last paper was published only five years later [18]—although received by the journal a couple of years before publication—and it has since often been popularized as *the proof* of Penrose conjecture. In Sect. 6.2.2 we will get back to the influence of this work in relation to cosmic censorship.

The methods pioneered by Christodoulou were refined for a number of more recent results, especially concerning the Einstein–Maxwell–scalar field model, see Sect. 6.2.4 below. These techniques, sharing a common approach involving the analysis of an initial value problem for the PDE governing system's evolution, led to a number of considerable result and advances in the study of problems in Mathematical Relativity, under very general assumptions. Some literature can be traced back from [67, Footnote 297]

On the other side, since the dynamical system picture was not fully resolved by the theorems in [15–18], a significant effort was devoted to many analytical and/or numerical studies of the'critical' case found out by Choptuik. The aim was to determine the extent to which Christodoulou's non-genericity result was essentially describing the same situation as the Choptuik black hole threshold. Despite some initial enthusiasm, it must now be concluded that the link between the two pictures, while displaying some overlapping features, has not been fully resolved to date.

On quite a different pathway, a few years later, stimulated by connections with extended gravity theories and string theory, a growing interest arose about models self interacting with a potential. Dealing with a cosmological setting, in this case one usually restricts to homogeneous models where the background metric interacting with the scalar field is a Robertson–Walker (RW) spacetime—this gives the obvious advantage to reduce the model's equations to an ODEs system. A huge variety of potentials $V(\phi)$ has been discussed within this framework, finding conditions under which the universe collapse to a future singularity, and global models without the formation of an apparent horizon has been discussed also in this situation. Of course, this study interacted with those aiming to find quantum mechanisms to avoid recollapsing universe, see e.g. [49]—a quite recent and complete review on this subject is [84].

The review is organized as follows. The basic results from Christodoulou and PDE approach are described and commented in Sect. 6.2. The following Sect. 6.3 is devoted to the advances in the study of the black hole threshold and the critical case, and the homogeneous case is described in Sect. 6.4. One last Sect. 6.5 is dedicated to some hints and studies that are not completely related to the topics covered in the previous section, and to sketch conclusions.

6.2 Scalar Field Collapse as a PDE Initial Value Problem

In this section we review some of the most influential and important results regarding the scalar field collapse, when the evolutionary PDE approach is embraced. As said in the Introduction, this field can be traced back to the earlier works of Christodoulou on the subject, but it must be also said that these were pioneering works, with a notation often invented specifically to solve the problem at hand, and therefore not yet consolidated in the literature at the time and sometimes not fully optimized, so the less accustomed reader may find it challenging at first glance. For example, in [15, p. 345] a geometric Bondi coordinate system is used to describe the metric,

$$\mathrm{d}s^2 = -a(u,s)\,\mathrm{d}u^2 - 2\,\mathrm{d}u\,\mathrm{d}s + r^2(u,s)\,\gamma_{\mathbb{S}^2}, \tag{6.5}$$

(we refer to $\gamma_{\mathbb{S}^2}$ as the Euclidean metric induced by $\mathbb{S}^2 \hookrightarrow \mathbb{R}^3$) but many of the calculations in the same work are done in a double null frame. The use of double null coordinates is rather effective in this PDE approach:

$$\mathrm{d}s^2 = -\Omega^2(u,v)\,\mathrm{d}u\,\mathrm{d}v + r^2(u,v)\,\gamma_{\mathbb{S}^2}. \tag{6.6}$$

Coherently with [16] and other works in this thread, we normalize the scalar field $\phi \to \frac{1}{\sqrt{4\pi}}\phi$ and the potential $V \to \sqrt{2\pi}V$ to get rid of a 4π factor in the coupling constant, thereby obtaining from (6.2)

$$r_{,uv} = -\frac{1}{r}r_{,u}\,r_{,v} - \frac{1}{4r}\Omega^2 + \frac{1}{2}r\Omega^2\,V(\phi), \tag{6.7a}$$

$$(\log\Omega)_{,uv} = \left(\frac{\Omega}{2r}\right)^2 + \frac{r_{,u}\,r_{,v}}{r^2} - \phi_{,u}\,\phi_{,v}\,, \tag{6.7b}$$

$$\phi_{,uv} = -\frac{\phi_{,u}\,r_{,v} + \phi_{,v}\,r_{,u}}{r} - \frac{1}{4}\Omega^2\,V'(\phi), \tag{6.7c}$$

as a system of independent PDEs. Moreover, the two relations

$$\left(\frac{r_{,u}}{\Omega^2}\right)_{,u} = -r\left(\frac{\phi_{,u}}{\Omega}\right)^2, \tag{6.7d}$$

$$\left(\frac{r_{,v}}{\Omega^2}\right)_{,v} = -r\left(\frac{\phi_{,v}}{\Omega}\right)^2, \tag{6.7e}$$

are consequence of (6.7a)–(6.7c).

It will also be useful to consider the Misner–Sharp mass[1] $m(u, v)$ defined by the relation

$$1 - \frac{2m}{r} = g(\nabla r, \nabla r), \tag{6.8}$$

that in view of (6.6) becomes

$$m(u, v) = \frac{r}{2}\left(1 + 4\frac{r_{,u}\,r_{,v}}{\Omega^2}\right). \tag{6.9}$$

6.2.1 *Trapped Region Formation in the Free Massless Case*

As already mentioned, Christodoulou considers the free massless case, $V(\phi) = 0$, that implies (see (6.4)) than the scalar field satisfies the homogeneous wave equation. The equations (6.7a)–(6.7e) can be transformed in a first order (overdetermined) system with the positions

$$\nu = r_{,u}\,, \quad \lambda = r_{,v}\,, \quad \zeta = r\phi_{,u}\,, \quad \theta = r\phi_{,v}\,, \tag{6.10}$$

in the unknown functions $(r, \lambda, \nu, m, \zeta, \theta)$ as follows (see e.g. [31, Eqs. (12)–(19)]):

[1] Often referred to as *Hawking* mass too, see e.g. [27].

$$\lambda_{,u} = \lambda \left(\frac{2m\,\nu}{r^2 - 2m\,r} \right), \qquad \nu_{,v} = \nu \left(\frac{2m\,\lambda}{r^2 - 2m\,r} \right), \tag{6.11a}$$

$$m_{,u} = \frac{(r-2m)\zeta^2}{2r\,\nu}, \qquad m_{,v} = \frac{(r-2m)\theta^2}{2r\,\lambda}, \tag{6.11b}$$

$$\theta_{,u} = -\frac{\zeta\lambda}{r}, \qquad \zeta_{,v} = -\frac{\theta\nu}{r}. \tag{6.11c}$$

The first paper on the subject, as already recalled, is [14], where a condition on the initial data is derived to have global forward–in–time solutions. This result was recently refined in [77, 78], where it is proved that initial data can be chosen such that their L^1 norm is infinite but they evolve again to a forward–in–time global solution. The construction can be extended in order to prescribe initial data at past null infinity. Moreover these refinements contain estimates useful for proving recent stability results of dispersive spherical solutions with respect to non–spherical perturbations [65, 81].

On the other side, on [15] a sufficient condition for trapped surface formation starting from collapsing initial data is formulated. The theorem is stated—and re–proved—in terms of the double null framework sketched above in the recent paper [1], that here we follow.

First of all, notice that spherical symmetry allows to consider a timelike curve Γ that will play the role of the symmetry centre. Characteristic initial data along $u = u_0$ and $v = v_1$ are prescribed (see Fig. 6.1), and $v_2 > v_1$ is taken to define the quantities

$$\eta_0 := \frac{m(u_0, v_2) - m(u_0, v_1)}{r(u_0, v_2)}, \qquad \delta_0 := 2\frac{r(u_0, v_2) - r(u_0, v_1)}{r(u_0, v_2)},$$

Now, a collapsing situation amounts to assume $r_{,u} < 0$ and then, recalling (6.9), trapped surface forms when also the other partial derivative becomes negative somewhere, $r_{,v} < 0$.

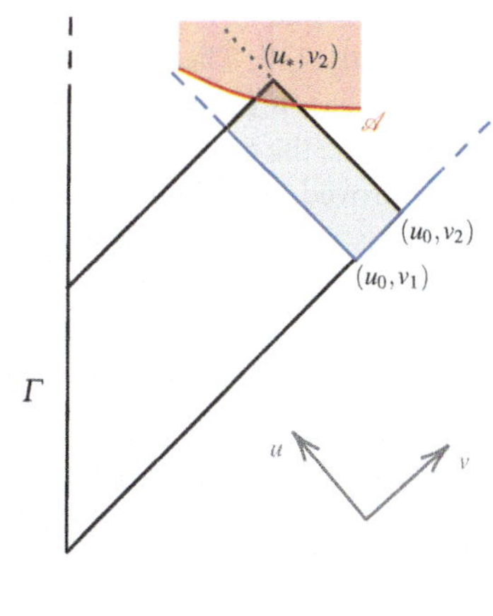

Fig. 6.1 The collapse theorem in [15]. The centre of symmetry is Γ, and initial data are prescribed in $u = u_0$ and $v = v_1$ (blue lines). A trapped region (in red), bounded by the apparent horizon $\mathscr{A}$, develops in the (grey) shaded region

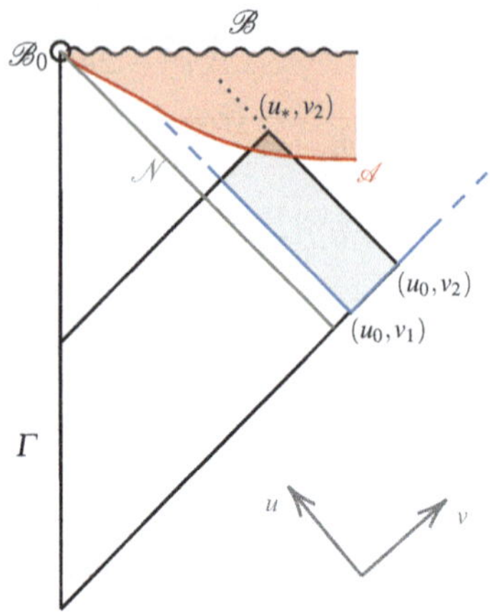

Fig. 6.2 The existence of a future singular boundary shown in [15]. The non-central boundary is strictly spacelike and terminates in the central boundary, where it is intersected by the apparent horizon $\mathscr{A}$. An incoming null hypersurface $\mathscr{N}$ also intersects the (singular) central boundary

Theorem 6.1 ([15], Theorem 5.1) *There exists a constant $c_1 \geq 1$ such that if*

$$\delta_0 \leq \frac{1}{e} \quad \textit{and} \quad \eta_0 > c_1 \delta_0 \log\left(\frac{1}{\delta_0}\right)$$

then a trapped surface forms in the region $[u_0, u_*] \times [v_1, v_2]$, where u_* is such that $r(u_*, v_2) = \frac{3\delta_0}{1+\delta_0}$.

It is worth mentioning here that the techniques pioneered in [15] to prove the above Theorem lie at the foundations of many other important results in the gravitational collapse mathematical theory, let me cite the celebrated [20] proof of trapped surface development from a packet of waves emitted at past infinity. There, the technique is refined to assume that the incoming radiation per unit solid angle is uniformly bounded from below. The original idea is contained in the above Theorem: the conditions on η_0 and δ_0 in Theorem 6.1 statement[2] are roughly equivalent to assuming that a sufficient amount of energy is emitted by the scalar field, at a sufficiently short amount of time.

Actually, [15, Theorem 5.1] contains other important results for the present context, that haven't been reported in the above statement of Theorem 6.1. In particular, see Fig. 6.2, the existence of a (non-central) strictly spacelike singular future boundary is proved. This boundary terminates at the boundary of the centre of symmetry where also the apparent horizon $\mathscr{A}$ arrives. In other words, the centre of symmetry becomes trapped exactly when it becomes singular.[3]

In a more recent paper [31] Mihalis Dafermos shows that these solutions can be regularly extended backwards up to past null infinity.

---

[2] In the original formulation of [15], the conditions were stated as $\delta_0 < \frac{1}{2}$ and $\eta_0 > E(\delta_0)$, with $E(x) = \frac{x}{(x+1)^2}(5 - x - \log(2x))$. The formulation shown here is introduced by the Author since [18], where the result is used for the proof of naked singularity instability (see below).

[3] This is a common feature shared by other models of gravitational collapse with matter, for instance collapsing stars with dust [86, 97] or anisotropic [41] matter. See also [42] and references therein.

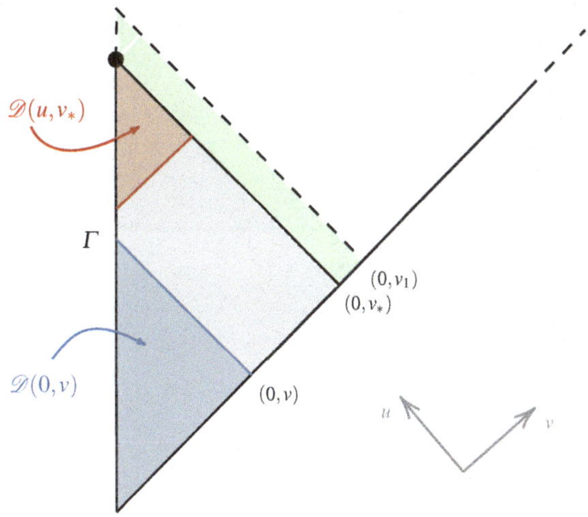

Fig. 6.3 The extension theorem from [16]. Condition (6.12) allows to extend the solution to a region (in green) across the future null cone of $(0, v_*)$

The paper [16] studies solutions with bounded variation (BV) initial data. With this notation it is meant that the functions ϕ, $r_{,u}$, $r_{,v}$, $(r\phi)_{,u}$, $(r\phi)_{,v}$ are BV with respect to one null variable, uniformly in the other, and vice versa; plus some technical regularity conditions on the centre of symmetry Γ. The paper considers both regular solutions and those developing singularities.

A regularity condition is stated in terms on the behaviour at the centre (see Fig. 6.3): if $v_* > 0$ is such that, $\forall v < v_*$, a solution exists in $\mathscr{D}(0, v)$, the domain of dependence of $(0, v)$, then $\exists \varepsilon > 0$ such that the condition

$$\lim_{u \to v_*} \sup_{\mathscr{D}(u, v_*)} \frac{2m}{r} < \varepsilon \tag{6.12}$$

implies the regular extension of the solution on $\mathscr{D}(0, v_1)$ for some $v_1 > v_*$.

In other words, in order to extend the regular centre at some point, $2m/r$ must tend to zero as this point is approached from every direction coming from $\mathscr{D}(u, v_*)$. If this condition isn't satisfied then a central singularity develops, and the existence of a non-central boundary, preceded by a trapped region as in Fig. 6.2, is proved.

6.2.2 *Non-genericity of Naked Singularity*

As said in the introduction, the solution exhibiting a naked singularity is built in the subsequent paper [17]. It is better to recall here that the naked singularity here corresponds to a situation where horizon does not exist at all[4]. We will discuss this

[4] In this sense this situation is very different from the naked singularity developing in the collapse of some matter models, see previous footnote. In those contexts there may be situations where outgoing

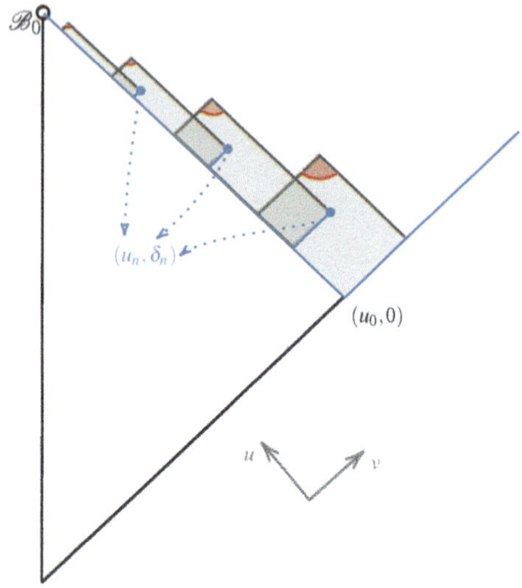

Fig. 6.4 The instability theorem in [18], in the double null setting argument proved in [75]. Here $\mathscr{B}_0$ has coordinates (0, 0). The sequence (u_n, δ_n) tends to $(0^-, 0^+)$ and each point is taken to satisfy hypotheses of Theorem 6.1 (on the blue lines) and find out a trapped region (in grey) such that a trapped surface (in red) develops in the upper corner at each step

example in next Section, since it is in some way connected to the critical solution numerically found by Choptuik [11]. Here we only stress that this example exhibits a pointwise singularity on the centre of symmetry. This fact is relevant because, in [18], it is shown that if one has bounded variation initial data on $u = 0$ developing such a solution, or more in general a solution that is neither future regular nor develops an event horizon, then a suitable perturbation of this initial data restores the apparent horizon and, hence, trapped surface formation.

More specifically, assigned BV initial data such that a naked pointwise singularity develops at the centre, then two BV functions $f_1(v)$, $f_2(v)$ exist, such that *any* (non-trivial) perturbation of the initial data with a linear combination of $f_1(v)$ and $f_2(v)$ results in a trapped surface formation. Let us discuss in brief this important result.

First of all, it must be said that the proof in [18] is carried out again in a Bondi coordinate setting similar to (6.5). However, in recent times a new proof appeared [75] where the argument is adapted in double null coordinates. The rough idea of the argument is sketched in Fig. 6.4.

Coordinates are shifted in such a way that (0, 0) is an isolated singularity in the (u, v)–plane as a development of some particular initial data $\alpha_0(v) = r\phi_{,v}(u_0, v)$.

The idea follows these steps:

1. finding proper estimates of the quantities involved (not satisfied with the initial data $\alpha_0(v)$), in order to determine a sequence (u_n, δ_n) approaching (0, 0), in such a way that one can apply Theorem 6.1 with $v_1 = 0$ and $v_2 = \delta_n$, and determine trapped surfaces existence. At every step of the sequence one is able to determine a piece of the apparent horizon. Since the sequence approaches the singular point, so does the apparent horizon, and the singularity is no more naked.

null geodesics are "emitted" from the central singularity and detected by a faraway observer, even when a horizon forms.

2. finding conditions on $r\phi_{,v}(0, v)$ such that estimates cited on step 1 hold. This is done by considering two separate cases, see [75, Theorems 5 and 6], corresponding to [18, Theorem 2.1 and Theorem 3.1].
3. determining—see [18, Sect. 4] and, in its new version, [75, Proof of Theorem 3]—two suitable functions $f_1(v)$ and $f_2(v)$ and proving that the initial data $\alpha_0(v) + \lambda_1 f_1(v) + \lambda_2 f_2(v)$ ($\forall \lambda_1, \lambda_2 \in \mathbb{R}$) evolves in such a way that $r\phi_{,v}(0, v)$ necessarily satisfies the conditions stated on step 2.

The instability result in [18] marks an important point in the history of Penrose cosmic censorship conjecture, and many still regard that paper as "the proof" of this conjecture, or at least of a re–designed version of it, allowing for naked singularities at most for an *exceptional* set of zero measure in the set of initial data for the evolution problem. The result shows that the codimension of the space of initial data is at least 2, being at least two "directions" where an initial data developing a point singularity can be perturbed, to restore trapped surface formation.

This and other results in this context seems to highlight a non-genericity feature of pointlike singularities developing without an horizon, but a central question may be the choice of the suitable—under a physical point of view—regularity requested for the initial data, because demanding a mild regularity property—as bounded variation solution may be, as commented in [85, Section 3.4]—can result in a too large space of admissible initial data.

In this sense, if we embrace the evolutionary PDE setting that is the object of the present section, and therefore identify the double null as the natural coordinate setting to work with, taking initial data on a null hypersurface, it must be said, regarding the instability theorem proved in [18] or in its new version of [75], that at least the function $f_2(v)$ is absolutely continuous in $[0, +\infty[$ (whereas f_1 has a jump discontinuity). This means that non-genericity (with codimension at least 1) is granted when one restricts to absolutely continuous functions.

Of course, the scalar field model treated so far is not the most general one (recall that potential is been set to zero so far), and then in the following subsection we wish to explore some more recent result that aim to extend the knowledge on the subject through analytical studies.

6.2.3 *Cosmological Constant and Potential*

Theorem 6.1 has been proved in other contexts. For example, in [25] the proof is made where the addition of a positive cosmological constant is considered, thereby obtaining the modified system with respect to (6.2)

$$R_{\mu\nu} - \frac{1}{2} g_{\mu\nu} R + \Lambda g_{\mu\nu} = 8\pi T_{\mu\nu}. \tag{6.13}$$

This amounts to work with system given by (6.2), where the potential $V(\phi)$ is assumed to be constant and positive. Therefore, with the same normalization of the scalar field considered above (see right after (6.6)) we arrive to the system (6.7a)–(6.7c) with $V(\phi) = \Lambda/2$. The conditions are similar to those appearing in 6.1, paying attention to the fact that the introduction of Λ affects the definition of mass (6.9) that in this context must be normalized as $\varpi = m - \frac{\Lambda}{6}r^3$.

The case with non-zero potential has in general been less covered in the PDE approach. Introducing a potential in principle adds a source term to Klein–Gordon equation and may significantly complicate the evolution in that, just to mention, scale invariance does not hold anymore. However some results have been proved all the same. Among these it must be certainly mentioned one of Dafermos' papers on the subject, i.e. [27]. At the time, paper [62] appeared that—heuristically trying to approximate the model with the anti–deSitter homogeneous spacetime, corresponding to the asymptotic state of a negative minima for the potential $V(\phi)$, was conjecturing the appearance of naked singularities generically arising but not from the centre of symmetry of the system. The paper [27] shows that this situation cannot happen for a collapsing scalar field with bounded from below potential, regardless of the sign of $V(\phi)$ at the minima.

In particular, the situation described in [27] is the following. We consider the future evolution $\mathcal{Q}^+$ from a spacelike Cauchy hypersurface S under the collapsing assumption $r_{,u} < 0$ in $\mathcal{Q}^+$. The notion of *first singularity* is introduced as follows:

Definition 6.1 A boundary point $p \in \partial\mathcal{Q}^+$ is a *first singularity* if

1. $J^+(p) \cap \mathcal{Q}^+$ is *eventually compactly generated* (i.e. $\exists X \subset \mathcal{Q}^+$ compact such that $J^-(p) \subset D^+(X) \cup J^-(X)$);
2. if $Y \subsetneq J^-(p) \cap \mathcal{Q}^+$ is eventually compactly generated $\Rightarrow \exists q \in \mathcal{Q}^+ : Y = J^-(q)$.

The result proved in [27] is that (see Fig. 6.5) *non central* first singularities are such that

$$J^-(p) \cap \mathcal{Q}^+ \cap D^+(X) \cap \mathcal{T} \neq \emptyset,$$

where $\mathcal{T}$ is the trapped surface defined by $r_{,v} < 0$.

This means that the horizon must form prior to the appearance of a non central singularity. The argument lies in careful estimates on the quantities involved in the "diamond" $J^-(p) \cap D^+(X)$, and that is why non centrality of p is crucial for this proof.

The paper [28] of the same period states general sufficient conditions on energy momentum tensor for the future null infinity I^+ to be future complete when $\mathcal{Q}^+ \setminus J(I^+) \neq \emptyset$ (i.e., when a trapped region arises) and in the scalar field collapse the result can be applied when $V(\phi) \geq 0$. Indeed, one of the sufficient conditions stated in [28] is the dominant energy condition that in this case requires $V(\phi) \geq 0$ to hold. In the evolutionary PDE approach, this corresponds to a reformulation of weak cosmic censorship given in [19], see also [69, Section 10.5]. It is also worth mentioning that the results proved in [27, 28] have been extended to higher dimensional spacetimes in [70].

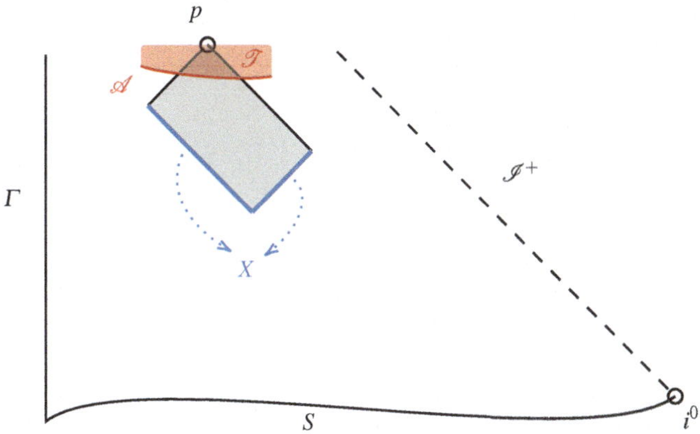

Fig. 6.5 The result in [27]. The grey region is $J^{-}(p) \cap D^{+}(X)$. If $V(\phi)$ is bounded from below—and then possibly attaining negative values—then a non-central singularity is necessarily hidden inside a trapped surface $\mathscr{T}$, bounded by an apparent horizon $\mathscr{A}$

6.2.4 Interaction with an Electromagnetic Field

Recalling the dualism between the (spherically symmetric) Reissner–Nordström spacetime and the (rotating) Kerr spacetime, in an effort to step out of the spherically symmetric case, the first attempt is to consider a spherical situation when an electromagnetic field is added to the system. This leads to modify the Lagrangian (6.1) as happens in the Einstein–Maxwell–scalar model, where the energy momentum tensor $T = T^{SF} + T^{EM}$ is made by the sum of two components T^{SF} given by (6.3) and an electromagnetic field

$$T^{EM}_{\mu\nu} = F_{\mu\lambda} F_{\nu\rho} g^{\lambda\rho} - \frac{1}{4} g_{\mu\nu} F_{\lambda\rho} F_{\sigma\tau} g^{\lambda\sigma} g^{\rho\tau}. \tag{6.14}$$

In this situations the scalar field and the electromagnetic field are uncoupled, so that each component of T separately satisfies Bianchi equations, i.e. the divergence of each T^{EM} and T^{SF} separately vanishes. Consequently, the Klein–Gordon equation still holds, together with Maxwell equations

$$F^{\mu\nu}{}_{;\nu} = 0, \qquad F_{[\mu\nu,\rho]} = 0, \tag{6.15}$$

that in view of the sphericity of the metric, gives a quite simplified form for T^{EM}:

$$(T^{EM})^{\mu}_{\nu} = \frac{q^2}{2r^4} \cdot \mathrm{diag}\{-1, -1, 1, 1\}, \tag{6.16}$$

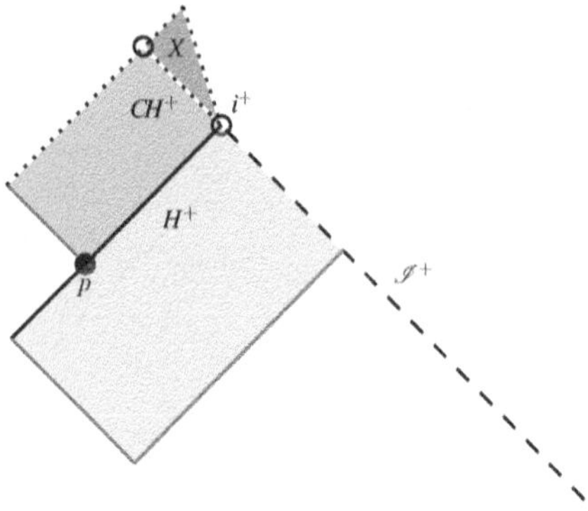

Fig. 6.6 The extension through a Cauchy horizon for the Einstein–Maxwell-scalar field studied in [26, 29, 30]. The light grey region extends to the event horizon $\mathscr{H}^+$ of the solution. The upper region is trapped and the evolution maximally extends to the Cauchy horizon CH^+. The metric is C^0 extendable across the Cauchy horizon in a region X. The results in [79, 80] exclude C^2 extendibility

where the constant q is the *charge* of the EM field. The mass m given in (6.8) is correspondingly renormalized as $\varpi = m + \frac{q^2}{2r^2}$.

The addition of this Maxwell field in the model, as studied by Dafermos in a series of his earlier papers [26, 29, 30], results in a quite different evolution of the collapse. Indeed, the future maximal development of the Cauchy problems ends at least partially in a Cauchy horizon that is not necessarily a singularity in the classical sense[5] (Fig. 6.6). Near this region the function r is positive and bounded away from zero [26, Theorem 1], but the renormalized mass ϖ blows up the at the Cauchy horizon [26, Theorem 2]. These facts imply [29, Theorems 1.1] that the metric can be (not uniquely) C^0 extended across the Cauchy horizon. In the same paper [29, Theorems 1.2] a sufficient condition, based on the growth estimate of $\phi_{,v}$ along the event horizon, is shown to imply that C^1 extension is not allowed. Actually Jonathan Luk and Sung–Jin Oh observe in papers [79, 80] that this pointwise estimate condition is difficult to be verified; in the same works, the C^2 inextendibility across the Cauchy horizon is proved, supporting a C^2–version of strong cosmic censorship.

These results and the techniques employed have been constantly refined in order to tackle the stability problem of the Cauchy horizon in Reissner–Nordstrom and later in Kerr spacetimes. This long programme has produced a huge literature in this sense and fostered a great interest on the study of stability properties of these features even outside the realm of spherical symmetry, see [32] and more recently [66] and again [69] with therein references.

[5] Interestingly enough, adding a charge to other matter models also results in weak singularity formation such as the shell crossing developing from the collapse of charged anisotropic models studied in [21].

As we have already said, Theorem 6.1 has been proved again in the paper [1], to carry on the argument in double null coordinates that are useful to study a generalization of the system that we have considered so far, the so called Einstein–Maxwell–*charged* scalar field. In this situation the scalar field is coupled to the electromagnetic field, which results in a modification of Maxwell equations because Bianchi identity does not hold anymore separately for the two components of T. We refer the reader to [100, Section 1.2.1] of Maxime Van de Moortel for some mathematical and physical aspects of this model. In the same paper, the existence of a Cauchy horizon is proved, together with C^0–extendibility across this horizon. In [1], with suitable modifications of the sufficient conditions from the original theorem in [15], the existence of trapped surface is proved for charged scalar field, confirming previous numerical insights on this model [99].

The above cited paper [100] actually also tackles the *massive* case $m \neq 0$ (recall (6.1)), where Cauchy horizon existence is proved also where mass of the scalar field is taken into account. In a later work [101] it is proved that the C^2–extension is not possible since Ricci curvature diverges. Results about continuous extendibility have recently been found too [64].

6.3 Scalar Field Collapse and the Numerical Approach

We have already commented in the Introduction that the "numerical counterpart" to Christodoulou for the scalar field collapse is represented by Choptuik and mostly by his celebrated[6] paper [11] (but see also [10] for some more details on the numerical scheme). His work is important because it represents a crucial advance to understand the scalar field's dynamics, but also because it revealed a behaviour that was surprisingly found to be satisfied by many other relativistic models along the years, as explained in the reviews [53, 55]. Gundlach and his collaborators gave a huge contribution in understanding this surprising behaviour—most of the material in the present section is taken from their works.

Scalar field spherical model for numerical analysis purposes is usually cast into a Schwarzschild gauge:

$$-\alpha^2(t,r)\,\mathrm{d}t^2 + a^2(t,r)\,\mathrm{d}r^2 + r^2(\mathrm{d}\theta^2 + \sin^2\theta\,\mathrm{d}\varphi^2). \tag{6.17}$$

Einstein equation (6.2) in the free massless case become[7]

---

[6] Almost 1000 citations received at the time the present work was written, Scopus data.

[7] Of course, one can always consider redefining Φ and Π to absorb the factor π in the Einstein equations, but this normalization, which is common in the analytical approach, is typically not utilized in the numerical approach to the problem.

$$\frac{a_{,r}}{a} + \frac{a^2 - 1}{2r} = 2\pi r(\Pi^2 + \Phi^2), \tag{6.18a}$$

$$\frac{\alpha_{,r}}{\alpha} - \frac{a_{,r}}{a} - \frac{a^2 - 1}{r} = 0, \tag{6.18b}$$

$$\frac{a_{,t}}{\alpha} = 4\pi r \Phi \Pi, \tag{6.18c}$$

where

$$\Phi = \phi_{,r}, \qquad \Pi = \frac{a}{\alpha}\phi_{,t}.$$

One advantage of using this gauge is that the first two equations contain only r–derivatives of the metric coefficients, and in principle they can be integrated in r once the dynamics of the scalar field is known. The latter is captured by the wave equation $\Box\phi = 0$ which takes the form

$$\Phi_{,t} = \left(\frac{\alpha}{a}\Pi\right)_{,r}, \tag{6.18d}$$

$$\Pi_{,t} = \frac{1}{r^2}\left(r^2\frac{\alpha}{a}\Phi\right)_{,r}. \tag{6.18e}$$

Also notice that the mass in this formulation obeys the relation

$$m_{,r} = 2\pi \left(\frac{r}{a}\right)^2 (\Phi^2 + \Pi^2), \tag{6.19}$$

which allows quite simply to calculate the total mass of the spacetime at each time just by integration,

$$M(t) = 2\pi \int_0^\infty \left(\frac{r}{a}\right)^2 (\Phi^2 + \Pi^2)\, \mathrm{d}r. \tag{6.20}$$

The numerical evaluation is performed on a typical finite difference grid. Φ and Π are specified at some initial time $t = 0$, and this completely defines $\alpha(0, r)$ and $a(0, r)$. The numerical scheme ignores (6.18c) and uses (6.18a)–(6.18b) together with (6.18d)–(6.18e) to determine completely the evolution of the system from the initial data—for more details reader is referred to [10], especially concerning how the discretization scale varies both in space and in time to keep the local truncation error below a fixed limit. Noticeably enough, the system is invariant under the rescaling

$$(t, r) \to (kt, kr), \qquad \forall k > 0;$$

we'll get back on the symmetry features of this model in a while.

6.3.1 The Black Hole Threshold

To describe the nature of the critical behaviour found out by Choptuik, let us alternatively assign $\phi_0(r) = \phi(0, r)$ instead of $\Phi(0, r)$ and $\Pi(0, r)$, for instance

$$\phi_0(r) = \phi_0 r^3 e^{-\left(\frac{r-r_0}{\delta}\right)^q}, \tag{6.21}$$

together with a condition that the scalar field is purely ingoing at initial time. Recalling the discussion in Sect. 6.2.1, if the mass $2M \ll L$, where L is the pulse thickness (controlled in (6.21) by δ) then the wave packet disperses, whereas when $2M \approx L$ then the strong gravity regime allows for trapped surfaces formation. What Choptuik did was to fix all the parameters of the initial data (6.21) but one, call it p, and vary it in order to obtain the critical value p^* by a simple bisection technique—that is, find two parameter values p^{HI} and p^{LO} such that $p > p^{HI}$ certainly implies black hole and $p < p^{LO}$ certainly implies dispersion, and then proceed by subsequent bisections of the interval $[p^{LO}, p^{HI}]$ to approximately determine p^*.

The study of this 1-parameter family of numerical solutions revealed that the mass of the black hole was obeying a scaling form depending on the free parameter p when $p \to (p^*)^+$:

$$M \simeq C|p - p^*|^\gamma, \tag{6.22}$$

where C and, of course p^*, depend on the parameter chosen to describe the family, but γ is a universal constant ($\gamma \approx 0.37$) at least for this matter model.

Moreover, the critical solution is *discretely self-similar*, that is there exists a conformal isometry of the spacetime with conformal factor equal to $e^{-2\Delta}$ for some Δ. This scale–periodicity (Christodoulou's terminology) or "echoing" (Choptuik's terminology) behaviour can be formalized as follows. Let us introduce the variable change

$$\tau = \log \frac{L}{t_* - t}, \qquad x = \log \frac{r}{t_* - t},$$

then for values t_* and L depending on the model, the unknown functions α, a, ϕ as a function of (τ, x) are periodic with respect to τ (for instance $\phi(\tau, x) = \phi(\tau + \Delta, x)$). For massless scalar field the period Δ is approximatively equal to 3.447.

Therefore, representing the space of initial data as the phase space associated to the dynamics of scalar field collapse, the situation is roughly explained in Fig. 6.7. The stable manifold for the critical solution is known as the *black hole threshold*, which separates data that evolves into a black hole from data that disperses towards future infinity. This is a rough description for many reasons: one of these is that the unstable mode for the black hole is not unique, as pointed out in [36]. Moreover, it's the numerical analysis made by considering several 1-parameter families of initial data, all exhibiting the same behaviour, that supports this universality, towards the conclusion that the black hole threshold is a codimension one submanifold.

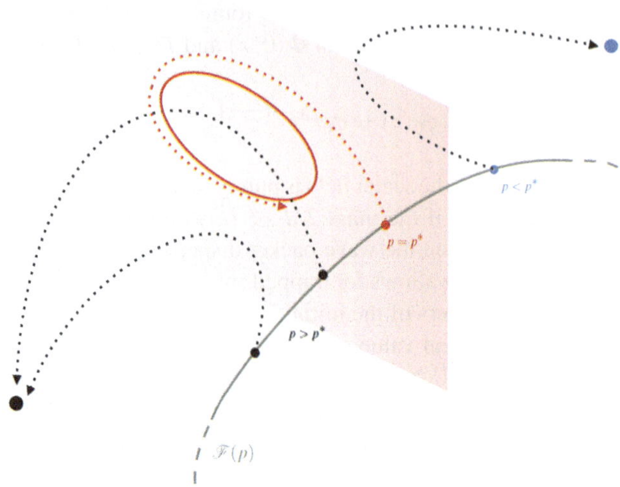

Fig. 6.7 The schematic picture of the critical behaviour [53, Figure 2]. Points in the space represent initial data for the scalar collapse evolution, dotted lines represent the dynamical evolution of the data. The red surface is the black hole threshold. Every one parameter family $\mathscr{F}(p)$ of initial data (grey curve) intersects the black hole threshold for a value of the parameter $p = p^*$ (the red point on $\mathscr{F}(p)$) that evolves near the critical solution, represented by the red limit cycle because of its periodicity property. When $p < p^*$ (blue points) then the spacetime evolves to a flat configuration where scalar field is dispersed at infinity. When $p > p^*$, then trapped surface forms and a black hole forms

It is to remember that the scalar field case is the first example exhibiting features of *universality*,—solutions' endstates depend on the positions of their data with respect to a critical submanifold only—*black hole mass scaling* like (6.22), and critical solution's *scale–invariance* of some kind. Attempts to study the critical behaviour have been made [8, 102] with both quantitative and qualitative results not completely overlapping with those from numerical studies. All these hints are probably the signature of profound phenomena that have yet to be entirely understood in their full generality.

What's for sure, is that numerical analysis has confirmed critical behaviour existence in many cases. In the realm of scalar field collapse, when considering a massive scalar field [9], in the non-minimally coupled case with a potential [63], and in other gravity theories like Brans–Dicke theory [12]. The latter case is relevant since in this context a proof of the existence of naked singularity, extending to this case the example of Christodoulou [17] that we have already cited and that we will discuss in next Subsection too, is given [6]. Further examples of critical behaviour in other matter models are given for instance in the case of the Einstein cluster studied in [58, 83]. For a much broader overview on critical behaviour we refer the reader to [55] once again and to the more recent review [13] also.

6.3.2 *Naked Singularity in the Critical Case*

The picture sketched in the above subsection shows that there is a so–called *critical solution* that lives in the threshold black hole and is neither regular nor contains a black hole covered by a horizon. In Fig. 6.7 this solution is represented as a limit cycle in the phase space since, as we have observed, it has periodicity features. This critical solution, as mentioned in the previous sections, is important also because it possesses a singularity that yet is not covered by a event horizon, and is visible a future null infinity. It is therefore a naked singularity.

The global structure of this spacetime has been the object of several works [51, 52, 54, 85]. However it must be remembered that its existence basically comes from numerical inference, although quite recently a computer–assisted proof appeared [95] that, starting from and approximate solution shows the existence of a real analytic *true* solution, precisely interpreted as the Choptuik critical case.

The picture emerging from the analysis of Gundlach and Martín–García (see in particular [85]) shows that it is a spacetime that can be uniquely continued up to the future outgoing null cone of the pointlike singularity, that therefore is a Cauchy horizon. This analysis seems confirmed also using other numerical approaches, see e.g. [37].

The solution can be extended in a non unique way beyond the Cauchy horizon, but the only discrete self–similar possible extensions are either with a regular centre or with a timelike central singularity (see Fig. 6.8), excluding null singularity followed by a spacelike dynamical singularity, similar to other collapsing matter model such as spherical dust. The past light cone is a self similar horizon—a null surface invariant by the self–similarity.

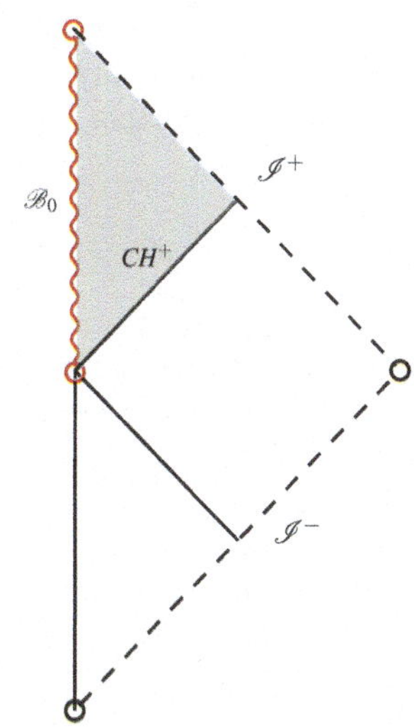

Fig. 6.8 The critical solution with a possible C^0 extension [85, Fig. 2]. The grey region represent a possible C^0 extension of the maximal analytical extension of Choptuik critical spacetime across the Cauchy horizon CH^+. In the maximal analytical extension, the naked central singularity is the lower end of the red vertical line, where the regular centre (black vertical line) terminates

Without considering the possible extension across the Cauchy horizon, the model has then a pointlike singularity, in that resembling Christodoulou counterexample built in [17]. The two solutions anyway cannot coincide, because the latter is actually a *continuous* self–similar solution, with metric

$$g = -e^{2\nu}\,\mathrm{d}u^2 - 2e^{\lambda+\nu}\,\mathrm{d}u\,\mathrm{d}r + r^2\gamma_{\mathbb{S}^2}, \tag{6.23}$$

where λ and ν are function of

$$x = -\frac{r}{u}.$$

In this way, defined the vector field $S = u\partial_u + r\partial_r$, one has $\mathscr{L}_S g = 2g$, $Sr = r$. Moreover the vector field acts on the scalar field ϕ so that $S\phi = -k \in \mathbb{R}$.

Self–similarity reduces Einstein equations to an ODE system, and in [17] it is shown that a solution without a horizon exists in the half–plane $u < 0$, such that the scalar field along two null directions terminating in $(u = 0, r = 0)$ diverges like $\log r$, and this makes $(0, 0)$ a singular point not covered by a horizon. In this example the solution is extended to the half plane $u > 0$ isometrically, so that one can obtain a solution that is regular everywhere except one single point.

At this stage, one may argue that Christodoulou analysis described in Sect. 6.2.1 and the picture emerging from critical behaviour don't overlap, since one may apply the instability argument described in Sect. 6.2.2 above to an initial data in the black hole threshold, obtaining a 2-parameter family of initial data always leading to black hole formation. As pointed out in [53], one possible explanation for this apparent contradiction lies in that the picture emerging from numerical evidence concerns solutions starting from *smooth* functions, unlike the argument presented in Sect. 6.2.2.

Indeed (see Fig. 6.9) let us consider initial data α_0 corresponding to a solution in the black hole threshold. In view of the above discussion we can suppose this data at past null infinity I^-, in order to be consistent with the setup of Sect. 6.2. The perturbative argument of Christodoulou considers two functions $f_1(v)$ and $f_2(v)$

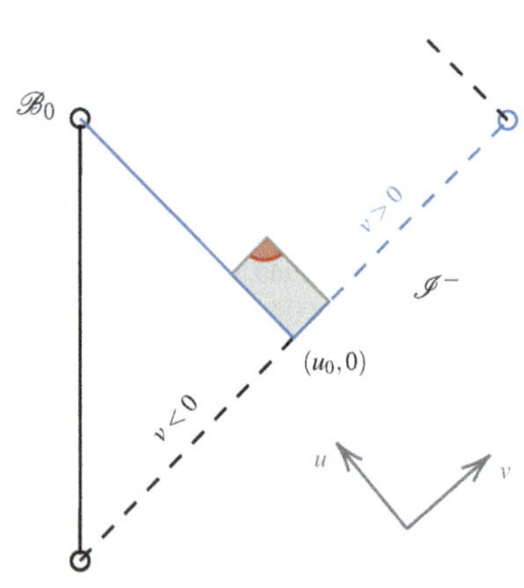

Fig. 6.9 Non-smoothness of data in Christodoulou instability argument. The argument is developed in the exterior of the past null cone of $\mathscr{B}_0$, defining suitable data for $v > 0$ (blue dashed line) in order to determine an apparent horizon (in the picture only the first element of the sequence of trapped surfaces is shown, see also Fig. 6.4). The data are then *non-smoothly* extended for $v < 0$ (black dashed line)

(see step 3. in Sect. 6.2.2). These functions will be set to zero for $v < 0$, so that the evolution is left unchanged in the interior of the past null cone of $\mathscr{B}_0$. Then they are defined for $v > 0$ in order to obtain an apparent horizon extendable until $\mathscr{B}_0$, as explained in Sect. 6.2.2. This results in that $f_1(v)$ has a jump discontinuity at $v = 0$, while $f_2(v)$ is absolutely continuous on the real line, but not smooth.

Therefore, although the whole 2-parameters family of initial data

$$\mathscr{F}(p_1, p_2) = \{\alpha_0(v) + p_1 f_1(v) + p_2 f_2(v) \; : \; p_1, p_2 \in \mathscr{R}\}$$

evolves to a black hole, this is not a family of smooth initial data, and then lies outside the range covered by numerical studies.

6.4 Homogeneous Scalar Fields

The scalar field collapsing model that we have discussed so far has no constraint imposed on the evolution of the scalar field, excluding the initial assumption that we have made at the very beginning that ϕ is defined on the quotient space $\mathscr{Q}$ for symmetry reasons. In a cosmological setting, however, isotropy arguments impose the further constraint that ϕ is a function of the cosmological time only. Of course this considerably reduce the mathematical complexity of the model, that will be now ruled by a system of ODEs. On the other side, this allows to include without significant drawbacks the self–interaction potential $V(\phi)$, and to study solutions' evolution by means of dynamical systems techniques [56], retaining information on the qualitative behaviour of the scalar field depending on assumptions made on $V(\phi)$.

The physical significance of scalar fields with potential in a cosmological setting is related to the interpretation of inflationary expansion phase of the universe [48]. But we find them also in string theory or in higher dimensional gravity, using Kaluza–Klein reduction [103]. Moreover, they arise also in nonlinear theories of gravitation [34, 82] such as $f(R)$ theories, since as is well known they can be reduced, up to a conformal transformation, to the ordinary gravity theory where the energy momentum tensor has a scalar field component, where $V(\phi)$ depends on the form of $f(R)$. Scalar field and spacetime singularities are also connected of course within the study of primordial gravitation, for instance to suggest alternative model to the standard ΛCDM paradigm [76] or within loop quantum cosmology [3].

Perhaps the best known example of homogeneous collapsing solutions is the Oppenheimer–Snyder [92] homogeneous dust ball collapsing to a black hole. Similarly, one may consider a scalar field—possibly coupled with other matter types—as interior of a star, that must be matched with a reasonable exterior to get a global model where one can, once again, study the collapse evolution and see whether singularities form with or without a horizon. In this context one can build homogeneous "scalar field stars" matching a Robertson–Walker (RW) interior with a—possibly non homogeneous—exterior solution, and the cosmological time will play the role of the co-moving time for this collapsing model.

All in all, there is a huge variety of situations that have been studied, and a wide account on the existing literature on the subject can be found in the two companion reviews [71, 72].

Einstein's field equations for the RW model

$$ds^2 = -dt^2 + a(t)^2 \left[\frac{1}{1-kr^2}\, dr^2 + r^2 \left(d\theta^2 + \sin^2\theta\, d\varphi^2 \right) \right], \tag{6.24}$$

with the stress–energy tensor (6.3) are

$$(G^0_0 = 8\pi T^0_0): \qquad -\frac{3(k+\dot{a}^2)}{a^2} = -\left(\frac{1}{2}\dot{\phi}^2 + V(\phi)\right), \tag{6.25}$$

$$(G^1_1 = 8\pi T^1_1): \qquad -\frac{(k+\dot{a}^2)+2a\ddot{a}}{a^2} = \left(\frac{1}{2}\dot{\phi}^2 - V(\phi)\right). \tag{6.26}$$

in the unknown functions $a(t)$, $\phi(t)$.[8] These equations imply as usual Bianchi identity that in this context becomes

$$T^{\mu}_{\ 0;\mu} = -\dot{\phi}\left(\ddot{\phi} + V'(\phi) + 3\frac{\dot{a}}{a}\dot{\phi}\right) = 0. \tag{6.27}$$

The above second order system can be translated into a first order system introducing the *Hubble function*

$$h = \frac{\dot{a}}{a},$$

and $v = \dot{\phi}$. We first briefly consider solutions such that $\phi(t) = \phi_0$ constant on some interval so (6.27) holds, but $V'(\phi_0) \neq 0$ (so the term in round bracket in (6.27) is non zero). These evolutions are obtained by (6.26):

$$\dot{a}^2 = \frac{V(\phi_0)}{3}a^2 - k,$$

corresponding to a (anti)deSitter solution (the regular value of the potential in ϕ_0 plays the role of the cosmological constant). Excluding this case we have that (ϕ, v, h) solves the regular system

$$\dot{\phi} = v, \tag{6.28a}$$

$$\dot{v} = -V'(\phi) - 3h\, v, \tag{6.28b}$$

$$\dot{h} = -h^2 - \tfrac{1}{3}(v^2 - V(\phi)). \tag{6.28c}$$

[8] As in the previous situations, the scalar field and the potential are normalized in order to get rid of the π factor in Einstein field equations. Unfortunately, this normalization changes from paper to paper in literature—here we will follow the normalization made (for the flat $k = 0$ case) in [44], that will possibly differ from those used in other papers that we will refer to.

Observe that the function

$$W(\phi, v, h) = h^2 - \frac{1}{3}\left(\frac{1}{2}v^2 + V(\phi)\right) \tag{6.29}$$

satisfies by (6.25) the identity

$$W(t) = -\frac{k}{a(t)^2}, \tag{6.30}$$

where $k = -\text{sgn}(W(0))$. The sign of W is invariant by the flow ($\dot{W} = -2hW$)—in other words the system (6.28a)–(6.28c) represents at once *all* RW cosmologies (6.24).

The equilibrium points are given by $(\phi_*, 0, h_*)$ such that $V'(\phi_*) = 0$, $h_*^2 = \frac{1}{3}V(\phi_*)$ (notice that $W(\phi_*, 0, h_*) = 0$), that physically correspond to (anti)deSitter universe. The sign of the Hubble function h tells us whether the solution is expanding ($h > 0$) or collapsing ($h < 0$).

6.4.1 (Re)collapsing Solutions and Nature of the Singularity

Let us first briefly discuss the evolution from expanding initial conditions, i.e. $h(0) > 0$ [87]. Figure 6.10 sketches some of the integral curves of system (6.28a)–(6.28c) in this case. A minimum of the potential $V(\phi)$ with positive critical value is a stable equilibrium point, and nearby cosmologies approach the corresponding deSitter state asymptotically: no collapse will take place here. These are the curves on the right of Fig. 6.10.

For the open topologies $k \leq 0$, corresponding to the portion of the phace space $W \geq 0$, the upper branch of $W = 0$ near this local minimum acts like a constraint forcing the solutions to approach the minimum. The deSitter space is attractive also for the portion ($W < 0$) of the phase space between the two branches of $W = 0$, corresponding to $k = 1$ cosmologies, at least when the minimum of $V(\phi)$ is *strictly* positive. When $V(\phi_0) = 0$ then recollapse is possible in $k = 1$ case.

On the left of Fig. 6.10 we have a situation near a local minimum for $V(\phi)$ with negative critical value. In this case the two branches of $W > 0$ intersect somewhere at $h = 0$, determining a sort of "tunnel" crossed by solutions in the open topology $k = -1$, that in this way reach the $h < 0$ portion of the space and then recollapse. A fortiori, that happens also for solutions with $k = 0$ and $k = 1$.

Once we have established that homogeneous scalar fields may collapse, even if they start expanding, we can ask whether the evolution ends into a singularity and if a horizon forms. Since the dependence on h in the right-hand side of (6.28c) is quadratic, a singularity will likely form in a finite amount of co-moving time. In [39] the flat case $k = 0$ is approached by considering that in a collapse situation $\dot{a}(t) < 0$, which allows to use the scale factor a as an independent variable, in order to express

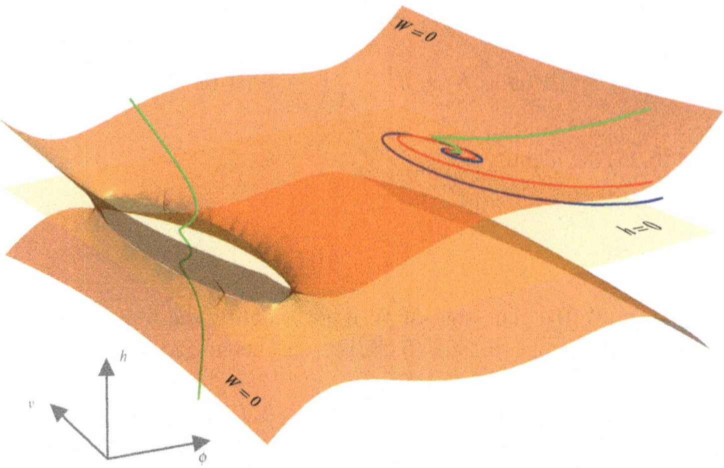

Fig. 6.10 The phase space of system (6.28a)–(6.28c), in a case where $V(\phi)$ has two local minima, one with negative critical value, the other with positive critical value (see [43, Fig. 1]). Example of cosmologies for the cases $k = -1$ (green curves), $k = 0$ (red curve) and $k = 1$ (blue curve) starting in the half space $h > 0$ (expanding initial conditions) are sketched

the energy density ρ_{SF} of the model, given by (minus) the right-hand side in (6.25), as

$$\rho_{SF}(t) := \frac{1}{2}\dot{\phi}(t) + V(\phi(t)) = 3\left(\frac{\psi(a)}{a}\right)^2. \tag{6.31}$$

Using (6.31) we obtain

$$\dot{a} = -\psi(a), \tag{6.32}$$

and then the singularity form in a finite amount of co-moving time if and only if [39, Theorem 3.1]

$$\int_0^{a_0} \frac{1}{\psi(a)}\, \mathrm{d}a < +\infty \tag{6.33}$$

for some $a_0 > 0$ corresponding to the initial data at $t = 0$.

Moreover, all the dynamical quantities involved can be expressed as a function of a, for example using (6.25)–(6.26) we have

$$\left(\frac{\mathrm{d}\phi}{\mathrm{d}a}\right)^2 = \frac{2}{a^2}\left(1 - \frac{a\,\psi'(a)}{\psi(a)}\right), \tag{6.34}$$

$$V(a) = \frac{2\psi^2(a)}{a^2}\left(1 + \frac{a\,\psi'(a)}{2\psi(a)}\right), \tag{6.35}$$

whereas the mass in (6.8) takes the form

$$m = \frac{r^3}{2} a(t)\dot{a}^2(t) = \frac{r^3}{2} a\, \psi(a)^2, \tag{6.36}$$

which allows to write the trapped region $R < 2m$ as

$$\psi(a)^2 r^2 > 1, \tag{6.37}$$

and in this way one can find out sufficient conditions on the function $\psi(a)$ to develop a horizon or a naked singularity [39, Theorem 5.2].

The idea sketched above extends the approach used by Rituparno Goswami and Pankaj S. Joshi [50], where $\rho_{SF} = \rho_0\, a^{-\nu}$ with $\nu > 0$ is considered. In the language of (6.31) that amounts to postulate

$$\psi(a) = \psi_0\, a^{1-\nu/2},$$

and in view of (6.33) a singularity always forms in a finite amount of time if $\nu > 0$. On the other side, if the inequality

$$r < \frac{1}{\psi_0} a^{\frac{\nu}{2}-1} \tag{6.38}$$

is satisfied until $a = 0$, then (6.37) cannot hold and then an apparent horizon with a trapped region cannot form. Then, if $\nu < 2$, the right-hand side of (6.38) diverges in the approach to the singularity, and then $\forall r > 0$ there is a region near the singular boundary that is untrapped, see Fig. 6.11.

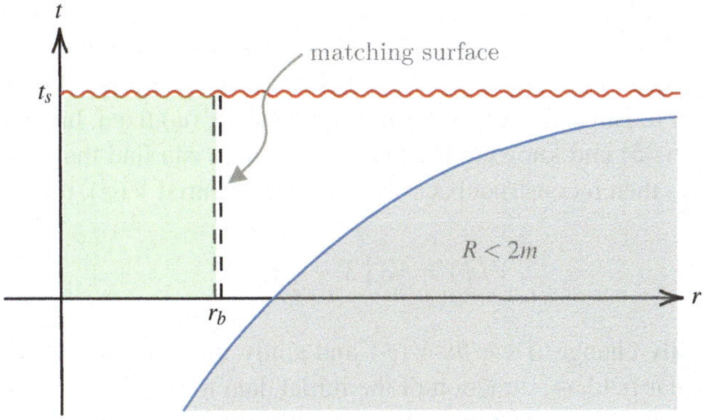

Fig. 6.11 Homogeneous scalar field with a naked singularity. The region nearby the singularity $t = t_s$ is such that $R > 2m$. In this way one can consider a region (in green) in a right neighbourhood $[0, r_b]$ of $r = 0$ for the scalar field solution and use Israel–Darmois condition to match this region with an exterior spacetime

To build a global model one can consider a RW metric for $r \leq r_b$ for some r_b, and get initial data $a(t) = a_0$ such that

$$r_b < \frac{1}{\psi_0} a_0^{\frac{\nu}{2}-1}.$$

In this way the solution will live in the untrapped region $\forall r \in [0, r_b]$, $\forall t : a(t) \in]0, a_0]$. Then a matching at $r = r_b$ using Israel–Darmois junction conditions can be performed with an exterior spacetime. This is made in [49] where generalized Vaidya spacetimes are considered, while in [39] an anisotropic generalization of deSitter—that is observed to be a subclass of the former—is used. The choice of the external solution is important also because it allows to evaluate the *strength* of the singularity. Many definitions in this sense exists in literature, and we refer to [39, Section 5.1] for a discussion about it, where a sufficient condition for the singularity discussed here to be strong is given. As a consequence, in the example above sketched, there exists timelike geodesics terminating into a strong naked singularity.

6.4.2 *Instability of the Naked Singularity*

The example discussed above has been further developed in [49] where is shown that loop quantum gravity effects prevent the formation of a singularity. But yet at a classical level, the fact that they develop a naked singularity for values of the parameter ν in an open interval of the real line, and therefore are stable with respect to that parameter's perturbation, could in principle be interpreted as a non-generic violation of cosmic censorship for these models. In the more recent [4] the problem is cast into a classical setting but adding a cosmological constant to (6.2), and finding once again naked singularities for a non-generic choice of a free parameter.

However, as already remarked, the genericity question crucially relies on what we set as fixed and what we consider as "variable". For instance, in the above example, the free parameter is ν, expressing the dependence of the energy density ρ_{SF} from the scale factor a. But this does not keep the potential $V(\phi)$ fixed. Indeed notice that, from (6.34)–(6.35) and knowing the profile $\psi(a)$ one can find that ϕ diverges like $-\sqrt{\nu}\log a$ and then reconstruct back the original potential $V(\phi)$, that is given by

$$V(\phi) = V_0 \left(3 - \frac{\nu}{2}\right) e^{\sqrt{\nu}\phi}.$$

Things radically change if we *fix* $V(\phi)$ and study the end state of homogeneous collapsing scalar field as a function of the initial data for $t = 0$.

Let us examine again the flat case $k = 0$ and start by observing that the choice of $(\phi_0, v_0) = (\phi(0), \dot{\phi}(0))$ fully determines the data at $t = 0$, because in system (6.28a)–(6.28c) the initial condition $h_0 = h(0)$ is determined by the constraint $W(\phi_0, v_0, h_0) = 0$ and the fact that we consider a collapse scenario. In the paper [44] a class of C^2 potentials $V(\phi)$ is considered, that includes quite general polynomial potentials such as those asymptotically diverging like ϕ^{2n} as a subclass. In particular, $V(\phi)$ is supposed to satisfy (see [44, Section II]) the following properties: (i) it must be bounded from below, (ii) its critical points must be isolated and possible local maxima must be non-degenerate, (iii) a suitable compact sublevel set $\{V \leq V_*\}$ must exist and (iv) a growth condition on the first and second derivative of $\log V(\phi)$ at $\pm\infty$ is assumed to hold. Other class of potentials were considered in [45].

Under the above assumptions is proved [44, Theorem 3.7] that—except at most for an exceptional zero–measured set of initial data (ϕ_0, v_0)— the scalar field diverges together with its derivative and develops a singularity at a finite amount of co-moving time. Moreover [44, Theorem 4.1] the scale factor $a(t)$ vanishes in the approach to the singularity but generically $\dot{a}(t) \to -\infty$. This means, recalling (6.32), that (6.37) is always satisfied in a right neighbourhood of $a = 0$, $\forall r > 0$, and then an horizon is developing before the singularity. We stress that this happens up to a non-generic choice of initial data, once $V(\phi)$ is fixed.

The proof of the above results is carried on by observing first that, introduced the function

$$y := \frac{V(\phi)}{\frac{1}{2}\dot{\phi}^2},$$

then the scalar field energy ρ_{SF} and the scale factor $a(t)$ satisfy the differential relations

$$\dot{\rho}_{SF}(t) = 2\sqrt{3}\rho_{SF}^{3/2}(t)\frac{1}{1+y(t)}, \tag{6.39a}$$

$$\ddot{a}(t) = -\frac{\dot{a}^2(t)}{a(t)}\left(\frac{2-y(t)}{1+y(t)}\right) \tag{6.39b}$$

Now the proof determines the asymptotic behaviour of $y(t)$, as follows:

1. first, it is shown that the scalar field $\phi(t)$ with collapsing initial data diverges and its derivative $\dot{\phi}(t)$ is eventually non-zero; then we can express—at least eventually the dynamical quantities with respect to ϕ;
2. the function y, seen as a function of ϕ, satisfies a differential relation that is studied using the growth assumption on $V(\phi)$, finding that, up a non-generic choice of initial data, $y(\phi) \xrightarrow{\phi\to\infty} 0$;
3. using steps 1. and 2. in (6.39a)–(6.39b), one finds that, for these non-generic data, $\rho_{SF}(t)$ diverges in a finite amount of co-moving time, and then a singularity develops, but $\dot{a}(t)$ is unbounded, which as observed above results generically in the formation of the horizon.

The function $y(t)$ is basically a modified way to "normalize" the dynamical system, that is a classical approach to the system under suitable condition on the potential $V(\phi)$, see [35] where, instead of considering the unknown functions and (6.28a)–(6.28c), takes their ratios to the Hubble function h, that in the flat $k = 0$ case is connected, by (6.29), to the energy density ρ_{SF}.

It follows that, the cases depicted in [50] leading to a naked singularity, necessarily correspond to the non-generic case excluded by the analysis of step 2 above, see [44, Section V.]). Same argument applies to the examples shown in [90]. In this case the potential considered is $V(\phi) = \lambda\phi^2(\phi^2 - \frac{8}{3})$, and it is prescribed $\rho_{SF}(a) = 64\lambda(k - \log a)^2$ for $k > 0$. Using the above relations it can be actually proved that the spacetime is regular for all $t > 0$—although in [90] it is observed that an ultra high density region is reached in a finite amount of time. However, once again, it can be seen that this situation corresponds to the non-generic choice of initial data excluded by step 2 described above. In fact, as observed in [44, Example 3.9], if a potential $V(\phi)$ in this class is fixed, then initial data generically lead to black hole formation, and prescribing the dependence of $\rho(a)$ as done in [90] precisely corresponds to the non-generic data $(\phi(0), \dot{\phi}(0))$ that possibly may not lead to the formation of an horizon.

In conclusion, once a potential $V(\phi)$ obeying suitable—yet rather inclusive—conditions is prescribed, the collapse may end into a naked singularity only for a non-generic choice of the initial data on the scalar field. Nevertheless, the non-generic naked singularity can be gravitationally strong, which interestingly enough is the same conclusion reached in [57] for the critical collapse described before in Sect. 6.3.2.

6.5 Other Developments and Conclusions

The literature discussed in this review, although covering some of the undoubtedly recognized as cornerstones in the analytical and numerical studies on scalar field collapse, is necessarily a selection in the huge amount of contributions on this subject.

Concerning the analytical studies, we have already cited some recent developments aiming to prove stability of spherically results with respect to general—hence possibly nonspherical—perturbations [65, 78, 79] in the free massless case. Just to name a few other analytical results, when a potential is added, collapse in cylindrical symmetry under self–similarity assumptions has been discussed in [22–24]. With aim to study the big bang singularity, backwards evolution of a free massless scalar field is considered in [96], where the initial data are close to the homogeneous case and given on a 3-dimensional torus $\mathbb{T}^3$. In [103], a scalar field self–interacting with a negative exponential potential is Kaluza–Klein lifted to a vacuum solution in $\mathbb{R}^{1+3} \times S^2$, to find singular solutions without trapped surface formation developing from regular data. Notice that the family of 4-dimensional scalar field solutions considered contains an example that extends to a case with potential the self–similar naked singularity of [17]. The stability of this important example, as we have said, remains a problem with some issues to be fully understood, although [18] represent of course a crucial step. When confining to the self–similar collapse, the situation has been studied e.g. in [5, 91]. It is worth citing also [73] that attempts a first important generalization of the naked singularity instability result when no symmetry on the spacetime geometry is assumed.

The PDE approach that has been presented here is not the only one used, even in an analytical context. Some results based on Fuchsian techniques can be found in [2, 33], while an interesting approach involving inverse reconstruction methods recently appeared [68] where no particular geometric symmetries are required. Also numerical studies have tried to cover solutions not necessarily spherically symmetric, in the attempt to reproduce the critical behaviour also in more general cases, see e.g. [60].

As already said, the study of scalar field cosmologies via dynamical system techniques also are a huge part of this subject, although many are understandably focused on expanding solutions. Having recalled once again reviews [71, 72] for extensive treatises in this context, let us cite the paper [61] where back–reaction terms are considered to find collapsing solutions that do not develop singularities. A great deal of attention has been also paid to (re)collapsing models where a matter contribution, typically a perfect fluid—either coupled or uncoupled to the scalar field—is added, mostly arising from higher order gravity theory [88, 89]. The problem has been studied in [98] for the case of positive exponential potentials and in [40] for the same class of potentials studied in [44], where again black hole formation is shown to generically form. As one may expect, big crunch singularities naturally arise also when one deals with negative potentials [46, 47]. Coupling effect may be mimicked also using deformation of the phase space, producing corrections to the dynamical system also in absence of a perfect fluid, as studied in [94], where anyway the deformation is shown to prevent, in some cases, the formation of singularities from gravitational collapse.

There are other approaches to the problem worth to be cited, for instance [7] where the scalar field is assumed to be timelike, in such a way that co-moving coordinates can be built considering a possibly inhomogeneous spacetime, although $\phi = \phi(t)$ only.

In summary, the collapse of a scalar field has been a valuable tool for investigating various aspects of gravitational physics. The scalar field's multiplicity and rich physical interpretations, both alone and when coupled with other matter models, have contributed to its continued relevance. Moreover, the associated mathematical model is relatively simple, particularly in the context of spherical symmetry, making it a useful test-bed for exploring classical and nonlinear gravity theories. As such, the collapse of a scalar field has been and continues to be an important means of probing physics in the theory of gravitation.

References

1. X. An, Z.F. Lim, Ann. Henri Poincare **23**(9), 3159–3190 (2022). https://doi.org/10.1007/s00023-022-01168-y. [arXiv:2005.04090 [math.AP]]
2. L. Andersson, [arXiv:gr-qc/0011104 [gr-qc]]
3. A. Ashtekar, P. Singh, Class. Quant. Grav. **28**, 213001 (2011). https://doi.org/10.1088/0264-9381/28/21/213001. ([arXiv:1108.0893 [gr-qc]].)

4. R. Baier, H. Nishimura, S.A. Stricker, Class. Quant. Grav. **32**(13), 135021 (2015). https://doi.org/10.1088/0264-9381/32/13/135021. [arXiv:1410.3495 [gr-qc]]
5. N. Banerjee, S. Chakrabarti, Phys. Rev. D **95**(2), 024015 (2017). https://doi.org/10.1103/PhysRevD.95.024015. [arXiv:1701.04235 [gr-qc]]
6. N. Bedjaoui, P.G. LeFloch, J.M. Martin-Garcia, J. Novak, Class. Quant. Grav. **27**, 245010 (2010). https://doi.org/10.1088/0264-9381/27/24/245010. ([arXiv:1008.4238 [gr-qc]].)
7. S. Bhattacharya, P.S. Joshi, K.I. Nakao, Phys. Rev. D **81**, 064032 (2010). https://doi.org/10.1103/PhysRevD.81.064032. [arXiv:0911.2297 [gr-qc]]
8. S. Bhattacharya, R. Goswami, P.S. Joshi, Int. J. Mod. Phys. D **20**, 1123–1133 (2011). https://doi.org/10.1142/S021827181101930X. [arXiv:1010.1757 [gr-qc]]
9. P.R. Brady, C.M. Chambers, S.M.C.V. Goncalves, Phys. Rev. D **56**, R6057-R6061 (1997). https://doi.org/10.1103/PhysRevD.56.R6057. [arXiv:gr-qc/9709014 [gr-qc]]
10. M. Choptuik, in ed. by R. D'Inverno, Approaches to Numerical Relativity (Cambridge University Press, Cambridge, 1992), pp. 202–222. https://doi.org/10.1017/CBO9780511524639.019
11. M.W. Choptuik, Phys. Rev. Lett. **70**, 9–12 (1993). https://doi.org/10.1103/PhysRevLett.70.9
12. M.W. Choptuik, E.W. Hirschmann, S.L. Liebling, Phys. Rev. D **55**, 6014–6018 (1997). https://doi.org/10.1103/PhysRevD.55.6014. [arXiv:gr-qc/9701011 [gr-qc]]
13. M.W. Choptuik, L. Lehner, F. Pretorius, Chapter 7, in *General Relativity and Gravitation: A Centennial Perspective*, ed. by A. Ashtekhar, B.K. Berger, J. Isenberg, M. MacCallum (Cambridge University Press, 2015) [arXiv:1502.06853 [gr-qc]]
14. D. Christodoulou, Commun. Math. Phys. **105**, 337–361 (1986). https://doi.org/10.1007/BF01205930
15. D. Christodoulou, Commun. Pure Appl. Math. **44**(3), 339–373 (1991). https://doi.org/10.1002/cpa.3160440305
16. D. Christodoulou, D., Comm. Pure Appl. Math. **46** 1131–1220 (1993). https://doi.org/10.1002/cpa.3160460803
17. D. Christodoulou, Ann. Math. **140**, 607–653 (1994). https://doi.org/10.2307/2118619
18. D. Christodoulou, Ann. Math. **149**, 183–217 (1999)
19. D. Christodoulou, Class. Quantum Grav. **16**, A23 (1999)
20. D. Christodoulou, The formation of black holes in general relativity. EMS Monographs Math. (2009). https://doi.org/10.4171/068. [arXiv:0805.3880 [gr-qc]]
21. F. Cipolletta, R. Giambò, Class. Quant. Grav. **29**, 245008 (2012). https://doi.org/10.1088/0264-9381/29/24/245008. ([arXiv:1208.0804 [gr-qc]].)
22. E. Condron, B.C. Nolan, [arXiv:1305.4866 [gr-qc]]
23. E. Condron, B.C. Nolan, Class. Quant. Grav. **31**(16), 165018 (2014). https://doi.org/10.1088/0264-9381/31/16/165018. [arXiv:1309.3977 [gr-qc]]
24. E. Condron, B.C. Nolan, Class. Quant. Grav. **31**(1), 015015 (2014). https://doi.org/10.1088/0264-9381/31/1/015015
25. J.L. Costa, Class. Quant. Grav. **37**(19), 195022 (2020). https://doi.org/10.1088/1361-6382/abb075. [arXiv:2005.03434 [gr-qc]]
26. M. Dafermos, Ann. Math **158**(3), 875–928 (2003)
27. M. Dafermos, Adv. Theor. Math. Phys. **9**(4), 575–591 (2005). https://doi.org/10.4310/ATMP.2005.v9.n4.a3. [arXiv:gr-qc/0403033 [gr-qc]]
28. M. Dafermos, Class. Quant. Grav. **22**, 2221–2232 (2005). https://doi.org/10.1088/0264-9381/22/11/019. [arXiv:gr-qc/0403032 [gr-qc]]
29. M. Dafermos, Commun. Pure Appl. Math. **58**, 0445–0504 (2005). [arXiv:gr-qc/0307013 [gr-qc]]
30. M. Dafermos, I. Rodnianski, Invent. math. **162**, 381–457 (2005). https://doi.org/10.1007/s00222-005-0450-3. [arXiv:gr-qc/0309115]
31. M. Dafermos, Commun. Math. Phys. **289**, 579–596 (2009). https://doi.org/10.1007/s00220-009-0775-7. [arXiv:gr-qc/0310040 [gr-qc]]
32. M. Dafermos, Proceedings of the ICM, Seoul 2014, vol. III, p. 747 (2014)

33. T. Damour, M. Henneaux, A.D. Rendall, M. Weaver, Ann. Henri Poincare **3**, 1049–1111 (2002). https://doi.org/10.1007/s000230200000. [arXiv:gr-qc/0202069 [gr-qc]]
34. M. Ferraris, M. Francaviglia, G. Magnano, Class. Quant. Grav. **5**, L95 (1988). https://doi.org/10.1088/0264-9381/5/6/002
35. S. Foster, Class. Quant. Grav. **15**, 3485–3504 (1998). https://doi.org/10.1088/0264-9381/15/11/014. [arXiv:gr-qc/9806098 [gr-qc]]
36. A.V. Frolov, Phys. Rev. D **61**, 084006 (2000). https://doi.org/10.1103/PhysRevD.61.084006. ([arXiv:gr-qc/9908046 [gr-qc]].)
37. A.V. Frolov, U.L. Pen, Phys. Rev. D **68**, 124024 (2003). https://doi.org/10.1103/PhysRevD.68.124024. ([arXiv:gr-qc/0307081 [gr-qc]].)
38. R.P. Geroch, Annals Phys. **48**, 526–540 (1968). https://doi.org/10.1016/0003-4916(68)90144-9
39. R. Giambò, Class. Quant. Grav. **22**, 2295–2305 (2005). https://doi.org/10.1088/0264-9381/22/11/023. [arXiv:gr-qc/0501013 [gr-qc]]
40. R. Giambò, J. Math. Phys. **50**, 012501 (2009). https://doi.org/10.1063/1.3032755. ([arXiv:0811.4570 [gr-qc]].)
41. R. Giambò, F. Giannoni, G. Magli, P. Piccione, Commun. Math. Phys. **235**, 545–563 (2003). https://doi.org/10.1007/s00220-003-0793-9. [arXiv:gr-qc/0204030 [gr-qc]]
42. R. Giambò, F. Giannoni, G. Magli, P. Piccione, Class. Quant. Grav. **20**, L75 (2003). https://doi.org/10.1088/0264-9381/20/6/102. ([arXiv:gr-qc/0212082 [gr-qc]].)
43. R. Giambò, F. Giannoni, G. Magli, Gen. Rel. Grav. **41**, 21–30 (2009). https://doi.org/10.1007/s10714-008-0647-z. [arXiv:0802.0157 [gr-qc]]
44. R. Giambò, F. Giannoni, G. Magli, J. Math. Phys. **49**, 042504 (2008). https://doi.org/10.1063/1.2907949. ([arXiv:0802.0992 [gr-qc]].)
45. R. Giambò, A. Stimilli, J. Geom. Phys. **59**, 400–408 (2009). https://doi.org/10.1016/j.geomphys.2008.11.014
46. R. Giambò, J. Miritzis, K. Tzanni, Class. Quant. Grav. **32**(3), 035009 (2015). https://doi.org/10.1088/0264-9381/32/3/035009. [arXiv:1411.0218 [gr-qc]]
47. R. Giambò, J. Miritzis and K. Tzanni, Class. Quant. Grav. **32**(16), 165017 (2015). https://doi.org/10.1088/0264-9381/32/16/165017. [arXiv:1506.08162 [gr-qc]]
48. D.S. Goldwirth, T. Piran, Phys. Rept. **214**, 223–291 (1992). https://doi.org/10.1016/0370-1573(92)90073-9
49. R. Goswami, P.S. Joshi, P. Singh, Phys. Rev. Lett. **96**, 031302 (2006). https://doi.org/10.1103/PhysRevLett.96.031302. ([arXiv:gr-qc/0506129 [gr-qc]].)
50. R. Goswami, P.S. Joshi, Mod. Phys. Lett. A **22**, 65–74 (2007). https://doi.org/10.1142/S0217732307020701. [arXiv:gr-qc/0410144]
51. C. Gundlach, Phys. Rev. Lett. **75**, 3214–3217 (1995). https://doi.org/10.1103/PhysRevLett.75.3214. [arXiv:gr-qc/9507054 [gr-qc]]
52. C. Gundlach, Phys. Rev. D **55**, 695–713 (1997). https://doi.org/10.1103/PhysRevD.55.695. [arXiv:gr-qc/9604019 [gr-qc]]
53. C. Gundlach, Phys. Rep. **376**(6), 339–405 (2003). https://doi.org/10.1016/S0370-1573(02)00560-4. [arXiv:gr-qc/0210101 [gr-qc]]
54. C. Gundlach, J.M. Martin-Garcia, Phys. Rev. D **68**, 064019 (2003). https://doi.org/10.1103/PhysRevD.68.064019. ([arXiv:gr-qc/0306001 [gr-qc]].)
55. C. Gundlach, J.M. Martin-Garcia, Living Rev. Rel. **10**, 5 (2007). https://doi.org/10.12942/lrr-2007-5. ([arXiv:0711.4620 [gr-qc]].)
56. E. Gunzig, V. Faraoni, A. Figueiredo, T.M. Rocha, L. Brenig, Class. Quant. Grav. **17**, 1783–1814 (2000). https://doi.org/10.1088/0264-9381/17/8/304
57. J.Q. Guo, L. Zhang, Y. Chen, P.S. Joshi, H. Zhang, Eur. Phys. J. C **80**(10), 924 (2020). https://doi.org/10.1140/epjc/s10052-020-08486-7. [arXiv:2011.06792 [gr-qc]]
58. T. Harada, A. Mahajan, Gen. Rel. Grav. **39**, 1847–1854 (2007). https://doi.org/10.1007/s10714-007-0493-4. [arXiv:0707.3000 [gr-qc]]
59. S.W. Hawking, G.F.R. Ellis (Cambridge University Press, 2023). ISBN 978-1-00-925316-1, 978-1-00-925315-4, 978-0-521-20016-5, 978-0-521-09906-6, 978-0-511-82630-6, 978-0-521-09906-6 https://doi.org/10.1017/9781009253161

60. J. Healy, P. Laguna, Gen. Rel. Grav. **46**, 1722 (2014). https://doi.org/10.1007/s10714-014-1722-2. ([arXiv:1310.1955 [gr-qc]].)
61. J.M. Hernández, M. Bellini, C. Moreno, Phys. Dark Univ. **23**, 100251 (2019). https://doi.org/10.1016/j.dark.2018.100251. ([arXiv:1810.07385 [gr-qc]].)
62. T. Hertog, G.T. Horowitz, K. Maeda, Phys. Rev. Lett. **92**, 131101 (2004). https://doi.org/10.1103/PhysRevLett.92.131101. ([arXiv:gr-qc/0307102 [gr-qc]].)
63. E. Jiménez-Vázquez, M. Alcubierre, Phys. Rev. D **105**(6), 064071 (2022). https://doi.org/10.1103/PhysRevD.105.064071. [arXiv:2112.04065 [gr-qc]]
64. C. Kehle, M. Van de Moortel, [arXiv:2105.04604 [gr-qc]], to appear on Analysis & PDE
65. E. Kilgore, Ann. Henri Poincare **23**(9), 3093–3157 (2022). https://doi.org/10.1007/s00023-022-01162-4. [arXiv:2108.13377 [gr-qc]]
66. S. Kleinerman, J. Szeftel (2022). arXiv:2210.14400 [math.AP]
67. K. Landsman, *Foundations of General Relativity: From Einstein to Black Holes* (Radboud University Press, 2021). https://doi.org/10.54195/EFVF4478
68. Y. Kurylev, M. Lassas, L. Oksanen, G. Uhlmann, Duke Math. J. **171**(16), 3215–3282 (2022). https://doi.org/10.1215/00127094-2022-0064
69. K. Landsman, Gen. Rel. Grav. **54**(10), 115 (2022). https://doi.org/10.1007/s10714-022-02973-w. [arXiv:2205.01680 [physics.hist-ph]]
70. P. Langfelder, R.B. Mann, Class. Quant. Grav. **22**, 1917–1932 (2005). https://doi.org/10.1088/0264-9381/22/11/002. [arXiv:gr-qc/0412087 [gr-qc]]
71. G. Leon, F.O.F. Silva, Class. Quant. Grav. **37**(24), 245005 (2020). https://doi.org/10.1088/1361-6382/abbd5a. [arXiv:2007.11140 [gr-qc]]
72. G. Leon, F.O.F. Silva, Class. Quant. Grav. **38**(1), 015004 (2021). https://doi.org/10.1088/1361-6382/abc095. [arXiv:2007.11990 [gr-qc]]
73. J. Li, J. Liu, J. Diff. Geom. **120**(1), 97–197 (2022). https://doi.org/10.4310/jdg/1641413698. [arXiv:1710.02422 [gr-qc]]
74. S.L. Liebling, C. Palenzuela, Living Rev. Rel. **15**, 6 (2012). https://doi.org/10.1007/s41114-023-00043-4. ([arXiv:1202.5809 [gr-qc]].)
75. J. Liu, J. Li, Commun. Math. Phys. **363**(2), 561–578 (2018). https://doi.org/10.1007/s00220-018-3157-1. [arXiv:1710.02922 [gr-qc]]
76. O. Luongo, M. Muccino, Phys. Rev. D **98**(10), 103520 (2018). https://doi.org/10.1103/PhysRevD.98.103520. [arXiv:1807.00180 [gr-qc]]
77. J. Luk, S.J. Oh, Anal. Part. Diff. Eq. **8**(7), 1603–1674 (2015). https://doi.org/10.2140/apde.2015.8.1603. [arXiv:1402.2984 [gr-qc]]
78. J. Luk, S.J. Oh, S. Yang, Ann. PDE **4**, 3 (2018). https://doi.org/10.1007/s40818-017-0038-4. [arXiv:1605.03893 [gr-qc]]
79. J. Luk, S.J. Oh, Ann. Math. **190**(1) (2019). https://doi.org/10.4007/annals.2019.190.1.1. [arXiv:1702.05715 [gr-qc]]
80. J. Luk, S.J. Oh, Ann. PDE **5**(6) (2019). https://doi.org/10.1007/s40818-019-0062-7. [arXiv:1702.05716 [gr-qc]]
81. J. Luk, S.J. Oh, Ann. Henri Poincare **23**(7), 2391–2521 (2022). https://doi.org/10.1007/s00023-021-01148-8. [arXiv:2108.13379 [gr-qc]]
82. G. Magnano, M. Ferraris, M. Francaviglia, Gen. Rel. Grav. **19**, 465 (1987). https://doi.org/10.1007/BF00760651
83. A. Mahajan, T. Harada, P.S. Joshi, K.I. Nakao, Prog. Theor. Phys. **118**, 865–878 (2007). https://doi.org/10.1143/PTP.118.865. [arXiv:0710.4315 [gr-qc]]
84. D. Malafarina, Universe **3**(2), 48 (2017). https://doi.org/10.3390/universe3020048. [arXiv:1703.04138 [gr-qc]]
85. J.M. Martin-Garcia, C. Gundlach, Phys. Rev. D **68**, 024011 (2003). https://doi.org/10.1103/PhysRevD.68.024011. ([arXiv:gr-qc/0304070 [gr-qc]].)
86. F.C. Mena, B.C. Nolan, Class. Quant. Grav. **18**, 4531–4548 (2001). https://doi.org/10.1088/0264-9381/18/21/310. [arXiv:gr-qc/0108008 [gr-qc]]
87. J. Miritzis, Class. Quant. Grav. **20**, 2981–2990 (2003). https://doi.org/10.1088/0264-9381/20/14/301. [arXiv:gr-qc/0303014 [gr-qc]]

88. J. Miritzis, J. Math. Phys. **44**, 3900–3910 (2003). https://doi.org/10.1063/1.1602161. [arXiv:gr-qc/0305062 [gr-qc]]
89. J. Miritzis, J. Math. Phys. **46**, 082502 (2005). https://doi.org/10.1063/1.2009648. ([arXiv:gr-qc/0505139 [gr-qc]].)
90. K. Mosani, D. Dey, K. Bhattacharya, P.S. Joshi, Phys. Rev. D **105**(6), 064048 (2022). https://doi.org/10.1103/PhysRevD.105.064048. [arXiv:2110.07343 [gr-qc]]
91. B.C. Nolan, Class. Quant. Grav. **23**, 4523–4538 (2006). https://doi.org/10.1088/0264-9381/23/13/015. [arXiv:gr-qc/0603105 [gr-qc]]
92. J.R. Oppenheimer, H. Snyder, Phys. Rev. **56**, 455–459 (1939). https://doi.org/10.1103/PhysRev.56.455
93. R. Penrose, Riv. Nuovo Cim. **1**, 252–276 (1969). https://doi.org/10.1023/A:1016578408204
94. S.M.M. Rasouli, A.H. Ziaie, J. Marto, P.V. Moniz, Phys. Rev. D **89**(4), 044028 (2014). https://doi.org/10.1103/PhysRevD.89.044028. [arXiv:1309.6622 [gr-qc]]
95. M. Reiterer, E. Trubowitz, Commun. Math. Phys. **368**(1), 143–186 (2019). https://doi.org/10.1007/s00220-019-03413-8. [arXiv:1203.3766 [gr-qc]]
96. I. Rodnianski, J. Speck, Ann. Math. **2**(1), 65–156 (2018). https://doi.org/10.4007/annals.2018.187.1.2. [arXiv:1407.6293 [math.AP]]
97. T.P. Singh, P.S. Joshi, Class. Quant. Grav. **13**, 559–572 (1996). https://doi.org/10.1088/0264-9381/13/3/019. [arXiv:gr-qc/9409062 [gr-qc]]
98. Y. Tavakoli, J. Marto, A.H. Ziaie, P. Vargas Moniz, Gen. Rel. Grav. **45**, 819–844 (2013). https://doi.org/10.1007/s10714-013-1503-3. [arXiv:1105.0445 [gr-qc]]
99. J.M. Torres, M. Alcubierre, Gen. Rel. Grav. **46**, 1773 (2014). https://doi.org/10.1007/s10714-014-1773-4. ([arXiv:1407.7885 [gr-qc]].)
100. M. Van de Moortel, Commun. Math. Phys. **360**, 103–168 (2018). https://doi.org/10.1007/s00220-017-3079-3. [arXiv:1704.05790 [gr-qc]]
101. M. Van de Moortel, Commun. Math. Phys. **382**(2), 1263–1341 (2021). https://doi.org/10.1007/s00220-020-03923-w. [arXiv:2001.11156 [gr-qc]]
102. X. Zhang, H. Lü, Phys. Rev. D **91**(4), 044046 (2015). https://doi.org/10.1103/PhysRevD.91.044046. [arXiv:1410.8337 [gr-qc]]
103. X. Zhang, X. An, Ann. Henri Poincare **19**(2), 619–651 (2018). https://doi.org/10.1007/s00023-017-0631-9. [arXiv:1509.07956 [gr-qc]]

Chapter 7
Cylindrically Symmetric Fields in General Relativity

N. O. Santos and Anzhong Wang

7.1 Introduction

Globally cylindrically symmetric solutions of the Einstein field equations, at first sight, may not seem to be physically relevant since they impose infinitely long sources. Nonetheless, under controlled circumstances, they can provide fairly accurate descriptions of different physical phenomena. Important features of cylindrical systems that can help relativistic astrophysics are, for instance, exact models for rotation and the dragging of spacetime, models of extragalactic jets, gravitational radiation, the effect to the cosmological constant describing dark energy, translating fluids that might model beams of light produced by stars, and many others. For references concerning these features see [1, 2].

Cylindrically symmetric fields started to be studied in the realm of General Relativity dating back to 1919, when Levi-Civita (LC) [3] obtained the vacuum solution of a static cylindrical vacuum spacetime. Ever since much has been written by researchers trying to grasp its physical and geometrical interpretations. However, this endeavour proved to be difficult and uncertain. Only in 1958 Marder [4] established that the LC solution has two independent parameters usually called σ and a. Understanding the origin, geometry and physics that lies behind these two parameters is a big challenge in understanding the solution. For small values of σ, noted by LC himself, the corresponding Newtonian field is the external gravitational field produced by an infinitely long homogeneous line mass with σ representing the mass

N. O. Santos
Observatoire de Paris/Meudon, LERMA, UPMC Université Paris 06, Sorbonne Université, 5 place Jules Janssen, 92195 Meudon Cedex, France
e-mail: nilton.santos@obspm.fr

A. Wang (✉)
GCAP-CASPER, Physics Department, Baylor University, Waco, TX 76798-7316, USA
e-mail: anzhong_wang@baylor.edu

D. Malafarina and P. S. Joshi (eds.), *New Frontiers in Gravitational Collapse and Spacetime Singularities*, Springer Series in Astrophysics and Cosmology,
https://doi.org/10.1007/978-981-97-1172-7_7

per unit length. In this approximation the parameter a is also associated with the constant arbitrary potential that exists in the Newtonian solution.

In 1979 Bonnor [5] observed that a is also dressed with a relevant global topological meaning, and cannot be removed by coordinate transformations. There is a series of obstacles and apparently contradictory properties of σ to allow possible interpretations. Unexpectedly, the parameter σ is the most difficult and elusive to be interpreted and there is a long list of articles dedicated to unveil its meaning. In spite of that, the Newtonian limit in cylindrical models agrees well with observations (see [6–8]).

Research extended LC to cylindrically stationary vacuum spacetimes, obtained independently, by Lanczos [10] in 1924 and Lewis [11] in 1932. Its physical interpretation is taking too a long time. In its usual form it is presented by two families, one with four real parameters and the other with complex parameters. Only not long ago (in 1998), it was proved by MacCallum and Santos [12] that the four parameters can be reduced to three independent ones. However, the general interpretation of these parameters still remain unclear, and the constraints relating them are unknown.

The vacuum spacetime produced by a translating source has been obtained by Griffiths and Santos in 2009 [13]. Its extension possessing a cosmological constant has not yet been considered.

In 1925 Beck was the first to study cylindrical time-dependent vacuum spacetime [14]. The interpretation of these solutions as representing cylindrical gravitational waves was given by Einstein. These same solutions were later reobtained by Einstein and Rosen [15] in 1937. It is still unknown a collapsing source that can be matched to this solution [16].

Our aim here is to present a brief review of the exact gravitational fields produced by different cylindrical perfect fluid sources which are static, rotating, under translation and under gravitational collapse. Some important unsolved questions are addressed too.

7.2 Einstein's Field Equations

The world constructed by Newton to describe his theory of gravitation had much simplicity. Space is always equal to itself where particles move and act upon each other. Furthermore, the gravitational effects propagate with an infinite large velocity. This concept of space sounded not plausible to Einstein and he came with the brilliant idea that space is the gravitational field meaning the Newtonian space itself [49]. This needed a better formulation of the field equations. Einstein, after a long struggle, found that the Riemann geometry, first proposed by Gauss and generalized to any dimensions by Riemann, could properly describe the curvature of space produced by different matter distributions. By doing this Einstein managed to express the gravitational field equations in an arbitrary coordinate system.

In Riemann's geometry the parallel transport of a vector is proportional to its curvature which is described by a quantity called the Riemann curvature tensor

$R_{\alpha\beta\gamma\delta}$. The justification of calling it curvature lies in the fact that it vanishes if and only if the space is flat.

Einstein [50] derived the field equations in the year 1915 which are given by

$$R_{\alpha\beta} - \frac{1}{2}Rg_{\alpha\beta} = \kappa T_{\alpha\beta}. \tag{7.1}$$

They show how the space curvature, represented by the Ricci tensor $R_{\alpha\beta} = R^{\gamma}_{\alpha\gamma\beta}$, and its scalar $R = g^{\alpha\beta}R_{\alpha\beta}$ are brought about by the source of curvature, the matter distribution given by the energy momentum tensor $T_{\alpha\beta}$. In fact not only space curves but time as well and for this reason we call spacetime where Einstein field equations dwell. The coupling constant κ in normalized units, the velocity of light $c = 1$ and the Newton's constant $G = 1$, values $\kappa = 8\pi G/c^4 = 8\pi$. This system of equations, called Einstein's field equations, constitutes a set of 10 partial differential equations with respect to the metric $g_{\alpha\beta}$. In the vacuum case where $T_{\alpha\beta} = 0$, these equations reduce to

$$R_{\alpha\beta} = 0. \tag{7.2}$$

When spacetime is deprived of sources producing gravitational fields, and thereby producing curvature, the spacetime is flat and given by the metric $\eta_{\alpha\beta}$, called the Minkowski metric,

$$ds^2 = \eta_{\alpha\beta}dx^{\alpha}dx^{\beta} = dt^2 - dx^2 - dy^2 - dz^2, \tag{7.3}$$

with t being the time coordinate and x, y and z the Cartesian coordinates. Hence, the Newtonian limit, which one expects to be obtained from Einstein's field equations is produced when the metric is given by

$$g_{\alpha\beta} = \eta_{\alpha\beta} + f_{\alpha\beta}, \tag{7.4}$$

with $f_{\alpha\beta}$ being a small deviation from the Minkowski spacetime.

7.3 LC Vacuum Spacetime

Einstein's field equations (7.1) are highly nonlinear thus imposing huge difficulties in finding their solutions. Furthermore, after one found their solutions, even more difficulties arise in interpreting them. In spite of great efforts to grasp their physical meanings as is well shown in [51], the majority or even almost the totality of these solutions are still not well understood given their high nonlinearity. We will consider these difficulties in the following by studying cylindrically symmetric gravitational fields.

Schwarzschild [52] in 1916 obtained the first vacuum solution to (7.2) describing the spherically symmetric vacuum field with the following line element

$$ds^2 = g_{\alpha\beta} dx^\alpha dx^\beta$$
$$= \left(1 - \frac{2M}{r}\right) dt^2 - \left(1 - \frac{2M}{r}\right)^{-1} dr^2 - r^2(d\theta^2 + \sin^2\theta d\phi^2), \quad (7.5)$$

where the spherical coordinates are numbered $x^0 = t$, $x^1 = r$, $x^2 = \theta$ and $x^3 = \phi$. In the Newtonian approximation (7.5) becomes

$$g_{00} = 1 + 2U, \quad (7.6)$$

where U is the Newtonian potential and comparing to (7.5) we have

$$U = -\frac{M}{r}, \quad (7.7)$$

hence the only parameter stemming from the integration of the field equations for a spherically symmetric vacuum spacetime is the Newtonian mass M.

The second solution of Einstein's field equations (7.2) was obtained a few years later, 1919, after Schwarzschild presented his in 1916. It corresponds to a cylindrical vacuum spacetime, and can be cast in the form

$$ds^2 = g_{\alpha\beta} dx^\alpha dx^\beta = \rho^{4\sigma} dt^2 - \rho^{4\sigma(2\sigma-1)}(d\rho^2 + dz^2) - \frac{1}{a}\rho^{2(1-2\sigma)} d\phi^2, \quad (7.8)$$

where the cylindrical coordinates are numbered $x^0 = t$, $x^1 = \rho$, $x^2 = z$ and $x^3 = \phi$. In the Newtonian approximation we have corresponding to (7.6),

$$g_{00} = e^{2U}, \quad (7.9)$$

and compared to (7.8) for this case we have the Newtonian potential

$$U = 2\sigma \ln \rho. \quad (7.10)$$

Hence, from this expression we clearly see that for small values of σ it is the Newtonian mass per unit length as produced by an infinitely long homogeneous line mass, as observed by LC himself. Ever since much has been written by researchers trying to unveil its physical and geometrical interpretations. However, the endeavour proved to be difficult and uncertain.

Only in 1958 did Marder [4] establish that the solution (7.8) contains two arbitrary independent parameters, σ and a, differently from the Newtonian fields in which there is only one independent parameter σ.

We call the attention that for spherical vacuum case, the Schwarzschild metric, in the relativistic and Newtonian theories there appears just one parameter. This fact already suggests some harder difficulties in understanding cylindrically symmetric fields, which is indeed the case.

In the following we review the main properties and physics that lies behind the LC spacetime, which so far has been grasped up to the present time in a large number of papers. We are aware that these results sometimes appear contradictory since some interpretations collide with others. In fact this is one of our main motivations to deepen in Einstein's theory.

7.3.1 Nature of the Coordinates of the LC Spacetime

The LC metric given by (7.8) can be written in the form [9]

$$ds^2 = \rho^{4\sigma} dt^2 - \rho^{4\sigma(2\sigma-1)}\left(d\rho^2 + \frac{1}{a_m} dm^2\right) - \frac{1}{a_n}\rho^{2(1-2\sigma)} dn^2, \tag{7.11}$$

where $-\infty < t < \infty$ is the time and $0 \le \rho < \infty$ the radial coordinates, and σ, a_m and a_n are constants. The nature of the coordinates m and n, so far unspecified, depends upon the behaviour of the metric coefficients. Either a_m or a_n can be transformed away by a scale transformation depending upon the behaviour of the coordinates m and n, thus leaving the metric with only two independent parameters. In order to find that behaviour we transform the radius ρ into a proper length radial coordinate r by defining

$$\rho^{2\sigma(2\sigma-1)} d\rho = dr, \tag{7.12}$$

thus obtaining

$$\rho = R^{1/\Sigma}, \quad R = \Sigma r, \quad \Sigma \equiv 4\sigma^2 - 2\sigma + 1, \tag{7.13}$$

and metric (7.11) becomes

$$ds^2 = f(r)dt^2 - dr^2 - h(r)dm^2 - l(r)dn^2, \tag{7.14}$$

with

$$f(r) = R^{4\sigma/\Sigma}, \quad h(r) = \frac{1}{a_m} R^{4\sigma(2\sigma-1)/\Sigma}, \quad l(r) = \frac{1}{a_n} R^{2(1-2\sigma)/\Sigma}. \tag{7.15}$$

Consider $0 < \sigma < 1/2$, which implies for this range $h(r)$ diverging when $r \to 0$ and $l(0) = 0$. Then we can interpret m as the axial coordinate $-\infty < z < \infty$ with $a_m = 1$ by rescaling z and n as the angular coordinate ϕ with the topological identification of every ϕ with $\phi + 2\pi$, and the metric (7.14) becomes

$$ds^2 = R^{4\sigma/\Sigma} dt^2 - dr^2 - R^{4\sigma(2\sigma-1)/\Sigma} dz^2 - \frac{1}{a} R^{2(1-2\sigma)/\Sigma} d\phi^2, \tag{7.16}$$

where a_n is replaced by a.

Now consider $1/2 < \sigma < \infty$, which implies $h(0) = 0$ and $l(r)$ diverging when $r \to 0$. Then we can interpret m as the angular coordinate ϕ with topological identification of every ϕ with $\phi + 2\pi$, and n as an axial coordinate $-\infty < z < \infty$ with $a_n = 1$ by again rescaling z, and metric (7.14) becomes

$$ds^2 = R^{4\sigma/\Sigma}dt^2 - dr^2 - R^{2(1-2\sigma)/\Sigma}dz^2 - \frac{1}{a}R^{4\sigma(2\sigma-1)/\Sigma}d\phi^2, \tag{7.17}$$

where we replaced a_m for a.

In both cases, (7.16) and (7.17), where $\sigma > 0$ we have $g_{00} \to 0$ as $r \to 0$, indicating an attractive singularity. While assuming $\sigma < 0$, we obtain $g_{00} \to \infty$ as $r \to 0$, indicating a repulsive singularity.

The invariant quantity, under coordinate transformations, built out of the Riemann curvature tensor given by $\mathscr{R} = R_{\alpha\beta\gamma\delta}R^{\alpha\beta\gamma\delta}$, called the Kretschmann scalar, is a good indicator of singularities. Calculating $\mathscr{R}$ for metrics (7.16) and (7.17) we obtain

$$\mathscr{R} = \frac{64\sigma^2(2\sigma - 1)^2}{\Sigma^3 r^4}. \tag{7.18}$$

From (7.18) we see that $\mathscr{R} \to \infty$ as $r \to 0$, and $\mathscr{R} \to 0$ for $\sigma = 0, 1/2$ and ∞.

Metric (7.16) for $\sigma = 0$ becomes

$$ds^2 = dt^2 - dr^2 - dz^2 - \frac{1}{a}r^2 d\phi^2, \tag{7.19}$$

representing the Minkowski spacetime when $a = 1$ in cylindrical polar coordinates (r, z, ϕ). However if (7.19) has $a > 0$ there is an angle deficit of $2\pi\delta$ given by $\delta = 1 - 1/\sqrt{a}$, producing flat spacetime everywhere except along the axis $r = 0$, which is interpreted as a cosmic string. The deficit can represent the tension of the string with mass per unit length $\mu = \delta/4$. If there is an angle excess $a < 1$ it would represent a cosmic string under compression with $\mu < 0$. Hence the constant a is directly linked to the gravitational analogue of the Aharonov–Bohm effect [17]. This effect shows that gravitation depends on the topological structure of spacetime giving rise to an angular deficit δ as in the electromagnetic Aharonov–Bohm effect, where a (classical) non-observable quantity (the vector potential) becomes observable (part of it) through a quantum non-local effect. Its gravitational analogue allows a (Newtonian) non-observable quantity (the additional constant to the Newtonian potential) to become observable in the relativistic theory through the angular deficit in strings. For a review in cosmic strings see [18].

In the case $\sigma = 1/2$ the two metric coefficients h and l in (7.15) are constants, so both a_m and a_n can be set to unity. Then neither m nor n is entitled to be an angular coordinate, and the three coordinates (r, m, n) are better visualized as Cartesian coordinates (x, y, z). Hence metric (7.11) can be written as

$$ds^2 = z^2 dt^2 - dx^2 - dy^2 - dz^2, \tag{7.20}$$

which is the static plane symmetric vacuum spacetime obtained by Rindler [13, 19, 20].

In the next section we calculate the circular geodesics for the above spacetime to try to get further understanding of the properties of different values that σ can attain for $\sigma \geq 0$, since $\sigma < 0$ would in some way correspond to negative mass densities.

7.3.2 *Geodesics of LC Spacetime*

For the metrics (7.16) and (7.17) the circular geodesics [21] have $\dot{r} = \dot{z} = 0$ and $g_{tt,r}\dot{t}^2 + g_{\phi\phi,r}\dot{\phi}^2 = 0$, where the dot stands for differentiation with respect to the affine parameter s. The geodesic angular velocity is defined by $\omega = \dot{\phi}/\dot{t}$, and the only non zero component of its velocity is the tangential one, given by $W^{\phi} = \omega\sqrt{g_{tt}}$, with it modulus defined by $W^2 = W^{\alpha}W_{\alpha}$ [22].

In the case $0 \leq \sigma < 1/2$, from (7.16) we obtain

$$\omega^2 = \frac{2\sigma}{1-2\sigma} a R^{2(4\sigma-1)/\Sigma}, \tag{7.21}$$

$$W^2 = \frac{2\sigma}{1-2\sigma}. \tag{7.22}$$

We note that for a given σ the velocity W in (7.22) is the same for all circular geodesics, in agreement with the corresponding Newtonian gravitation. Furthermore, we see that W increases monotonically with σ, that is from $\sigma = 0$ producing $W = 0$, to $\sigma = 1/4$ attaining $W = 1$, the speed of light, and finally $\sigma = 1/2$ producing geodesics with $W = \infty$. For small σ and $a = 1$ from (7.21) and (7.22) we obtain the Newtonian limit $W = \omega r$.

In the case $1/2 < \sigma < \infty$ from (7.17) we obtain

$$\omega^2 = \frac{1}{2\sigma - 1} a R^{8\sigma(1-\sigma)/\Sigma}, \tag{7.23}$$

$$W^2 = \frac{1}{2\sigma - 1}. \tag{7.24}$$

With σ increasing beyond $1/2$, we note from (7.23) that W diminishes, attaining $W = 1$ for $\sigma = 1$ and $W = 0$ for $\sigma = \infty$.

In other words, the circular geodesics are timelike when either $0 < \sigma < 1/4$ or $\sigma > 1$, are lightlike when $\sigma = 1/4$ or 1, and are spacelike when $1/4 < \sigma < 1$.

If we redefine σ by

$$\sigma = \frac{1}{4\bar{\sigma}}, \tag{7.25}$$

then metric (7.17) with a rescaling of its coordinates becomes (7.16), hence (7.22) and (7.23) reduce to (7.20) and (7.21). This means that the parameter range $1/2 < \sigma < \infty$

is equivalent to the range $0 < \sigma < 1/2$ and the coordinates z and ϕ switching their nature.

Hence we might have the following picture for the different values of σ. For small values of σ the metric (7.16) with t and r constants describes cylindrical surfaces with ϕ as the periodic coordinate. As σ increases the cylindrical surfaces open out and become an infinite plane for $\sigma = 1/2$. For values of σ bigger than $1/2$ the coordinate z becomes periodic forming new cylindrical surfaces perpendicular to previous ones for $0 < \sigma < 1/2$.

Another interesting geodesic is the one that describes the motion of the particle along the axis of symmetry z. These geodesics calculated with (7.16) (we restrict the calculation of this metric since (7.17) is equivalent) produce

$$\ddot{z} = \frac{4\sigma(1-2\sigma)\dot{r}\dot{z}}{\Sigma r}. \tag{7.26}$$

It means that particles increase their speed along z when distancing radially from the axis, while diminishing their axial speed when moving radially towards the axis. This result indicates that a force parallel to the z axis appears. In the flat case $\sigma = 0$ such an effect vanishes, bringing out its non-Newtonian nature. We further discuss this weird geodesic property in the section containing the Lewis spacetime.

For radial geodesics it has been shown [22] that there exist timelike particles approaching z that are reflected at $r = r_{min}$ and move outwards until attaining $\dot{r} = 0$ at $r = r_{max}$ repeating endlessly this trajectory. This motion is called confinement of test particles.

In the next section we see some further limits that LC metric satisfies.

7.3.3 LC Spacetime as a Limit of the γ Spacetime

In cylindrical coordinates, static axially symmetric solutions of Einstein's vacuum equations are given by the Weyl metric [23]

$$ds^2 = e^{2\lambda}dt^2 - e^{-2\lambda}[e^{2\mu}(d\rho^2 + dz^2) + \rho^2 d\phi^2], \tag{7.27}$$

with $\lambda(\rho, z)$ and $\mu(\rho, z)$ satisfying

$$\lambda_{,\rho\rho} + \frac{1}{\rho}\lambda_{,\rho} + \lambda_{zz} = 0, \tag{7.28}$$

and

$$\mu_{,\rho} = \rho(\lambda_{,\rho}^2 - \lambda_{,zz}^2), \tag{7.29}$$

$$\mu_{,z} = 2\rho\lambda_{,\rho}\lambda_{,z}, \tag{7.30}$$

where the comma stands for partial derivation. Synge writes [24], as the most amazing fact, that (7.28) is just the Laplace equation for λ in the Euclidean space. Metric (7.27) with the solution of (7.28) corresponding to a Newtonian source of constant density γ uniformly distributed on the z axis between $z = -m$ and $z = m$ is usually referred to as the γ metric, and the corresponding functions λ and μ are given by

$$e^{2\lambda} = \left(\frac{R_+ + R_- - 2m}{R_+ + R_- + 2m}\right)^{\gamma}, \tag{7.31}$$

$$e^{2\mu} = \left[\frac{(R_+ + R_- + 2m)(R_+ + R_- - 2m)}{4R_+R_-}\right]^{\gamma^2}, \tag{7.32}$$

where

$$R_{\pm}^2 = \rho^2 + (z \pm m)^2, \tag{7.33}$$

and γ and m are two integration constants. These solutions were first found by Bach and Weyl in 1922 [25]. Calculating its Newtonian potential (7.10) we obtain

$$U = \gamma \ln\left(\frac{R_- + z - m}{R_+ + z + m}\right), \tag{7.34}$$

which corresponds to a potential due to a line segment of length $2m$ and mass per unit length $\gamma/2$ symmetrically distributed along the z axis. Hence the total mass M of the line segment is $M = \gamma m$. The particular case $\gamma = 1$ corresponds to the Schwarzschild metric. This can be seen by taking Erez–Rosen coordinates [26] defined by

$$\rho^2 = (r^2 - 2mr)\sin^2\theta, \quad z = (r - m)\cos\theta, \tag{7.35}$$

and the γ metric becomes

$$ds^2 = -F dt^2 - \frac{1}{F}[G dr^2 + H d\theta^2 + (r^2 - 2mr)\sin^2\theta d\phi^2], \tag{7.36}$$

where

$$F = \left(1 - \frac{2m}{r}\right)^{\gamma}, \tag{7.37}$$

$$G = \left(\frac{r^2 - 2mr}{r^2 - 2mr + m^2\sin^2\theta}\right)^{\gamma^2 - 1}, \tag{7.38}$$

$$H = \frac{(r^2 - 2mr)^{\gamma^2}}{(r^2 - 2mr + m^2\sin^2\theta)^{\gamma^2 - 1}}. \tag{7.39}$$

Now we can easily check that for $\gamma = 1$ the metric (7.36) reduces to the Schwarzschild metric.

If we want to compare the γ metric to the LC metric, in the limit when its length segment $m \to \infty$, one notices that by taking this limit in (7.31) and (7.32) the metric diverges. For this reason one can use the Cartan scalars approach to obtain a finite limit. These scalars are the components of the Riemann tensor and its covariant derivatives calculated in a constant frame. Two metrics are equivalent if and only if there exist coordinate and Lorentz transformations which transform the Cartan scalars of one of the metrics into the Cartan scalars of the other. Although the Cartan scalars provide a local characterization of the spacetime, global properties such as topological defects do not appear in them. By doing all this procedure one can prove that locally in the limit $m \to \infty$ the γ metric is the same as the LC metric. Details of these long calculations are given in [27]. Here we come to an interesting and, so far, weird result showing how apparently unexpected results can stem from long known results like the Schwarzschild and the LC solutions. When the density per unit length of the rod $\sigma = \gamma/2$ has the value $\gamma = 1$, or the mass density per unit length $\sigma = 1/2$, it becomes the Schwarzschild spherically symmetric spacetime, and in the limit $m \to \infty$, it becomes the Rindler static plane symmetric vacuum spacetime. This is a remarkable result.

For all the different limiting metrics that LC spacetime can undergo see [27]. The limits for the circular geodesics of the γ spacetime to the LC spacetime are well studied in [22].

In the next section we make a brief review of possible sources to the LC spacetime.

7.3.4 Sources Producing LC Spacetime

The LC spacetime, as we saw, contains two essential constants denoted by a and σ. The constant a refers to an angle deficit or excess that produces cosmological strings. However, it is σ that presents the most serious obtacles to its interpretation. Indeed, for small values $0 < \sigma < 1/4$, LC describes the spacetime generated by an infinite line mass, with σ as mass per unit coordinate length. When $\sigma = 0$ the spacetime is flat. However, circular timelike geodesics only exist for $0 < \sigma < 1/4$, becoming null for $\sigma = 1/4$ and being spacelike for $\sigma > 1/4$. Furthermore, as the value of σ increases from 1/4 to 1/2 the corresponding Kretchmann scalar (7.18) diminishes monotonically and vanishes when $\sigma = 1/2$, implying that the spacetime is flat.

Thus, the question is what does the LC metric represent outside the range $0 \leq \sigma \leq 1/4$?

First, we observe that there are known physically satisfactory fluid sources for the LC spacetime satisfying boundary conditions for both ranges of σ (see for example [28]). On the other hand the fact that the scalar Kretschmann decreases with increasing σ may not be associated with the strength of the gravitational field. Instead it could be associated with the acceleration of a test particle that would measure more suitably the strength of the gravitational field, which is the case for $1/4 < \sigma < 1/2$.

A possible interpretation of the LC solution is a spacetime generated by a cylinder whose radius increases with σ and tends to infinity as σ approaches $1/2$. This interpretation suggest that when $\sigma = 1/2$ the cylinder becomes a plane.

It might be instructive to consider a static cylinder filled with anisotropic perfect fluid and calculate its mass per unit length by using the junction conditions to its exterior LC spacetime. The Whittaker formula [29] for the active gravitational mass per unit length ν of the static distribution of perfect fluid with energy density μ and principal stresses P_r, P_z and P_ϕ inside a cylinder of surface S is

$$\nu = 2\pi \int_0^{r_S} (\mu + P_r + P_z + P_\phi)\sqrt{-g}dr. \tag{7.40}$$

Considering a static cylindrically symmetric metric

$$ds^2 = Adt^2 - dr^2 - Cdz^2 - Dr^2 d\phi^2, \tag{7.41}$$

in which A, C and D are only functions of r, and from Einstein's field equations we obtain

$$\frac{A_{,rr}}{A} - \frac{1}{2}\frac{A_{,r}}{A}\left(\frac{A_{,r}}{A} - \frac{2}{r} - \frac{C_{,r}}{C} - \frac{D_{,r}}{D}\right) = \kappa(\mu + P_r + P_z + P_\phi). \tag{7.42}$$

Substituting (7.42) into (7.40) we have the simple expression

$$\nu = \frac{1}{4}\left(\frac{A_{,r}}{A}\sqrt{-g}\right)_S, \tag{7.43}$$

where regularity on the axis of symmetry [28] has been assumed. Now taking the LC metric (7.16) for the exterior spacetime of the cylindrical surface S, and imposing that it satisfies Darmois' junction conditions, [28, 30], with its interior spacetime described by (7.41) it amounts to impose the continuity of the metric functions and its derivatives on S. By doing so, from (7.43) we obtain

$$\nu = \frac{\sigma}{\sqrt{a}}, \tag{7.44}$$

where a is the constant defined in (7.16). From (7.44) we have that when $a > 1$, there is a topological angle deficit, then $\nu < \sigma$, and if $a < 1$ there is an angle excess producing $\nu > \sigma$. However, when there is no topological defect, $a = 1$, then ν can in fact be interpreted as mass per unit length of its source. Furthermore, since with cylindrical sources no black holes are formed, one might conclude that the minimum mass per unit length to form a black hole satisfies the constraint $\nu > 1/2$. This result would fulfil the present knowledge of black holes that a lower mass limit is required for its formation. We are aware that ν is model depended and cannot be given a general meaning, but nonetheless it fulfils some of the expected properties.

A last comment on sources for the LC spacetime is the fact that conformally flat static sources do not admit Darmois' matching conditions satisfied for an exterior LC spacetime. This result is proved in [31]. Conformal flatness of a metric means that its corresponding Weyl tensor vanishes. The interpretation of the Weyl tensor is that it describes the purely gravitational characteristics of the source. This interesting result means that a static cylindrical source deprived of its purely gravitational character cannot be smoothly matched to the exterior LC spacetime. It is conspicuous that for spherical symmetry this result does not stand, since there are conformally flat static spherical sources matched to the Schwarzschild spacetime [32].

7.4 Lewis Vacuum Spacetime

The extension of the LC static cylindrically symmetric vacuum spacetime to a stationary cylindrically symmetric vacuum spacetime was obtained independently by Lanczos in 1924 [10] and Lewis in 1932 [11]. We consider the spacetime described by the cylindrically symmetric stationary metric

$$ds^2 = f dt^2 - 2k dt d\phi - e^{\mu}(dr^2 + dz^2) - l d\phi^2, \tag{7.45}$$

where f, k, μ and l are functions only of r. The ranges of the coordinates are $-\infty < t < \infty$ for the time coordinate, $0 \le r < \infty$ for the radial coordinate, $-\infty < z < \infty$ for the axial coordinate and $0 \le \phi \le 2\pi$ for the angular coordinate with the hypersurfaces $\phi = 0$ and $\phi = 2\pi$ being identified. The coordinates are numbered $x^0 = t, x^1 = r, x^2 = z$ and $x^3 = \phi$. The general vacuum solution $R_{\alpha\beta} = 0$ for metric (7.45), in the notation given by [19, 21], is

$$f = ar^{1-n} - a\delta^2 r^{1+n}, \tag{7.46}$$
$$k = -(1 - ab\delta)\delta r^{1+n} - abr^{1-n}, \tag{7.47}$$
$$l = \frac{(1 - ab\delta)^2}{a} r^{1+n} - ab^2 r^{1-n}, \tag{7.48}$$
$$e^{\mu} = r^{(n^2-1)/2}, \tag{7.49}$$

with

$$\delta = \frac{c}{an}. \tag{7.50}$$

The constants n, a, b and c can be either real or complex, and the corresponding solutions belong to the Weyl class or Lewis class, respectively. For the Lewis class these constants are given by

$$n = im, \tag{7.51}$$

$$a = \frac{1}{2}(a_1 + b_1)^2, \tag{7.52}$$

$$b = \frac{a_2 + ib_2}{a_1 + ib_1}, \tag{7.53}$$

$$c = \frac{m}{2}(a_1^2 + b_1^2), \tag{7.54}$$

where m, a_1, a_2, b_1 and b_2 are real constants satisfying

$$a_1 b_2 - a_2 b_1 = 1. \tag{7.55}$$

The metric coefficients (7.46)–(7.49) with (7.51)–(7.54) become

$$f = r(a_1^2 - b_1^2)\cos(m \ln r) + 2ra_1b_1 \sin(m \ln r), , \tag{7.56}$$

$$k = -r(a_1a_2 - b_1b_2)\cos(m \ln r) - r(a_1b_2 + a_2b_1)\sin(m \ln r), \tag{7.57}$$

$$l = -r(a_2^2 - b_2^2)\cos(m \ln r) - 2ra_2b_2 \sin(m \ln r), \tag{7.58}$$

$$e^{\mu} = r^{-(m^2+1)/2}. \tag{7.59}$$

A simple deduction of the Lewis metric, where one does not need to consider complex parameters to obtain the Lewis class is given in [33]. There a physical interpretation of the field equations is also provided, which permits to have some understanding of the four parameters appearing in the Lewis solution (7.46)–(7.49). Another derivation of the Lewis metric is given in [12]. It is found that three parameters are essential, of which one characterizes the local gravitational field, while the remaining two provide information about the topological identification made to produce cylindrical symmetry.

In Newtonian physics the potential due to cylindrical matter source, being static or stationary, has the same dependence, that is, it depends only on one parameter, the mass per unit length. For the static vacuum cylindrical field in General Relativity the solution is the LC metric, the one we studied in the previous section, revealing two essential parameters, while for the stationary cylindrical rotating source it has, in its usual form four parameters reducible to three essential parameters [12].

In the next section we study the meaning of the parameters appearing in the Lewis metric.

7.4.1 *The Parameters of the Weyl Class*

The transformation

$$d\tau = \sqrt{a}(dt + bd\phi), \tag{7.60}$$

$$d\bar{\phi} = -\frac{1}{n}[cdt - (n - bc)d\phi], , \tag{7.61}$$

casts the metric (7.45) with (7.46)–(7.49) in the form

$$ds^2 = r^{1-n} d\tau^2 - r^{(n^2-1)/2}(dr^2 + dz^2) - \frac{r^{n+1}}{a} d\bar{\phi}^2. \quad (7.62)$$

This is locally the LC metric. Nevertheless, since $\phi = 0$ and $\phi = 2\pi$ are identified, τ defined in (7.60) attains a periodic nature unless $b = 0$. On the other hand, the new coordinate $\bar{\phi}$ ranges from $-\infty$ to ∞. A more detailed account of this subject can be found in [34]. In order to globally transform the Weyl class of the Lewis metric into the static LC metric we have to make $b = 0$ and $c = 0$. Note that in this case, from the transformations (7.60) and (7.61), τ and $\bar{\phi}$ become true time and angular coordinates. Hence we can say that b and c are responsible for the non-staticity of this family of solutions of the Lewis metric.

As mention previously, the Cartan scalars provide the local characteristics of a metric. They are obtained through the components of the Riemann tensor and its covariant derivatives calculated in a constant frame. Two metrics are equivalent if and only if there is a coordinate and Lorentz transformations which transform the Cartan scalars of one of the metrics into the Cartan scalars of the other. By performing these calculations for both metrics, the LC metric and the Weyl class of the Lewis metric we find that both are equivalent locally and indistinguishable, which confirms the coordinate analysis made in the beginning of this subsection. Furthermore, we showed that only the parameter n curves spacetime for both static and stationary Weyl class metrics. However, we shall see that the two metrics possess very different topological behaviour.

Details of the calculations of the Cartan scalars are given in [21].

Considering a cylindrical matter source for the Weyl class metric consisting of a rigidly rotating anisotropic fluid, one of Einstein's field equations can be integrated. This integration produces

$$fk_{,r} - kf_{,r} = \xi r, \quad (7.63)$$

where ξ is an integration constant. Calculating the rotation of the source as given in [21] produces the rotation magnitude given by $\xi/(2fe^{\mu}/2)$. Now using the matching conditions on the surface of the source cylinder as given by Darmois [30] we find

$$c = -\frac{\xi}{2}. \quad (7.64)$$

Note that this constant c is different from the speed of light used in other places of this chapter. In fact, now it measures the rotation of the cylindrical source, as can be seen from (7.64).

In the Newtonian limit, the velocity term is negligible, then from (7.64) $c \approx 0$, and recalling (7.46), we find that

$$f = e^{2U}. \quad (7.65)$$

Then the Newtonian potential is

$$U = 2\sigma \ln r + \frac{\ln a}{2}, \tag{7.66}$$

where σ is given by

$$\sigma = \frac{1-n}{4}. \tag{7.67}$$

In Newtonian theory, (7.66) is the gravitational potential of an infinite uniform line mass with mass per unit length σ. The constant $(\ln a)/2$ represents the constant arbitrary potential that exists in the Newtonian solution. The metric (7.46)–(7.49) has infinite curvature, according to its Cartan scalars, only at $r = 0$ for all n except $n = \pm 1$, i.e., $\sigma = 0$ and $1/2$. Thus the Weyl class metric has a singularity along the axis $r = 0$, then we can say that this spacetime is generated by an infinite uniform line source for $0 < \sigma < 1/2$.

Considering the static limit of the Weyl class metric when $n = 1$ ($\sigma = 0$) and $b = c = 0$, we have from (7.46)–(7.49)

$$ds^2 = d\tau^2 - dr^2 - dz^2 - \frac{r^2}{a} d\phi^2, \tag{7.68}$$

which is the limit of the LC metric when $\sigma = 0$. In the previous section it has been pointed that it generates strings when $a > 1$ with mass per unit length $\mu = \delta/4$ and how a is directly linked to the gravitational analog of the Aharonov–Bohm effect [17].

Considering $c = 0$ and $n = 1$ ($\sigma = 0$) in (7.46)–(7.49) we have

$$ds^2 = d\tau^2 + 2b\sqrt{a} d\tau d\phi - dr^2 - dz^2 - \left(\frac{r^2}{a} - b^2 a\right) d\phi^2, \tag{7.69}$$

producing a locally flat spacetime. In this case (7.69) represents the exterior spacetime of a spinning string along the axis of symmetry [35] with the same mass per unit length $\mu = \delta/4$ but with an angular momentum J given by

$$J = -\frac{b\sqrt{a}}{4}, \tag{7.70}$$

for $a > 1$.

It has been shown [35] that a quantum scalar particle moving around a spinning cosmic string as given by (7.69), exhibits a phase factor proportional to J in its angular momentum. It is a reminiscence of the Aharonov–Bohm effect. It is also worth mentioning that even if $b = 0$, an Aharonov–Bohm like effect exists, thought of a different kind, appears (as commented in the static case), since the angular momentum spectrum differs from the usual one, if only $a > 1$.

7.4.2 Geodesics of Weyl Class Spacetime

The Weyl class of solutions is given by (7.45) with (7.46)–(7.49) and all parameters being real, as shown above. Then for circular geodesics [36] $\dot{r} = \dot{z} = 0$, we find that $f_{,r}\dot{t}^2 - 2k_{,r}\dot{t}\dot{\phi} - l_{,r}\dot{\phi}^2 = 0$, where the dot stands for differentiation with respect to an affine parameter s. The geodesic angular velocity is defined by $\omega = \dot{\phi}/\dot{t}$, and the velocity of the test particle has only two non zero components, $W^t = k\omega/(f^{3/2} - \sqrt{f}k\omega)$ and $W^\phi = \sqrt{f}\omega/(f - k\omega)$, with

$$\omega = \frac{c \pm n\omega_0}{n - b(c \pm n\omega_0)}, \quad W = \frac{\delta r^n \pm W_0}{1 \pm \delta r^n W_0}, \tag{7.71}$$

where ω_0 is the LC angular velocity and W_0 is the LC tangential velocity,

$$\omega_0^2 = \left(\frac{1-n}{1+n}\right)\frac{a^2}{r^{2n}}, \quad W_0^2 = \frac{1-n}{1+n}. \tag{7.72}$$

We note that ω and W vanish for $\omega_0 = \mp c/n$ and $W_0 = \mp\delta r^n$, respectively, which are equivalent to say that the free particle in the present case is simply static. This could come about if the "centrifugal repulsion" balances the gravitational attraction.

The geodesic motion of a particle along the axis of symmetry z for metric (7.45) produces

$$\ddot{z} = \frac{(1-n^2)\dot{r}\dot{z}}{2r}. \tag{7.73}$$

It is interesting to note that for this geodesic the parameters b and c, due to the stationarity of spacetime, do not appear in (7.73) and in fact it is indistinguishable from the static limit, the LC spacetime. There is a force that tends to damp the motion along the axis, $\ddot{z} < 0$, whenever the particle approaches the axis, $\dot{r} < 0$, and reverses this tendency in the opposite case. In the flat case, $n = 1$ or $\sigma = 0$, such an effect vanishes, exposing its non-Newtonian nature.

For quasi-spherical objects it has been shown [37] that positive radial acceleration can be produced along its axis of symmetry.

It is also worth noticing that the non-Newtonian forces parallel to the z axis, also appear in the field of axially symmetric rotating bodies [38]. However the force parallel to z found in [38], unlike the current case, is directly related to the spin of the source. For the Kerr black hole, it is shown that particles produced by the Penrose process, can be ejected from the ergosphere surface, covering the the black hole through repulsive gravitational fields. In this case too, unlike in the Lewis spacetime, the gravitational repulsion is created by the spin of the Kerr black hole. These ejected particles are highly collimated and might be a mechanism for the observed extragalactic jets [39–41].

7.4.3 The Parameters of the Lewis Class

Using the transformation

$$d\phi = d\bar{\phi} + \omega dt, \quad \omega = -\frac{k}{l}, \tag{7.74}$$

the metric (7.56)–(7.59) can be diagonalized. In order to have an integral coordinate transformation ω must be a constant, therefore, from (7.56)–(7.59), $m = 0$. This implies, from (7.51)–(7.54) that $n = 0$ and $c = 0$. Thus the line element becomes

$$ds^2 = -\frac{rdt^2}{a_2^2 - b_2^2} - \frac{(dr^2 + dz^2)}{\sqrt{r}} - (a_2^2 - b_2^2)rd\bar{\phi}^2. \tag{7.75}$$

This is a particular case of the static LC metric with the energy density per unit length σ given by (7.67) being equal to $1/4$. Nevertheless, the transformation (7.74) is not global, since the new coordinate $\bar{\phi}$ ranges from $-\infty$ to ∞, instead of ranging from 0 to 2π [34].

Considering, as in the case of the Weyl class, a rigidly rotating anisotropic fluid, one of the Einstein's field equations can be integrated producing (7.63). With (7.56)–(7.59) the matching conditions given by Darmois [30] yield

$$\xi = -m(a_1^2 + b_1^2), \tag{7.76}$$

and so

$$c = -\frac{\xi}{2}. \tag{7.77}$$

Hence, in order to have the rotation equal to zero, i.e. $\xi = 0$, we need $m = 0$ since $a_1^2 + b_1^2 \neq 0$.

Observe the difference, at this point, between the Weyl class and the Lewis class. In the latter the vanishing of the rotation yields a locally LC spacetime, whereas in the former the vanishing of the rotation does not necessarily imply that the metric can be reduced either globally or locally to a static spacetime.

For the Cartan scalars that produce the local characteristics of a metric, we have the following results for the Lewis class. In the Lewis class as in the Weyl class, only the constant n appears in the Cartan scalars. Nevertheless, now n must be substituted by its complex value im (7.51). However, contrary to the Weyl class, the Cartan scalar for the Lewis class are distinguishable from the LC metric, except for $m = 0$. Furthermore, there is no value of m for which the Cartan scalars are all zero, implying at once that the Lewis class does not include the Minkowski spcetime as a special case. This fact implies too that there must be a lower limit to the energy per unit length of its source. The Cartan scalars impose a upper bound on the parameter m, given by

$$m \leq \sqrt{3}, \tag{7.78}$$

since for larger values of m than this, the singularity is at $r = \infty$ and not at $r = 0$.

7.4.4 Sources Producing Lewis Spacetime

In a fine paper, well ahead of his time, as observed by Bill Bonnor [42], in 1937 van Stockum [43] completely solved the problem of a rigidly rotating infinitely long cylinder of dust, including the application of adequate boundary conditions. The solution is a remarkable one. The metric for the interior is simple and unique depending on one parameter w in our notation. But, for the vacuum exterior, $r > R$, where R is the coordinate radius of the cylinder, there are three cases depending on the mass per unit length of the interior. For the metric (7.45) we have the following results.

Case $wR < 1/2$:

$$\begin{aligned} f &= -r\left[2\beta\cosh(2N\ln r) + \frac{\alpha^2+\beta^2}{\alpha}\sinh(2N\ln r)\right], \\ k &= -r\left[\cosh(2N\ln r) + \frac{\beta}{\alpha}\sinh(2N\ln r)\right], \\ l &= \frac{r}{\alpha}\sinh(2N\ln r), \\ e^{\mu} &= \lambda\left(\frac{r}{R}\right)^{(4N^2-1)/2}, \end{aligned} \tag{7.79}$$

with

$$\begin{aligned} N &= \frac{1}{2}(1-4w^2R^2)^{1/2}, \quad \alpha = \frac{(1-4w^2R^2)^{1/2}}{2w^3R^4}, \\ \beta &= \frac{2w^2R^2-1}{2w^3R^4}, \quad \lambda = e^{-w^2R^2}. \end{aligned} \tag{7.80}$$

Case $wR > 1/2$:

$$\begin{aligned} f &= r\left[2\beta\sin(2N\ln r) + \frac{\alpha^2-\beta^2}{\alpha}\cos(2N\ln r)\right], \\ k &= r\left[\sin(2N\ln r) - \frac{\beta}{\alpha}\cos(2N\ln r)\right],, \\ l &= \frac{r}{\alpha}\cos(2N\ln r), \\ e^{\mu} &= \lambda\left(\frac{r}{R}\right)^{(1-4N^2)/2}, \end{aligned} \tag{7.81}$$

with

$$N = \frac{1}{2}(4w^2R^2 - 1)^{1/2}, \quad \alpha = \frac{(4w^2R^2 - 1)^{1/2}}{2w^3R^4},$$
$$\beta = \frac{2w^2R^2 - 1}{2w^3R^4}, \quad \lambda = e^{-w^2R^2}. \tag{7.82}$$

For the case $wR = 1/2$ one obtains the relations from the limits either from $wR < 1/2$ or $wR > 1/2$ which are equal. The solution for $wR < 1/2$ belongs to the Weyl class and its real parameters n, a, b and c assume the values

$$n = (1 - w^2R^2)^{1/2}, \quad a = \frac{(\alpha - \beta)^2}{2\alpha},$$
$$b = \pm\frac{1}{\alpha - \beta}, \quad c = \frac{\alpha^2 - \beta^2}{\alpha}N. \tag{7.83}$$

The solution $wR > 1/2$ belong to the Lewis class and its real parameters m, a_1, a_2, b_1 and b_2 assume the values

$$m = (4w^2R^2 - 1)^{1/2}, \quad a_1 = \frac{\beta}{b_1}, \quad a_2 = -\frac{1}{b_1},$$
$$b_1 = (-\alpha)^{1/2}, \quad b_2 = 0. \tag{7.84}$$

For the Weyl class we have the Newtonian mass per unit length given by $\sigma = (1 - n)/4$ which implies, for Case $wR < 1/2$, from (7.83),

$$\sigma = \frac{1}{4}\left[1 - (1 - 4w^2R^2)^{1/2}\right]. \tag{7.85}$$

Hence (7.85) establishes a lower limit for σ in the Lewis class and being $\sigma = 1/4$. This value is the frontier between the Weyl class metric and the Lewis class metric, at least for the rotating dust solution obtained by van Stockum [43].

For the Lewis class metric the Cartan scalars, as it was remarked, do not admit the Minkovski spacetime. This is in accordance with the existence of a lower limit for σ in the van Sockum solution [43] $wR > 1/2$, since with a lower limit the source cannot be made a vacuum, and therefore the exterior solution cannot be Minkovski.

The Cartan scalars also impose an upper bound on the parameter m, given by

$$m \leq \sqrt{3}, \tag{7.86}$$

since for values of m larger than this, the singularity is at $r = \infty$ and not in $r = 0$. When we substitute this value in (7.84), considering the equality, we have $wR = 1$.

Van Stockum solution is studied at length in [42], its properties concerning gravitoelectric and gravitomagnetic fields in [38, 45], its confinement properties in [44], its extension to non-rigid rotation in [46]. A range of stationary cylindrically symmetric perfect fluid sources are presented in [47] and an anisotropic cylindrical stationary source can be found in [48].

7.5 Translation of Cylinders

It has been proved [13] that the vacuum field produced by a rotating mass cylinder is mathematically closely related to the field produced by a translating mass cylinder along its axis of symmetry. Nonetheless, its physical and geometrical properties differ significantly since the relativistic frame dragging for rotation and translation physically differ considerably.

We assume the general cylindrically symmetric metric with its source translating parallel to its axis of symmetry given by

$$ds^2 = Adt^2 - 2Kdtdz - Bd\rho^2 - Cdz^2 - D\rho^2 d\phi^2, \tag{7.87}$$

with the usual properties of its coordinates and A, K, B, C and D functions only of ρ. Making $\rho = e^r$ and after rescaling, the metric (7.87) can be written as

$$ds^2 = fdt^2 - 2kdtdz - e^{\mu}(dr^2 + d\phi^2) - dz^2. \tag{7.88}$$

The general vacuum solution for (7.88) is the stationary Lewis metric. So, the vacuum solution corresponding to (7.87) is simply the Lewis solution with the coordinates z and ϕ interchanged. Hence the metric coefficients in (7.88) are the same as in (7.46)–(7.49).

In spite of the mathematical similarity between the vacuum solutions for the fields produced by rotating and translating cylinder filled with perfect fluid they differ substantially. Unlike the rotating case, the translating cylinder cannot be filled with pressure free dust, as there is nothing here equivalent to a centrifugal force that would prevent the matter collapsing to the axis. The pressure must therefore be nonzero. Furthermore, unlike the rotating case where matter can be rigidly rotating, which means with shear free rotation, the translating matter case if translating rigidly can always be transformed to a frame where the system is static [13].

The field of a cylinder of matter that is in translational motion along its length has not been studied in detail, and any differences with the static case are unknown. It is therefore of interest to determine whether or not frames are dragged by motion along the cylinder in a way similar to that in which they are dragged around it.

7.6 Gravitational Collapse of Cylinders

Gravitational collapse and the emission of gravitational waves has been one of the most important problems in Einstein's theory. However, due to the complexity of Einstein's field equations, even in simple cases it is not well understood. It is well known, due to the Birkhoff's theorem it is not possible to have gravitational radiation in spherically symmetric spacetimes. The next simplest symmetry assumption is cylindrical symmetry. In this sense gravitational collapse of a cylindrical distribution of perfect fluid mass matched to an exterior time depended vacuum field might be important since it can be well stated and might give us hope to obtain exact solutions satisfying matching conditions with parameters that can help us to understand the mechanism of the eventual production of gravitational waves.

For the exterior of a cylindrically anisotropic perfect fluid under gravitational collapse one considers the Beck–Einstein–Rosen metric, [14, 15], for vacuum (7.2) given by

$$ds^2 = e^{2(\gamma-\psi)}(dT^2 - dR^2) - e^{2\psi}dZ^2 - e^{-2\psi}R^2d\phi^2, \tag{7.89}$$

where γ and ψ are functions of T and R, and vacuum field equations become

$$\psi_{,TT} - \psi_{,RR} - \frac{\psi_{,R}}{R} = 0, \tag{7.90}$$

and

$$\gamma_{,T} = 2R\psi_{,T}\psi_{,R}, \quad \gamma_{,R} = R(\psi_{,T}^2 + \psi_{,R}^2). \tag{7.91}$$

Equation (7.90) is the cylindrically symmetric wave equation in an Euclidean spacetime, suggesting the presence of a gravitational wave field. All the set of equations that (7.89) has to satisfy in order to be matched, following Darmois' conditions, to an anisotropic perfect fluid under gravitational collapse are given in [16]. There it is demonstrated that for a shear free cylindrically symmetric anisotropic fluids if the exterior spacetime is static, i.e. the LC spacetime, the cylindrical source must be static too. However, if the fluid is isotropic and shear free, only a Robertson-Walker dust interior is possible. It is concluded as well that there is no energy transport in its exterior.

It is still an open question the more general behaviour of a cylindrically collapsing anisotropic fluid producing shear.

Further studies with collapsing dust have been undertaken by considering nonzero shear but with zero expansion [53] and other simplifying assumptions see references in [2, 53].

7.7 Conclusions

With the beginning of the era of the gravitational wave astronomy, the strong gravitational field regime will be soon explored observationally in various aspects. Theoretical studies can be carried out analytically and/or numerically. In the former, due to the complexity of Einstein's field equations symmetries of spacetimes are often imposed such as spherical, plane and cylindrical [54]. Although they are all ideal models, and in realistic situations any of these may not exist, they do provide solvable problems from which some fundamental issues of physics can be addressed. A good example is the Schwarzschild solution which plays a central role in the studies of black hole physics.

In this brief review concerning different cylindrically symmetrical perfect fluid sources, static, stationary, translating and collapsing, we present what is so far achieved in understanding the physics and geometry. We hope that this brief adventure into cylindrical systems in General Relativity helps to justify its importance and motivate further studies on unsolved issues like the ones that we called the attention in this review.

References

1. K.A. Bronnikov, N.O. Santos, A. Wang, Class. Quantum Grav. **37**, 113002 (2020)
2. I. Brito, M.F.A. da Silva, F.C. Mena, N.O. Santos, Class. Quantum Grav. **32**, 185015 (2015)
3. T. Levi-Civita, Rend. Acad. Lincei **28**, 101 (1915)
4. L. Marder, Proc. Roy. Soc. A **244**, 524 (1958)
5. W.B. Bonnor, J. Phys. A **12**, 843 (1979)
6. M. Fujimoto, Astron. Soc. Japan **15**, 107 (1963)
7. R.W. Hockney, Astrophys. J. **150**, 797 (1967)
8. S. Song, C.W. Jung, S.Y. Kim, Y. Choi, Astrophys. Space Sci. **193**, 217 (1992)
9. L. Herrera, N.O. Santos, A.F.F. Teixeira, A.Z. Wang, Class. Quantum Grav. **18**, 3847 (2001)
10. C. Lanczos, Z. Physik **21**, 73 (1924)
11. T. Lewis, Proc. Roy. Soc. A **136**, 176 (1932)
12. M.A.H. MacCallum, N.O. Santos, Class. Quantum Grav. **15**, 1627 (1998)
13. J. Griffiths, N.O. Santos, Int. J. Mod. Phys. D **19**, 79 (2010)
14. G. Beck, Z. Physik **33**, 713 (1925)
15. A. Einstein, R. Rosen, J. Franklin Inst. **223**, 43 (1937)
16. A. di Prisco, L. Herrera, M.A.H. MacCallum, N.O. Santos, Phys. Rev. D **80**, 064031 (2009)
17. J.S. Dowker, Nuovo Cimento B **52**, 129 (1967)
18. M.B. Hindmarsh, T.W.B. Kibble, Rep. Prog. Phys. **58**, 477 (1995)
19. M.F.A. da Silva, A. Wang, N.O. Santos, Phys. Lett. A **244**, 462 (1998)
20. W. Rindler, *Essential Relativity*, 2nd edn. (Springer, Berlin, 1977), pp.156–164
21. M.F.A. da Silva, L. Herrera, F.M. Paiva, N.O. Santos, Gen. Rel. Grav. **27**, 859 (1995)
22. L. Herrera, F.M. Paiva, N.O. Santos, Int. J. Mod. Phys. D **9**, 649 (2000)
23. H. Weyl, Ann. Physik **59**(364), 185 (1919)
24. J.L. Synge, *Relativity: The General Theory* (North-Holland Publishing Company, Amsterdam, 1960), p.313
25. R. Bach, H. Weyl, Math. Zeits. **13**, 134 (1922)
26. G. Erez, N. Rosen, Bull. Res. Counc. Israel **8F**, 47 (1959)

27. L. Herrera, F.M. Paiva, N.O. Santos, J. Math. Phys. **40**, 4064 (1999)
28. L. Herrera, J. Ruifernández, N.O. Santos, Gen. Rel. Grav. **33**, 515 (2001)
29. E.T. Whittaker, Proc. Roy. Soc. Lond. A **149**, 384 (1935)
30. E. Darmois, *Mémorial des sciences mathématiques Facicule 25 Gauttier-Villars* (Paris, 1927)
31. L. Herrera, G. Le Denmat, G. Marcilhacy, N.O. Santos, J. Mod. Phys. D **14**, 657 (2005)
32. L. Herrera, A. Di Prisco, J. Ospino, E. Fuenmayor, J. Math. Phys. **42**, 2129 (2001)
33. J. Gariel, G. Marcilhacy, N.O. Santos, J. Math. Phys. **41**, 7535 (2000)
34. J. Stachel, Phys. Rev. D **26**, 1281 (1982)
35. B. Jensen, J. Kucera, J. Math. Phys. **34**, 4975 (1993)
36. L. Herrera, N.O. Santos, J. Math. Phys. **39**, 3817 (1998)
37. L. Herrera, N.O. Santos, Astrophys. Space Sci. **310**, 251 (2007)
38. W.B. Bonnor, Class. Quantum Grav. **12**, 1483 (1995)
39. J. Gariel, G. Marcilhacy, N.O. Snatos, Astrophys. J. **774**, 109 (2013)
40. J. Gariel, N.O. Santos, A. Wang, Gen. Rel. Grav. **49**, 43 (2017)
41. J.A. Pacheco, J. Gariel, G. Marcilhacy, N.O. Santos, Astrophys. J. **759**, 125 (2012)
42. W.B. Bonnor, J. Phys. A: Maths. Gen. **13**, 2121 (1980)
43. W.J. van Stockum, Proc. Roy. Soc. Edinburgh A **57**, 135 (1937)
44. R. Opher, N.O. Santos, A. Wang, J. Math. Phys. **37**, 1982 (1996)
45. L.F. Costa, J. Natário, N.O. Santos, Class. Quantum Grav. **38**, 055003 (2021)
46. W.B. Bonnor, B.R. Steadman, Gen. Rel. Grav. **41**, 1381 (2009)
47. M.-N. Célérier, N.O. Santos, Phys. Rev. D **102**, 044026 (2020)
48. F. Debbasch, L. Herrera, P.R.C.T. Pereira, N.O. Santos, Gen. Rel. Grav. **38**, 1825 (2006)
49. C. Rovelli, *Reality Is Not What It Seems* (Penguin Random House, UK, 2017)
50. A. Einstein, Sitz. Preuss. Akad. Wis 844 (1915)
51. J.B. Griffiths, J. Podolský, *Exact Space-times in Einstein's General Relativity*. Cambridge Monographs on Mathematical Physics (Cambridge University Press, 2009)
52. K. Schwarzschild, Sitz. Preuss. Akad. Wiss. 180 (1916)
53. I. Brito, M.F.A. da Silva, F.C. Mena, N.O. Santos, Class. Quantum Grav. **34**, 205005 (2017)
54. H. Stephani, D. Kramer, M.A.H. MacCallum, C. Hoenselaers, E. Herlt, *Exact Solutions of Einstein's Field Equations*. Cambridge Monographs on Mathematical Physics (Cambridge University Press, 2009)

Chapter 8
Complexity and Simplicity of Self-Gravitating Fluids

L. Herrera

8.1 Introduction

In recent decades many efforts have been made towards a rigorous definition of complexity in different branches of science (see [5, 19, 21, 26, 30, 35, 81, 84, 86, 94, 95, 100] and references therein). However, despite all the work done so far, there is not yet a consensus on a precise definition.

The reason behind such interest stems from the fact that at least at an intuitive level, complexity, no matter how we define it, is a physical concept deeply intertwined with fundamental aspects of the system. In other words, we expect that a suitable definition of complexity of the system could allow us to infer relevant conclusions about its behavior.

Therefore, it is of utmost relevance to provide a precise definition of an observable quantity which allows measurement of such an important property of the system. Thus, when dealing with a situation that intuitively is judged as "complex", we have to be able to quantify this complexity by defining an observable measuring it.

Among the many definitions that have been proposed so far, most of them resort to concepts such as information and entropy (see for example [21, 86]), and are based on the intuitive idea that complexity should, somehow, measure a basic property related to the structural characteristics of the system.

This Chapter is devoted to the review of the concept of complexity introduced in [54] for self-gravitating systems, and its applications under variety of circumstances.

An extension of the definition of complexity based on the work developed by López-Ruiz and collaborators [21, 86] has already been proposed for self-gravitating systems in [22, 27, 28, 101].

L. Herrera (✉)
Instituto Universitario de Física Fundamental y Matemáticas, Universidad de Salamanca, Salamanca, Spain
e-mail: lherrera@usal.es

D. Malafarina and P. S. Joshi (eds.), *New Frontiers in Gravitational Collapse and Spacetime Singularities*, Springer Series in Astrophysics and Cosmology,
https://doi.org/10.1007/978-981-97-1172-7_8

However, such a poposal suffers from two drawbacks, the most important of which is the fact that it only involves the energy density, ignoring the role of stresses. This motivated the introduction of a quite different definition which was proposed in [54] for the static spherically symmetric case, and extended further in [55] to the general full dynamic case.

The definition given in [54], although intuitively associated with the very concept of "structure" within the fluid distribution, is not related (at least directly) to information or disequilibrium; rather it stems from the basic assumption that the simplest system (or at least one of them) is represented by the homogeneous fluid with isotropic pressure. Having assumed this conjecture for a vanishing complexity system, the very definition of complexity emerges in the development of the fundamental theory of self-gravitating compact objects, in the context of general relativity.

The variable responsible for measuring complexity, which we call the complexity factor, appears in the orthogonal splitting of the Riemann tensor, and the justification for such a proposition, roughly speaking, is as follows (details are given in the next section).

Once the point of view that for a static fluid distribution, the simplest system is represented by a homogeneous (in the energy density), locally isotropic fluid (principal stresses equal), then it is reasonable to assign zero value of the complexity factor to such a distribution. Next, let us recall the concept of Tolman mass [117], which may be interpreted as the "active" gravitational mass, and may be expressed, for an arbitrary distribution, through its value for the zero-complexity case plus two terms depending on the energy density inhomogeneity and pressure anisotropy, respectively. These latter terms in turn may be expressed through a single scalar function that we call the complexity factor. It obviously vanishes when the fluid is homogeneous in the energy density, and isotropic in pressure, but also may vanish when the two terms containing density inhomogeneity and anisotropic pressure cancel each other out. Thus, as in [86], vanishing complexity may correspond to very different systems.

When dealing with time-dependent systems, we face two different problems; on the one hand, we have to generalize the concept of complexity of the structure of the fluid distribution to time-dependent dissipative fluids, and on the other hand we also have to evaluate the complexity of the patterns of evolution and propose what we consider to be the simplest of them.

In [55] it was shown that is reasonable to assume that the complexity factor for the structure of the fluid distribution is the same scalar function as for the static case, which now includes the dissipative variables. As for the simplest pattern of evolution, it was shown that the homologous condition characterizes the simplest possible mode. However, as was shown later on, it may be useful to relax this last condition to enlarge the set of possible solutions, by adopting the so-called quasi-homologous condition [60].

The axially symmetric static case has been considered in [58], while some particular cases of cylindrically symmetric fluid distributions have been studied in [103, 104]. Further applications of the concept of complexity as defined in [54] may be found in [20, 61, 76, 105, 109]. Always within the context of general relativity, exact solutions for static fluid distributions endowed with hyperbolical symmetry and satisfying the condition of minimal complexity were presented in [24, 61, 136], while dynamic solutions endowed with hyperbolical symmetry and satisfying the condition of minimal complexity were obtained in [62], such solutions evolve in the so-called quasi-homologous regime.

The concept of complexity as defined in [54] has also been extended to other theories of gravity in [1–4, 93, 106–108, 110, 120–128, 134, 135]. More recent developments of this subject within a variety of contexts may be found in [6–10, 13–15, 23, 29, 34, 36, 79, 82, 83, 89, 90, 96, 97, 99, 111–114, 129–132, 137] and references therein.

All the above-mentioned cases concern fluid distributions (which eventually may be charged); however, the vacuum case has barely been treated. The only known example so far is the extension of the complexity factor as defined in [54] to the vacuum solutions of the Einstein equations represented by the Bondi metric [59]. A complexity hierarchy may be established in this case, ranging from the Minkowski spacetime (the simplest one) to gravitationally radiating systems (the most complex).

8.2 Defining Complexity for Static Self–gravitating Systems

Let us start by introducing the notion of complexity of a self–gravitating fluid as represented by a variable (more than one for non–spherical configurations) aimed to measure the degree of "complexity" of the fluid distribution. At this level of generality, due to the vagueness of the concept of "complexity", the problem appears to be quite difficult to solve, accordingly let's consider instead a much simpler question, namely: what is the simplest fluid configuration we can envisage?

In order to answer to the above question, let us first restrict the discussion to static and spherically symmetric distributions. Under these conditions, the fluid is described by the energy density and the pressure. The simplest energy density distribution corresponds to an incompressible fluid (a constant energy density) whereas the simplest pressure distribution corresponds to an isotropic fluid (principal stresses equal). Indeed, it should be noticed that since for a bounded configuration the pressure vanishes at the boundary surface, it cannot be homogeneous.

Thus, based on these very simple considerations we shall assume that the simplest structure (at least one of them) of a static spherically symmetric fluid corresponds to a homogeneous (in the energy density) and isotropic (in the pressure) fluid. We shall now see how this assumption leads us to a reliable definition of complexity.

The Static Spherically Symmetric Case

We consider spherically symmetric distributions of static fluid, which for the sake of completeness we assume to be locally anisotropic and bounded by a spherical surface Σ.

The line element is given in Schwarzschild–like coordinates by

$$ds^2 = e^\nu dt^2 - e^\lambda dr^2 - r^2\left(d\theta^2 + \sin^2\theta d\phi^2\right), \tag{8.1}$$

where $\nu(r)$ and $\lambda(r)$ are functions of their arguments. We number the coordinates: $x^0 = t;\ x^1 = r;\ x^2 = \theta;\ x^3 = \phi$.

The energy–momentum tensor corresponding to our fluid distribution may be written as

$$T^\mu_\nu = \mu u^\mu u_\nu - P h^\mu_\nu + \Pi^\mu_\nu, \tag{8.2}$$

with

$$\Pi^\mu_\nu = \Pi\left(s^\mu s_\nu + \frac{1}{3}h^\mu_\nu\right); \quad P = \frac{P_r + 2P_\perp}{3}$$
$$\Pi = P_r - P_\perp; \quad h^\mu_\nu = \delta^\mu_\nu - u^\mu u_\nu, \tag{8.3}$$

where μ, P_r, $P_\perp$, $\Pi_{\mu\nu}$, u^μ denote the energy density, the radial pressure, the tangential pressure, the anisotropic tensor and the four–velocity respectively.

The vector s^μ is defined by

$$s^\mu = (0, e^{-\frac{\lambda}{2}}, 0, 0), \tag{8.4}$$

whereas the four–velocity vector is given by:

$$u^\mu = (e^{-\frac{\nu}{2}}, 0, 0, 0), \tag{8.5}$$

with the properties $s^\mu u_\mu = 0$, $s^\mu s_\mu = -1$.

From (8.5) we can calculate the four acceleration, $a^\alpha = u^\alpha_{;\beta}u^\beta$, whose only non–vanishing component is:

$$a_1 = -\frac{\nu'}{2}, \tag{8.6}$$

where prime denotes derivative with respect to r.

At the exterior of the fluid distribution the spacetime is described by the Schwarzschild metric

$$ds^2 = \left(1 - \frac{2M}{r}\right)dt^2 - \frac{dr^2}{\left(1 - \frac{2M}{r}\right)} - r^2\left(d\theta^2 + \sin^2\theta d\phi^2\right). \tag{8.7}$$

In order to match smoothly the two metrics above on the boundary surface $r = r_\Sigma = constant$, we require the continuity of the first and the second fundamental forms across that surface, producing

$$e^{\nu_\Sigma} = 1 - \frac{2M}{r_\Sigma}, \tag{8.8}$$

$$e^{-\lambda_\Sigma} = 1 - \frac{2M}{r_\Sigma}, \tag{8.9}$$

$$[P_r]_\Sigma = 0, \tag{8.10}$$

where, from now on, subscript Σ indicates that the quantity is evaluated on the boundary surface Σ.

Equations (8.8), (8.9), and (8.10) are the necessary and sufficient conditions for a smooth matching of the two metrics (8.1) and (8.7) on Σ.

Next, let us recall that the Tolman mass for a spherically symmetric static distribution of matter is given by [117]

$$m_T = 4\pi \int_0^{r_\Sigma} r^2 e^{(\nu+\lambda)/2} \left(T_0^0 - T_1^1 - 2\,T_2^2\right) dr, \tag{8.11}$$

which we shall extend for any sphere of radius r, completely inside Σ, as

$$m_T = 4\pi \int_0^{r} \tilde{r}^2 e^{(\nu+\lambda)/2} \left(T_0^0 - T_1^1 - 2\,T_2^2\right) d\tilde{r}.$$

This extension of the global concept of energy to a local level is suggested by the conspicuous role played by m_T as the "active gravitational mass".

Indeed, as it follows from (8.6), the gravitational acceleration ($a = -s^\nu a_\nu$) of a test particle, instantaneously at rest in a static gravitational field, is given by

$$a = \frac{e^{-\lambda/2}\,\nu'}{2} = \frac{e^{-\nu/2} m_T}{r^2}. \tag{8.12}$$

Another expression for m_T, which appears to be more suitable for the discussion is (see [39, 40] for details, but notice slight changes in notation)

$$m_T = (m_T)_\Sigma \left(\frac{r}{r_\Sigma}\right)^3 - r^3 \int_r^{r_\Sigma} e^{(\nu+\lambda)/2} \left[\frac{8\pi}{\tilde{r}} (P_\perp - P_r) + \frac{1}{\tilde{r}^4} \int_0^{\tilde{r}} 4\pi \tilde{r}^3 \mu' d\tilde{r}\right] d\tilde{r}, \tag{8.13}$$

The important point to keep in mind here is that the second integral in (8.13) describes the contribution of density inhomogeneity and local anisotropy of pressure to the Tolman mass.

Next, using the orthogonal splitting of the Riemann tensor first considered by Bel [11] (see also [31]), let us introduce the tensor

$$Y_{\alpha\beta} = R_{\alpha\gamma\beta\delta} u^{\gamma} u^{\delta}, \tag{8.14}$$

which can be splitted in terms of its trace and the corresponding trace–free tensor, as (see [45] for details)

$$TrY \equiv Y_T = 4\pi(\rho + 3P_r - 2\Pi), \tag{8.15}$$

and

$$Y_{<\alpha\beta>} = Y_{TF}(s_\alpha s_\beta + \frac{h_{\alpha\beta}}{3}), \tag{8.16}$$

with

$$Y_{TF} \equiv (4\pi\Pi + E) = 8\pi\Pi - \frac{4\pi}{r^3}\int_0^r \tilde{r}^3 \mu' d\tilde{r}, \tag{8.17}$$

where E is the scalar defining the electric part of the Weyl tensor $E_{\alpha\beta} = C_{\alpha\gamma\beta\delta} u^{\gamma} u^{\delta}$, (the magnetic part vanishes identically), which may be written as

$$E_{\alpha\beta} = E(s_\alpha s_\beta + \frac{1}{3} h_{\alpha\beta}), \tag{8.18}$$

with

$$E = -\frac{e^{-\lambda}}{4}\left[\nu'' + \frac{\nu'^2 - \lambda'\nu'}{2} - \frac{\nu' - \lambda'}{r} + \frac{2(1 - e^{\lambda})}{r^2}\right], \tag{8.19}$$

satisfying the following properties:

$$E^{\alpha}_{\ \alpha} = 0, \quad E_{\alpha\gamma} = E_{(\alpha\gamma)}, \quad E_{\alpha\gamma} u^{\gamma} = 0. \tag{8.20}$$

Using (8.17) in (8.13) we get

$$m_T = (m_T)_\Sigma \left(\frac{r}{r_\Sigma}\right)^3 + r^3 \int_r^{r_\Sigma} \frac{e^{(\nu+\lambda)/2}}{\tilde{r}} Y_{TF} d\tilde{r}. \tag{8.21}$$

Thus we see that this single scalar function, Y_{TF}, encompasses all the modifications produced by the energy density inhomogeneity and the anisotropy of the pressure, on the active gravitational (Tolman) mass. More specifically, it describes how these two factors modify the value of the Tolman mass, with respect to its value for the homogeneous isotropic fluid.

Then, following our starting conjecture stating that at least one of the simplest configurations corresponds to an incompressible fluid (a constant energy density) with isotropic pressure, and noticing that for such a system, the scalar Y_{TF} vanishes, it appears well justified to identify the complexity factor with Y_{TF}.

The following remarks are in order at this point:

- The scalar Y_{TF} belongs to a family of functions known as structure scalars, defined in [45] from a detailed analysis of the orthogonal splitting of the Riemann tensor.
- As is apparent from (8.12) the Toman mass is a measure of the strength of the gravitational interaction.
- The complexity factor defined as above, not only vanishes for the homogeneous, isotropic fluid, where the two terms in (8.17) vanish identically, but also for all configurations where the two terms in (8.17) cancel each other, implying that there are a wealth of configurations satisfying the vanishing complexity condition.
- It is worth noticing that whereas the contribution of the pressure anisotropy to Y_{TF} is local, the contribution of the density energy inhomogeneity is not.
- For a charged fluid Y_{TF} includes contributions from the electric charge as well (see Eq. (25) in [48]).

In the next section, we shall present two examples of inhomogeneous and anisotropic fluid configurations, satisfying the vanishing complexity factor condition.

8.2.1 *Fluid Distributions with Vanishing Complexity Factor*

The vanishing complexity factor condition, according to (8.17) reads

$$\Pi = \frac{1}{2r^3}\int_0^r \tilde{r}^3 \mu' d\tilde{r}. \tag{8.22}$$

However, since the Einstein equations for a spherically symmetric static, anisotropic fluid form a system of three ordinary differential equations for five unknown functions $(\nu, \lambda, \mu, P_r, P_\perp)$, the condition $Y_{TF} = 0$ is not enough to close the system and we need still one condition in order to solve it. Just for the sake of illustration, we shall propose two examples in the next subsections.

8.2.1.1 The Gokhroo and Mehra Ansatz

A family of anisotropic spheres has been found in [33], which leads to physically satisfactory models for compact objects.

These models are obtained from an ansatz on the form of the metric function λ which reads

$$e^{-\lambda} = 1 - \alpha r^2 + \frac{3K\alpha r^4}{5r_\Sigma^2}, \tag{8.23}$$

producing,

$$\mu = \mu_0\left(1 - \frac{Kr^2}{r_\Sigma^2}\right), \tag{8.24}$$

and

$$m(r) = \frac{4\pi\mu_0 r^3}{3}\left(1 - \frac{3Kr^2}{5r_\Sigma^2}\right), \tag{8.25}$$

where K is a constant in the range (0, 1), $\alpha = \frac{8\pi\mu_0}{3}$, and $m(r)$ is the mass function, defined by

$$1 - e^{-\lambda} = \frac{2m}{r}, \tag{8.26}$$

or, equivalently

$$m = 4\pi\int_0^r \tilde{r}^2\mu d\tilde{r}. \tag{8.27}$$

Then using the field equations it can be shown that the line element becomes (see [54] for details)

$$ds^2 = -e^{\int(2z(r)-2/r)dr}dt^2 + \frac{z^2(r)e^{\int(\frac{4}{r^2z(r)}+2z(r))dr}}{\Omega}dr^2$$
$$+ r^2d\theta^2 + r^2sin^2\theta d\phi^2, \tag{8.28}$$

where C is a constant of integration, $\Omega \equiv r^6(-2\int\frac{z(r)(1+\frac{\Pi(r)r^2}{8\pi})e^{\int(\frac{4}{r^2z(r)}+2z(r))dr}}{r^8}dr + C)$, and

$$e^{\nu(r)} = e^{\int(2z(r)-2/r)dr}. \tag{8.29}$$

For the physical variables we have

$$4\pi P_r = \frac{z(r-2m) + m/r - 1}{r^2}, \tag{8.30}$$

$$4\pi\mu = \frac{m'}{r^2}, \tag{8.31}$$

and

$$4\pi P_\perp = \left(1 - \frac{2m}{r}\right)\left(z' + z^2 - \frac{z}{r} + \frac{1}{r^2}\right) + z\left(\frac{m}{r^2} - \frac{m'}{r}\right). \tag{8.32}$$

The so obtained solution is regular at the origin, and satisfies the conditions $\mu > 0$, $\mu > P_r, P_\perp$.

Also, to avoid singular behaviour of physical variables on the boundary of the source (Σ), the solution should satisfy the Darmois conditions on the boundary (8.8), (8.9), (8.10).

8.2.1.2 The Polytrope with Vanishing Complexity Factor

The polytropic equation of state plays an important role in the study of self–gravitating systems, both, in Newtonian and general relativistic astrophysics. In the isotropic pressure case such an equation of state is sufficient to integrate the field equations. However in case of an anisotropic fluid additional information is required. The study of polytropes for anisotropic matter has been considered in detail in [50, 51, 53].

Once the polytropic equation of state is assumed, in the case of anisotropic matter, we still need an additional condition in order to solve the corresponding system of equations. We propose to assume the vanishing complexity factor condition as such a complementary information.

Thus our model is characterized by

$$P_r = K\mu^\gamma = K\mu^{1+1/n}; \qquad Y_{TF} = 0, \tag{8.33}$$

where constants K, γ, and n are usually called polytropic constant, polytropic exponent, and polytropic index, respectively.

It is worth mentioning that the generalization of the Newtonian polytrope to the general relativistic case admits two possibilities. One is the Eq. (8.33), the other is

$$P_r = K\mu_b^\gamma = K\mu_b^{1+1/n} \tag{8.34}$$

where μ_b denotes the baryonic (rest) mass density. The treatment of this latter case has been described in detail in [51].

From the polytropic equation of state (8.33) we obtain two equations which read:

$$\xi^2 \frac{d\Psi}{d\xi}\left[\frac{1-2(n+1)\alpha v/\xi}{1+\alpha\Psi}\right] + v + \alpha\xi^3\Psi^{n+1} + \frac{2\Pi\Psi^{-n}\xi}{P_{rc}(n+1)}\left[\frac{1-2\alpha(n+1)v/\xi}{1+\alpha\Psi}\right] = 0, \tag{8.35}$$

and

$$\frac{dv}{d\xi} = \xi^2\Psi^n, \tag{8.36}$$

where

$$\alpha = P_{rc}/\mu_c, \quad r = \xi/A, \quad A^2 = 4\pi\mu_c/\alpha(n+1), \tag{8.37}$$

$$\Psi^n = \mu/\mu_c, \quad v(\xi) = m(r)A^3/(4\pi\mu_c), \tag{8.38}$$

where subscript c indicates that the quantity is evaluated at the center. At the boundary surface $r = r_\Sigma$ ($\xi = \xi_\Sigma$) we have $\Psi(\xi_\Sigma) = 0$ (see [53] for details).

Equations (8.35), (8.36), form a system of two first order ordinary differential equations for the three unknown functions: Ψ, v, Π, depending on a duplet of parameters n, α. In order to proceed further with the modeling of a compact object, we shall assume the vanishing complexity factor condition, which with the notation above, reads

$$\frac{6\Pi}{n\mu_c} + \frac{2\xi}{n\mu_c}\frac{d\Pi}{d\xi} = \Psi^{n-1}\xi\frac{d\Psi}{d\xi}. \tag{8.39}$$

Now we have a system of three ordinary differential equations (8.35), (8.36), (8.39) for the three unknown functions Ψ, v, Π, which may be integrated for any arbitrary duplet of values of the parameters n, α, only constrained by the physical conditions (see [51] for details)

$$\mu > 0, \qquad \alpha\Psi \leq 1, \qquad \frac{3v}{\xi^3\Psi^n} + \alpha\Psi - 1 \leq 1. \tag{8.40}$$

The equations equivalent to (8.35) and (8.39), for the equation of state (8.34) read

$$\xi^2\frac{d\Psi_b}{d\xi}\left[\frac{1-2(n+1)\alpha v/\xi}{1+\alpha\Psi_b}\right] + v + \alpha\xi^3\Psi_b^{n+1} + \frac{2\Pi\Psi_b^{-n}\xi}{P_{rc}(n+1)}\left[\frac{1-2\alpha(n+1)v/\xi}{1+\alpha\Psi_b}\right] = 0, \tag{8.41}$$

$$\frac{6\Pi}{n\mu_{bc}} + \frac{2\xi}{n\mu_{bc}}\frac{d\Pi}{d\xi} = \Psi_b^{n-1}\xi\frac{d\Psi_b}{d\xi}\left[1 + K(n+1)\mu_{bc}^{1/n}\Psi_b\right], \tag{8.42}$$

with $\Psi_b^n = \mu_b/\mu_{bc}$.

Further research on the application of the concept of complexity in the study of polytropes may be found in [75–79, 82].

So far we have been concerned with spherically symmetric systems, we shall next consider the axially symmetric (static) case.

The Static Axially–Symmetric Case

Let us now extend the concept of complexity as defined for the spherically symmetric case to the most general axially symmetric static fluid distributions.

The reason to undertake such a task is based on the well known fact that (close to the horizon) there is a bifurcation between any finite perturbation of Schwarzschild spacetime and any Weyl solution, even when the latter is characterized by parameters arbitrarily close to those corresponding to spherical symmetry (see [12, 25, 43, 44, 73, 119] and references therein for a discussion on this point). This fact in turn is related to the well known result, usually referred to as Israel theorem [66], stating that the only regular static and asymptotically flat vacuum spacetime possessing a regular horizon is the Schwarzchild solution, while all the others Weyl exterior solutions exhibit singularities in the curvature invariants (as the boundary of the source approaches the horizon).

Thus, even though observational evidence suggest that deviations from spherical symmetry in compact self-gravitating objects (white dwarfs, neutron stars), are likely to be incidental rather than basic features of these systems, it is clear that for very compact objects, deviations from spherical symmetry (no matter how small) should be studied resorting to exact solution of the Einstein equations, instead as perturbations of spherically symmetric systems.

As shown in the previous sections, the scalar intended to measure the degree of complexity for a spherically symmetric fluid distribution (the complexity factor), is identified as one of the scalar functions (structure scalars) which appears in the orthogonal splitting of the Riemann tensor. More specifically, it is related to one of the scalar functions appearing in the splitting of the electric part of the Riemann tensor.

In spite of the fact that in the axially symmetric case the situation is much more complicated, and the number of structure scalars is larger than in the spherically symmetric case, we shall proceed in a similar way to fix the general criterium to define the variable(s) measuring the complexity of the fluid distribution.

Thus, we start by asking ourselves the same question as in the previous case, namely: which is the simplest fluid configuration? As in the spherically symmetric case we shall assume that such a configuration corresponds to the incompressible (constant energy density), isotropic (in the pressure) spheroid. From this simple assumption, we shall see that as the obvious candidates to measure the degree of complexity of the fluid distribution, appear three of the eight structure scalars corresponding to the axially symmetric static fluid distribution. Explicit forms of these structure scalars as well as some useful differential equations relating the inhomogeneities of the energy density to some of the structure scalars were already found in [52].

As in the spherically symmetric case, the vanishing of the three complexity factors corresponds not only to the incompressible, isotropic spheroid, but also to a large family of solutions where the density inhomogeneity terms cancel the pressure anisotropic terms in the equations relating these variables to the complexity factors. Some of these solutions will be exhibited.

Thus, let us consider static and axially symmetric sources. For such a system the line element may be written in "Weyl spherical coordinates" (please notice that in this section we are using signature +2 instead of −2 as in the previous case, which leads to some changes in the sign of some variables), as

$$ds^2 = -A^2 dt^2 + B^2 \left(dr^2 + r^2 d\theta^2\right) + D^2 d\phi^2, \tag{8.43}$$

where the coordinates t and ϕ are adapted to the two Killing vectors admitted by our line element, and therefore the metric functions depend only on r and θ.

We recall that, unlike the vacuum case, the assumption of the Weyl gauge ($R^3_3 + R^0_0 = 0$, where R^α_β denotes the Ricci tensor), reducing the number of independent metric functions to two, cannot be used without loss of generality in the interior, which explains why our line element is described in terms of three independent functions.

In a purely locally Minkowski frame (hereafter referred to as l.M.f.) where the first derivatives of the metric vanish (locally) [17], the most general energy–momentum tensor is given by:

$$\widehat{T}_{\alpha\beta} = \begin{pmatrix} \mu & 0 & 0 & 0 \\ 0 & P_{xx} & P_{xy} & 0 \\ 0 & P_{yx} & P_{yy} & 0 \\ 0 & 0 & 0 & P_{zz} \end{pmatrix}, \tag{8.44}$$

where μ, P_{xy}, P_{xx}, P_{yy}, P_{zz} denote the energy density and different stresses, respectively, as measured by our locally defined Minkowskian observer.

Also observe that $P_{xy} = P_{yx}$ and, in general $P_{xx} \neq P_{yy} \neq P_{zz}$.

Then transforming back to our coordinates, we obtain the components of the energy momentum tensor in terms of the physical variables as defined in the l.M.f.

$$\begin{aligned} T_{\alpha\beta} &= (\mu + P_{zz}) V_\alpha V_\beta + P_{zz} g_{\alpha\beta} + (P_{xx} - P_{zz}) K_\alpha K_\beta \\ &+ (P_{yy} - P_{zz}) L_\alpha L_\beta + 2P_{xy} K_{(\alpha} L_{\beta)}, \end{aligned} \tag{8.45}$$

with

$$\begin{aligned} V_\alpha &= (-A, 0, 0, 0); \quad K_\alpha = (0, B, 0, 0); \\ L_\alpha &= (0, 0, Br, 0); \quad S_\alpha = (0, 0, 0, D), \end{aligned} \tag{8.46}$$

where we are considering observers at rest with respect to the fluid distribution.

Alternatively we may write the energy momentum tensor in the "canonical" form

$$T_{\alpha\beta} = (\mu + P) V_\alpha V_\beta + P g_{\alpha\beta} + \Pi_{\alpha\beta}, \tag{8.47}$$

with

$$\begin{aligned} \Pi_{\alpha\beta} &= (P_{xx} - P_{zz}) \left(K_\alpha K_\beta - \frac{h_{\alpha\beta}}{3} \right) \\ &+ (P_{yy} - P_{zz}) \left(L_\alpha L_\beta - \frac{h_{\alpha\beta}}{3} \right) + 2P_{xy} K_{(\alpha} L_{\beta)}, \end{aligned} \tag{8.48}$$

$$P = \frac{P_{xx} + P_{yy} + P_{zz}}{3}, \quad h_{\mu\nu} = g_{\mu\nu} + V_\nu V_\mu. \tag{8.49}$$

The anisotropic tensor may also be written as

$$\Pi_{\alpha\beta} = \frac{1}{3}(2\Pi_I + \Pi_{II})\left(K_\alpha K_\beta - \frac{h_{\alpha\beta}}{3}\right) + \frac{1}{3}(2\Pi_{II} + \Pi_I)\left(L_\alpha L_\beta - \frac{h_{\alpha\beta}}{3}\right) + \Pi_{KL}\left(K_\alpha L_\beta + K_\beta L_\alpha\right), \tag{8.50}$$

with

$$\Pi_{KL} = K^\alpha L^\beta T_{\alpha\beta}, \tag{8.51}$$

$$\Pi_I = \left(2K^\alpha K^\beta - L^\alpha L^\beta - S^\alpha S^\beta\right) T_{\alpha\beta}, \tag{8.52}$$

$$\Pi_{II} = \left(2L^\alpha L^\beta - K^\alpha K^\beta - S^\alpha S^\beta\right) T_{\alpha\beta}. \tag{8.53}$$

The relationships between the above scalars and the variables P_{xy}, P_{xx}, P_{yy}, P_{zz} are (besides (8.49)),

$$\Pi_2 \equiv \frac{1}{3}(2\Pi_I + \Pi_{II}) = P_{xx} - P_{zz}, \tag{8.54}$$

$$\Pi_3 \equiv \frac{1}{3}(2\Pi_{II} + \Pi_I) = P_{yy} - P_{zz}, \tag{8.55}$$

$$\Pi_{KL} = P_{xy}. \tag{8.56}$$

or, inversely:

$$P_{zz} = P - \frac{1}{3}(\Pi_2 + \Pi_3), \tag{8.57}$$

$$P_{xx} = P + \frac{1}{3}(2\Pi_2 - \Pi_3), \tag{8.58}$$

$$P_{yy} = P + \frac{1}{3}(2\Pi_3 - \Pi_2). \tag{8.59}$$

The explicit form of the Einstein equations as well as the conservation equations, for the line element (8.43) and the energy–momentum tensor (8.47), are given in the Appendix 8.6.

8.2.2 The Structure Scalars

The structure scalars for our problem were calculated in [52]. For their definition we need first to obtain the electric part of the Weyl tensor (the magnetic part vanishes identically), whose components can be obtained directly from its definition,

$$E_{\mu\nu} = C_{\mu\alpha\nu\beta}\, V^{\alpha}\, V^{\beta}, \tag{8.60}$$

where $C_{\mu\alpha\nu\beta}$ denotes the Weyl tensor.

Equivalently, the electric part of the Weyl tensor may also be written as

$$E_{\alpha\beta} = \mathscr{E}_1 \left(K_\alpha L_\beta + L_\alpha K_\beta\right) + \mathscr{E}_2 \left(K_\alpha K_\beta - \frac{1}{3}h_{\alpha\beta}\right) + \mathscr{E}_3 \left(L_\alpha L_\beta - \frac{1}{3}h_{\alpha\beta}\right), \tag{8.61}$$

where explicit expressions for the three scalars $\mathscr{E}_1$, $\mathscr{E}_2$, $\mathscr{E}_3$ are given in the Appendix 8.7.

Next, let us calculate the electric part of the Riemann tensor (the magnetic part vanishes identically), which is defined by

$$Y^{\rho}_{\beta} = V^{\alpha} V^{\mu} R^{\rho}_{\alpha\beta\mu}. \tag{8.62}$$

After some lengthy calculations we find;

$$\begin{aligned} Y_{\alpha\beta} &= Y_{TF1} \left(K_\alpha L_\beta + K_\beta L_\alpha\right) + Y_{TF2} \left(K_\alpha K_\beta - \frac{1}{3}h_{\alpha\beta}\right) \\ &+ Y_{TF3} \left(L_\alpha L_\beta - \frac{1}{3}h_{\alpha\beta}\right) + \frac{1}{3} Y_T h_{\alpha\beta}, \end{aligned} \tag{8.63}$$

where

$$Y_T = 4\pi(\mu + 3P), \tag{8.64}$$

$$Y_{TF1} = \mathscr{E}_1 - 4\pi \Pi_{KL}, \tag{8.65}$$

$$Y_{TF2} = \mathscr{E}_2 - 4\pi \Pi_2, \tag{8.66}$$

$$Y_{TF3} = \mathscr{E}_3 - 4\pi \Pi_3. \tag{8.67}$$

Finally, we shall find the tensor associated with the double dual of Riemann tensor, defined as

$$X_{\alpha\beta} = {}^{*}R^{*}_{\alpha\gamma\beta\delta} V^{\gamma} V^{\delta} = \frac{1}{2}\eta_{\alpha\gamma}{}^{\epsilon\rho} R^{*}_{\epsilon\rho\beta\delta} V^{\gamma} V^{\delta}, \tag{8.68}$$

with $R^*_{\alpha\beta\gamma\delta} = \frac{1}{2}\eta_{\epsilon\rho\gamma\delta} R_{\alpha\beta}{}^{\epsilon\rho}$, where $\eta_{\epsilon\rho\gamma\delta}$ denotes the permutation symbol.
Thus, we find

$$X_{\alpha\beta} = X_{TF1}\left(K_\alpha L_\beta + K_\beta L_\alpha\right) + X_{TF2}\left(K_\alpha K_\beta - \frac{1}{3}h_{\alpha\beta}\right) + X_{TF3}\left(L_\alpha L_\beta - \frac{1}{3}h_{\alpha\beta}\right) + \frac{1}{3}X_T h_{\alpha\beta}, \tag{8.69}$$

where

$$X_T = 8\pi\mu, \tag{8.70}$$

$$X_{TF1} = -(\mathcal{E}_1 + 4\pi \Pi_{KL}), \tag{8.71}$$

$$X_{TF2} = -\left(\mathcal{E}_2 + 4\pi \Pi_2\right), \tag{8.72}$$

$$X_{TF3} = -\left(\mathcal{E}_3 + 4\pi \Pi_3\right). \tag{8.73}$$

The scalars $Y_T, Y_{TF1}, Y_{TF2}, Y_{TF3}, X_T, X_{TF1}, X_{TF2}, X_{TF3}$, are the structure scalars for our system.

Next, we shall need two differential equations which relate the spatial derivatives of the physical variables and the Weyl tensor, obtained from Bianchi identities, they have been found before for the spherically symmetric and the cylindrically symmetric cases (see [45, 49] and references therein). For our case they have been calculated in [52], and read

$$\frac{\mathcal{E}_{1\theta}}{r} + \frac{1}{3}(2\mathcal{E}_2 - \mathcal{E}_3)' + \frac{\mathcal{E}_1}{r}\left(\frac{2B_\theta}{B} + \frac{D_\theta}{D}\right) + \mathcal{E}_2\left(\frac{B'}{B} + \frac{D'}{D} + \frac{1}{r}\right) - \mathcal{E}_3\left(\frac{B'}{B} + \frac{1}{r}\right) = \frac{4\pi}{3}(2\mu + 3P)' + 4\pi\left[\mu + P + \frac{1}{3}(2\Pi_2 - \Pi_3)\right]\frac{A'}{A} + 4\pi \Pi_{KL}\frac{A_\theta}{Ar}, \tag{8.74}$$

$$\mathcal{E}_1' + \frac{1}{3r}(2\mathcal{E}_3 - \mathcal{E}_2)_\theta + \mathcal{E}_1\left(\frac{2B'}{B} + \frac{D'}{D} + \frac{2}{r}\right) - \frac{\mathcal{E}_2 B_\theta}{Br} + \frac{\mathcal{E}_3}{r}\left(\frac{B_\theta}{B} + \frac{D_\theta}{D}\right) = \frac{4\pi}{3r}(2\mu + 3P)_\theta + 4\pi\left[\mu + P + \frac{1}{3}(2\Pi_3 - \Pi_2)\frac{A_\theta}{Ar}\right] + 4\pi \Pi_{KL}\frac{A'}{A}, \tag{8.75}$$

which, using (8.64)–(8.67) and (8.70)–(8.73), may be written in terms of structure scalars, producing

$$\frac{8\pi\mu'}{3} = \frac{1}{r}\left[Y_{TF1\theta} + 8\pi\Pi_{KL\theta} + (Y_{TF1} + 8\pi\Pi_{KL})(\ln B^2 D)_\theta\right]$$
$$+ \left[\frac{2}{3}(Y'_{TF2} + 8\pi\Pi'_2) + (Y_{TF2} + 8\pi\Pi_2)(\ln BDr)'\right]$$
$$- \left[\frac{1}{3}(Y'_{TF3} + 8\pi\Pi'_3) + (Y_{TF3} + 8\pi\Pi_3)(\ln Br)'\right], \qquad (8.76)$$

$$\frac{8\pi\mu_\theta}{3r} = -\frac{1}{r}\left[\frac{1}{3}(Y_{TF2\theta} + 8\pi\Pi_{2\theta}) + (Y_{TF2} + 8\pi\Pi_2)(\ln B)_\theta\right]$$
$$+ \frac{1}{r}\left[\frac{2}{3}(Y_{TF3\theta} + 8\pi\Pi_{3\theta}) + (Y_{TF3} + 8\pi\Pi_3)(\ln BD)_\theta\right]$$
$$+ \left[Y'_{TF1} + 8\pi\Pi'_{KL} + (Y_{TF1} + 8\pi\Pi_{KL})(\ln B^2 D r^2)'\right], \qquad (8.77)$$

where prime and subscript θ denote derivatives with respect to r and θ respectively.

We have now available all the elements necessary to identify the complexity factors for the fluid distribution under consideration. For doing so we recall our basic ansatz consisting in assuming that the simplest possible fluid (or at least one of them) is the incompressible (constant energy density) fluid with isotropic pressure.

Now, in [52] it has been shown that the necessary and sufficient conditions for the vanishing of the (invariantly defined) spatial derivatives of the energy density are $X_{TF1} = X_{TF2} = X_{TF3} = 0$. In other words

$$X_{TF1} = X_{TF2} = X_{TF3} = 0 \Leftrightarrow \mu' = \mu_\theta = 0. \qquad (8.78)$$

Therefore the homogeneous energy–density condition implies $X_{TF1} = X_{TF2} = X_{TF3} = 0$, which in turn produces

$$Y_{TF1} = -8\pi\Pi_{KL}; \quad Y_{TF2} = -8\pi\Pi_2; \quad Y_{TF3} = -8\pi\Pi_3. \qquad (8.79)$$

From the above it follows that the isotropic pressure condition would imply $Y_{TF1} = Y_{TF2} = Y_{TF3} = 0$.

In other words, following the rationale exposed in the spherically symmetric case, it is reasonable to identify the three structure scalars Y_{TF} (more precisely, their absolute values) with the complexity factors. As in the previous case, we notice that they vanish for the incompressible (constant energy density) fluid with isotropic pressure, but may also vanish for inhomogeneous, anisotropic fluids, provided these two factors combine in such a way that they cancel the three complexity factors.

In the next subsections, just to illustrate the way by means of which such models may be obtained, we shall present two solutions with vanishing complexity factors.

8.2.3 The Incompressible, Isotropic Spheroid

The first solution we shall present corresponds to the case where the complexity factors vanish because the energy density is homogeneous and the pressure is isotropic. This solution was previously obtained and analyzed in [52]. Here we just present it without details.

Let us first notice that, from (8.78), (8.71)–(8.73) and $P_{xx} = P_{yy} = P_{zz} = P$, $P_{xy} = 0$, $\mu = \mu_0 = constant$, it follows that such a solution is also conformally flat.

Next, for simplicity we shall assume the boundary surface Σ to be defined by the equation:

$$r = r_1 = constant, \tag{8.80}$$

which is not the most general form of a possible boundary surface.

Then, from the above and (8.228) and (8.231) it follows that

$$P \stackrel{\Sigma}{=} 0, \tag{8.81}$$

where $\stackrel{\Sigma}{=}$ means that both sides of the equation are evaluated on Σ

Under the conditions above, (8.233) and (8.234) can be integrated to obtain:

$$P + \mu_0 = \frac{\zeta}{A}, \tag{8.82}$$

and

$$P + \mu_0 = \frac{\xi(r)}{A}, \tag{8.83}$$

where ξ is an arbitrary function of its argument. Using boundary conditions (8.81) in (8.82) (8.83) it follows that:

$$A(r_1, \theta) = const. = \frac{\alpha}{\mu_0}, \qquad \zeta = constant. \tag{8.84}$$

Finally, the metric for this model can be written as follows

$$ds^2 = \frac{1}{(\gamma r^2 + \delta + br\cos\theta)^2}\left[-(\alpha r^2 + \beta + ar\cos\theta)^2 dt^2 + dr^2 + r^2 d\Omega^2\right], \tag{8.85}$$

from which, the physical variables can be easily calculated, producing

$$8\pi\mu = 12\gamma\delta - 3b^2, \tag{8.86}$$

$$8\pi P = (3b^2 - 12\gamma\delta)\left[1 - \frac{\alpha r_1^2 + \beta}{\gamma r_1^2 + \delta}\frac{\gamma r^2 + \delta + br\cos\theta}{\alpha r^2 + \beta + ar\cos\theta}\right], \tag{8.87}$$

where b, γ, δ are constants, and

$$\zeta = \mu_0 \frac{\alpha r_1^2 + \beta}{\gamma r_1^2 + \delta}, \quad a = \frac{\alpha r_1^2 + \beta}{\gamma r_1^2 + \delta} b, \tag{8.88}$$

in order to satisfy the junction condition (8.81).

It is important to stress the fact that this solution cannot be matched to any Weyl exterior, except in the spherically symmetric case, even though it has a surface of vanishing pressure (see [52] for details). As shown in [52] this result is a consequence of the energy density homogeneity and the pressure isotropy. So, to find matchable solutions we should relax these two conditions. Also, it is worth mentioning that the above result is in agreement with previous works indicating that static, perfect fluid (isotropic in pressure) sources are spherical (see [91] and references therein).

8.2.4 Anisotropic Inhomogeneous Spheroids

In order to obtain a metric smoothly matchable to any Weyl space–time we shall consider a solution with vanishing complexity factors but inhomogeneous energy density and anisotropic pressure (see details in [58]).

The metric variables of the solution are

$$A(r,\theta) = \frac{a_1 r \sin\theta}{b_1 r^2 + b_2}, \tag{8.89}$$

$$B(r,\theta) = \frac{1}{b_1 r^2 + b_2}, \tag{8.90}$$

$$D(r,\theta) = \frac{b_1 r^2 - b_2}{b_1 r^2 + b_2} F\left(\frac{r\cos\theta}{b_1 r^2 - b_2}\right). \tag{8.91}$$

It is a simple matter to check that the vanishing complexity factors conditions (8.246)–(8.248) in Appendix 8.8, are satisfied for (8.89)–(8.91).

From the above, using the Einstein equations (8.227)–(8.231), one obtains for the physical variables.

$$8\pi\mu = 12b_1b_2 - \frac{(b_1r^2+b_2)^2}{(b_1r^2-b_2)^2}\left[\frac{4b_1b_2r^2cos^2\theta}{(b_1r^2-b_2)^2}+1\right]\frac{F_{zz}}{F}, \tag{8.92}$$

$$8\pi P = -12b_1b_2 + \frac{(b_1r^2+b_2)^2}{3(b_1r^2-b_2)^2}\left[\frac{4b_1b_2r^2cos^2\theta}{(b_1r^2-b_2)^2}+1\right]\frac{F_{zz}}{F}, \tag{8.93}$$

$$8\pi\Pi_2 \equiv 8\pi(P_{xx}-P_{zz}) = \frac{F_{zz}}{4F}\frac{(b_1r^2+b_2)^2}{(b_1r^2-b_2)^2}\sin^2\theta, \tag{8.94}$$

$$8\pi\Pi_3 \equiv 8\pi(P_{yy}-P_{zz}) = \frac{F_{zz}}{4F}\frac{(b_1r^2+b_2)^4cos^2\theta}{(b_1r^2-b_2)^4}, \tag{8.95}$$

$$8\pi\Pi_{KL} \equiv 8\pi P_{xy} = -\frac{F_{zz}}{2F}\frac{r(b_1r^2+b_2)^3}{(b_1r^2-b_2)^3}\sin 2\theta, \tag{8.96}$$

where a_1, b_1, b_2 are constant, and

$$F(z) \equiv F\left(\frac{r\cos\theta}{b_1r^2-b_2}\right). \tag{8.97}$$

It is not difficult to find a range of values of the parameters, for which the physical behavior of physical variables is acceptable and the metric may be matched smoothly on the boundary surface to a Weyl solution.

The Static Hyperbolically Symmetric Case

Motivated by a new version of the Schwarzschild black hole proposed in [56, 57], there has been a renewed interest in self–gravitating systems admitting hyperbolical symmetry.

In this picture, the space–time outside the horizon is represented by the usual Schwarzschild metric, whereas the region inner to the horizon is described by the line element

$$ds^2 = \left(\frac{2M}{R}-1\right)dt^2 - \frac{dr^2}{\left(\frac{2M}{R}-1\right)} - R^2d\Omega^2,$$
$$d\Omega^2 = d\theta^2 + \sinh^2\theta d\phi^2, \tag{8.98}$$

which is a static solution with the (θ, ϕ) space describing a positive Gaussian curvature, admitting the four Killing vectors

$$\chi_{(0)} = \partial_{\mathbf{t}}, \quad \chi_{(2)} = -\cos\phi\partial_\theta + \coth\theta\sin\phi\partial_\phi$$
$$\chi_{(1)} = \partial_\phi \quad \chi_{(3)} = \sin\phi\partial_\theta + \coth\theta\cos\phi\partial_\phi. \tag{8.99}$$

A solution to the Einstein equations of the form given by (8.98), defined by the hyperbolic symmetry (8.99), was first considered by Harrison [38], and has been more recently the subject of research in different contexts (see [32, 74, 85, 87, 88, 98] and references therein).

Our purpose here is to present some exact solutions to Einstein equations endowed with the symmetry given by (8.99), satisfying the vanishing complexity factor condition and which might serve as the source of (8.98), see [61] for details.

Thus, let us consider hyperbolically symmetric distributions of static fluid, which for the sake of completeness we assume to be locally anisotropic and which may be (or may be not) bounded from the exterior by a surface Σ^e whose equation is $r = r_{\Sigma^e} =$ constant. On the other hand as it appears from the study of such fluids (see [57] for details) the fluid distribution cannot fill the central region, in which case we may assume that such a region is represented by an empty vacuole, implying that the fluid distribution is also bounded from the inside by a surface Σ^i whose equation is $r = r_{\Sigma^i} =$ constant.

The line element is given in polar coordinates by

$$ds^2 = e^{\nu} dt^2 - e^{\lambda} dr^2 - r^2 \left(d\theta^2 + \sinh^2\theta d\phi^2\right), \tag{8.100}$$

where, due to the imposed symmetry, $\nu(r)$ and $\lambda(r)$ are exclusively functions of r. We number the coordinates: $x^0 = t$; $x^1 = r$; $x^2 = \theta$; $x^3 = \phi$.

The metric (8.100) has to satisfy Einstein field equations

$$G^{\nu}_{\mu} = 8\pi T^{\nu}_{\mu}. \tag{8.101}$$

We may write for the energy momentum tensor

$$T_{\alpha\beta} = (\mu + P)V_\alpha V_\beta - P g_{\alpha\beta} + \Pi_{\alpha\beta}, \tag{8.102}$$

or

$$T_{\alpha\beta} = (\mu + P_{zz})V_\alpha V_\beta - P_{zz} g_{\alpha\beta} + (P_{xx} - P_{zz})K_\alpha K_\beta. \tag{8.103}$$

Since we choose the fluid to be comoving in our coordinates, then

$$V^\alpha = (e^{-\nu/2}, 0, 0, 0); \quad V_\alpha = (e^{\nu/2}, 0, 0, 0), \tag{8.104}$$

and

$$K_\alpha = (0, -e^{\lambda/2}, 0, 0), \tag{8.105}$$

It would be useful to express the anisotropic tensor in the form

$$\Pi_{\alpha\beta} = \Pi \left(K_\alpha K_\beta + \frac{h_{\alpha\beta}}{3}\right), \tag{8.106}$$

with $h_{\mu\nu} = g_{\mu\nu} - V_\nu V_\mu$,

$$\Pi = P_{xx} - P_{zz}, \tag{8.107}$$

and

$$P = \frac{P_{xx} + 2P_{zz}}{3}. \tag{8.108}$$

Since the Lie derivative and the partial derivative commute, then

$$\mathscr{L}_\chi G_{\alpha\beta} = 8\pi \mathscr{L}_\chi T_{\alpha\beta} = 0, \tag{8.109}$$

for any χ defined by (8.99), implying that all physical variables only depend on r.

If the fluid is bounded from the exterior by a hypersurface Σ^e described by the equation $r = r_{\Sigma^e} = constant$, then the smooth matching of (8.98) and (8.100) on Σ^e requires the fulfillment of the Darmois conditions, imposing the continuity of the first and the second fundamental forms, which imply

$$e^{\nu_{\Sigma^e}} = \frac{2M}{r_{\Sigma^e}} - 1, \qquad e^{\lambda_{\Sigma^e}} = \frac{1}{\frac{2M}{r_{\Sigma^e}} - 1}, \qquad P_{xx}(r_{\Sigma^e}) = 0, \tag{8.110}$$

and the continuity of the mass function $m(r)$ defined below. If we assume that the central region is surrounded by an empty cavity whose delimiting surface is $r = r_{\Sigma^i} = constant$, then the fulfillment of Darmois conditions on Σ^i implies

$$e^{\nu_{\Sigma^i}} = 1, \qquad e^{\lambda_{\Sigma^i}} = 1, \qquad P_{xx}(r_{\Sigma^i}) = 0, \tag{8.111}$$

and $m(r_{\Sigma^i}) = 0$.

The non–vanishing components of the Einstein equations for the metric (8.100) and the energy momentum tensor (8.103) are

$$8\pi\mu = -\frac{(e^{-\lambda} + 1)}{r^2} + \frac{\lambda'}{r} e^{-\lambda}, \tag{8.112}$$

$$8\pi P_r = \frac{(e^{-\lambda} + 1)}{r^2} + \frac{\nu'}{r} e^{-\lambda}, \tag{8.113}$$

$$8\pi P_\perp = \frac{e^{-\lambda}}{2}\left(\nu'' + \frac{\nu'^2}{2} - \frac{\lambda'\nu'}{2} + \frac{\nu'}{r} - \frac{\lambda'}{r}\right), \tag{8.114}$$

where we have used the standard notation $P_{xx} \equiv P_r$ and $P_{zz} = P_{yy} \equiv P_\perp$, and primes denote derivatives with respect to r.

It is worth stressing the differences between these equations and the corresponding to the spherically symmetric case.

From the equations above or using the conservation laws $T^\alpha_{\beta;\alpha} = 0$ we obtain, besides the identity $\dot{\mu} = 0$ (where dot denotes derivative with respect to t), the corresponding hydrostatic equilibrium equation (the generalized Tolman–Oppenheimer–Volkoff equation)

$$P_r' + (\mu + P_r)\frac{\nu'}{2} + \frac{2}{r}\Pi = 0. \tag{8.115}$$

Let us now define the mass function $m = m(r)$. For doing so, let us notice that using (8.98) we have that outside the fluid distribution (but inside the horizon)

$$M = -\left(\frac{R}{2}\right) R^3_{232}, \tag{8.116}$$

where the Riemann tensor component R^3_{232}, has been calculated with (8.98).

Then generalizing the above definition of mass for the interior of the fluid distribution we may write

$$m(r) = -\left(\frac{r}{2}\right) R^3_{232} = \frac{r(1 + e^{-\lambda})}{2} \tag{8.117}$$

where now the Riemann tensor component is calculated with (8.100).

Feeding back (8.117) into (8.112) we obtain

$$m'(r) = -4\pi r^2 \mu \Rightarrow m = -4\pi \int_0^r \mu r^2 dr. \tag{8.118}$$

Since m as defined by (8.117) is a positive quantity, then μ should be negative and therefore the weak energy condition is violated, a result already obtained in [87]. However it is important to stress that our definition of mass function differs from the one introduced in [87]. In particular our m is positive defined whereas the expression used in [87] is negative (for the hyperbolically symmetric fluid).

The following comments are in order at this point.

- If the energy density is regular everywhere then the mass function must vanish at the center as $m \sim r^3$, this implies (as it follows from (8.117)) that the fluid cannot fill the space in the neighborhood of the center, i.e. there is a cavity around the center which may be, either empty, or filled with a fluid distribution non endowed with hyperbolical symmetry. Thus the hyperbolically symmetric fluid spans from a minimal value of the coordinate r until its external boundary. For the extreme case $\mu = constant$, this minimal value $r_{min.}$ is defined by $-\frac{8\pi}{3}\mu r^2_{min.} > 1$. Obviously, if the energy density is singular in the neighborhood of the center, then this region must also be excluded by physical reasons.

From the above it follows that, strictly speaking, we should write instead of (8.118)

$$m = 4\pi \int_{r_{min}}^r |\mu| r^2 dr, \tag{8.119}$$

where due to the fact that μ is negative, we have replaced it by $-|\mu|$ (as we shall do from now on).

The situation described above is fully consistent with the results obtained in [57] where it was shown that test particles cannot reach the center for any finite value of its energy.

Next, using (8.113) and (8.117) we obtain

$$\nu' = 2\frac{4\pi r^3 P_r - m}{r(2m - r)}, \tag{8.120}$$

from which we may write (8.115) as

$$P_r' + (P_r - |\mu|)\frac{4\pi r^3 P_r - m}{r(2m - r)} + \frac{2}{r}\Pi = 0. \tag{8.121}$$

This is the hydrostatic equilibrium equation for our fluid. Let us analyze in some detail the physical meaning of its different terms.

The first term is just the gradient of pressure, which is usually negative and opposing gravity. The second term describes the gravitational "force" and contains two different contributions: on the one hand the term $P_r - |\mu|$ which we expect to be negative (or zero for the stiff equation of state) and is usually interpreted as the "passive gravitational mass density" (p.g.m.d.), and on the other hand the term $4\pi r^3 P_r - m$ that is proportional to the "active gravitational mass" (a.g.m.), and which is negative if $4\pi r^3 P_r < m$. Finally the third term describes the effect of the pressure anisotropy, whose sign depends on the difference between principal stresses. Two important remarks are in order at this point:

- It is worth stressing that while the self–regenerative pressure effect (described by the $4\pi r^3 P_r$ term in (8.121)) has the same sign as in the spherically symmetric case, the mass function contribution in the second term has the opposite sign with respect to the latter case. This of course is due to the fact that the energy density is negative.
- If, both, the p.g.m.d. and the a.g.m. are negative, the final effect of the gravitational interaction would be as usual, to oppose the negative pressure gradient. However, because of the equivalence principle, a negative p.g.m.d. implies a negative inertial mass, which in turn implies that the hydrostatic force term (the pressure gradient and the anisotropic term), and the gravitational force term, switch their roles with respect to the positive energy density case.

In this case the complexity factor Y_{TF} is given by

$$Y_{TF} = 4\pi\Pi + \mathcal{E}, \tag{8.122}$$

or

$$Y_{TF} = 8\pi\Pi + \frac{4\pi}{r^3}\int_0^r \tilde{r}^3|\mu|'d\tilde{r}, \tag{8.123}$$

where $\mathscr{E}$ is the scalar defining the electric part of the Weyl tensor (the magnetic part vanishes identically as in the spherically symmetric case).

Also, as in the spherically symetric case Y_{TF} encompasses the influence of the local anisotropy of pressure and density inhomogeneity on the Tolman mass. Or, in other words, Y_{TF} describes how these two factors modify the value of the Tolman mass, with respect to its value for the homogeneous isotropic fluid.

Indeed, in this case it can be shown that the following expressions for the Tolman mass can be obtained (see [61] for details)

$$m_T = \frac{(cosh\pi - 1)}{4}e^{(\nu-\lambda)/2}r^2\nu', \tag{8.124}$$

and

$$m_T = (m_T)_{\Sigma^e}\left(\frac{r}{r_{\Sigma^e}}\right)^3 + \frac{(cosh\pi - 1)}{2}r^3\int_r^{r_{\Sigma^e}}\frac{e^{(\nu+\lambda)/2}}{\tilde{r}}Y_{TF}d\tilde{r}. \tag{8.125}$$

We shall next present two models endowed with hyperbolical symmetry and satisfying the vanishing complexity factor condition.

8.2.4.1 A Model with Vanishing Complexity Factor and Vanishing Radial Pressure

Since the vanishing complexity factor is not enough to close the full system of Einstein equations we have to impose an additional restriction in order to obtain a specific model. Here we shall assume (besides the vanishing complexity factor), the condition $P_r = 0$.

Thus, assuming $P_r = 0$, we obtain from (8.113)

$$\nu' = -\frac{2g}{(2g-1)r}, \tag{8.126}$$

where g is defined by

$$e^{-\lambda} = 2g - 1. \tag{8.127}$$

Next, imposing $Y_{TF} = 0$ in (8.125) it follows that

$$m_T = (m_T)_{\Sigma^e}\frac{r^3}{r^3_{\Sigma}}. \tag{8.128}$$

The combination of (8.124), (8.126), (8.127) and (8.128) produces

$$e^{\nu} = \frac{4(m_T^2)_{\Sigma^e} r^4}{r_{\Sigma}^6 (cosh\pi - 1)^2} \frac{(2g - 1)}{g^2}. \tag{8.129}$$

On the other hand the condition $Y_{TF} = 0$ may be written as

$$g' r(1 - g) + g(5g - 2) = 0, \tag{8.130}$$

whose solution reads

$$C_2 r^{10} = \frac{g^5}{(5g - 2)^3}, \tag{8.131}$$

where C_2 is a constant of integration.

Then for the physical variables we obtain

$$|\mu| = \frac{3}{4\pi r^2} \frac{g(2g - 1)}{(g - 1)}, \tag{8.132}$$

$$P_{\perp} = \frac{3}{8\pi r^2} \frac{g^2}{(g - 1)}. \tag{8.133}$$

In this case, the fluid distribution is restricted by a minimal value of the r coordinate, satisfying $g(r_{min}) > 1$. The specific value of r_{min} is obtained from (8.131). For $0 < r < r_{min}$ we may assume, as in precedent models, an empty cavity surrounding the center. Also as in precedent models, the discontinuity of the mass function across Σ^i implies that a thin shell appears on it. Finally, since the radial pressure is assumed to be zero, both the a.g.m. and the p.g.m.d. are negative.

8.2.4.2 A Model with the Stiff Equation of State and Vanishing Complexity Factor

Finally we shall consider a solution satisfying the so called stiff equation of state proposed by Zeldovich [133] and which is thought to be suitable to describe ultradense matter, and the vanishing complexity factor.

In its original form the stiff equation of state assumes that energy density equals pressure (in relativistic units). In our case we shall assume

$$|\mu| = P_r. \tag{8.134}$$

then (8.121) becomes

$$P_r' + \frac{2}{r}\Pi = 0. \tag{8.135}$$

Using this latter condition in (8.123) with $Y_{TF} = 0$, and feeding back the resulting expression into (8.135) one obtains

$$P_r'' + \frac{3}{r} P_r' = 0, \tag{8.136}$$

whose solution reads

$$P_r = \frac{b}{r^2} - a, \tag{8.137}$$

where a and b are two positive constants of integration.

Then from (8.117) and (8.118) it follows at once

$$m = 4\pi r \left(b - \frac{ar^2}{3} \right), \tag{8.138}$$

from which we easily obtain λ. Finally, feeding back these expressions into (8.120) we may obtain ν.

Assuming the fluid distribution to be bounded from the exterior by the surface Σ^e described by $r = r_{\Sigma^e} = constant$, then we may write

$$P_r = b \left(\frac{1}{r^2} - \frac{1}{r^2_{\Sigma^e}} \right). \tag{8.139}$$

So far we have only considered static configurations. In the next section we shall tackle the problem of defining complexity for dynamic dissipative fluids.

8.3 Dynamic Spherically Symmetric Fluids

As mentioned before, when dealing with non–static dissipative fluids we encounter two additional problems to define complexity. On the one hand we have to include the dissipative flux in the variable measuring the complexity of the structure. On the other hand we have also to describe the complexity of the pattern of evolution.

In other words, in the dynamic case we have to ask ourselves two questions instead of one, namely, what is the simplest fluid? and, what is the simplest mode of evolution of the fluid? These two questions, although related in some way, are completely different. A "simple" fluid may evolve exhibiting a very complex pattern of evolution, while a fluid with a high degree of complexity may evolve through a simple pattern of evolution.

This problem was studied in detail in [55], here we shall sketch the main steps leading to the appropriate definition of complexity for dynamic fluids and discuss about the answer to the question, what is the simplest mode of evolution of the fluid? Afterwards we shall present some solutions.

We shall restrict the discussion to the spherically symmetric case.

8.3.1 The Dynamic Spherically Symmetric Case

So, let us consider a spherically symmetric distribution of collapsing fluid, which may be bounded by a spherical surface Σ, or not. The fluid is assumed to be locally anisotropic (principal stresses unequal) and undergoing dissipation in the form of heat flow (diffusion approximation).

Choosing comoving coordinates, the general interior metric can be written

$$ds^2 = -A^2dt^2 + B^2dr^2 + R^2(d\theta^2 + \sin^2\theta d\phi^2), \tag{8.140}$$

where A, B and R are functions of t and r and are assumed positive. We number the coordinates $x^0 = t$, $x^1 = r$, $x^2 = \theta$ and $x^3 = \phi$. Observe that A and B are dimensionless, whereas R has the same dimension as r.

The energy–momentum tensor $T_{\alpha\beta}$ of the fluid distribution has the form

$$\begin{aligned} T_{\alpha\beta} &= (\mu + P_\perp)V_\alpha V_\beta + P_\perp g_{\alpha\beta} + (P_r - P_\perp)\chi_\alpha\chi_\beta \\ &+ q_\alpha V_\beta + V_\alpha q_\beta, \end{aligned} \tag{8.141}$$

where μ is the energy density, P_r the radial pressure, $P_\perp$ the tangential pressure, q^α the heat flux, V^α the four velocity of the fluid, and χ^α a unit four vector along the radial direction. These quantities satisfy

$$V^\alpha V_\alpha = -1, \quad V^\alpha q_\alpha = 0, \quad \chi^\alpha\chi_\alpha = 1, \quad \chi^\alpha V_\alpha = 0. \tag{8.142}$$

Or in the equivalent (canonical) form

$$T_{\alpha\beta} = \mu V_\alpha V_\beta + P h_{\alpha\beta} + \Pi_{\alpha\beta} + q\left(V_\alpha\chi_\beta + \chi_\alpha V_\beta\right) \tag{8.143}$$

with

$$P = \frac{P_r + 2P_\perp}{3}, \qquad h_{\alpha\beta} = g_{\alpha\beta} + V_\alpha V_\beta,$$

$$\Pi_{\alpha\beta} = \Pi\left(\chi_\alpha\chi_\beta - \frac{1}{3}h_{\alpha\beta}\right), \qquad \Pi = P_r - P_\perp,$$

where q is a function of t and r.

Since we are considering comoving observers, we have

$$V^\alpha = A^{-1}\delta_0^\alpha, \quad q^\alpha = qB^{-1}\delta_1^\alpha, \quad \chi^\alpha = B^{-1}\delta_1^\alpha, \tag{8.144}$$

It is worth noticing that, both, bulk and shear viscosity could be easily introduced to the system through a redefinition of the radial and tangential pressures, P_r and $P_\perp$. Also, dissipation in the free streaming approximation could be introduced by redefining μ, P_r and q.

The Einstein equations for (8.140) and (8.143), are explicitly written in Appendix 8.9.

The acceleration a_α and the expansion Θ of the fluid are given by

$$a_\alpha = V_{\alpha;\beta} V^\beta, \quad \Theta = V^\alpha{}_{;\alpha}. \tag{8.145}$$

and its shear $\sigma_{\alpha\beta}$ by

$$\sigma_{\alpha\beta} = V_{(\alpha;\beta)} + a_{(\alpha} V_{\beta)} - \frac{1}{3}\Theta h_{\alpha\beta}, \tag{8.146}$$

from which we easily obtain

$$a_1 = \frac{A'}{A}, \quad a = \sqrt{a^\alpha a_\alpha} = \frac{A'}{AB}, \tag{8.147}$$

$$\Theta = \frac{1}{A}\left(\frac{\dot{B}}{B} + 2\frac{\dot{R}}{R}\right), \tag{8.148}$$

$$\sigma_{11} = \frac{2}{3}B^2\sigma, \quad \sigma_{22} = \frac{\sigma_{33}}{\sin^2\theta} = -\frac{1}{3}R^2\sigma, \tag{8.149}$$

where

$$\sigma^{\alpha\beta}\sigma_{\alpha\beta} = \frac{2}{3}\sigma^2, \tag{8.150}$$

with

$$\sigma = \frac{1}{A}\left(\frac{\dot{B}}{B} - \frac{\dot{R}}{R}\right), \tag{8.151}$$

where the prime stands for r differentiation and the dot stands for differentiation with respect to t.

Next, the mass function $m(t, r)$ introduced by Misner and Sharp [92] reads

$$m = \frac{R^3}{2} R_{23}{}^{23} = \frac{R}{2}\left[\left(\frac{\dot{R}}{A}\right)^2 - \left(\frac{R'}{B}\right)^2 + 1\right], \tag{8.152}$$

and introducing the proper time derivative D_T given by

$$D_T = \frac{1}{A}\frac{\partial}{\partial t}, \tag{8.153}$$

we can define the velocity U of the collapsing fluid as the variation of the areal radius with respect to proper time, i.e.

$$U = D_T R, \tag{8.154}$$

where R defines the areal radius of a spherical surface inside the fluid distribution (as measured from its area).

Then (8.152) can be rewritten as

$$E \equiv \frac{R'}{B} = \left(1 + U^2 - \frac{2m}{R}\right)^{1/2}. \tag{8.155}$$

Using (8.155) we can express (8.254) as

$$4\pi q = E\left[\frac{1}{3}D_R(\Theta - \sigma) - \frac{\sigma}{R}\right], \tag{8.156}$$

where D_R denotes the proper radial derivative,

$$D_R = \frac{1}{R'}\frac{\partial}{\partial r}. \tag{8.157}$$

Using (8.250)–(8.252) with (8.157) we obtain from (8.152)

$$D_R m = 4\pi\left(\mu + q\frac{U}{E}\right)R^2, \tag{8.158}$$

which implies

$$m = 4\pi\int_0^r\left(\mu + q\frac{U}{E}\right)R^2 R' dr, \tag{8.159}$$

satisfying the regular condition $m(t, 0) = 0$.

Integrating (8.159) we find

$$\frac{3m}{R^3} = 4\pi\mu - \frac{4\pi}{R^3}\int_0^r R^3\left(D_R\mu - 3q\frac{U}{RE}\right)R' dr. \tag{8.160}$$

8.3.2 *Defining Complexity for the Dynamic Fluid*

As we have already mentioned, in the dynamic case the definition of a quantity measuring the complexity of the system poses two additional problems with respect to the static case.

On the one hand, the definition of the complexity of the structure of the fluid, which in this case also involves dissipative variables, and on the other hand the problem of defining the complexity of the pattern of evolution of the system.

For the static fluid distribution it was assumed that the scalar function Y_{TF} is an appropriate measure of the complexity of the fluid, and therefore was identified as the complexity factor.

We shall assume in the dynamic case that Y_{TF} still measures the complexity of the system, in what corresponds to the structure of the object, and we shall adopt initially an assumption about the simplest possible pattern of evolution. Specifically, we shall assume that the simplest evolution pattern (one of them at least) is described by the homologous evolution. However, as we shall see below, this last condition might be too stringent, ruling out many interesting scenarios from the astrophysical point of view and therefore we shall consider also other possible (less restrictive) mode of evolution which also could be used to describe the simplest mode of evolution, and which we call quasi–homologous.

In order to provide the necessary mathematical expressions for carrying out our task, let us start by finding the expression for the Weyl tensor.

As is well known, in the spherically symmetric case the Weyl tensor ($C^{\rho}_{\alpha\beta\mu}$) is defined by its "electric" part $E_{\gamma\nu}$ alone, since its "magnetic" part vanishes, with

$$E_{\alpha\beta} = C_{\alpha\mu\beta\nu} V^{\mu} V^{\nu}, \tag{8.161}$$

where the electric part of Weyl tensor may also be written as

$$E_{\alpha\beta} = \mathscr{E}(\chi_{\alpha}\chi_{\beta} - \frac{1}{3}h_{\alpha\beta}). \tag{8.162}$$

with

$$\begin{aligned} \mathscr{E} = & \frac{1}{2A^2}\left[\frac{\ddot{R}}{R} - \frac{\ddot{B}}{B} - \left(\frac{\dot{R}}{R} - \frac{\dot{B}}{B}\right)\left(\frac{\dot{A}}{A} + \frac{\dot{R}}{R}\right)\right] \\ & + \frac{1}{2B^2}\left[\frac{A''}{A} - \frac{R''}{R} + \left(\frac{B'}{B} + \frac{R'}{R}\right)\left(\frac{R'}{R} - \frac{A'}{A}\right)\right] - \frac{1}{2R^2}. \end{aligned} \tag{8.163}$$

Then, proceeding as in the previous cases (see [55] for details) we obtain

$$Y_T = 4\pi(\mu + 3P_r - 2\Pi), \qquad Y_{TF} = \mathscr{E} - 4\pi\Pi. \tag{8.164}$$

Next, using (8.250), (8.252), (8.253) with (8.152) and (8.163) we obtain

$$\frac{3m}{R^3} = 4\pi\,(\mu - \Pi) - \mathscr{E}, \tag{8.165}$$

which combined with (8.160) and (8.164) produces

$$Y_{TF} = -8\pi\Pi + \frac{4\pi}{R^3}\int_0^r R^3\left(D_R\mu - 3q\frac{U}{RE}\right)R'dr. \tag{8.166}$$

Again, we notice that due to a different signature, the sign of Y_{TF} in the above equation differs from the sign of the Y_{TF} used in [54] for the static case.

Thus the scalar Y_{TF} may be expressed through the Weyl tensor and the anisotropy of pressure or in terms of the anisotropy of pressure, the density inhomogeneity and the dissipative variables.

Once the complexity factor for the structure of the fluid distribution has been established, it remains to elucidate what is the simplest pattern of evolution. Based on purely intuitive thoughts we shall first identify the homologous evolution as the simplest mode of evolution.

In order to obtain a mathematical description of the homologous evolution, we shall proceed as follows

First of all observe that we can write (8.156) as

$$D_R\left(\frac{U}{R}\right)=\frac{4\pi}{E}q+\frac{\sigma}{R}, \tag{8.167}$$

which after integration becomes

$$U=\tilde{a}(t)R+R\int_0^r\left(\frac{4\pi}{E}q+\frac{\sigma}{R}\right)R'dr, \tag{8.168}$$

where $\tilde{a}$ is an integration function, or,

$$U=\frac{U_\Sigma}{R_\Sigma}R-R\int_r^{r_\Sigma}\left(\frac{4\pi}{E}q+\frac{\sigma}{R}\right)R'dr. \tag{8.169}$$

If the integrand in the above equations vanishes we have from (8.168) or (8.169)

$$U=\tilde{a}(t)R. \tag{8.170}$$

This relationship is characteristic of the homologous evolution in Newtonian hydrodynamics [37, 80, 102]. In our case this may occur if the fluid is shear–free and non dissipative, or if the two terms in the integral cancel each other.

In [55], the term "homologous evolution" was used to characterize relativistic systems satisfying, besides (8.170), the condition

$$\frac{R_I}{R_{II}}=\text{constant}, \tag{8.171}$$

where R_I and R_{II} denote the areal radii of two concentric shells (I, II) described by $r=r_I=$ constant, and $r=r_{II}=$ constant, respectively.

Now, it is very important to be aware of the fact that conditions (8.170) and (8.171), in the general relativistic case, are different and more specifically, that (8.170) does not imply (8.171).

Indeed, (8.170) implies that for the two shells of fluids I, II we have

$$\frac{U_I}{U_{II}} = \frac{A_{II}\dot{R}_I}{A_I\dot{R}_{II}} = \frac{R_I}{R_{II}}, \tag{8.172}$$

that implies (8.171) only if $A = A(t)$, which by a simple coordinate transformation becomes $A =$ constant. Thus in the non–relativistic regime, (8.171) always follows from the condition that the radial velocity is proportional to the radial distance, whereas in the relativistic regime the condition (8.170) implies (8.171), only if the fluid is geodesic.

We shall define quasi–homologous evolution as that restricted only by condition (8.170), implying

$$\frac{4\pi}{R'}Bq + \frac{\sigma}{R} = 0. \tag{8.173}$$

Let us first consider the homologous evolution (both (8.170) and (8.171) are satisfied).

From the Eq. (8.171) it follows that R is a separable function, i.e. we can write

$$R = R_1(t)R_2(r). \tag{8.174}$$

To summarize, the homologous condition implies (8.174), and (8.173).

Feeding back (8.173) into (8.254), we obtain

$$(\Theta - \sigma)' = 0, \tag{8.175}$$

whereas, using (8.148) and (8.151) we get

$$(\Theta - \sigma)' = \left(\frac{3}{A}\frac{\dot{R}}{R}\right)' = 0. \tag{8.176}$$

Then using (8.174) it follows at once that

$$A' = 0, \tag{8.177}$$

implying that the fluid is geodesic, as it follows from (8.147). Also, by reparametrizing the coordinate t, we may put, without loss of generality, $A = 1$.

The inverse is in general untrue, e.g. the Lemaitre–Tolman–Bondi case, unless we assume that $(\Theta - \sigma)'$ is of class C^{ω}.

In the non–dissipative case, the homologous condition not only implies that the fluid is geodesic, but also that it is shear–free, as it follows at once from (8.173).

An important point to mention here is that as it has been shown in [46], an initially shear–free geodesic fluid remains shear–free during the evolution iff $Y_{TF} = 0$. This implies that a system which starts its evolution from the rest ($\sigma = 0$), will remain shear–free if the fluid is geodesic (or equivalently, homologous) and $Y_{TF} = 0$. This is an additional argument supporting our choice of Y_{TF} as the complexity factor.

If we impose the homologous condition, the Eq. (8.259) in Appendix 8.10 becomes

$$D_T U = -\frac{m}{R^2} - 4\pi P_r R, \tag{8.178}$$

or in terms of Y_{TF}

$$\frac{3D_T U}{R} = -4\pi\,(\mu + 3P_r - 2\Pi) + Y_{TF}, \tag{8.179}$$

where (8.164) has been used.

Next from the field equations we obtain

$$4\pi\,(\mu + 3P_r - 2\Pi) = -\frac{2\ddot{R}}{R} - \frac{\ddot{B}}{B}, \tag{8.180}$$

and from the definition of U

$$\frac{3D_T U}{R} = \frac{3\ddot{R}}{R}, \tag{8.181}$$

feeding back the two equations above into (8.179), it follows that

$$\frac{\ddot{R}}{R} - \frac{\ddot{B}}{B} = Y_{TF}. \tag{8.182}$$

Now, the vanishing complexity factor condition ($Y_{TF} = 0$), produces after the integration of (8.182)

$$B = R_1(t)\left(b_1(r)\int \frac{dt}{R_1(t)^2} + b_2(r)\right), \tag{8.183}$$

where $b_1(r)$ and $b_2(r)$ are two functions of integration, or

$$B = R_1(t)R_2'(r)\left(\tilde{b}_1(r)\int \frac{dt}{R_1(t)^2} + \tilde{b}_2(r)\right), \tag{8.184}$$

with $b_1(r) = \tilde{b}_1(r)R_2'$ and $b_2(r) = \tilde{b}_2(r)R_2'$.

Then introducing the variable

$$Z = \tilde{b}_1(r)\int \frac{dt}{R_1(t)^2} + \tilde{b}_2(r), \tag{8.185}$$

we may write

$$B = ZR'. \tag{8.186}$$

Let us now use the expressions above to analyze first the non–dissipative case.

Thus, if we further assume the fluid to be non–dissipative, recalling that in this case the homologous condition implies the vanishing of the shear, we obtain because of (8.151)

$$\frac{\ddot{R}}{R} - \frac{\ddot{B}}{B} = 0 \quad \Rightarrow Y_{TF} = 0. \tag{8.187}$$

In other words, in this particular case, the homologous condition already implies the vanishing complexity factor condition.

More so, since the fluid is shear–free, we have because of (8.151) and (8.183)

$$b_1(r) = 0 \Rightarrow B = b_2(r)R_1(t) = \tilde{b}_2(r)R_1(t)R_2'. \tag{8.188}$$

Then reparametrising r as $\tilde{b}_2(r)dr \Rightarrow dr$, we may put without loss of generality $B = R_1(t)R_2(r)'$, or equivalently $Z = 1$, implying that all non–dissipative configurations evolving homologously (and thereby satisfying $Y_{TF} = 0$), belong to what are known as "Euclidean stars" [47], characterized by the condition $Z = 1 \Rightarrow B = R'$. However among all possible solutions satisfying the "Euclidean condition", only one evolves homologously and satisfies the condition $Y_{TF} = 0$.

Indeed, from the field Eqs. (8.252) and (8.253) we may write

$$8\pi(P_r - P_\perp) = \frac{\dot{Z}\dot{R}}{ZR} + \frac{1}{Z^2R^2}\left(\frac{Z'R}{ZR'} + 1 - Z^2\right). \tag{8.189}$$

Since in this case we have $Z = 1$ then $\Pi = P_r - P_\perp = 0$ which implies because of the $Y_{TF} = 0$ condition, that $\mu' = 0$.

However it is known that a shear–free, geodesic (non–dissipative) fluid with isotropic pressure is necessarily dust with homogeneous energy density and vanishing Weyl tensor (see [41, 45]). It goes without saying that this kind of system represents the simplest conceivable configuration (Friedman–Robertson–Walker).

Thus for the non–dissipative case, the homologous condition implies $Y_{TF} = 0$ and produces the simplest configuration. This configuration is the only one evolving homologously and satisfying $Y_{TF} = 0$.

Of course, solutions satisfying $Y_{TF} = 0$ but not evolving homologously do exist. They only require

$$8\pi\Pi = \frac{4\pi}{R^3}\int_0^r R^3\mu' dr. \tag{8.190}$$

In such a case the solutions are shearing, and neither conformally flat nor geodesic.

Based on all the precedent comments, it seems reasonable to consider the homologous condition as a good candidate to describe the simplest mode of evolution.

8.3.3 The Dissipative Case

In the dissipative case, we may obtain from (8.151) and (8.187),

$$\dot{\sigma} = -Y_{TF} + \left(\frac{\dot{R}}{R}\right)^2 - \left(\frac{\dot{B}}{B}\right)^2. \tag{8.191}$$

Then, taking the t-derivative of (8.173) and using (8.191) we obtain

$$Y_{TF}\frac{R'}{R} = 4\pi Bq\left(\frac{\dot{q}}{q} + 2\frac{\dot{B}}{B} + \frac{\dot{R}}{R}\right). \tag{8.192}$$

If we assume $Y_{TF} = 0$, then we obtain

$$q = \frac{f(r)}{B^2 R}, \tag{8.193}$$

implying

$$\dot{q} = -q(\Theta + \sigma), \tag{8.194}$$

where f is an arbitrary integration function. Solutions of this kind might be found by using the general methods presented in [70–72, 115, 116].

In the dissipative case we need to provide a transport equation to describe the evolution and distribution of temperature. Assuming a causal dissipative theory (e.g. the Israel–Stewart theory [67–69]), the transport equation for the heat flux reads

$$\tau h^{\alpha\beta} V^{\gamma} q_{\beta;\gamma} + q^{\alpha} = -\kappa h^{\alpha\beta}\left(T_{,\beta} + T a_{\beta}\right) - \frac{1}{2}\kappa T^2 \left(\frac{\tau V^{\beta}}{\kappa T^2}\right)_{;\beta} q^{\alpha}, \tag{8.195}$$

where κ denotes the thermal conductivity, and T and τ denote temperature and relaxation time, respectively.

In the spherically symmetric case under consideration, the transport equation has only one independent component, which may be obtained from (8.195) by contracting with the unit spacelike vector χ^{α}, producing

$$\tau V^{\alpha} q_{,\alpha} + q = -\kappa\left(\chi^{\alpha} T_{,\alpha} + T a\right) - \frac{1}{2}\kappa T^2 \left(\frac{\tau V^{\alpha}}{\kappa T^2}\right)_{;\alpha} q. \tag{8.196}$$

Sometimes, it is possible to simplify the equation above, assuming the so-called truncated transport equation when the last term in (8.195) may be neglected [118], producing

$$\tau V^{\alpha} q_{,\alpha} + q = -\kappa\left(\chi^{\alpha} T_{,\alpha} + T a\right). \tag{8.197}$$

Now, in a dissipative process, it appears reasonable to consider the stationary state, prevailing once the system has relaxed and transient phenomena have vanished, as an example of the simplest dissipative regime. Thus, if we assume the stationary state (neglecting the relaxation time), then the transport Eq. (8.195) reads

$$q = -\frac{\kappa T'}{B}. \tag{8.198}$$

Combining the above equation with (8.193) we obtain

$$T' = -\frac{f(r)}{\kappa B R}. \tag{8.199}$$

At this point, however, neither can we provide solid arguments to support further the assumption about the vanishing of the relaxation time as an indicator of minimum complexity of the dissipative regime, nor can we prove that exact solutions of this kind exist.

So far we have assumed the homologous condition in order to describe the simplest mode of evolution, however as indicated in the lines above, the resulting models are perhaps too restrictive, and it could be wise to consider less stringent conditions. One possible example could be the quasi–homologous condition (8.173). In [60, 63–65] the reader may find a long list of exact solutions satisfying the vanishing complexity factor and the quasi–homologous condition for spherically symmetric and hyperbolically symmetric dissipative fluids.

Finally we shall tackle the problem of defining complexity of vacuum solutions to Einstein equations. We shall restrict our discussion to the Bondi metric.

8.4 Complexity of the Bondi Space–Time

Let us now consider the extension of our concept of complexity to vacuum space-times. More specifically we shall consider the Bondi metric [16], which encompasses a vast numbers of spacetimes, including the Minkowski spacetime, the static Weyl metrics, non–radiative non–static metrics and gravitationally radiating metrics. Furthermore, the Bondi approach has the virtue of providing a clear and precise criterion for the existence of gravitational radiation. Namely, if the news function (defined by Bondi in [16]) is zero over a time interval, then there is no radiation over that interval.

As we have seen, in the case of fluid distributions, the variable(s) measuring the complexity of the fluid (the complexity factor(s)) appear in the trace free part of the orthogonal splitting of the electric Riemann tensor. In vacuum the Riemann tensor and the Weyl tensor are the same, so if we extrapolate to the vacuum case the same definition of complexity as for the fluid self–gravitating system, we shall need the scalar functions defining the electric part of the Weyl tensor for the Bondi metric.

The general form of an axially and reflection symmetric asymptotically flat metric given by Bondi [16] is

$$ds^2 = \left(\frac{V}{r}e^{2\beta} - U^2 r^2 e^{2\gamma}\right) du^2 + 2e^{2\beta} du dr + 2Ur^2 e^{2\gamma} du d\theta - r^2 \left(e^{2\gamma} d\theta^2 + e^{-2\gamma} \sin^2\theta d\phi^2\right), \tag{8.200}$$

where V, β, U and γ are functions of u, r and θ.

We number the coordinates $x^{0,1,2,3} = u, r, \theta, \phi$ respectively. u is a timelike coordinate ($g_{uu} > 0$) converging to the retarded time as $r \to \infty$. The hypersurfaces $u = constant$ define null surfaces (their normal vectors are null vectors), which at null infinity ($r \to \infty$) coincides with the Minkowski null light cone open to the future. r is a null coordinate ($g_{rr} = 0$) and θ and ϕ are two angle coordinates (see [16] for details).

Regularity conditions in the neighborhood of the polar axis ($\sin\theta = 0$), imply that as $\sin\theta \to 0$, then $V, \beta, U/\sin\theta, \gamma/\sin^2\theta$ each equals a function of $\cos\theta$ regular on the polar axis.

The four metric functions are assumed to be expanded in series of $1/r$, then using the field equations, Bondi gets

$$\gamma = cr^{-1} + \left(C - \frac{1}{6}c^3\right) r^{-3} + \cdots, \tag{8.201}$$

$$U = -\left(c_\theta + 2c\cot\theta\right) r^{-2} + \left[2N + 3cc_\theta + 4c^2\cot\theta\right] r^{-3} \cdots, \tag{8.202}$$

$$V = r - 2M - \left(N_\theta + N\cot\theta - c_\theta^2 - 4cc_\theta\cot\theta - \frac{1}{2}c^2(1 + 8\cot^2\theta)\right) r^{-1} \cdots, \tag{8.203}$$

$$\beta = -\frac{1}{4}c^2 r^{-2} + \cdots, \tag{8.204}$$

where c, C, N and M are functions of u and θ satisfying the constraint

$$4C_u = 2c^2 c_u + 2cM + N\cot\theta - N_\theta, \tag{8.205}$$

and letters as subscripts denote derivatives. The three functions c, M and N are further related by the supplementary conditions

$$M_u = -c_u^2 + \frac{1}{2}\left(c_{\theta\theta} + 3c_\theta\cot\theta - 2c\right)_u, \tag{8.206}$$

$$-3N_u = M_\theta + 3cc_{u\theta} + 4cc_u\cot\theta + c_u c_\theta. \tag{8.207}$$

In the static case M equals the mass of the system and is called by Bondi the "mass aspect", whereas N and C are closely related to the dipole and quadrupole moments respectively.

Next, Bondi defines the mass $m(u)$ of the system as

$$m(u) = \frac{1}{2}\int_0^\pi M \sin\theta d\theta, \tag{8.208}$$

which by virtue of (8.206) yields

$$m_u = -\frac{1}{2}\int_0^\pi c_u^2 \sin\theta d\theta. \tag{8.209}$$

Arriving at this point let us summarize the main conclusions emerging from Bondi's approach.

1. If γ, M and N are known for some $u = a$(constant) and c_u (the news function) is known for all u in the interval $a \leq u \leq b$, then the system is fully determined in that interval. In other words, whatever happens at the source, leading to changes in the field, it can only do so by affecting c_u and vice versa. In the light of this comment the relationship between news function and the occurrence of radiation becomes clear.
2. As it follows from (8.209), the mass of a system is constant if and only if there is no news.

Now, for an observer at rest in the frame of (8.200), the four-velocity vector has components

$$V^\alpha = \left(\frac{1}{A}, 0, 0, 0\right), \tag{8.210}$$

with

$$A \equiv \left(\frac{V}{r}e^{2\beta} - U^2 r^2 e^{2\gamma}\right)^{1/2}. \tag{8.211}$$

Next, let us introduce the unit, spacelike vectors **K**, **L**, **S**, with components

$$K^\alpha = \left(\frac{1}{A}, -e^{-2\beta}A, 0, 0\right), \tag{8.212}$$

$$L^\alpha = \left(0, Ure^\gamma e^{-2\beta}, -\frac{e^{-\gamma}}{r}, 0\right) \tag{8.213}$$

$$S^\alpha = \left(0, 0, 0, -\frac{e^\gamma}{r\sin\theta}\right), \tag{8.214}$$

For the observer defined by (8.210) the vorticity vector may be written as (see [42] for details)

$$\omega^\alpha = \left(0, 0, 0, \omega^\phi\right). \tag{8.215}$$

The explicit expressions for ω^ϕ and its absolute value $\Omega \equiv (-\omega_\alpha \omega^\alpha)^{1/2}$ are given in the Appendix 8.13.

The electric and magnetic parts of Weyl tensor, $E_{\alpha\beta}$ and $H_{\alpha\beta}$, respectively, are formed from the Weyl tensor $C_{\alpha\beta\gamma\delta}$ and its dual $\tilde{C}_{\alpha\beta\gamma\delta}$ by contraction with the four velocity vector given by (8.210)

$$E_{\alpha\beta} = C_{\alpha\gamma\beta\delta} V^\gamma V^\delta, \tag{8.216}$$

$$H_{\alpha\beta} = \tilde{C}_{\alpha\gamma\beta\delta} V^\gamma V^\delta = \frac{1}{2}\epsilon_{\alpha\gamma\epsilon\delta} C^{\epsilon\delta}{}_{\beta\rho} V^\gamma V^\rho,$$

$$\epsilon_{\alpha\beta\gamma\delta} \equiv \sqrt{-g}\ \eta_{\alpha\beta\gamma\delta}, \tag{8.217}$$

where $\eta_{\alpha\beta\gamma\delta}$ is the permutation symbol.

The electric part of the Weyl tensor has only three independent non-vanishing components, whereas only two components define the magnetic part. Thus we may write

$$E_{\alpha\beta} = \mathcal{E}_1 \left(K_\alpha L_\beta + L_\alpha K_\beta\right) + \mathcal{E}_2 \left(K_\alpha K_\beta + \frac{1}{3}h_{\alpha\beta}\right) + \mathcal{E}_3 \left(L_\alpha L_\beta + \frac{1}{3}h_{\alpha\beta}\right), \tag{8.218}$$

and

$$H_{\alpha\beta} = H_1(S_\alpha K_\beta + S_\beta K_\alpha) + H_2(S_\alpha L_\beta + S_\beta L_\alpha). \tag{8.219}$$

with $h_{\mu\nu} = g_{\mu\nu} - V_\nu V_\mu$, and

$$\mathcal{E}_1 = L^\alpha K^\beta E_{\alpha\beta}, \tag{8.220}$$

$$\mathcal{E}_2 = (2K^\alpha K^\beta + L^\alpha L^\beta) E_{\alpha\beta}, \tag{8.221}$$

$$\mathcal{E}_3 = (2L^\alpha L^\beta + K^\alpha K^\beta) E_{\alpha\beta}, \tag{8.222}$$

these three scalars represent the complexity factors of our solutions.

For the magnetic part we have

$$H_2 = S^\alpha L^\beta H_{\alpha\beta}, \tag{8.223}$$

$$H_1 = S^\alpha K^\beta H_{\alpha\beta}. \tag{8.224}$$

Explicit expressions for these scalars are given in the Appendixes 8.11 and 8.12.

In [42] it was obtained that if we put $H^{\alpha}_{\beta} = 0$ then the field is non–radiative and up to order $1/r^3$ in γ, the metric is static, and the mass, the "dipole" (N) and the "quadrupole" (C) moments correspond to a static situation. However, the time dependence might enter through coefficients of higher order in γ, giving rise to what Bondi calls "non–natural–non–radiative moving system" (NNNRS). In this latter case, the system keeps the first three moments independent of time, but allows for time dependence of higher moments. This class of solutions is characterized by $M_\theta = 0$.

A second family of time dependent non–radiative solutions exists for which $M_\theta \neq 0$. These are called natural non–radiative moving system" (NNRS), and their magnetic Weyl tensor is non–vanishing.

Let us now discuss the hierarchy of different spacetimes belonging to the Bondi family, according to their complexity.

The simplest spacetime corresponds to the vanishing of the three complexity factors, and this is just Minkowski.

Indeed, as it was shown in [42], if we assume $E^{\alpha}_{\beta} = 0$ and use regularity conditions, we find that the spacetime must be Minkowski, giving further support to the conjecture that there are no purely magnetic vacuum space–times [18].

At the other end (maximal complexity) we have a gravitationally radiating system which requires all three complexity factors to be different from zero.

Indeed, if we assume that $\mathcal{E}_1 = 0$, then it follows at once from (8.261) that $c_u = 0$ (otherwise c_u would be a non–regular function of θ on the symmetry axis). Thus $\mathcal{E}_1 = 0$ implies that the system is non–radiative.

If instead we assume that $\mathcal{E}_2 = 0$, then from the first order in (8.262) we obtain that $c_{uu} = 0$, this implies that either $c_u = 0$ or $c_u \sim u$. Bondi refers to this latter case as "mass loss without radiative Riemann tensor" and dismisses it as being of little physical significance. As a matter of fact, in this latter case the system would be radiating "forever", which according to (8.209) requires an unbounded source, incompatible with an asymptotically flat spacetime. Thus in this case too, we have $c_u = 0$, and the system is non–radiative.

Finally, if we assume $\mathcal{E}_3 = 0$ it follows at once from the first order in (8.263), that $c_{uu} = 0$, leading to $c_u = 0$, according to the argument above.

Thus, a radiative system requires all three complexity factors to be nonvanishing, implying a maximal complexity.

In the middle of the two extreme cases we have, on the one hand the spherically symmetric spacetime (Schwarzschild), characterized by a single complexity factor (the same applies for any static metric), $\mathcal{E}_1 = \mathcal{E}_3 = 0$, and $\mathcal{E}_2 = \frac{3M}{r^3}$. On the other hand, we have the non–static non–radiative case.

Let us now analyze in detail this latter case. This group of solutions encompasses two subclasses, which using Bondi notation are:

1. Natural–non–radiative systems (NNRS) characterized by $M_\theta \neq 0$.
2. Non–natural–non–radiative systems (NNNRS) characterized by $M_\theta = 0$.

Let us first consider the NNNRS subcase. Using (8.261) we obtain $\mathscr{E}_1 = 0$, (up to order $1/r^3$), while the first non–vanishing terms in $\mathscr{E}_2$ and $\mathscr{E}_3$ are respectively, $3M$ and 0, where (8.8), (8.206), (8.207), (8.262) and (8.263) have been used.

Thus, the NNNRS are characterized by only one non–vanishing complexity factor ($\mathscr{E}_2$). Furthermore, as it follows from (8.268) the vorticity of the congruence of observers at rest with respect to the frame of (8.200) vanishes, and the field is purely electric. However as mentioned before we cannot conclude that the field is static, since the u dependence might appear through coefficients of higher order in y.

Let us now consider the "natural–non–radiative system" (NNRS). In this subcase, using (8.261) we obtain $\mathscr{E}_1 = 0$, (up to order $1/r^3$) as for the NNNRS subcase, while the first non–vanishing term in $\mathscr{E}_2$ and $\mathscr{E}_3$ (up to order $1/r^3$) are respectively $3M + M_{\theta\theta}/4 - M_\theta \cot\theta/4$ and $M_{\theta\theta}/2 - M_\theta \cot\theta/2$.

Also, up to the same order, it follows from (8.264) and (8.265) that $H_1 = 0$ for both subcases, while the corresponding term in H_2 is (for NNRS)

$$H_2 = -\frac{1}{4r^3}(M_{\theta\theta} - M_\theta \cot\theta), \tag{8.225}$$

which of course vanishes for the NNNRS subcase.

It should be observed that if we assume $\mathscr{E}_3 = 0$ or $H_2 = 0$ then it follows at once from the above that

$$M_{\theta\theta} - M_\theta \cot\theta = 0 \Rightarrow M = a\cos\theta, \quad a = constant. \tag{8.226}$$

But this implies because of (8.208) that the Bondi mass function of the system vanishes. Therefore, the only physically meaningful NNRS requires $\mathscr{E}_3 \neq 0$, $\Omega \neq 0$ and $H_2 \neq 0$ implying that the complexity is characterized by two complexity factors ($\mathscr{E}_2, \mathscr{E}_3$).

Thus a hierarchy of spacetimes according to their complexity has been established. This allows us to discriminate between two classes of spacetimes that depend on time but are not radiative (vanishing of the news function). These two classes were called by Bondi [16] natural and non–natural non–radiative moving systems, and are characterized by different forms of the mass aspect. They exhibit different degree of complexity.

Unfortunately, though, up to the leading order of the complexity factors analyzed here, it is impossible to discriminate between different radiative systems according to their complexity. Higher order terms would be necessary for that purpose, although it is not clear at this point if it is possible to establish such a hierarchy of radiative systems after all.

The simplest system (Minkowski) is characterized by the vanishing of all the complexity factors. Next, the static case (including Schwarzchild) is described by a single complexity factor.

The time dependent non–radiative solutions split in two subgroups depending on the form of the mass aspect M. If $M_\theta = 0$ which corresponds to the NNNRS the complexity is similar to the static case. Also, in this case, as in the static situation,

Table 8.1 Complexity factors for different spacetimes of the Bondi metric ($\mathscr{E}^{(n)}_{1,2,3}$ are the coefficients of order $\mathscr{O}(r^{-n})$)

Complexity Hierarchy

$Complex.fac.\diagdown$ $spacetimes$	Minkowski	Static	NNNRS	NNRS	Radiative
$\mathscr{E}_1$	0	0	0	0	$\mathscr{E}^{(n)}_1 \neq 0$, $n \geq 1$
$\mathscr{E}_2$	0	$\mathscr{E}^{(3)}_2 = 3M$	$\mathscr{E}^{(3)}_2 = 3M$	$\mathscr{E}^{(3)}_2 = 3M + \frac{M_{\theta\theta}}{4} - \frac{M_\theta \cot\theta}{4}$	$\mathscr{E}^{(n)}_2 \neq 0$, $n \geq 1$
$\mathscr{E}_3$	0	0	0	$\mathscr{E}^{(3)}_3 = \frac{1}{2}(M_{\theta\theta} - M_\theta \cot\theta)$	$\mathscr{E}^{(n)}_3 \neq 0$, $n \geq 1$

Table 8.2 The magnetic parts of the Weyl tensor and the vorticity for different spacetimes of the Bondi metric

Magnetic parts and vorticity

$Mag.Weyl; \Omega.\diagdown$ $spacetimes$	Minkowski	Static	NNNRS	NNRS	Radiative
H_1	0	0	0	0	$H^{(n)}_1 \neq 0$, $n \geq 1$
H_2	0	0	0	$H^{(3)}_2 = -\frac{1}{4}(M_{\theta\theta} - M_\theta \cot\theta)$	$H^{(n)}_2 \neq 0$, $n \geq 1$
Ω	0	0	0	$\Omega^{(2)} = M_\theta$	$\Omega^{(n)} \neq 0$, $n \geq 1$

the vorticity vanishes and the field is purely electric. This result could suggest that in fact NNNRS are just static, and no time dependence appears in the coefficients of higher order in y. On the contrary for the NNRS there are two complexity factors, the vorticity is non–vanishing and the field is not purely electric.

All these results are summarized in Tables 8.1 and 8.2. Thus, NNNRS and NNRS are clearly differentiated through their degree of complexity, as measured by the complexity factors considered here.

The fact that radiative systems necessarily decay into NNRS, NNNRS or static systems, since the Bondi mass function must be finite, suggests that higher degrees of complexity might be associated with stronger stability. Of course a proof of this conjecture requires a much more detailed analysis.

It is also worth mentioning the conspicuous link between vorticity and complexity factors. Indeed vorticity appears only in NNRS and radiative systems, which are the most complex systems, while it is absent in the simplest systems (Minkowski, static, NNNRS). In the radiative case there are contributions at order $\mathscr{O}(r^{-1})$ related to the news function, and at order $\mathscr{O}(r^{-2})$, while for the NNRS there are only contributions at order $\mathscr{O}(r^{-2})$, these describe the effect of the tail of the wave.

8.5 Conclusions

We have discussed about the concept of complexity of self–gravitating relativistic fluid distributions. For the static case we were concerned exclusively with the notion of complexity of the structure of the fluid. However in the dynamic case we have also tackled the question about the complexity of the pattern of evolution. These two questions, although related, refer to different aspects of the definition of complexity. However, it is remarkable that in the non–dissipative case the homologous condition implies the vanishing complexity factor.

As a measure of complexity of the structure of the fluid (the complexity factor) we have chosen the scalar function(s) Y_{TF} defining the trace–free part of the electric Riemann tensor.

Next, we discussed about the complexity of the pattern of evolution. Two possibilities appear as the more obvious candidates: the homologous condition and the quasi–homologous condition. The latter, being less restricted than the former allows to consider larger number of models.

All this having been said, we are well aware of the fact that the above mentioned candidate for measuring the complexity of a self-gravitating system is by no means unique and many different alternatives may be proposed. In the same line of arguments, the simplest patterns of evolution assumed so far are the homologous and the quasi-homologous regimes. But, it is not clear whether or not other patterns of evolution could also fit the role of the simplest pattern of evolution.

Additionally, we believe that alternative definitions of complexity for vacuum space–time is worth considering. Finally, new exact solutions to the field equations in the context of the Einstein theory or any alternative one, would serve as a test-bed for the definition of complexity.

These remarks suggest a list of questions and open issues which we believe that deserve to be considered further.

- Are there alternative definitions of complexity, different from the one proposed in [54]?
- Are there other ways to extend the definition of complexity for vacuum space–time?
- Besides the homologous and the quasi-homologous regime, could we define another pattern of evolution that could qualify as the simplest one?
- Do physically meaningful dissipative models satisfying (8.193) exist?
- If the answer to the above question is positive then, is there a unique solution or are there a large number of them?
- What is the physical meaning of such solution(s)?
- Is it physically reasonable to neglect transient effects when considering the simplest dissipative system, and assume that the relaxation time vanishes?
- To summarize the four points above: is there a specific dissipative regime that could be considered as the simplest one?
- Can we relate the complexity factor(s) in the non-spherically symmetric case to the active gravitational mass, as in the spherically symmetric case?

- Can we single out a specific family of exact axially symmetric static solutions satisfying the vanishing complexity factor(s) condition?
- Can any of the above solutions be matched smoothly to any vacuum Weyl solution?
- The definition of complexity proposed in [54] is not directly related to entropy or disequilibrium, although it is possible that such a link might exist after all. If so, how could such relationship be brought out?
- What is the simplest mode of evolution of axially–symmetric fluids? How is such a mode related to the emission (or not) of gravitational radiation?
- Could it be possible to provide a definition of the arrow of time in terms of the complexity factor?
- How is the complexity factor related to physical relevant properties of the source, in terms of stability or maximal degree of compactness?
- How does the complexity factor evolve? Do physically meaningful systems prefer vanishing complexity factors?
- As we have seen the FRW model satisfies the vanishing complexity factor and evolves homologously. Should any other, physically sound, cosmological model have a vanishing complexity factor? Should it evolve in the homologous or quasi-homologous regime?
- The complexity factor for a charged fluid is known, but what is the complexity factor for a different type of field (e.g., scalar field?).
- How should we define the complexity factor in the context of other alternative theories of gravity that have not been considered so far?
- How can we find new solutions satisfying the vanishing complexity factor? Could we use the general methods described in [70–72, 115, 116] to obtain such solutions?
- What relevant physical features share solutions satisfying the vanishing complexity factor?
- Is there a link between the concept of complexity and some kind of symmetry (e.g. motions, conformal motions, affine collineations, curvature collineations, matter collineations, etc.)?
- We have extended the concept of complexity, adopted for fluids, to the vacuum case for the Bondi metric, the three complexity factors corresponding to the three scalars defining the electric part of the Weyl tensor. Could it be possible to further refine the scheme proposed here so as to discriminate between different radiative systems according to their complexity? Or, in other words, among radiative systems is there a simplest one ?
- Also in the vacuum case, could it be possible to discriminate between different static spacetimes of the Weyl family?

Acknowledgements This work was partially supported by Ministerio de Ciencia, Innovacion y Universidades. Grant number: PID2021-122938NB-I00. I also wish to thank Universitat de les Illes Balears for financial support and hospitality.

Appendix

8.6 The Einstein and the Conservation Equations for the Axially Symmetric Static Case

For the line element (8.43) and the energy momentum (8.47), the Einstein equations read:

$$8\pi\mu = -\frac{1}{B^2}\left\{\frac{B''}{B} + \frac{D''}{D} + \frac{1}{r}\left(\frac{B'}{B} + \frac{D'}{D}\right) - \left(\frac{B'}{B}\right)^2 + \frac{1}{r^2}\left[\frac{B_{\theta\theta}}{B} + \frac{D_{\theta\theta}}{D} - \left(\frac{B_\theta}{B}\right)^2\right]\right\}, \tag{8.227}$$

$$8\pi P_{xx} = \frac{1}{B^2}\left[\frac{A'B'}{AB} + \frac{A'D'}{AD} + \frac{B'D'}{BD} + \frac{1}{r}\left(\frac{A'}{A} + \frac{D'}{D}\right) + \frac{1}{r^2}\left(\frac{A_{\theta\theta}}{A} + \frac{D_{\theta\theta}}{D} - \frac{A_\theta B_\theta}{AB} + \frac{A_\theta D_\theta}{AD} - \frac{B_\theta D_\theta}{BD}\right)\right], \tag{8.228}$$

$$8\pi P_{yy} = \frac{1}{B^2}\left[\frac{A''}{A} + \frac{D''}{D} - \frac{A'B'}{AB} + \frac{A'D'}{AD} - \frac{B'D'}{BD} + \frac{1}{r^2}\left(\frac{A_\theta B_\theta}{AB} + \frac{A_\theta D_\theta}{AD} + \frac{B_\theta D_\theta}{BD}\right)\right], \tag{8.229}$$

$$8\pi P_{zz} = \frac{1}{B^2}\left\{\frac{A''}{A} + \frac{B''}{B} - \left(\frac{B'}{B}\right)^2 + \frac{1}{r}\left(\frac{A'}{A} + \frac{B'}{B}\right) + \frac{1}{r^2}\left[\frac{A_{\theta\theta}}{A} + \frac{B_{\theta\theta}}{B} - \left(\frac{B_\theta}{B}\right)^2\right]\right\}, \tag{8.230}$$

$$8\pi P_{xy} = \frac{1}{B^2}\left\{\frac{1}{r}\left[-\frac{A'_\theta}{A} - \frac{D'_\theta}{D} + \frac{B_\theta}{B}\left(\frac{A'}{A} + \frac{D'}{D}\right) + \frac{B'}{B}\frac{A_\theta}{A} + \frac{B'}{B}\frac{D_\theta}{D}\right] + \frac{1}{r^2}\left(\frac{A_\theta}{A} + \frac{D_\theta}{D}\right)\right\}. \tag{8.231}$$

The nonvanishing components of the conservation equations $T^{\alpha\beta}_{;\beta} = 0$ yield: the trivial equation

$$\dot{\mu} = 0, \tag{8.232}$$

where the overdot denotes derivative with respect to t, and the two hydrostatic equilibrium equations

$$\left[P + \frac{1}{3}(2\Pi_2 - \Pi_3)\right]' + \frac{A'}{A}\left[\mu + P + \frac{1}{3}(2\Pi_2 - \Pi_3)\right] + \frac{B'}{B}(\Pi_2 - \Pi_3) + \frac{D'}{D}\Pi_2$$
$$+ \frac{1}{r}\left[\left(\frac{A_\theta}{A} + 2\frac{B_\theta}{B} + \frac{D_\theta}{D}\right)\Pi_{KL} + \Pi_{KL\theta} + \Pi_2 - \Pi_3\right] = 0, \tag{8.233}$$

$$\left[P + \frac{1}{3}(2\Pi_3 - \Pi_2)\right]_{,\theta} + \frac{A_\theta}{A}\left[\mu + P + \frac{1}{3}(2\Pi_3 - \Pi_2)\right] + \frac{B_\theta}{B}(\Pi_3 - \Pi_2)$$
$$+ \frac{D_\theta}{D}\Pi_3 + r\left[\left(\frac{A'}{A} + 2\frac{B'}{B} + \frac{D'}{D}\right)\Pi_{KL} + \Pi'_{KL}\right] + 2\Pi_{KL} = 0. \tag{8.234}$$

8.7 Expression for the Components of the Electric Weyl Tensor

There are four nonvanishing components as calculated from (8.60), however they are not independent since they satisfy the relationship:

$$E_{11} + \frac{1}{r^2}E_{22} + \frac{B^2}{D^2}E_{33} = 0, \tag{8.235}$$

implying that the Weyl tensor may be expressed through three independent scalar functions $\mathcal{E}_1, \mathcal{E}_2, \mathcal{E}_3$.

These four components are

$$E_{11} = \frac{1}{6}\left[\frac{2A''}{A} - \frac{B''}{B} - \frac{D''}{D} - \frac{3A'B'}{AB} - \frac{A'D'}{AD} + \left(\frac{B'}{B}\right)^2 + \frac{3B'D'}{BD} + \frac{1}{r}\left(2\frac{D'}{D} - \frac{B'}{B} - \frac{A'}{A}\right)\right]$$
$$+ \frac{1}{6r^2}\left[-\frac{A_{\theta\theta}}{A} - \frac{B_{\theta\theta}}{B} + \frac{2D_{\theta\theta}}{D} + \frac{3A_\theta B_\theta}{AB} - \frac{A_\theta D_\theta}{AD} + \left(\frac{B_\theta}{B}\right)^2 - \frac{3B_\theta D_\theta}{BD}\right], \tag{8.236}$$

$$E_{22} = -\frac{r^2}{6}\left[\frac{A''}{A} + \frac{B''}{B} - \frac{2D''}{D} - \frac{3A'B'}{AB} + \frac{A'D'}{AD} - \left(\frac{B'}{B}\right)^2 + \frac{3B'D'}{BD} + \frac{1}{r}\left(\frac{D'}{D} + \frac{B'}{B} - \frac{2A'}{A}\right)\right]$$
$$- \frac{1}{6}\left[-\frac{2A_{\theta\theta}}{A} + \frac{B_{\theta\theta}}{B} + \frac{D_{\theta\theta}}{D} + \frac{3A_\theta B_\theta}{AB} + \frac{A_\theta D_\theta}{AD} - \left(\frac{B_\theta}{B}\right)^2 - \frac{3B_\theta D_\theta}{BD}\right], \quad (8.237)$$

$$E_{33} = -\frac{D^2}{6B^2}\left[\frac{A''}{A} - \frac{2B''}{B} + \frac{D''}{D} - \frac{2A'D'}{AD} + 2\left(\frac{B'}{B}\right)^2 + \frac{1}{r}\left(\frac{D'}{D} - \frac{2B'}{B} + \frac{A'}{A}\right)\right]$$
$$- \frac{D^2}{6B^2r^2}\left[\frac{A_{\theta\theta}}{A} - \frac{2B_{\theta\theta}}{B} + \frac{D_{\theta\theta}}{D} - \frac{2A_\theta D_\theta}{AD} + 2\left(\frac{B_\theta}{B}\right)^2\right], \quad (8.238)$$

$$E_{12} = \frac{1}{2}\left[\frac{A'_\theta}{A} - \frac{D'_\theta}{D} + \frac{B_\theta}{B}\frac{D'}{D} - \frac{A'B_\theta}{AB} - \frac{B'A_\theta}{AB} + \frac{D_\theta}{D}\frac{B'}{B}\right.$$
$$\left. - \frac{1}{r}\left(\frac{A_\theta}{A} - \frac{D_\theta}{D}\right)\right]. \quad (8.239)$$

For the three scalars $\mathscr{E}_1, \mathscr{E}_2, \mathscr{E}_3$ we obtain

$$\mathscr{E}_1 = \frac{1}{2B^2}\left[\frac{1}{r}\left(\frac{A'_\theta}{A} - \frac{D'_\theta}{D} - \frac{B_\theta}{B}\frac{A'}{A} + \frac{D'}{D}\frac{B_\theta}{B} - \frac{B'}{B}\frac{A_\theta}{A} + \frac{D_\theta}{D}\frac{B'}{B}\right)\right.$$
$$\left. + \frac{1}{r^2}\left(\frac{D_\theta}{D} - \frac{A_\theta}{A}\right)\right], \quad (8.240)$$

$$\mathscr{E}_2 = -\frac{1}{2B^2}\left[-\frac{A''}{A} + \frac{B''}{B} + \frac{A'B'}{AB} + \frac{A'D'}{AD} - \left(\frac{B'}{B}\right)^2 - \frac{B'D'}{BD} + \frac{1}{r}\left(\frac{B'}{B} - \frac{D'}{D}\right)\right]$$
$$- \frac{1}{2B^2r^2}\left[\frac{B_{\theta\theta}}{B} - \frac{D_{\theta\theta}}{D} - \frac{A_\theta B_\theta}{AB} + \frac{A_\theta D_\theta}{AD} - \left(\frac{B_\theta}{B}\right)^2 + \frac{B_\theta D_\theta}{BD}\right], \quad (8.241)$$

$$\mathscr{E}_3 = -\frac{1}{2B^2}\left[\frac{B''}{B} - \frac{D''}{D} - \frac{A'B'}{AB} + \frac{A'D'}{AD} - \left(\frac{B'}{B}\right)^2 + \frac{B'D'}{BD} + \frac{1}{r}\left(\frac{B'}{B} - \frac{A'}{A}\right)\right]$$
$$- \frac{1}{2B^2r^2}\left[\frac{B_{\theta\theta}}{B} - \frac{A_{\theta\theta}}{A} + \frac{A_\theta B_\theta}{AB} + \frac{A_\theta D_\theta}{AD} - \left(\frac{B_\theta}{B}\right)^2 - \frac{B_\theta D_\theta}{BD}\right]. \quad (8.242)$$

Or, using Einstein equations we may also write

$$\mathcal{E}_1 = \frac{E_{12}}{B^2 r} = 4\pi \Pi_{KL} + \frac{1}{B^2 r}\left[\frac{A'_\theta}{A} - \frac{A' B_\theta}{AB} - \frac{A_\theta}{A}\left(\frac{B'}{B} + \frac{1}{r}\right)\right], \quad (8.243)$$

$$\mathcal{E}_2 = -\frac{2E_{33}}{D^2} - \frac{E_{22}}{B^2 r^2} = 4\pi(\mu + 3P + \Pi_2) - \frac{A'}{B^2 A}\left(\frac{2D'}{D} + \frac{B'}{B} + \frac{1}{r}\right)$$
$$+ \frac{A_\theta}{AB^2 r^2}\left(\frac{B_\theta}{B} - \frac{2D_\theta}{D}\right) - \frac{1}{B^2 r^2}\frac{A_{\theta\theta}}{A}, \quad (8.244)$$

$$\mathcal{E}_3 = -\frac{E_{33}}{D^2} + \frac{E_{22}}{B^2 r^2} = 4\pi \Pi_3 - \frac{A'}{B^2 A}\left(\frac{D'}{D} - \frac{B'}{B} - \frac{1}{r}\right) - \frac{A_\theta}{AB^2 r^2}\left(\frac{D_\theta}{D} + \frac{B_\theta}{B}\right)$$
$$+ \frac{1}{B^2 r^2}\frac{A_{\theta\theta}}{A}. \quad (8.245)$$

8.8 Vanishing Complexity Factor Conditions

$$Y_{TF1} = \frac{1}{B^2 r}\left[\frac{A'_\theta}{A} - \frac{A' B_\theta}{AB} - \frac{A_\theta}{A}\left(\frac{B'}{B} + \frac{1}{r}\right)\right] = 0, \quad (8.246)$$

$$Y_{TF2} = \frac{A''}{B^2 A} - \frac{A'}{B^2 A}\left(\frac{D'}{D} + \frac{B'}{B}\right) + \frac{A_\theta}{AB^2 r^2}\left(\frac{B_\theta}{B} - \frac{D_\theta}{D}\right) = 0, \quad (8.247)$$

$$Y_{TF3} = -\frac{A'}{B^2 A}\left(\frac{D'}{D} - \frac{B'}{B} - \frac{1}{r}\right) - \frac{A_\theta}{AB^2 r^2}\left(\frac{D_\theta}{D} + \frac{B_\theta}{B}\right) + \frac{1}{B^2 r^2}\frac{A_{\theta\theta}}{A} = 0. \quad (8.248)$$

8.9 Einstein Equations for the Dynamical Spherically Symmetric Case

Einstein's field equations

$$G_{\alpha\beta} = 8\pi T_{\alpha\beta}, \tag{8.249}$$

for the interior spacetime (8.140) read

$$8\pi T_{00} = 8\pi\mu A^2 = \left(2\frac{\dot{B}}{B} + \frac{\dot{R}}{R}\right)\frac{\dot{R}}{R} - \left(\frac{A}{B}\right)^2 \left[2\frac{R''}{R} + \left(\frac{R'}{R}\right)^2 - 2\frac{B'}{B}\frac{R'}{R} - \left(\frac{B}{R}\right)^2\right], \tag{8.250}$$

$$8\pi T_{01} = -8\pi q A B = -2\left(\frac{\dot{R}'}{R} - \frac{\dot{B}}{B}\frac{R'}{R} - \frac{\dot{R}}{R}\frac{A'}{A}\right), \tag{8.251}$$

$$8\pi T_{11} = 8\pi P_r B^2 = -\left(\frac{B}{A}\right)^2 \left[2\frac{\ddot{R}}{R} - \left(2\frac{\dot{A}}{A} - \frac{\dot{R}}{R}\right)\frac{\dot{R}}{R}\right] + \left(2\frac{A'}{A} + \frac{R'}{R}\right)\frac{R'}{R} - \left(\frac{B}{R}\right)^2, \tag{8.252}$$

$$\begin{aligned} 8\pi T_{22} = \frac{8\pi}{\sin^2\theta} T_{33} = 8\pi P_\perp R^2 = &-\left(\frac{R}{A}\right)^2 \left[\frac{\ddot{B}}{B} + \frac{\ddot{R}}{R} - \frac{\dot{A}}{A}\left(\frac{\dot{B}}{B} + \frac{\dot{R}}{R}\right) + \frac{\dot{B}}{B}\frac{\dot{R}}{R}\right] \\ &+ \left(\frac{R}{B}\right)^2 \left[\frac{A''}{A} + \frac{R''}{R} - \frac{A'}{A}\frac{B'}{B} + \left(\frac{A'}{A} - \frac{B'}{B}\right)\frac{R'}{R}\right]. \end{aligned} \tag{8.253}$$

The component (8.251) can be rewritten with (8.148) and (8.151) as

$$4\pi q B = \frac{1}{3}(\Theta - \sigma)' - \sigma\frac{R'}{R}. \tag{8.254}$$

8.10 Dynamical Equations

The non trivial components of the Bianchi identities, $T^{\alpha\beta}_{;\beta} = 0$, from (8.249) yield

$$T^{\alpha\beta}_{;\beta} V_\alpha = -\frac{1}{A}\left[\dot{\mu} + (\mu + P_r)\frac{\dot{B}}{B} + 2(\mu + P_\perp)\frac{\dot{R}}{R}\right] - \frac{1}{B}\left[q' + 2q\frac{(AR)'}{AR}\right] = 0, \tag{8.255}$$

$$T^{\alpha\beta}_{;\beta} \chi_\alpha = \frac{1}{A}\left[\dot{q} + 2q\left(\frac{\dot{B}}{B} + \frac{\dot{R}}{R}\right)\right] + \frac{1}{B}\left[P_r' + (\mu + P_r)\frac{A'}{A} + 2(P_r - P_\perp)\frac{R'}{R}\right] = 0, \tag{8.256}$$

or, by using (8.147), (8.148), (8.157) and (8.155), they become, respectively,

$$D_T\mu + \frac{1}{3}(3\mu + P_r + 2P_\perp)\Theta + \frac{2}{3}(P_r - P_\perp)\sigma + ED_R q + 2q\left(a + \frac{E}{R}\right) = 0, \tag{8.257}$$

$$D_T q + \frac{2}{3}q(2\Theta + \sigma) + ED_R P_r + (\mu + P_r)a + 2(P_r - P_\perp)\frac{E}{R} = 0. \tag{8.258}$$

This last equation may be further transformed as follows, the acceleration $D_T U$ of an infalling particle can be obtained by using (8.147), (8.252), (8.152) and (8.155), producing

$$D_T U = -\frac{m}{R^2} - 4\pi P_r R + Ea, \tag{8.259}$$

and then, substituting a from (8.259) into (8.258), we obtain

$$\begin{aligned}(\mu + P_r)D_T U = &-(\mu + P_r)\left[\frac{m}{R^2} + 4\pi P_r R\right] - E^2\left[D_R P_r + 2(P_r - P_\perp)\frac{1}{R}\right] \\ &- E\left[D_T q + 2q\left(2\frac{U}{R} + \sigma\right)\right].\end{aligned} \tag{8.260}$$

8.11 The Complexity Factors for the Bondi Metric

$$\mathscr{E}_1 = \frac{1}{r^2}(2c_u \cot\theta + c_{\theta u}) + \mathscr{O}(r^{-n}), \quad n \geq 4, \tag{8.261}$$

$$\begin{aligned}\mathscr{E}_2 = &\frac{1}{r}c_{uu} - \frac{1}{2r^2}\left(c_{\theta\theta u} - 4Mc_{uu} + 2c_u + c_{\theta u}\cot\theta - \frac{4c_u}{\sin^2\theta}\right) \\ &+\frac{1}{r^3}\left[cc_u + 2c_\theta c_{\theta u} + 3M + \frac{\cot\theta}{2}(3c_u c_\theta + 5cc_{\theta u}) - M_u c + \frac{1}{2}M_{\theta\theta} + N_{\theta u} + P_{uu}\right. \\ &- \cot\theta\left(Mc_{\theta u} + \frac{1}{2}M_\theta + N_u - Nc_{uu}\right) - Mc_u\left(1 - \frac{4}{\sin^2\theta}\right) + c_u\left(cc_u + \frac{1}{2}c_{\theta\theta}\right) \\ &\left. + c_{uu}(4M^2 + N_\theta) - c_{\theta\theta u}\left(M - \frac{3}{2}c\right)\right] \\ &+\mathscr{O}(r^{-n}),\end{aligned} \tag{8.262}$$

$$\mathscr{E}_3 = \frac{2}{r}c_{uu} - \frac{1}{r^2}\left(c_{\theta\theta u} - 4Mc_{uu} + 2c_u + c_{\theta u}\cot\theta - \frac{4c_u}{\sin^2\theta}\right)$$
$$+\frac{1}{r^3}\left[-4cc_u + 4c_\theta c_{\theta u} + \cot\theta\left(3c_u c_\theta + 5cc_{\theta u}\right) - 2M_u c + M_{\theta\theta} + 2N_{\theta u} + 2P_{uu}\right.$$
$$-\cot\theta\left(2Mc_{\theta u} + M_\theta + 2N_u - 2Nc_{uu}\right) - 2Mc_u\left(1 - \frac{4}{\sin^2\theta}\right) + c_u\left(2cc_u + c_{\theta\theta}\right)$$
$$\left.+2c_{uu}\left(4M^2 + N_\theta\right) - c_{\theta\theta u}\left(2M - 3c\right)\right]$$
$$+\mathscr{O}(r^{-n}), \tag{8.263}$$

8.12 The Magnetic Part of the Weyl Tensor

$$H_1 = -\frac{1}{r^2}\left(2c_u\cot\theta + c_{\theta u}\right) + \mathscr{O}(r^{-n}), \quad n \geq 4, \tag{8.264}$$

$$H_2 = -\frac{1}{r}c_{uu} - \frac{1}{r^2}\left[-c_u\left(1 - \frac{2}{\sin^2\theta}\right) - \frac{\cot\theta}{2}c_{\theta u} + \right.$$
$$\left.2c_{uu}(M - c) - \frac{1}{2}c_{\theta\theta u}\right] - \frac{1}{r^3}\left\{-Mc_u\left(1 - \frac{4}{\sin^2\theta}\right) - \frac{4cc_u}{\sin^2\theta} + \left(\frac{5}{2}c - M\right)c_{\theta\theta u}\right.$$
$$+\cot\theta\left[\frac{3}{2}c_u c_\theta - N_u - \frac{1}{2}M_\theta + Nc_{uu} + \left(\frac{7}{2}c - M\right)c_{\theta u}\right] + \left(\frac{5}{2}c - M\right)c_{\theta\theta u}$$
$$+\frac{1}{2}c_{\theta\theta}c_u + 2c_\theta c_{\theta u} + cc_u^2 + \frac{1}{2}M_{\theta\theta} - cM_u + N_{\theta u} +$$
$$\left.P_{uu} + c_{uu}\left(4c^2 + 4M^2 - 4Mc + N_\theta\right)\right\} + \mathscr{O}(r^{-n}), \quad n \geq 4 \tag{8.265}$$

where $P = C - \frac{c^3}{6}$.

8.13 The Vorticity

$$\omega^\phi = -\frac{e^{-2\beta}}{2r^2\sin\theta}\left[2\beta_\theta e^{2\beta} - \frac{2e^{2\beta}A_\theta}{A} - \left(Ur^2e^{2\gamma}\right)_r + \frac{2Ur^2e^{2\gamma}}{A}A_r\right. \tag{8.266}$$
$$\left.+\frac{e^{2\beta}\left(Ur^2e^{2\gamma}\right)_u}{A^2} - \frac{Ur^2e^{2\gamma}}{A^2}2\beta_u e^{2\beta}\right],$$

and for the absolute value of ω^α we get

$$\Omega \equiv (-\omega_\alpha \omega^\alpha)^{1/2} = \frac{e^{-2\beta-\gamma}}{2r}\left[2\beta_\theta e^{2\beta} - 2e^{2\beta}\frac{A_\theta}{A} - \left(Ur^2e^{2\gamma}\right)_r \right. \tag{8.267}$$
$$\left. +2Ur^2e^{2\gamma}\frac{A_r}{A} + \frac{e^{2\beta}}{A^2}\left(Ur^2e^{2\gamma}\right)_u - 2\beta_u\frac{e^{2\beta}}{A^2}Ur^2e^{2\gamma}\right]$$

Feeding back (8.201)–(8.204) into (8.268) and keeping only the two leading terms, we obtain

$$\Omega = -\frac{1}{2r}(c_{u\theta} + 2c_u \cot\theta) + \frac{1}{r^2}\left[M_\theta - M(c_{u\theta} + 2c_u\cot\theta) - cc_{u\theta}\right. \tag{8.268}$$
$$\left. +6cc_u\cot\theta + 2c_uc_\theta\right].$$

References

1. G. Abbas, H. Nazar, Eur. Phys. J. C **78**, 510 (2018)
2. G. Abbas, H. Nazar, Eur. Phys. J. C **78**, 957 (2018)
3. G. Abbas, R. Ahmed, Astron. Space Sci. **364**, 194 (2019)
4. G. Abbas, H. Nazar, Int. J. Geo Meth. Mod. Phys. **17**, 2050043 (2020)
5. P.W. Anderson, Phys. Today **44**, 54 (1991)
6. J. Andrade et al., Int. J. Mod. Phys. D **31**, 2250093 (2022)
7. C. Arias et al., Ann. Phys. **436**, 168671 (2022)
8. I. Aslam et al., Chin. J. P.hys. **77**, 2136 (2022)
9. R. Avalos et al., Eur. Phys. J. C **82**, 420 (2022)
10. P. Bargueno et al., Ann. Phys. **443**, 169012 (2022)
11. L. Bel, Ann. Inst. H Poincaré **17**, 37 (1961)
12. L. Bel, Gen. Relativ. Gravit. **1**, 337 (1971)
13. M. Bhatti et al., Eur. Phys. J. C **82**, 714 (2022)
14. R. Bogadi, M. Govender, Eur. Phys. J. C **82**, 475 (2022)
15. R. Bogadi et al., Eur. Phys. J. C **82**, 747 (2022)
16. H. Bondi, M.G.J. van der Burg, A.W.K. Metzner, Proc. R. Soc. **A269**, 21 (1962)
17. H. Bondi, Proc. R. Soc. London **A281**, 39 (1964)
18. W.B. Bonnor, Class. Quantum Gravity **12**, 1483 (1995)
19. X. Calbet, R. López-Ruiz, Phys. Rev. E **63**, 066116 (2001)
20. R. Casadio, E. Contreras, J. Ovalle, A. Sotomayor, Z. Stuchlik, Eur. Phys. J. C **79**, 826 (2019)
21. R.G. Catalán, J. Garay, R. López-Ruiz, Phys. Rev. E **66**, 011102 (2002)
22. K.C. Chatzisavvas, V.P. Psonis, C.P. Panos, C.C. Moustakidis, Phys. Lett. A **373**, 3901 (2009)
23. E. Contreras et al., Eur. Phys. J. C **82**, 187 (2022)
24. E. Contreras, Z. Stulchlik, Eur. Phys. J. C **82**, 706 (2022)
25. F.I. Cooperstock, G.J. Junevicus, Nuovo Cimento **16B**, 387 (1973)
26. J.P. Crutchfield, K. Young, Phys. Rev. Lett. **63**, 105 (1989)
27. M.G.B. de Avellar, J.E. Horvath, Phys. Lett. A **376**, 1085 (2012)
28. M.G.B. de Avellar, R.A. Souza, J.E. Horvath, D.M. Paret, Phys. Lett. A **378**, 3481 (2014)
29. U. Farwa et al., Phys. Scrip. **7**, 105307 (2022)

30. D.P. Feldman, J.P. Crutchfield, Phys. Lett. A **238**, 244 (1998)
31. A. García–Parrado Gomez Lobo, Class. Quantum Gravity **25**, 015006 (2008)
32. M. Gaudin, V. Gorini, A. Kamenshchik, U. Moschella, V. Pasquier, Int. J. Mod. Phys. D **15**, 1387 (2006)
33. M.K. Gokhroo, A.L. Mehra, Gen. Relativ. Gravit. **26**, 75 (1994)
34. M. Govender et al., Eur. Phys. J. C **82**, 832 (2022)
35. P. Grassberger, Int. J. Theor. Phys. **25**, 907 (1986)
36. S. Guemede et al., Entropy **24**, 645 (2022)
37. C. Hansen, S. Kawaler, *Stellar Interiors: Physical Principles* Structure and Evolution. (Springer, Berlin, 1994)
38. B.K. Harrison, Phys. Rev. **116**, 1285 (1959)
39. L. Herrera, N.O. Santos, Phys. Rep. **286**, 53 (1997)
40. L. Herrera, A. Di Prisco, J. Hernńdez-Pastora, N.O. Santos, Phys. Lett. **237**, 113 (1998)
41. L. Herrera, A. Di Prisco, J. Martin, J. Ospino, N.O. Santos, O. Troconis, Phys. Rev. D **69**, 084026 (2004)
42. L. Herrera, N.O. Santos, J. Carot, J. Math. Phys. **47**, 052502 (2006)
43. L. Herrera, Int. J. Mod. Phys. D **17**, 557 (2008)
44. L. Herrera, Int. J. Mod. Phys. D **17**, 2507 (2008)
45. L. Herrera, J. Ospino, A. Di Prisco, E. Fuenmayor, O. Troconis, Phys. Rev D **79**, 064025 (2009)
46. L. Herrera, A. Di Prisco, J. Ospino, Gen. Relativ. Gravit. **42**, 1585 (2010)
47. L. Herrera, N.O. Santos, Gen. Relativ. Gravit. **42**, 2383 (2010)
48. L. Herrera, A. Di Prisco, J. Ibañez, Phys. Rev. D **84**, 107501 (2011)
49. L. Herrera, A. Di Prisco, J. Ospino, Gen. Relativ. Gravit. **44**, 2645 (2012)
50. L. Herrera, W. Barreto, Phys. Rev. D **87**, 087303 (2013)
51. L. Herrera, W. Barreto, Phys. Rev. D **88**, 084022 (2013)
52. L. Herrera, A. Di Prisco, J. Ibáñez, J. Ospino, Phys. Rev. D **87**, 024014 (2013)
53. L. Herrera, E. Fuemayor, P. Leon, Phys. Rev. D **93**, 024047 (2016)
54. L. Herrera, Phys. Rev. D **97**, 044010 (2018)
55. L. Herrera, A. Di Prisco, J. Ospino, Phys. Rev. D **98**, 104059 (2018)
56. L. Herrera, L. Witten, Adv. High En. Phys. **2018**, 3839103 (2018)
57. L. Herrera, A. Di Prisco, J. Ospino, L. Witten, Phys. Rev. D **101**, 064071 (2020)
58. L. Herrera, A. Di Prisco, J. Ospino, Phys. Rev. D **99**, 044049 (2019)
59. L. Herrera, A. Di Prisco, J. Carot, Phys. Rev. D **99**, 124028 (2019)
60. L. Herrera, A. Di Prisco, J. Ospino, Eur. Phys. J. C **80**, 631 (2020)
61. L. Herrera, A. Di Prisco, J. Ospino, Phys. Rev. D **103**, 024037 (2021)
62. L. Herrera, A. Di Prisco, J. Ospino, Symmetry **13**, 1568 (2021)
63. L. Herrera, A. Di Prisco, J. Ospino, Entropy **23**, 1219 (2021)
64. L. Herrera, A. Di Prisco, J. Ospino, Universe **8**, 296 (2022)
65. L. Herrera, A. Di Prisco, J. Ospino, Symmetry **15**, 754 (2023)
66. W. Israel, Phys. Rev. **164**, 1776 (1967)
67. W. Israel, Ann. Phys. (NY) **100**, 310 (1976)
68. W. Israel, J. Stewart, Phys. Lett. A **58**, 213 (1976)
69. W. Israel, J. Stewart, Ann. Phys. (NY) **118**, 341 (1979)
70. B. Ivanov, Int. J. Mod. Phys. D **20**, 319 (2011)
71. B. Ivanov, Astrophys. Space Sci. **361**, 18 (2016)
72. B. Ivanov, Int. J. Mod. Phys. D **25**, 1650049 (2016)
73. A. Janis, E.T. Newman, J. Winicour, Phys. Rev. Lett. **20**, 878 (1968)
74. AYu. Kamenshchik, E.O. Pozdeeva, A.A. Starobinsky, A. Tronconi, T. Vardanyan, G. Venturi, SYu. Verno, Phys. Rev. D **98**, 124028 (2018)
75. S. Khan, S. Mardan, M. Rehman, Eur. Phys. J. C **79**, 779 (2019)
76. S. Khan, S. Mardan, M. Rehman, Eur. Phys. J. C **79**, 1037 (2019)
77. S. Khan, S. Mardan, M. Rehman, Eur. Phys. J. C **81**, 602 (2021)
78. S. Khan, S. Mardan, M. Rehman, Eur. Phys. J. C **82**, 620 (2022)
79. S. Khan et al., JCAP **2022**, 023 (2022)

80. R. Kippenhahn, A. Weigert, *Stellar Structure and Evolution* (Springer, Berlin, 1990)
81. A.N. Kolmogorov, Prob. Inform. Theory J. **1**, 3 (1965)
82. P. Leon, C. Las Heras, Gen. Relativ. Gravit. **54**, 138 (2022)
83. P. Leon, C. Las Heras, Phys. Rev. D **107**, 064010 (2023)
84. S. Lloyd, H. Pagels, Ann. Phys. **188**, 186 (1988)
85. F.S.N. Lobo, J.P. Mimoso, Phys. Rev. D **82**, 044034 (2010)
86. R. López-Ruiz, H.L. Mancini, X. Calbet, Phys. Lett. A **209**, 321 (1995)
87. A. Maciel, M. Le Delliou, J.P. Mimoso, Class. Quantum Gravity **37**, 125005 (2020)
88. T. Madler, Phys. Rev. D **99**, 104048 (2019)
89. S. Maurya et al., Eur. Phys. C **82**, 48 (2022)
90. S. Maurya, M. Govender, G. Mustafa, R. Nag, Eur. J. Phys. C **82**, 1006 (2022)
91. A.K.M. Masood-ul-Alama, Gen. Relativ. Gravit. **39**, 55 (2007)
92. C. Misner, D. Sharp, Phys. Rev. **136**, B571 (1964)
93. H. Nazar, G. Abbas, Int. J. Geo. Meth. Mod. Phys. **16**, 1950170 (2019)
94. C.P. Panos, N.S. Nikolaidis, K.C. Chatzisavvasand, C.C. Tsouros, Phys. Lett. A **373**, 2343 (2009)
95. G. Parisi, Phys. World **6**, 42 (1993)
96. A. Rincon, G. Panotopoulos, I. Lopez, Universe **9**, 72 (2023)
97. A. Rincon, G. Panotopoulos, I. Lopez, Eur. Phys. J. C **83**, 116 (2023)
98. L. Rizzi, S.L. Cacciatori, V. Gorini, A. Kamenshchik, O.F. Piattella, Phys. Rev. D **82**, 027301 (2010)
99. D. Santana et al., Eur. Phys. J. C **82**, 703 (2022)
100. J. Sañudo, R. López-Ruiz, Phys. Lett. A **372**, 5283 (2008)
101. J. Sañudo, A.F. Pacheco, Phys. Lett. A **373**, 807 (2009)
102. M. Schwarzschild, *Structure and Evolution of the Stars* (Dover, New York, 1958)
103. M. Sharif, I. Butt, I. Eur, Phys. J. C **78**, 688 (2018)
104. M. Sharif, I. Butt, I. Eur, Phys. J. C **78**, 850 (2018)
105. M. Sharif, I. Butt, Chin. J. Phys. **61**, 238 (2019)
106. M. Sharif, A. Majid, Chin. J. Phys. **61**, 38 (2019)
107. M. Sharif, Majid, Int. J. Geo. Meth. Mod. Phys. **16**, 1950174 (2019)
108. M. Sharif, A. Majid, M. Nasir, Int. J. Mod. Phys. A **2019**(34), 19502010 (2019)
109. M. Sharif, S. Tariq, Mod. Phys. Lett. A **35**, 2050231 (2020)
110. M. Sharif, A. Majid, Eur. Phys. J. C **80**, 1185 (2020)
111. M. Sharif, A. Anjum, Gen. Relativ. Gravit. **54**, 112 (2022)
112. M. Sharif, T. Naseer, Eur. Phys. J. P **137**, 947 (2022)
113. M. Sharif, T. Naseer, Class. Quantum Gravity **40**, 035004 (2023)
114. D. Suarez et al., Eur. Phys. J. C **82**, 177 (2022)
115. S. Thirukkanesh, S.D. Maharaj, J. Math. Phys. **50**, 022502 (2009)
116. S. Thirukkanesh, S.D. Maharaj, J. Math. Phys. **51**, 072502 (2010)
117. R. Tolman, Phys. Rev. **35**, 875 (1930)
118. J. Triginer, D. Pavón, Class. Quantum Gravity **12**, 689 (1995)
119. J. Winicour, A.I. Janis, E.T. Newman, Phys. Rev. **176**, 1507 (1968)
120. Z. Yousaf, Phys. Scrip. **95**, 075307 (2020)
121. Z. Yousaf, M. Bhatti, T. Naseer, Phys. Dark. Univ. **28**, 100535 (2020)
122. Z. Yousaf, M. Bhatti, K. Hassan, Eur. Phys. J. P **135**, 397 (2020)
123. Z. Yousaf, M. Bhatti, T. Naseer, I. Ahmad, Phys. Dark. Univ. **29**, 100581 (2020)
124. Z. Yousaf, M. Bhatti, T. Naseer, Ann. Phys. **420**, 168267 (2020)
125. Z. Yousaf, M. Bhatti, T. Naseer, Int. J. Mod. Phys. D **29**, 2050061 (2020)
126. Z. Yousaf M. Bhatti, T. Naseer, Eur. Phys. J. P **135**, 323 (2020)
127. Z. Yousaf, M.Y. Khlopov, M. Bhatti, T. Naseer, Month. Not. R. Astron. Soc. **495**, 4334 (2020)
128. Z. Yousaf, K. Bamba, M. Bhatti, K. Hassan, New Astron. **84**, 101541 (2021)
129. Z. Yousaf, Universe **8**, 131 (2022)
130. Z. Yousaf et al., Chin. J. Phys. **77**, 2078 (2022)
131. Z. Yousaf, M. Bhatti, S. Rehman, K. Bamba, Gen. Relativ. Gravit. **55**, 31 (2023)

132. Z. Yousaf, M. Bhatti, S. Rehman, Zeits. Natur. A **78**, 161 (2023)
133. Y.B. Zeldovich, Zh. Eksp. Teor. Fiz. **41**, 1609 (1969); Sov. Phys. JETP **14**, 1143 (1962)
134. M. Zubair, H. Azmat, Int. J. Mod. Phys. D **29**, 2050014 (2020)
135. M. Zubair, H. Azmat, Phys. Dark. Univ. **28**, 100531 (2020)
136. M. Zubair, Eur. Phys. J. C **82**, 984 (2022)
137. M. Zubair et al., New Astr. **100**, 101996 (2023)

Chapter 9
Weyl Curvature and Cosmic Censorship Conjecture: A Geometrical Perspective

Rituparno Goswami

9.1 Introduction

We are thus presented with what is perhaps the most fundamental unanswered question of general-relativistic collapse theory, namely: does there exist a "cosmic censor" who forbids the appearance of naked singularities, clothing each one in an absolute event horizon? In one sense, a "cosmic censor" can be shown not to exist. For it follows from a theorem of Hawking that the "big bang" singularity is, in principle, observable. It is not known whether singularities observable from outside will ever arise in a generic collapse which starts off from a perfectly reasonable nonsingular initial state [28]. Ever since Penrose proposed this famous *Cosmic Censorship Conjecture (CCC)* in 1969, there have been numerous attempts towards validating this conjecture by means of a rigorous mathematical proof. However, this conjecture still remains unproved, and it has been recognised as one of the most important open problems in gravitational physics.

The reasons for these attempts are twofold. On one hand, the validity of the above will confirm the already widely accepted and applied theory of black hole dynamics and thermodynamics, which has considerable amount of astrophysical applications, especially in this era of *precision cosmology*. On the other hand if this conjecture is proven wrong then we will face a mammoth task of reformulating the proofs of a number of important fundamental global theorems in black hole physics that assume that the spacetime manifold is *future asymptotically predictable*. In other words there should be no singularity to the future of the partial Cauchy surface which is 'naked' or visible from the future null infinity [20].

Although no conclusive proof or disproof of CCC could be formulated, extensive studies of various dynamical and spherically symmetric collapse models for a wide

R. Goswami (✉)
Astrophysics Research Centre and School of Mathematics, Statistics and Computer Science, University of KwaZulu-Natal, Private Bag X54001, Durban 4000, South Africa
e-mail: Goswami@ukzn.ac.za

D. Malafarina and P. S. Joshi (eds.), *New Frontiers in Gravitational Collapse and Spacetime Singularities*, Springer Series in Astrophysics and Cosmology,
https://doi.org/10.1007/978-981-97-1172-7_9

range of matter fields continued over the past half century, investigating the final outcome of gravitational collapse, in terms of naked or covered singularities. Please refer to [15, 21, 24] for a detailed description and references on this subject. The generic conclusions from these studies were extremely interesting as they conclusively indicated the following: *In spite of the collapse commencing from a physically realistic and non-singular initial epoch, trapped surfaces may not develop early enough to always shield the curvature generated fireballs, characterised by diverging densities and curvatures, from an outside observer*. Not just isolated trajectories but families of non-spacelike geodesics emerge from such a naked singularity, providing a non-zero measure set of trajectories escaping away.

The above conclusion immediately prompts us to ask this important question:

Question: *Can we identify the possible physical and geometrical factors that are responsible for this delay in the formation of trapped regions that cover the spacetime singularity?*

Answering the above question will definitely help us in obtaining a transparent understanding of the physics of black hole or naked singularity formation in gravitational collapse. Moreover, being geometrical in nature, such factors should be covariant as well as universal, that is independent of the collapsing matter models. To this end, the pioneering work was done by Joshi et al. [22] followed by [23], which illuminated the crucial role that spacetime shear plays in determining the end state of continual collapse. It was clearly shown that there exists a remarkable connection between the spacetime shear and the inhomogeneity of collapsing matter. The spacetime shear generated by the matter inhomogeneity can distort the geometry of the trapped region in such a way that future directed null geodesics from the vicinity of the central singularity can escape being trapped.

In this chapter, we continue with our quest by recasting a spherical collapse scenario as a dynamical model of geometrical and thermodynamic scalars, that have well defined physical and geometrical interpretations. We recall that spherically symmetric spacetimes are a subclass of non-rotating Locally Rotationally Symmetric (LRS) spacetimes, that contain a continuous non-trivial isotropy group at every point, consisting of spatial rotations [12]. Via this isotropy group, these spacetimes are equipped with a covariantly defined unique preferred spatial direction at every point. Obviously this preferred spatial congruence will be orthogonal to the timelike congruence of the fluid flow lines and hence we can use these congruences to decompose the spacetime locally in a semi-tetrad way. This local decomposition then beautifully brings out the definition of the boundary of the trapped region, the *marginally outer trapped surface* (MOTS) (or the *apparent horizon*) in terms of the geometrical scalars associated with the congruences [19]. Analysing the geometrical properties of the MOTS, we establish the importance of Weyl curvature, in a quantitative and transparent manner, in deforming and delaying the trapped region to make the central singularity locally naked. As we know, the Weyl tensor, which is the trace-free part of the Riemann curvature tensor, gives the measure of the 'pure' geometrical effect (or source free gravity) on the spacetime manifold, as this tensor can be non-zero even in the absence of any matter field. The Weyl tensor depicts the tidal forces, resulting in volume distortion and generating the spacetime shear. In fact, it is well known that the variation of the acceleration vector along with the electric part of the Weyl tensor act as the source of the shear evolution equation in general relativity [13]. Hence, in general, in a spacetime with a non-zero electric part of Weyl, shear will be generated even if it is zero at a given epoch. Further to this, Weyl curvature is directly related to the energy density of the *free gravitational field* [11], whose energy momentum tensor can be obtained as a 2-index square root of the Bell Robinson tensor [3–5, 26, 29]. Hence, relating Weyl curvature with the violation of censorship gives a nice geometrical and thermodynamic perspective of the gravitational collapse problem, which is the aim of this chapter.

Unless otherwise specified, we use natural units ($c = 8\pi G = 1$) throughout this chapter. Latin indices run from 0 to 3. The symbol ∇ represents the usual covariant derivative and ∂ corresponds to partial differentiation. We use the $(-,+,+,+)$ signature and the Riemann tensor is defined by

$$R^a{}_{bcd} = \Gamma^a{}_{bd,c} - \Gamma^a{}_{bc,d} + \Gamma^e{}_{bd}\Gamma^a{}_{ce} - \Gamma^e{}_{bc}\Gamma^a{}_{de} \,, \tag{9.1}$$

where the $\Gamma^a{}_{bd}$ are the Christoffel symbols (i.e. symmetric in the lower indices) defined in terms of the partial derivatives of the metric tensor g_{ab} and it's inverse g^{ab} as

$$\Gamma^a{}_{bd} = \frac{1}{2}g^{ae}\left(\partial_d g_{be} + \partial_b g_{ed} - \partial_e g_{bd}\right) \,. \tag{9.2}$$

The Ricci tensor is obtained by contracting the *first* and the *third* indices

$$R_{ab} = g^{cd}R_{cadb} \,, \tag{9.3}$$

while the Ricci scalar is defined to be the trace of the Ricci tensor ($R = g^{ab}R_{ab}$). The Hilbert–Einstein action in the presence of matter is given by

$$\mathscr{S} = \frac{1}{2}\int d^4x\sqrt{-g}\left[R - 2\Lambda - 2\mathscr{L}_m\right] \,, \tag{9.4}$$

the variation of which gives the Einstein field equations as

$$G_{ab} + \Lambda g_{ab} = T_{ab} \,, \tag{9.5}$$

where

$$G_{ab} \equiv R_{ab} - \frac{1}{2}g_{ab}R \,, \tag{9.6}$$

is the Einstein tensor, T_{ab} is the energy momentum tensor of the matter field and Λ is the *cosmological constant*, that arises as the constant of integration of the first integral of doubly contracted Bianchi identities: $\nabla_b G^{ab} = 0$.

9.2 Geometry of Self Gravitating Systems

In general relativity, the complete information about the spacetime curvature is encoded in the 20 independent components of the Riemann curvature tensor $R^a{}_{bcd}$. It is well known that this tensor can be Ricci-decomposed into its irreducible sub-tensors in the following way [6, 16]

$$R_{abcd} = G_{abcd} + E_{abcd} + C_{abcd} \tag{9.7}$$

where we have

$$G_{abcd} = \frac{1}{12}R(g_{ac}g_{bd} - g_{ad}g_{bc}) \,, \tag{9.8}$$

and

$$E_{abcd} = \frac{1}{2}(g_{ac}S_{bd} + g_{bd}S_{ac} - g_{ad}S_{bc} - g_{bc}S_{ad}) \,, \; S_{ab} = R_{ab} - \frac{1}{4}g_{ab}R \,. \tag{9.9}$$

From the Einstein equations (9.5), it is obvious that both the sub-tensors G_{abcd} and E_{abcd} are directly related to the matter energy momentum tensor T_{ab}. In other words, these will vanish in the absence of matter. However, the tensor C_{abcd}, which is the Weyl tensor, can still be non-zero in vacuum. That is the reason why this tensor gives the measure of the *'pure' geometrical effect* (or source free gravity) on the spacetime manifold. It can easily be seen that the Weyl tensor satisfies the same algebraic symmetries as the Riemann tensor, namely

$$C_{abcd} = C_{[ab][cd]} = C_{[cd][ab]} \,, \; C_{[abcd]} = 0 \,. \tag{9.10}$$

Furthermore this tensor is traceless, as any contraction of this tensor vanishes:

$$C^a{}_{bad} = 0 \,. \tag{9.11}$$

The properties (9.10) and (9.11) imply that the left dual of this tensor is equal to its right dual:

$$^*C_{abcd} \equiv \frac{1}{2}\eta_{absm}C^{sm}{}_{cd} = C^*_{abcd} \equiv \frac{1}{2}\eta_{cdsm}C_{ab}{}^{sm} \,, \tag{9.12}$$

where η_{abcd} is the 4-dimensional volume element.

9.2.1 *The Case of Spherical Symmetry*

As we discussed in the previous section, spherically symmetric spacetimes are a subclass of non-rotating *Locally Rotationally Symmetric spacetimes*, that contain a continuous non-trivial isotropy group of spatial rotations at every point. Hence, they are equipped with a preferred spatial direction that is covariantly defined at every point. Using the timelike unit vector u^a (defined along the fluid flow lines) and the unit vector along the preferred spatial direction e^a (orthogonal to u^a), we can decompose the spacetime as [9, 10, 18]

$$g_{ab} = -u_a u_b + e_a e_b + N_{ab}\,, \tag{9.13}$$

where N_{ab} is the 2-dimensional metric that spans the spherical 2-shells. The geometrical quantities defined for the timelike congruence are the expansion scalar Θ, acceleration 3-vector $\dot{u}^a$ and the shear 3-tensor σ_{ab}. The timelike congruence also uniquely defines the electric and magnetic parts of the Weyl tensor in the following way [13]

$$E_{ab} = C_{acbd}u^c u^d\,, \quad H_{ab} = {}^*C_{acbd}u^c u^d\,. \tag{9.14}$$

The spherical symmetry ensures that the magnetic part H_{ab}, which gives rise to gravitational waves, vanishes identically. Hence the only non-trivial part of the Weyl tensor is the electric part E_{ab}, which is responsible for tidal forces that leads to matter inhomogeneity. This timelike congruence u^a can further decompose the energy momentum tensor of the matter to give the energy density μ, isotropic pressure p, heat flux 3-vector q^a and anisotropic stress 3-tensor π_{ab}. The only non-vanishing geometrical quantity related to the preferred spacelike congruence is the volume expansion ϕ. Other geometrical quantities related to this congruence vanish identically due to the spherical symmetry. Using this preferred spatial congruence, we can then extract the set of covariant scalars from the aforementioned 3-vectors and 3-tensors that govern the dynamics of the system in the following way

$$\mathcal{A} = \dot{u}^a e_a\,, \ \Sigma = \sigma_{ab}e^a e^b\,, \ \mathcal{E} = E_{ab}e^a e^b\,, \ Q = q^a e_a\,, \ \Pi = \pi_{ab}e^a e^b\,. \tag{9.15}$$

Therefore, the set of quantities that fully describe the spherically symmetric class of spacetimes are

$$\mathcal{D} \equiv \{\Theta, \mathcal{A}, \Sigma, \mathcal{E}, \phi, \mu, p, \Pi, Q\}\,. \tag{9.16}$$

These geometrical and thermodynamical scalars, together with their directional derivatives along u^a (denoted by a dot) and along e^a (denoted by a hat) completely specify the corresponding Ricci identities of u^a and e^a and the Bianchi identities and thereby specify the dynamics completely. The full covariant derivatives of e_a and u_a can be written in terms of these variables as

$$\nabla_a e_b = -\mathcal{A} u_a u_b + \left(\Sigma + \frac{1}{3}\Theta\right) e_a u_b + \frac{1}{2}\phi N_{ab}, \tag{9.17}$$

$$\nabla_a u_b = -\mathcal{A} u_a e_b + e_a e_b \left(\Sigma + \frac{1}{3}\Theta\right) - \frac{1}{2}N_{ab}\left(\Sigma - \frac{2}{3}\Theta\right). \tag{9.18}$$

9.2.2 The Governing Evolution and Propagation Equations

We now present evolution equations (containing the directional derivatives of the scalars along the fluid flow lines) propagation equations (containing the directional derivatives of the scalars along the preferred spatial direction) that govern the complete dynamics of spherically symmetric spacetimes. We refer to [9, 10] for the derivations of these equations from the Ricci identities of the vectors u^a and e^a and the once and twice contracted Bianchi identities. We would like to emphasize here that these equations completely encompass the Einstein field equations governing a spherically symmetric system. However, since these are a system of first order equations in terms of well defined geometrical and thermodynamic scalars, they are relatively easier to analyse and physically interpret than the Einstein equations in the usual form (9.5).

Propagation:

$$\hat{\phi} = -\tfrac{1}{2}\phi^2 + \left(\tfrac{1}{3}\Theta + \Sigma\right)\left(\tfrac{2}{3}\Theta - \Sigma\right) - \tfrac{2}{3}(\mu + \Lambda) - \mathcal{E} - \tfrac{1}{2}\Pi, \quad (9.19)$$

$$\hat{\Sigma} - \tfrac{2}{3}\hat{\Theta} = -\tfrac{3}{2}\phi\Sigma - Q\,, \quad (9.20)$$

$$\hat{\mathcal{E}} - \tfrac{1}{3}\hat{\mu} + \tfrac{1}{2}\hat{\Pi} = -\tfrac{3}{2}\phi\left(\mathcal{E} + \tfrac{1}{2}\Pi\right) + \left(\tfrac{1}{2}\Sigma - \tfrac{1}{3}\Theta\right)Q. \quad (9.21)$$

Evolution:

$$\dot{\phi} = -\left(\Sigma - \tfrac{2}{3}\Theta\right)\left(\mathcal{A} - \tfrac{1}{2}\phi\right) + Q\,, \quad (9.22)$$

$$\dot{\Sigma} - \tfrac{2}{3}\dot{\Theta} = -\mathcal{A}\phi + 2\left(\tfrac{1}{3}\Theta - \tfrac{1}{2}\Sigma\right)^2 + \tfrac{1}{3}(\mu + 3p - 2\Lambda) - \mathcal{E} + \tfrac{1}{2}\Pi, \quad (9.23)$$

$$\dot{\mathcal{E}} - \frac{1}{3}\dot{\mu} + \frac{1}{2}\dot{\Pi} = \left(\frac{3}{2}\mathcal{E} + \frac{1}{4}\Pi - \frac{1}{2}(\mu + p)\right)\left(\Sigma - \frac{2}{3}\Theta\right) + \frac{1}{2}\phi Q. \quad (9.24)$$

Propagation/Evolution:

$$\hat{\mathcal{A}} - \dot{\Theta} = -(\mathcal{A} + \phi)\mathcal{A} + \tfrac{1}{3}\Theta^2 + \tfrac{3}{2}\Sigma^2 + \tfrac{1}{2}(\mu + 3p - 2\Lambda)\,, \quad (9.25)$$

$$\dot{\mu} + \hat{Q} = -\Theta(\mu + p) - (\phi + 2\mathcal{A})Q - \tfrac{3}{2}\Sigma\Pi, \quad (9.26)$$

$$\dot{Q} + \hat{p} + \hat{\Pi} = -\left(\tfrac{3}{2}\phi + \mathcal{A}\right)\Pi - \left(\tfrac{4}{3}\Theta + \Sigma\right)Q - (\mu + p)\mathcal{A}\,. \quad (9.27)$$

We note that Eq. (9.25) is the famous Raychaudhuri equation which gave the first important geometrical insight towards a generic result relating to spacetime singularities. Henceforth, throughout this chapter, we will consider the cosmological constant Λ to be zero, as this has no geometrical effect on the local violation of the cosmic censorship conjecture.

9.2.3 The Gaussian Curvature of Spherical Shells and Misner-Sharp Mass

Via the projection tensor $g_{ab} + u_a u_b$, we can compute, in a straightforward manner, the 3 dimensional Ricci scalar of the spacelike three surface orthogonal to the timelike matter flow congruence. This can be expressed as

$$^3R = -2\left[\hat{\phi} + \tfrac{3}{4}\phi^2 - K\right] . \tag{9.28}$$

Here K is the Gaussian curvature of spherical 2-shells that foliate the 3-space. In fact, the Gaussian curvature scalar is the only independent component of the 2-dimensional Riemann tensor on these spherical shells, which is given as $^2R_{abcd} = K(N_{ac}N_{bd} - N_{ad}N_{bc})$. This curvature gives the geometric notion of the area radius of the spherical shells $\mathscr{R}$, as by definition the Gaussian curvature is the inverse square of the area radius. In terms of the covariant scalars, we can write the Gaussian curvature K as

$$K \equiv \frac{1}{\mathscr{R}^2} = \frac{1}{3}\mu - \mathscr{E} - \frac{1}{2}\Pi + \frac{1}{4}\phi^2 - \left(\frac{1}{3}\Theta - \frac{1}{2}\Sigma\right)^2 , \tag{9.29}$$

with the associated directional derivatives along u^a and e^a given by

$$\dot{K} = \left(\Sigma - \frac{2}{3}\Theta\right)K , \ \hat{K} = -\phi K . \tag{9.30}$$

The Misner-Sharp mass ($\mathscr{M}$) is a measure of the amount of gravitational mass within a given spherical shell at any instant of time [27, 32]. This is a purely geometrical measure, defined locally, in terms of the area radius of the spherical shell ($\mathscr{R}$) in the following way

$$\mathscr{M} = \frac{\mathscr{R}}{2}\left(1 - \nabla_a\mathscr{R}\nabla^a\mathscr{R}\right) . \tag{9.31}$$

The above can be written in terms of the Gaussian curvature of the shell $K \equiv \mathscr{R}^{-2}$ as

$$\mathscr{M} = \frac{1}{2\sqrt{K}}\left(1 - \frac{1}{4K^3}\nabla_a K \nabla^a K\right) . \tag{9.32}$$

We know that the symmetry of the spacetime dictates that the 4-gradient of any scalar Φ can be written as $\nabla_a\Phi \equiv -\dot{\Phi}u^a + \hat{\Phi}e^a$. Substituting (9.29) and (9.30) into (9.32), we finally get the expression for the Misner-Sharp mass in terms of the covariant scalars [33]

$$\mathscr{M} = \frac{1}{2K^{\frac{3}{2}}}\left(\frac{1}{3}\mu - \mathscr{E} - \frac{1}{2}\Pi\right) . \tag{9.33}$$

We note that the above equation for the gravitational mass is an equivalent alternative representation without using the spatial integral of the component $T^0{}_0$ of the matter energy momentum tensor $T^a{}_b$. The present representation is interesting as it transparently relates the total gravitational mass with matter inhomogeneity (generated by the tidal effects of the Weyl scalar) and matter anisotropy (generated by the anisotropic stress scalar Π which is proportional to the difference between radial and tangential pressure). From the field equations (9.19)–(9.27), we can also write down the associated directional derivatives of this mass along u^a and e^a as

$$\hat{\mathcal{M}} = \frac{1}{4K^{\frac{3}{2}}}\left[\phi\mu - \left(\Sigma - \frac{2}{3}\Theta\right)Q\right], \tag{9.34}$$

$$\dot{\mathcal{M}} = \frac{1}{4K^{\frac{3}{2}}}\left[\left(\Sigma - \frac{2}{3}\Theta\right)(p+\Pi) - \phi Q\right]. \tag{9.35}$$

9.3 Radial Null Geodesics in a Spherical Self Gravitating System

On any spacetime $(\mathcal{M}, g)$, null geodesics (light rays) are characterised by the curves $x^a(\nu)$, where ν is an affine parameter along the geodesics. The tangent to these curves is defined by $k^a \equiv dx^a(\nu)/d\nu$, where k^a is a null vector obeying $k^a k_a = 0$. Also, since the tangent vector to the geodesic is parallelly propagated to itself, we can write

$$k^b\nabla_b k^a = \frac{\delta k^a}{\delta\nu} = 0\,, \tag{9.36}$$

where $\delta/\delta\nu = k^b\nabla_b$ is the derivative along the ray with respect to the affine parameter. If light rays move along the preferred spatial direction (that is, they are *radial null rays*), then the components of these null curves on the spherical shells are zero. Also, at this point, we may define the notion of locally *outgoing* and *ingoing* null geodesics with respect to the preferred spatial direction in the following way [14].

Definition 9.1 Consider any open subset $\mathcal{S}$ of $(\mathcal{M}, g)$ and let $x^a(\nu)$ be a null geodesic in $\mathcal{S}$. Let k^a be the tangent to this geodesic. If $e^a k_a > 0$ in $\mathcal{S}$ then the geodesic is considered to be outgoing with respect to the preferred direction in $\mathcal{S}$. Similarly $e^a k_a < 0$ denotes an ingoing geodesic.

Therefore for spherically symmetric spacetimes, the equation of the tangent to the outgoing null geodesics along the preferred spatial direction can be written as

$$k^a = \frac{E}{\sqrt{2}}\left(u^a + e^a\right)\,, \tag{9.37}$$

where $E : [u, e] \rightarrow \mathbb{R}$ denotes the energy of the light ray. The propagation of E along the null geodesic is given by the geodesic equation (9.36) (using (9.17) and (9.18)) as

$$E' \equiv \frac{\delta E}{\delta v} = \mp E^2 \mathscr{A} - E^2 \left(\Sigma + \tfrac{1}{3}\Theta\right) . \tag{9.38}$$

Now it is evident that the hypersurface orthogonal to null vector k^a, contains k^a. Therefore we need a different construction to define a locally orthogonal space with respect to the outgoing null geodesics. Let us now define the projection tensor $\tilde{h}_{ab}$, which projects tensors and vectors onto the 2-dimensional screen space orthogonal to k^a, as [8, 14]

$$\tilde{h}_{ab} \equiv g_{ab} + 2k_{(a}l_{b)}, \;\; \tilde{h}^a_a = 2, \;\; \tilde{h}_{ac}\tilde{h}^c_b = \tilde{h}_{ab}, \;\; \tilde{h}_{ab}k^b = 0, \tag{9.39}$$

where l_a is the null ingoing geodesic that obeys

$$l^a l_a = 0, \;\; k^a l_a = -1 \text{ and } \frac{\delta l^a}{\delta v} = k^b \nabla_b l^a = 0. \tag{9.40}$$

Using these definitions, the general form of l^a can be written as

$$l^a = \frac{1}{\sqrt{2}E}\left(u^a - e^a\right), \tag{9.41}$$

and substituting (9.41) into (9.39) the screen-space projection tensor is obtained as

$$\tilde{h}_{ab} = g_{ab} + u_a u_b - e_a e_b = N_{ab} . \tag{9.42}$$

Hence we note that the 2-dimensional screen projection tensor is absolutely the same as the projection tensor N_{ab}. For spherically symmetric spacetimes, the local semi-tetrad decomposition using the mutually orthogonal vector fields u^a and e^a, gives the covariant derivative of the null vector k^a as [8, 14]

$$\nabla_b k_a = \tfrac{1}{2}\tilde{h}_{ab}\tilde{\Theta}_{out} + \tilde{\sigma}_{ab} + \tilde{X}_a k_b + \tilde{Y}_b k_a \, , +\lambda k_a k_b, \tag{9.43}$$

where $\tilde{X}_a = \frac{1}{E}e^d\nabla_d k_a$, $\tilde{Y}_a = \frac{1}{E}e^d\nabla_a k_d$ and $\lambda = -\frac{1}{E^2}e^c e^d \nabla_d k_c$. Also $\tilde{\Theta}_{out}$ and $\tilde{\sigma}_{ab}$ represent the expansion and shear of the outgoing null congruence respectively. From the above definitions and the geometry of spherically symmetric spacetimes we can easily calculate that

$$\tilde{\Theta}_{out} \equiv \tilde{h}^{ab}\nabla_b k_a = \frac{E}{\sqrt{2}} N^{ab}\nabla_a \left(u_b + e_b\right) , \tag{9.44}$$

which using (9.17) and (9.18) results in

$$\tilde{\Theta}_{out} = \frac{E}{\sqrt{2}}\left(\tfrac{2}{3}\Theta - \Sigma + \phi\right) . \tag{9.45}$$

An absolutely similar analysis involving the ingoing null geodesic l^a can be performed, giving

$$\tilde{\Theta}_{in} = \frac{1}{\sqrt{2}E}\left(\tfrac{2}{3}\Theta - \Sigma - \phi\right) . \tag{9.46}$$

9.4 Geometry of Marginally Outer Trapped Surface (MOTS)

As described in [20], let us consider a spherical emitter of light, surrounding a massive body, emitting a flash of light. In a normal situation, (like in the Minkowski spacetime or Schwarzschild spacetime outside the horizon), the volume expansion of the outgoing null congruence orthogonal to the sphere is always positive ($\tilde{\Theta}_{out} > 0$) while that of the incoming congruence is always negative ($\tilde{\Theta}_{in} < 0$). However, if a sufficiently large amount of matter is present within the emitting sphere, the volume expansion of the outgoing null congruence orthogonal to the sphere becomes negative. This is the case where both outgoing and ingoing wavefronts collapse towards the centre of the emitting sphere. In this case, the emitting sphere is then called a *closed trapped 2-surface*. The collection of all closed trapped 2-surfaces in a four dimensional manifold constitutes a 4 dimensional trapped region. *Marginally outer trapped surfaces* (MOTS) are the 3 dimensional boundary of the trapped region. We will now derive the equations for a MOTS surface and its evolution in the $[u, e]$ plane. To define MOTS we generalise the definition of dynamical horizons by Ashtekar and Krishnan [2] by removing the restriction that it is a spacelike surface [14] in the following way.

Definition 9.2 A smooth, three-dimensional sub-manifold $\mathscr{H}$ in a spacetime $(\mathscr{M}, g)$ is said to be a **Marginally outer trapped surface** if it is foliated by a preferred family of 2-spheres such that, on each leaf $\mathscr{S}$, the expansion $\tilde{\Theta}_{out}$ of the outgoing null normal k^a vanishes and the expansion $\tilde{\Theta}_{in}$ of the other ingoing null normal l^a is strictly negative.

Consequently a MOTS is a 3-manifold which is foliated by marginally trapped 2-spheres. On this definition, such 3-surfaces can be timelike, spacelike, or null, with rather different properties. The essential point is that they are locally defined, and therefore are able to respond to local dynamic change. We don't need to know what is happening at infinity in order to determine a local physical effect.

At this point, let us consider the classes of spherically symmetric spacetimes, where at any given epoch, the local Gaussian curvature of the 2-shells are monotonically decreasing along the integral curves of the preferred spatial direction. This assumption has a strong physical basis, as by definition of the Gaussian curvature this will imply that the area radius of the 2-shells are monotonically increasing along the integral curves of the preferred spatial direction and this is a sufficient condition for *no shell-crossing*. We will discuss the concept of shell crossings in the next section. This will then immediately imply positive spatial volume expansion of the preferred spatial direction ($\phi > 0$). Furthermore, by spherical symmetry we can easily see that it suffices to study the one dimensional MOTS curve in the local $[u, e]$ plane to determine its local properties. Now using Eqs. (9.45) and (9.46) it is evident that the equation of the MOTS curve in the local $[u, e]$ plane is given by

$$\Psi \equiv \left(\tfrac{2}{3}\Theta - \Sigma + \phi\right) = 0\,. \tag{9.47}$$

Since a monotonically decreasing Gaussian curvature implies $\phi > 0$, on the MOTS curve we have $\tilde{\Theta}_{out} = 0$ and $\tilde{\Theta}_{in} < 0$, as it should be by the above definition. Here we would like to show another important geometrical property of MOTS. Let us calculate the quantity $\nabla_a K \nabla^a K$ for a spherically symmetric spacetime (where $K \neq 0$):

$$\nabla_a K \nabla^a K = -\dot{K}^2 + \hat{K}^2. \tag{9.48}$$

Now using (9.30) in (9.48) we get

$$\nabla_a K \nabla^a K = \left(\tfrac{2}{3}\Theta - \Sigma + \phi\right)\left(\tfrac{2}{3}\Theta - \Sigma - \phi\right) K^2\,. \tag{9.49}$$

Hence for any 2-shell on the MOTS, the 4-gradient of its local Gaussian curvature is null. This property allows us to easily locate the MOTS in a given spacetime in the following way. From the equation of the Misner-Sharp mass (9.32), it is clear that if the 4-gradient of the Gaussian curvature is null then

$$2\mathscr{M}\sqrt{K} = 1\,. \tag{9.50}$$

This gives an alternate definition of MOTS for the spacetimes with no shell crossing:

Definition 9.3 A smooth, three-dimensional sub-manifold $\mathscr{H}$ in a spacetime $(\mathscr{M}, g)$ is said to be a **Marginally outer trapped surface** if it is foliated by a preferred family of 2-spheres such that the gravitational mass enclosed by each leaf $\mathscr{S}$ is half of the area radius of the leaf.

Having defined the MOTS surface purely in terms of the geometrical scalars associated with the matter flow lines and preferred spatial direction, the next obvious question would be to describe how the MOTS evolves geometrically.

Question: *What are the local conditions in terms of the geometry and the matter energy momentum tensor that determine whether MOTS will be locally timelike, spacelike or null ?*

As we stated earlier, due to the spherical symmetry, it suffices to study the behaviour of the MOTS curve in the local $[u, e]$ plane. Let the vector $\Psi^a = \alpha u^a + \beta e^a$ be the tangent to the curve $\Psi = 0$ in the local $[u, e]$ plane. Then we must have $\Psi^a \nabla_a \Psi = 0$.

Since we know that $\nabla_a \Psi = -\dot{\Psi} u_a + \hat{\Psi} e_a$, we can immediately see that the slope of the tangent to the MOTS on the local $[u, e]$ plane is given by $\alpha/\beta = -\hat{\Psi}/\dot{\Psi}$. Now using this decomposition with the field equations (9.19) to (9.27) on the surface $\Psi = 0$, we obtain

$$\nabla_a \Psi = \left(\frac{1}{3}\mu + p - \mathcal{E} + \frac{1}{2}\Pi - Q\right) u_a + \left(-\frac{2}{3}\mu - \frac{1}{2}\Pi - \mathcal{E} + Q\right) e_a, \quad (9.51)$$

and hence

$$\frac{\alpha}{\beta} = \frac{\frac{2}{3}\mu + \frac{1}{2}\Pi + \mathcal{E} - Q}{-\frac{1}{3}\mu - p + \mathcal{E} - \frac{1}{2}\Pi + Q}. \quad (9.52)$$

We now define the notion of *Future Outgoing (Ingoing)* in the following way. If $\alpha/\beta > 0$ then the MOTS is said to be future outgoing and if $\alpha/\beta < 0$ then the MOTS is said to be future ingoing. The nature of the MOTS in terms of it being timelike, spacelike or null can also be determined by the ratio α/β. We can easily see that

$$\Psi^a \Psi_a = \beta^2 \left(1 - \frac{\alpha^2}{\beta^2}\right). \quad (9.53)$$

Therefore $\alpha^2/\beta^2 > 1 (< 1)$ denotes the MOTS as locally timelike (spacelike). If $\alpha^2/\beta^2 = 1$ (as in the case of vacuum spacetimes, Schwarzschild or Schwarzschild-de Sitter, where all the thermodynamical terms vanish), the MOTS are null.

9.5 Local Violation of CCC in Spherical Collapse and Effect of Weyl Curvature

Having developed the complete geometrical picture of spherically symmetric self gravitating systems, the radial null geodesics and MOTS, we are now in a position to answer the question posed in the introductory section. Let us consider a sufficiently massive star that is undergoing continual gravitational collapse to form a shell focussing strong curvature singularity at the centre, where all the curvature scalars blow up strongly. We would like to find out if it is possible to have future directed radial null geodesics (and hence families of null geodesics) reaching far away observers with the past end point arbitrarily close to the central singularity. In that case the censorship will be violated. So our key question here is, under what conditions on the geometrical scalars that describe the spherical self gravitating system, such a scenario will be possible.

In this regard it is important to emphasize that apart from the shell focussing strong curvature singularity at the centre of the collapsing star (where the entire stellar matter collapse to a zero physical radius), there may be other kinds of singularities present in the spacetime which are called "*Shell crossing singularities*". These singularities occur when nearby shells of matter intersect, creating momentary density singularities where some of the curvature scalars could blow up [21, 24]. This may happen at any stage of the collapse and the situation is very similar to the formation of *caustics* in Newtonian fluid dynamics where two nearby fluid layers cross each other and thereby create shock waves travelling through the medium. In the context of general relativity, these shell crossing singularities are known to be weak in nature which are resolved through a suitable extension of the spacetime (as opposed to Big Bang or strong curvature shell focusing singularities), However, as we have discussed in the previous section, the assumption that the Gaussian curvature of the spherical shells are to be decreasing along the integral curves of the preferred spatial direction is a sufficient condition to make the collapse devoid of any shell crossings. This implies a positive spatial volume expansion of the preferred spatial direction ($\phi > 0$) within the collapsing star. Hence we will only concentrate on the nature of the final shell focussing singularity.

Another important aspect of considering no shell crossing in the vicinity of the collapsing central shell is that the outer shells do not become singular (with vanishing area radius) before the central shell does. Hence this will make the singularity curve (the time taken for a given shell to reach the singularity from the commencement of continual collapse) a non-decreasing function of the affine parameter of the integral curve of the vector e^a. Secondly, we specifically focus on the final fate of the singularity of the central shell in terms of being naked or covered as it has been shown numerous times that if all the physically reasonable energy conditions are satisfied by the collapsing matter, then the non-central singularities are always covered [24]. This is due to the reason that non central shells always enclose a positive gravitational mass. Hence, when their area radius becomes zero, the ratio $2\mathscr{M}/\mathscr{R}$ goes to infinity. Which implies that these singularities are well within the trapped region whose boundary is the surface $2\mathscr{M}/\mathscr{R} = 1$. Also it was shown very transparently in [15] that the spacetime point on which the central shell becomes singular *must* lie on the MOTS. We now define the notion of local nakedness give the necessary and sufficient condition for the singularity of the central shell to be naked.

Definition 9.4 If there exists an open set $\mathscr{U}$ in $(\mathscr{M}, g)$ that contains the outgoing null rays with their past end points arbitrarily close to a spacetime singularity and the null rays do not hit the MOTS in $\mathscr{U}$, then the singularity is said to be locally naked in the open set $\mathscr{U}$.

Proposition 9.1 *Consider the continued collapse of a general spherically symmetric matter cloud that obeys the physically reasonable energy conditions. from a regular initial epoch. If the following conditions are satisfied:*

1. *The spacetime is free of shell crossing singularities,*
2. *Closed trapped surfaces exist,*

then the necessary and sufficient condition for the existence of an open set $\mathscr{U}$ in which the central singularity is locally naked, is that the MOTS at the central singularity should be future outgoing and non-spacelike.

Proof The key point of the proof of this proposition lies in picturing that if the null geodesic with the past end point arbitrarily close to the central singularity is future directed and outgoing at a given point P, then it must not hit the MOTS at any point in the causal past of P. Otherwise the ray will be trapped with $\tilde{\Theta}_{out} < 0$ and it will fall back to the singularity.

If the MOTS is future ingoing at the central singularity then the trapping is inevitable as the outer shells get trapped before the central singularity. The same argument applies if the MOTS at the central singularity is future outgoing and spacelike as any outgoing null direction from the central singularity will be necessarily within the trapped region. Therefore, the necessary condition for a central singularity to be locally naked in $\mathscr{U}$ is that the MOTS at the central singularity is future outgoing and non-spacelike ($\alpha/\beta \geq 1$).

Conversely, suppose there exists a future directed null geodesic that has reached any future non-trapped point P in $\mathscr{U}$ from the points arbitrarily close to the central singularity in the $[u, e]$ plane . Then that would imply that for this ray $\tilde{\Theta}_{out} > 0$ for every point on the geodesic that is in the causal past of P and hence the ray has not hit the MOTS in the causal past of P. This is possible if and only if the MOTS is future outgoing and non-spacelike at the central singularity. □

This result is extremely interesting as it transparently explains the role of the energy momentum tensor of the collapsing matter field as well as the Weyl curvature in making a spacetime singularity locally visible [19]. In a way, this result is universal in the realm of spherical symmetry, as we haven't a priori assumed any form of the collapsing matter field . Also, as shown in [21], if a null geodesic emerges from the singularity, then there exist families of future directed non-spacelike curves which also necessarily escape from the same. The existence of such families is crucial to the physical visibility of the singularity at least locally. Further to this, it is clear that if such geodesics cross the boundary of the collapsing star, then they can travel to the future null infinity through the exterior spacetime and hence the condition of the spacetime being future asymptotically simple will be violated.

We now look into the crucial role the Weyl curvature plays, in making the central singularity locally naked in the next proposition and its corollary. To make the picture more transparent, we consider the collapsing matter field to be a perfect fluid. This is a very physical assumption as most stellar models assume the perfect fluid form of matter. This then implies that the pressure anisotropy and heat flux of the collapsing matter vanish identically.

Proposition 9.2 *Consider the gravitational collapse of a spherically symmetric perfect fluid obeying all physically reasonable energy conditions. If the following conditions are satisfied :*

1. *The spacetime is free of shell crossing singularities,*
2. *Closed trapped surfaces exist,*
3. *The central singularity is marginally naked ($\frac{\alpha}{\beta} = 1$),*

then the limit of $\frac{|\mathscr{E}|}{\mu+p}$ at the central singularity along the apparent horizon curve diverges.

Proof We know that for a perfect fluid we have $Q = \Pi = 0$, and at the central singularity $\frac{\alpha}{\beta} = 1$ implies

$$\frac{\frac{2}{3}\mu + \mathscr{E}}{-\frac{1}{3}\mu - p + \mathscr{E}} = 1 \,. \tag{9.54}$$

This can have two consequences. The first one is, $\mathscr{E}$ is finite $\mu + p = 0$. But this doesn't relate to a dynamical collapse scenario and can be excluded. To get the second consequence, when $\mu + p \neq 0$ but diverges at the central singularity along with $\mathscr{E}$, we subtract 1 from both sides of the above equation, which gives

$$\left[\frac{\mathscr{E}}{\mu + p} - \frac{1}{3}\frac{\mu + 3p}{\mu + p}\right]^{-1} = 0. \tag{9.55}$$

For the perfect fluid satisfying the physically reasonable energy conditions, $\frac{\mu+3p}{\mu+p}$ is finite even when both μ and p diverge and hence $\frac{|\mathscr{E}|}{\mu+p}$ at the central singularity along the apparent horizon must tend to infinity. □

The above result clearly shows that the electric part of the Weyl scalar, which is the measure of free gravity in this spherically symmetric case and that manifests itself via the tidal deformations, must diverge faster than the total inertial energy density $\mu + p$, along the apparent horizon curve, for a singularity to be locally naked. In fact, this result closely relates to the result obtained in [22] that the square of the shear 3-tensor (σ^2) must diverge faster than the energy density at the central singularity of the collapsing dustlike matter. This clearly gives us a comprehensive geometrical picture of the whole process. The tidal effects diverging faster than the total inertial energy density would generate a diverging spacetime shear deformation (via the shear evolution Eq. (9.23)), which will in turn deform the geometry of MOTS in such a way that radially outgoing null rays can escape locally from points arbitrarily close to the central singularity without getting trapped. In fact the next corollary cements this idea further.

Corollary 9.1 *Consider the continued gravitational collapse of a spherically symmetric perfect fluid obeying all physically reasonable energy conditions. If the spacetime is conformally flat then the central singularity is necessarily covered,*

Proof A conformally flat spacetime implies the vanishing of the Weyl tensor. Hence we have $\mathscr{E} = 0$. Also for a perfect fluid $Q = \Pi = 0$. We therefore have

$$\frac{\alpha}{\beta} = \frac{-\frac{2}{3}\mu}{\frac{1}{3}\mu + p} \,. \tag{9.56}$$

Strong energy conditions immediately imply that $\alpha/\beta < 0$ and thus the MOTS is future ingoing at the central singularity and by earlier discussions, we can see that it is necessarily covered. In fact, one can explicitly calculate the norm of the tangent to show that

$$\Psi^a\Psi_a = -\beta^2 \frac{1}{3}(\mu + p)(\mu - 3p) . \tag{9.57}$$

If the weak energy condition is satisfied we have $\mu + p > 0$, then we have the following cases:

- If $\mu > 3p$, the Ψ^a is "ingoing" timelike.
- If $\mu = 3p$, the Ψ^a is "ingoing" null.
- If $\mu < 3p$, the Ψ^a is "ingoing" spacelike.

In all these cases, the region around the centre gets trapped before the central singularity. Hence the central singularity is always covered. □

9.6 How to Relate Local Violation of CCC with Thermodynamics of Gravity

Thus far we have identified the key geometrical mechanism that is responsible for producing a locally naked central singularity at the end state of a continual gravitational collapse, that commences from a regular initial epoch, and the physically reasonable energy conditions being obeyed throughout the process. Now the natural question that arises is the following:

Question: *Can we relate our analysis with the thermodynamics of gravitational field in a quantitative fashion?*

This is indeed a very difficult question. Since the birth of general relativity, the problem which perpetually eluded a satisfactory solution was the exact mechanism through which energy can be transferred from one body to another via gravitational interactions. In electromagnetism this problem is well understood as the electromagnetic energy can be transferred across empty space via induction or radiation, enabled by the electromagnetic field F_{ab}. One would then naturally expect the same in the case of gravity.

The biggest hurdle to adequately tackle the above problem is the absence of any satisfactory energy momentum tensor for free gravity (gravity in the absence of matter). One approach has been via stress-energy-momentum pseudotensors, for example the Landau-Lifshitz pseudotensor [25]; but as they are not tensors, they can be made to vanish by a suitable choice of coordinates. Roger Penrose suggested that it should be constructed entirely via the Weyl tensor, which describes the free gravity on the manifold, but did not give a specific formula. In this direction, Maartens and Basset [26] presented a very interesting result. They proved that the Bell–Robinson tensor [3–5, 29]

$$T_{abcd} \equiv \frac{1}{4}\left(C_{eabf}C^{e}{}_{cd}{}^{f} + C^{*}_{eabf}C^{*e}_{cd}{}^{f}\right) \tag{9.58}$$

is a unique Maxwellian tensor that can be constructed from the Weyl tensor so as to act as the 'super energy momentum' tensor for gravitational fields (see also [7, 30,

31]), with only one caveat, that the dimension of this tensor is L^{-4} and not L^{-2} like the energy momentum tensor should be. Based on this result, Clifton, Ellis and Tavakol [11] proposed that the symmetric 2-index square root $\mathscr{T}_{ab}$, of the Bell–Robinson tensor would act as the effective energy momentum tensor for free gravitational field. Although the square root may not exist for general spacetimes, it was shown that a unique square root exists (up to an arbitrary trace) for spacetimes that are Petrov type D which has a Coloumb like gravitational field, or N, that is a wave like gravitational field [32].

As described in [11] , for Coulomb-like gravitational fields that are described by a Petrov type D spacetime (of which spherical spacetimes are a subclass), the symmetric 2 index square root of the Bell–Robinson tensor $\mathscr{T}_{ab}^{G}$ can be written as

$$\mathscr{T}_{ab}^{G} = \alpha_0 \left[3|\Psi_2| \left(m_{(a}\bar{m}_{b)} + l_{(a}k_{b)}\right) + f g_{ab}\right], \tag{9.59}$$

where α_0 is a constant, l_a and n_a are real null vectors, and m_a and $\bar{m}_a$ are complex null vectors at a given point in the spacetime in the Newman-Penrose tetrad formalism. Ψ_2 is the only non zero complex Weyl scalar for Petrov type D spacetimes and 'f' is a free function that can be fixed via physical requirements. In terms of electric and magnetic parts of the Weyl tensor, the complex Weyl scalar can be written as

$$|\Psi_2| = \sqrt{\frac{1}{6}\left(E_{ab}E^{ab} + H_{ab}H^{ab}\right)}\,. \tag{9.60}$$

The Newman Penrose null tetrads are related to the timelike direction and preferred spatial direction of spherically symmetric spacetimes in the following way

$$k^a = \frac{1}{\sqrt{2}}\left(u^a + e^a\right),\; l^a = \frac{1}{\sqrt{2}}\left(u^a - e^a\right),\; N_{ab} = 2m_{(a}\bar{m}_{b)}\,. \tag{9.61}$$

Since for spherical symmetry we have $H_{ab} = 0$ and $E_{ab}E^{ab} = (3/2)\mathscr{E}^2$, we get

$$|\Psi_2| = \pm\frac{1}{2}\mathscr{E}\,. \tag{9.62}$$

Therefore, the effective energy momentum tensor for the free gravitational field in spherical symmetry takes the form

$$\mathscr{T}_{ab}^{G} = \alpha_0 \left[\frac{3}{2}\mathscr{E}\left(u_a u_b - e_a e_b\right) + \left(f + \frac{3}{4}\mathscr{E}\right) g_{ab}\right]. \tag{9.63}$$

To define the arbitrary function f, we assume the physically reasonable condition that this effective energy momentum tensor $\mathscr{T}_{ab}^{G}$ is conserved (divergence free) in vacuum. Then using the field equations (9.24) and (9.21) in vacuum, we can uniquely determine the function f as

$$f = -\frac{1}{4}\mathscr{E} + \lambda_1, \tag{9.64}$$

where λ_1 is another arbitrary constant. Therefore the energy momentum tensor for the free gravitational field, that is obtained via the 2-index square root of the Bel Robinson tensor, is given as [17]

$$\mathscr{T}^G_{ab} = \alpha_0 \left[\frac{3}{2}\mathscr{E}\left(u_a u_b - e_a e_b\right) + \left(\frac{1}{2}\mathscr{E} + \lambda_1\right) g_{ab}\right], \tag{9.65}$$

with the following thermodynamic quantities

$$\mu^G = \alpha_0(\mathscr{E} + \lambda_1),\;\; p^G = \alpha_0\lambda_1,\;\; \Pi^G = -\alpha_0\mathscr{E},\;\; Q^G = 0\,. \tag{9.66}$$

The above result is extremely interesting as now we can relate in a straightforward fashion, the conditions on the energy momentum tensor of the free gravity with the conditions on the energy momentum tensor of the collapsing matter for the scenarios where the CCC is locally violated in an open set $\mathscr{U}$. For example, in the case of the continual collapse of the perfect fluid form of matter, the main result in Proposition 9.2, translates to the following: If the central singularity has to be locally naked, at the vicinity of the central singularity we must have

$$\frac{\mu + p}{\mu^G} \to 0\,. \tag{9.67}$$

In other words, in the vicinity of the central singularity, the tidal effects on collapsing matter, which definitely give rise to the matter inhomogeneity, must absolutely overcome the inertial matter energy density if the central singularity is locally naked. This result is also corroborated from the fact that, under the same condition, the Misner-Sharp mass enclosed by the near central collapsing shells can be written as

$$\mathscr{M} \approx \frac{1}{2K^{\frac{3}{2}}}|\mathscr{E}|\,. \tag{9.68}$$

These observations allow us to state the following proposition

Proposition 9.3 *For the central singularity formed due to a continual collapse of a spherical self gravitating system from a regular initial epoch and all physically reasonable energy conditions being satisfied, to be locally naked in an open set $\mathscr{U}$, in the vicinity of the central singularity the total energy of the self gravitating system should be completely dominated by the the energy of the free gravitational field. In this case the trace-free part of the Riemann curvature will dominate over the other parts.*

9.7 The Story so Far and Open Questions to Ponder About

We began our journey by posing an important question in the introductory section. With the mission of finding a satisfactory answer to the question posed, we worked with a semi-tetrad covariant and model independent formalism within the realm of spherical symmetry. We successfully identified the physical and geometrical mechanisms responsible for delaying the trapped surface formation and making the central singularity locally naked during the continued gravitational collapse of a spherical massive star. By working out the geometry of the boundary of the trapped region we transparently and quantitatively identified the role of the Weyl curvature in deforming the boundary in such a way that the central singularity can be locally naked.

The complete physical and geometrical picture can be heuristically summed up in the following way. In the previous section we conclusively observed that if the central singularity is locally naked then the total energy in the vicinity of the singularity must be completely dominated by the energy of the free gravitational field, which is the Weyl curvature. Therefore, since the Weyl curvature is responsible for the tidal forces within the collapsing cloud, it generates the spacetime shear or the deformation. This deformation then naturally dominates over the collapsing matter density and deform the MOTS geometry in such a way that there will exist an open set where in principal we can sneak a peek into the singular point.

A word of caution here. Our analysis is necessarily local and we showed how the singularity can be locally naked in an open set $\mathscr{U}$ by making the MOTS outgoing null at the central singularity, so that the outgoing null rays from points that are arbitrarily close to the central singularity can escape. However, the MOTS will not continue to be non-spacelike in general. As more and more collapsing shells cross the MOTS, it will be dynamical and become spacelike as clearly shown by Ashtekar and Krishnan [2]. Hence, there will be a continuous race between the MOTS and the escaping rays and it is as well possible that these rays get trapped before they can cross the stellar surface. In that case we can avoid the global nakedness and save the future asymptotically simple nature of the global spacetime. Also this will then make all the concerns raised by Penrose against the naked singularities [28] irrelevant.

However, the key physical point to be noted here is this. Even if we cannot sneak a peek into an *infinite* spacetime curvature region, regions with *sufficiently* high curvature (way higher than the case of an Oppenheimer Snyder type collapse where the star gets trapped much earlier than the singularity) can still be visible to the outside observers. This immediately brings us to these open questions:

1. What will be the observational signatures of these *sufficiently* high curvature regions in a collapsing star? Surely at such high curvature the purely classical fluid picture of the collapsing fluid will break down. The fluid picture will then become more statistical in nature as Einstein-Vlasov kinetic matter [1] and semi-classical effects will start creeping in. Can these semiclassical effect be observed and identified via astrophysical observations?

2. Although our analysis is purely geometrical and independent of the matter models, still it is within the realm of spherical symmetry. No astrophysical stellar object is exactly spherical. What will happen to our results if we perturb the geometry?
3. And finally, how sensitive are these results to the theory of gravity? To explain our current observations of the universe, especially the dark sectors of the energy budget, there has been continuous proposals of various modifications of general relativity. What will be the impact of these modifications to the censorship conjecture?

Acknowledgements The author thanks Prof. Naresh Dadhich for a number of extremely important comments and suggestions. Also Dr. C. Hansraj for carefully proof reading the chapter. The author is supported by the National Research Foundation (South Africa) and University of KwaZulu-Natal.

References

1. H. Andréasson, Living Rev. Relativ. 14, 4 (2011). https://doi.org/10.12942/lrr-2011-4 https://doi.org/10.12942/lrr-2011-4
2. A. Ashtekar, B. Krishnan, Phys. Rev. Lett. **89**, 261101 (2002). arXiv:gr-qc/0207080
3. L. Bel, C. R. Acad. Sci. **247**, 1094 (1958)
4. L. Bel, C.R. Acad. Sci. **248**, 1297 (1959)
5. L. Bel, Cah. Phys. **138**, 59 (1962)
6. A.L. Besse, *Einstein Manifolds*. Results in Mathematics and Related Areas (3), vol. 10. (Springer, Berlin, New York, 1987), p. 510. ISBN 978-3-540-15279-8
7. M.A.G. Bonilla, J.M.M. Senovilla, Gen. Relativ. Gravit. **29**, 91 (1997)
8. B. de Swardt, P.K.S. Dunsby, C. Clarkson, arXiv:1002.2041 [gr-qc]
9. C. Clarkson, R.K. Barrett, Class. Quantum Grav. **20**, 3855 (2003)
10. C. Clarkson, Phys. Rev. D **76**, 104034 (2007)
11. T. Clifton, G.F.R. Ellis, R. Tavakol, Class. Quant. Grav. **30**, 125009 (2013). ([arXiv:1303.5612 [gr-qc]])
12. G.F.R. Ellis, J. Math. Phys. **8**, 1171–1194 (1967). H. van Elst, G.F.R. Ellis, Class. Quantum Grav. **13**, 1099 (1996). arXiv:gr-qc/9510044
13. G. F. R. Ellis & H. van Elst, *Cosmological Models*, Cargèse Lectures 1998, in Theoretical and Observational Cosmology, Ed. M Lachze-Rey, (Dordrecht: Kluwer 1999), 1. [arXiv:gr-qc/9812046]
14. G.F.R. Ellis, R. Goswami, A.I.M. Hamid, S.D. Maharaj, Phys. Rev. D **90**(8), 084013 (2014). https://doi.org/10.1103/PhysRevD.90.084013. arXiv:1407.3577 [gr-qc]
15. R. Goswami, P.S. Joshi, Phys. Rev. D **76**, 084026 (2007). https://doi.org/10.1103/PhysRevD.76.084026. arXiv:gr-qc/0608136 [gr-qc]
16. R. Goswami, G.F.R. Ellis, Class. Quant. Grav. **34**(5), 055009 (2017). https://doi.org/10.1088/1361-6382/aa5dc0. arXiv:1511.05700 [gr-qc]
17. R. Goswami, G.F.R. Ellis, Class. Quant. Grav. **35**(16), 165007 (2018). https://doi.org/10.1088/1361-6382/aad0ad. arXiv:1805.00927 [gr-qc]
18. R. Goswami, G.F.R. Ellis, Class. Quant. Grav. **38**(8), 085023 (2021). https://doi.org/10.1088/1361-6382/abdaf3. arXiv:1912.00591 [gr-qc]
19. A.I.M. Hamid, R. Goswami, S.D. Maharaj, Class. Quant. Grav. **31**, 135010 (2014). https://doi.org/10.1088/0264-9381/31/13/135010. arXiv:1402.4355 [gr-qc]
20. S.W. Hawking, G.F.R. Ellis, *The Large Scale Structure of Space-Time* (Cambridge University Press, Cambridge, 1973). ISBN 978-1-00-925316-1. https://doi.org/10.1017/9781009253161

21. P.S. Joshi, *Global Aspects in Gravitation and Cosmology*. International Series of Monographs on Physics (87) (Clarendon Press, Oxford, 1994)
22. P.S. Joshi, N. Dadhich, R. Maartens, Phys. Rev. D **65**, 101501 (2002). arXiv:gr-qc/0109051
23. P.S. Joshi, R. Goswami, N. Dadhich, Phys. Rev. D **70**, 087502 (2004)
24. P.S. Joshi, *Gravitational Collapse and Spacetime Singularities* (Cambridge University Press, Cambridge, 2008). ISBN 978-1-107-40536-3. https://doi.org/10.1017/CBO9780511536274
25. L.D. Landau, E.M. Lifshitz, *The Classical Theory of Fields* (Pergamon Press, 1951). Chapter 11, Section 96
26. R. Maartens, B.A. Bassett, Gravitoelectromagnetism. Class. Quant. Grav. **15**, 705 (1998). ([gr-qc/9704059])
27. C.W. Misner, D.H. Sharp, Phys. Rev. D **136**, B571 (1964)
28. R. Penrose, Riv. Nuovo Cim. **1**, 252–276 (1969). https://doi.org/10.1023/A:1016578408204
29. I. Robinson, *Lectures at Kings College (London)* (1958) (unpublished)
30. J.M.M. Senovilla, Gen. Relat. Gravit. **32**, 2043 (2000)
31. J.M.M. Senovilla, Class. Quantum Gravity **17**, 2779 (2000)
32. H. Stephani, E. Herlt, M. MacCullum, C. Hoenselaers, D. Kramer, *Exact Solutions to Einstein's Field Equations* (Cambridge University Press, Cambridge, UK, 2003)
33. S. Singh, D. Baboolal, R. Goswami, S.D. Maharaj, Class. Quant. Grav. **39**(23), 235010 (2022). https://doi.org/10.1088/1361-6382/ac9efe. arXiv:2206.03828 [gr-qc]

Chapter 10
Spherical and Aspherical Gravitational Collapse: Locally and Globally Naked Singularities

Dipanjan Dey and Karim Mosani

10.1 Introduction

For stars having mass greater than ten solar masses, the quantum pressures (electron and neutron degeneracy pressure) inside the collapsing matter become insufficient to hold it further. What one obtains, then is its continual gravitational collapse. The singularity theorem proposed by Hawking and Penrose predicts that under generic conditions of gravitational collapse, a spacetime singularity should always form. One version of the singularity theorem states: If (M, g) (a four-dimensional spacetime) is

1. Connected, globally hyperbolic spacetime with non-compact Cauchy surface Σ,
2. $Ric\ (\xi, \xi) \geq 0\ \forall\ \xi$ (Ric: Ricci tensor, ξ: null vector), and
3. Admits a trapped surface T,

then at least one inextendible future-directed orthogonal null geodesic from T exists that is future incomplete.

The above statement predicts that the gravitational collapse of a generic matter cloud inevitably leads to a spacetime singularity [1]. Once a singularity is formed due to gravitational collapse, it may or may not be naked (visible).

In the context of the visibility of a singularity, a singularity can be locally or globally visible. In both scenarios, the singularity is non-spacelike. A globally visible singularity is causally connected to the future null infinity; on the other hand, a locally visible singularity is only causally connected to the neighbouring spacetime points

D. Dey (✉)
Department of Mathematics and Statistics, Dalhousie University, Halifax, NS B3H 3J5, Canada
e-mail: deydipanjan7@gmail.com

K. Mosani
Department of Mathematics, University of Tuebingen, Auf der Morgenstelle 10, 72076 Tuebingen, Germany
e-mail: kmosani2014@gmail.com

D. Malafarina and P. S. Joshi (eds.), *New Frontiers in Gravitational Collapse and Spacetime Singularities*, Springer Series in Astrophysics and Cosmology,
https://doi.org/10.1007/978-981-97-1172-7_10

since they are always covered by an event horizon. The event horizon is the boundary of all non-spacelike geodesics reaching the future null infinity. In case a singularity is locally visible, singular null geodesics can escape from it but ultimately get trapped inside the event horizon. In contrast, in case a singularity is globally visible, singular null geodesics reach an asymptotic observer. We say that a singularity is spacelike if it is globally as well as locally not visible (hidden). A spacelike singularity is always covered by trapped surfaces. As an example of stationary spacetimes, in the case of Schwarzschild black hole, the central singularity is covered by trapped surfaces. In contrast, in the case of Reissner-Nordstrom's (RN) black hole, trapped surfaces do not exist in the neighbourhood of the singularity. Therefore, Schwarzschild singularity is spacelike, and RN singularity is timelike.

In Oppenheimer-Snyder-Datt (OSD) collapse model, the collapsing matter is considered as spherically symmetric, homogeneous (density $\rho \equiv \rho(t)$), dust (pressure $P = 0$). One can show that the spacetime singularity formed as an end state of the gravitational collapse of such matter cloud is locally and globally hidden (spacelike singularity). Such singularity is at par with the Cosmic Censorship Conjecture (CCC) proposed by Roger Penrose [2]. The two forms of CCC are as follows:

1. The weak cosmic censorship [3] states that a spacetime singularity should always be globally hidden,
2. The strong cosmic censorship [4, 5] states that the singularity formed during the gravitational collapse should always be spacelike, i.e., it can never be locally naked.

For cosmic censorship to hold true, spacetime should be globally hyperbolic. On the contrary, one can obtain initial conditions of gravitational collapse such that the endstate is a naked singularity. If we consider a collapsing inhomogeneous matter cloud, the evolutions of the event horizon and the apparent horizon are such that locally and/or globally visible singularities can be obtained [6–8].

There has been a keen interest in investigating the visibility or otherwise of spacetime singularities formed in the continuous gravitational collapse of matter in the realm of general relativity as well as modified gravity theories [9–27]. One of the reasons is that if the singularity formed due to gravitational collapse is visible to distant observers, we could, in principle, observe the realm of quantum gravity in the ultra-high-density region. At the same time, for the black hole case, since the singularity is spacelike, that information is hidden inside the event horizon.

Under various settings, it can be shown that the singularity formed during the gravitational collapse of an inhomogeneous spherical cloud can be visible to an asymptotic observer [8, 20, 22–27], and that there exists a nonzero measured set of initial parameters giving rise to such singularities. The existence or otherwise of a locally/globally visible singularity could give rise to various phenomenological outcomes that can be observed. Examples are shadow of the singularity [28], timelike and null trajectories in the vicinity of the singularity [29, 30], accretion disc properties [31], and gravitational lensing [32, 33].

Coming back to the singularity theorems, one of their interpretations is that general relativity predicts its breakdown. Therefore, the existence of singularities puts a question on the accuracy of the general relativity in the ultra-strong gravity regime, where singularities are likely to occur. A modified theory of gravity in the strong gravity regime is studied in [35–37]. In this context, in [38, 39], the authors argue that general relativistic predictions near the singularities should be investigated further since it may guide us into what one could expect from improved theories of gravity such as quantum gravity, which can apply to the strong gravity regime and the theory would be singularity free.

Even if the singularity formed in the gravitational collapse is visible to faraway observers, one could argue that the information about the physics of ultra-high gravity region carried by the outgoing singular null geodesics can get distorted due to the scattering of the null geodesic caused by the collapsing matter surrounding the singularity. Suppose a singularity formed during a spherical gravitational collapse becomes globally visible. In that case, the null geodesics must travel through the collapsing matter. Due to the spherical symmetry, it takes an equal amount of time to reach the boundary of collapsing matter along every angular direction. However, in the case of aspherical collapse, singularity can be highly shifted towards the boundary along a certain direction, and therefore, null geodesics coming out from the singularity can be less scattered around that certain angular direction. The closer the singularity is to the boundary, the lesser will be the scattering of the outgoing singular null geodesics in certain directions.

Here, we give a short review of the developments in the understanding of locally as well as globally visible singularities formed due to gravitational collapse. The chapter contains the following: first, we discuss the OSD collapse, which is a spherical homogeneous dust collapse and shows that the singularity formed in this type of collapse is always spacelike. Next, we discuss the spherical inhomogeneous dust collapse modelled by Lemaitre-Tolman-Bondi (LTB) spacetime. There, we show that the singularity, in this case, can be locally and globally visible. Lastly, we discuss the aspherical Szekeres collapse model [40] and show the influence of asphericity on the local and global visibility of the singularity. This chapter is representation mainly of the following papers: [6, 13, 22, 27, 51, 52].

10.2 Oppenheimer-Snyder-Datt (OSD) Collapse

Datt [49], and independently Oppenheimer and Snyder [48], studied the first model of gravitational collapse in the context of general relativity. The setup of the matter model is as follows: (1) Spatial homogeneity, (2) Spherical symmetry, and (3) Perfect fluid with zero pressure. The corresponding stress-energy tensor of such matter model has components in the comoving coordinate basis as $T^{\mu}_{\nu} = \text{diag}(\rho, 0, 0, 0)$. Such matter cloud collapses to a singularity. Such singularity is hidden behind an event horizon, and hence a black hole is obtained as an end-state. The previous statement will be justified by the end of this section. The spacetime metric for such a stress-energy tensor is the Friedmann-Lemaître-Robertson-Walker (FLRW) metric. In the comoving coordinate basis, we write it as

$$ds^2 = -dt^2 + \frac{v(t)^2}{1-kr^2}(dr^2 + r^2 d\Omega^2), \tag{10.1}$$

where

$$d\Omega = d\theta^2 + \sin^2\theta d\phi^2 \tag{10.2}$$

is the line element on the two-sphere. Here, $k \in \{-1, 0, 1\}$. It represents the spatial curvature. $v(t)$ is the scale factor of the FLRW metric. The first Einstein's equation gives an expression of the density of the collapsing cloud in terms of the Misner-Sharp mass function $F(t, r)$ as

$$\rho = \frac{F'}{R^2 R'}. \tag{10.3}$$

The Misner-Sharp mass function is the measurement of the mass inside a spherical collapsing shell of radial coordinate r and at time t. We use the notation: subscript prime and dot for the partial derivative of the function with respect to radial and time coordinates, respectively. The Misner-Sharp mass function in terms of the metric component R and its time derivative is written as

$$F = R\left(\dot{R}^2 + kr^2\right). \tag{10.4}$$

The second Einstein's field equation gives us the pressure evolution of the collapsing cloud with the function of metric components and its derivatives as

$$p = -\frac{\dot{F}}{r^3 v^2 \dot{v}}. \tag{10.5}$$

However, since the collapsing matter cloud under consideration has zero pressure, we obtain

$$\dot{F} = 0, \tag{10.6}$$

which means that $F = F(r)$ and does not depend on t. Thus, in the case of dust collapse, we say that the mass of the collapsing cloud inside a shell of fixed radial coordinate r is conserved. For maintaining spatial homogeneity, one can choose a suitable functional form of the Misner-Sharp mass function as

$$F = F_0 r^3, \tag{10.7}$$

where $F_0 > 0$. We now integrate Eq. (10.4) with respect to time and obtain

$$t - t_s = -\frac{v^{\frac{3}{2}}}{\sqrt{F_0}}\mathcal{G}\left(\frac{kv}{F_0 r^2}\right). \tag{10.8}$$

Here $\mathcal{G}(y)$ is defined as follows:

$$\mathcal{G}(y) = \left(\frac{\arcsin\sqrt{y}}{y^{\frac{3}{2}}} - \frac{\sqrt{1-y}}{y} \right) \quad \text{for} \quad 0 < y < 1,$$
$$\mathcal{G}(y) = \frac{2}{3} \quad \text{for} \quad y = 0, \tag{10.9}$$
$$\mathcal{G}(y) = \left(\frac{-\text{arcsinh}\sqrt{-y}}{(-y)^{\frac{3}{2}}} - \frac{\sqrt{1-y}}{y} \right) \quad \text{for} -\infty < y < 0.$$

We call Eq. (10.8) the time curve $t(r, v)$. It gives us the comoving time the shell (identified by radial coordinate r) takes to reach the epoch v from the initial epoch $v = 1$. t_s is the constant of integration and gives us the time of formation of the singularity. It is the time curve for epoch $v = 0$, i.e., it gives us the comoving time the shell (identified by radial coordinate r) takes to reach the epoch $v = 0$ from the initial epoch $v = 1$. The singularity curve is expressed as

$$t_s = \frac{1}{\sqrt{F_0}} \mathcal{G} \left(\frac{k}{F_0} \right). \tag{10.10}$$

From here, we learn that concentric spherical shells corresponding to different radial coordinates r, collapses at the same comoving time. We will see in the next section that dropping the assumption of spatial homogeneity corresponds to $t_s = t_s(r)$, i.e., the constant of integration is different for different r indicating that the shells of different radial coordinate contracts to a singularity at different comoving times.

The surrounding space-time $r > r_c$ for some $r_c \in \mathbb{R}^+$ is governed by the Schwarzschild metric and is written as

$$ds^2 = -(1 - \frac{r_s}{r})dt^2 + (1 - \frac{r_s}{r})^{-1} dr^2 + r^2 d\Omega^2. \tag{10.11}$$

Here r_s is the Schwarzschild radius and is twice the total mass of the collapsing cloud in the units $c = G = 1$.

We call the FLRW metric that governs the collapsing spatially homogeneous dust cloud as the internal metric ($g^-_{\mu\nu}$), and the Schwarzschild metric that governs the surrounding spacetime, as the external metric ($g^+_{\mu\nu}$).

It is desirable to make the union of internal and external metrics a valid solution of Einstein's field equation. To do so, certain conditions need to be imposed on the otherwise free functions of the metric components. These conditions are called junction conditions.

The first junction condition dictates that the induced metric, from the internal and the external metric, on the hypersurface Σ partitioning the two regions should be the same. This condition ensures the elimination of the singular term arising in the first derivative of the new metric $g_{\mu\nu}$ that incorporates both the internal and the external metric such that

$$g_{\mu\nu} = \Theta(m) g^+_{\mu\nu} + \Theta(-m) g^-_{\mu\nu}. \tag{10.12}$$

$\Theta(m)$ is called the Heaviside distribution function that is defined as

$$\begin{aligned} \Theta(m) &= +1, & m > 0, \\ &= 0, & m < 0, \\ &= \text{indeterminite}, & m = 0. \end{aligned} \tag{10.13}$$

The second junction condition dictates that the Lie derivative of both the induced metrics with respect to the vector field that is orthogonal to the hypersurface Σ (identified by $m = 0$) should be the same. This condition ensures the elimination of the surface stress-energy tensor, thereby ensuring that Σ is non-singular. However, it should be noted that the violation of the second junction condition has an acceptable physical interpretation that the Σ is a thin shell with a non-vanishing stress-energy tensor. These junction conditions can put constraints on the otherwise free functions in the metric components.

The evolution of the apparent horizon in the case of OSD collapse can be found from the expression of the expansion scalar of the outgoing singular null geodesic, which in the metric under consideration (Eq. (10.1)) becomes

$$\theta = 2\left(\frac{\dot{v}}{v} + \frac{\sqrt{1-k}}{rv}\right). \tag{10.14}$$

Using Eq. (10.4), this can be rewritten as

$$\theta = \frac{2}{rv}\left(\sqrt{1-k} - \sqrt{\frac{F}{rv} - k}\right). \tag{10.15}$$

θ is zero for all points in the apparent horizon. Therefore, the apparent horizon is represented by

$$F = rv. \tag{10.16}$$

For $k = 0$, we obtain the scale factor of the collapsing homogeneous dust cloud using Eqs. (10.8) and (10.10) as

$$v(t) = \left(1 - \frac{3}{2}\sqrt{F_0}t\right)^{\frac{2}{3}}. \tag{10.17}$$

In the time curve, if we substitute the condition that is satisfied on the apparent horizon, namely the identity (10.16), we obtain the evolution of the apparent horizon as

$$t_{AH}(r) = \frac{2}{3}\left(\frac{1}{\sqrt{F_0}} - r^3 F_0\right). \tag{10.18}$$

We call the function $t_{AH}(r)$ the apparent horizon curve. We now have the physical radius $R(t, r) = rv(t, r)$ (Eq. (10.17)) and the apparent horizon curve $t_{AH}(r)$. Now

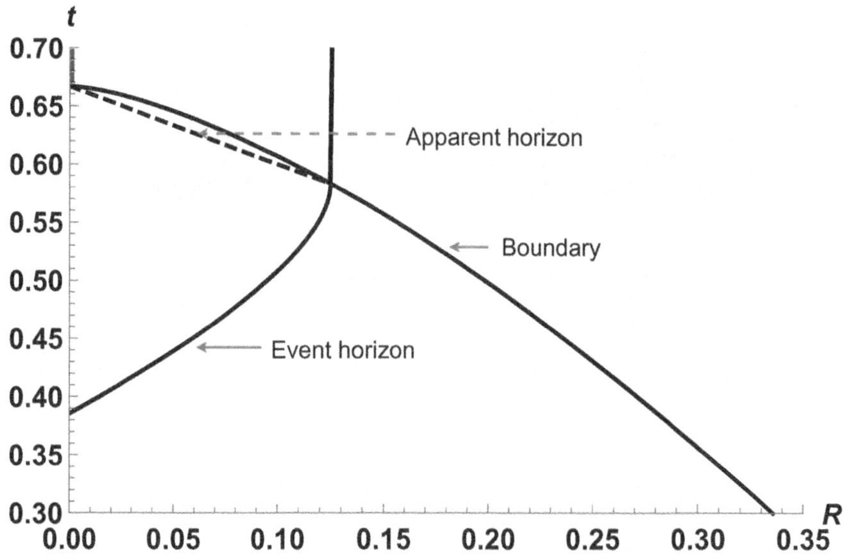

Fig. 10.1 Spacetime plot: Datt-Oppenheimer-Snyder collapse. R and t are the physical radius of the collapsing cloud and the comoving time, respectively. We set a certain numerical value to the comoving radius of the boundary of the collapsing cloud, beyond which the spacetime is governed by the Schwarzschild metric. The Misner-Sharp mass function is $F = F_0 r^3$, where $F_0 = 1$. We set $k = 0$ in the FLRW spacetime governing the collapsing cloud. The red segment on the vertical axis represents the worldline of the singularity

we can do a parametric plot $(F(r), t_{AH}(r))$ and obtain the apparent horizon curve in the (R, t) plane. We observe that in (R, t) plane, the curve representing the evolution of the apparent horizon is a monotone decreasing function of R. This behaviour is depicted in Fig. 10.1. This is not necessarily true if we drop the assumption of spatial homogeneity. We will see in the next section that introducing inhomogeneity in the collapsing cloud, such that the density is a monotone-decreasing function of the radial coordinate, the curve representing the evolution of the apparent horizon in (R, t) plane is a monotone increasing function of R. Hence, the apparent horizon evolves from the centre towards the boundary of the collapsing cloud. Such dynamics may make the central singularity naked, at least locally.

Coming back to the OSD collapse, the event horizon is a null hypersurface, and its evolution is obtained by substituting $ds^2 = d\Omega^2 = 0$ in the metric mentioned in Eq. (10.1). We then obtain:

$$\frac{dt}{dr} = \frac{rv}{\sqrt{1 - kr^2}}. \tag{10.19}$$

Here, v is obtained using Eq. (10.17). For $k = 0$, the general solution of this differential equation is obtained as

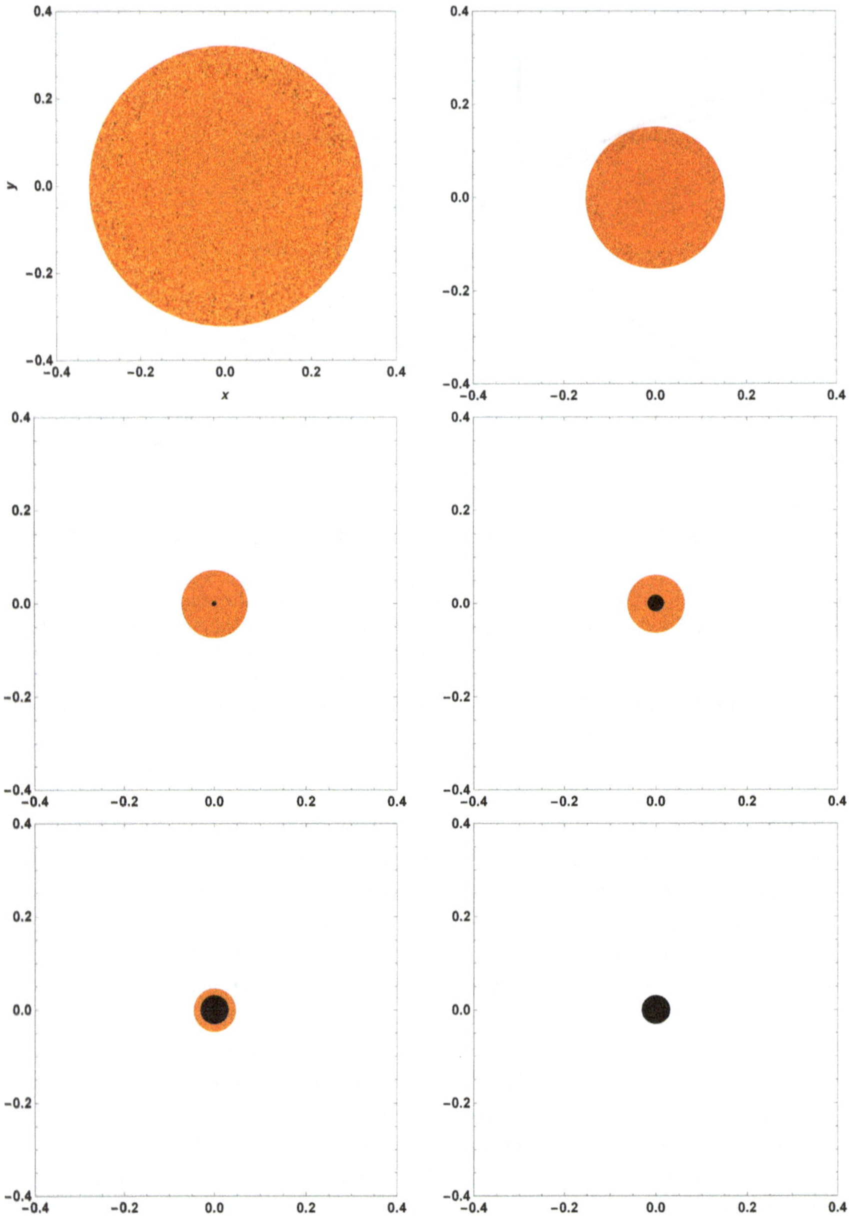

Fig. 10.2 Density plot: Datt-Oppenheimer-Snyder collapse. The central black disk depicts the evolution of the event horizon. We fix a numerical value for the comoving radius of the boundary. The Misner-Sharp mass function is $F = F_0 r^3$, where $F_0 = 1$. We set $k = 0$ in the FLRW spacetime governing the collapsing cloud

$$t_{NG}(r) = \frac{2}{3\sqrt{F_0}} + \frac{F_0 r^3}{12} + \frac{F_0 r^2 C}{16^{\frac{1}{3}}} + \frac{F_o r C^2}{4^{\frac{1}{3}}} + \frac{F_0 C^3}{3}. \tag{10.20}$$

This equation represents the time evolution of the null geodesic from the centre of the collapsing cloud. The centre is non-singular in the OSD collapse. Here C is a constant of integration. To obtain the evolution of the event horizon, we consider a particular curve that coincides with the apparent horizon at the boundary r_c of the collapsing cloud. Hence, we set an initial condition

$$t_{AH}(r_c) = t_{NG}(r_c). \tag{10.21}$$

Using Eqs. (10.18), (10.20) and (10.21), the value of C is obtained as

$$C = -\frac{3r_c}{4^{\frac{1}{3}}}. \tag{10.22}$$

Substituting C from Eq. (10.22) in Eq. (10.20), the event horizon curve $t_{EH}(r)$ is finally obtained as

$$t_{EH}(r) = \frac{2}{3\sqrt{F_0}} + \frac{F_0}{12}(r - 3r_c)^3 . \tag{10.23}$$

At $r = 0$,

$$t_{EH}(0) < t_s. \tag{10.24}$$

Hence, in the OSD collapse, the event horizon at the centre $r = 0$ forms before the formation of the singularity. This dynamics is represented in Fig. 10.2. The event horizon is a one-way membrane since there is no causal connection from inside the event horizon to the outside universe. Hence the singularity formed in the OSD collapse cannot be seen from the outside universe. It is worth mentioning here that the necessary criteria for a singularity to be visible in the global sense is to have the event horizon form at the same comoving time as that of the singularity at $r = 0$, i.e., $t_s(0) = t_{EH}(0)$.

As hinted previously, introducing inhomogeneity in the density of the collapsing cloud (in other words, dropping the assumption of spatial homogeneity) can alter the evolution of the apparent and event horizon. In the next section, we discuss such a model that has spatial inhomogeneity and see how it differs from the Datt-Oppenheimer-Snyder collapse.

10.3 Lemaitre-Tolman-Bondi (LTB) Collapse

A spherically symmetric perfect fluid with zero pressure can be modeled by the well-known Lemaitre-Tolman-Bondi (LTB) spacetime. The spacetime metric is given by

$$ds^2 = -dt^2 + \frac{R'^2}{1+f}dr^2 + R^2 d\Omega^2, \quad (10.25)$$

in the comoving coordinate basis corresponding to the spherical coordinates (t, r, θ, ϕ). R is the physical radius of the collapsing cloud that the above metric models, and the superscripts dot and prime denotes the partial derivative of the function with respect to the time and radial coordinates, respectively. f is called the velocity function. Einstein's field equations give us the expression of density and pressure, respectively as

$$\rho = \frac{F'^2}{R^2 R'}, \quad (10.26)$$

$$p = -\frac{\dot{F}}{R^2 \dot{R}}. \quad (10.27)$$

Here, F is the usual Misner-Sharp mass function, and is given by

$$F = R(\dot{R}^2 - f). \quad (10.28)$$

The Misner-Sharp mass function in the case of dust is a function of r only and independent of t. This can be seen from Eq. (10.27), which tells us that $\dot{F} = 0$ since $p = 0$ in the case of dust. Just like the case of the Datt-Oppenheimer-Snyder collapse, in the case of collapsing cloud governed by the LTB metric for zero pressure, this mass is conserved inside a fixed radial shell. We have to ensure the maintenance of certain restrictions on free functions. They are as follows:

1. The Misner-Sharp mass function should have the following expression:

$$F(r) = r^3 M(r). \quad (10.29)$$

   Here, $M > 0$ is a regular, at least C^2 function, having a finite value at the limit of approach to the centre, and is called the mass profile of the collapsing cloud. In the case F goes as r^2 or lower power, it could be seen from Einstein's field Eq. (10.26) that the density blows up at the center at the initial epoch itself, which is undesirable.
2. We avoid a cusp in the energy density; the function space of M is further restricted to follow the condition

$$M'(0) = 0. \quad (10.30)$$

3. We maintain the positivity of energy density by restricting $F' > 0$ and $R' > 0$. This is obtained by restricting the mass profile as follows:

$$3M + rM' > 0. \quad (10.31)$$

4. $R' > 0$ also does not let individual shells cross each other. The shell crossing [53] creates a weak momentary singularity, which is not so desirable.
5. We restrict the physical radius further such that $\dot{R} < 0$. Individual shells, each of fixed radial coordinate contracts to form a singularity, and the singularity corresponds to $R = 0$.
6. It should be noted that R also vanishes at the regular centre. Hence, to distinguish these two cases, we further express R as

$$R(t, r) = rv(t, r). \tag{10.32}$$

Now, a shell of radial coordinate r is said to form a singularity at time t_s when $v(t_s, r) = 0$. Rescaling of the physical radius is done using the coordinate freedom such that

$$R(t_i, r) = r, \tag{10.33}$$

where t_i is the initial time. This can be rewritten as $v(t_i, r) = 1$. Notice the analogy with the scale factor of the FLRW spacetime in the previous section, in which case it is just a function of t.
7. The functional form of the velocity function is restricted as

$$f = r^2 f_0(r), \tag{10.34}$$

where $f_0(r)$ is at least continuous at $r = 0$ and moreover, sufficiently differentiable function. This restriction is obtained by rewriting Eq. (10.28) as $\dot{v} = \sqrt{M + f/r^2}$.

Once we put the abovementioned restrictions, we proceed to integrate Eq. (10.28) and obtain

$$t - t_s(r) = -\frac{R^{\frac{3}{2}}\mathcal{G}(-fR/F)}{\sqrt{F}}. \tag{10.35}$$

Here $\mathcal{G}(y)$ is defined as in (10.9). The constant of integration in Eq. (10.35) can be obtained using Eq. (10.33) as

$$t_s(r) = \frac{r^{3/2}\mathcal{G}(-fr/F)}{\sqrt{F}}. \tag{10.36}$$

This is then the singularity curve. Equation (10.35) is a generalization of Eq. (10.8), while the Eq. (10.36) is a generalization of Eq. (10.10). Notice that in the case of the OSD collapse, all the concentric shells collapses to the singularity at the same comoving time, while more generally, we have a singularity curve that depicts the time at which a shell identified by a radial coordinate r contracts to a singularity.

A singularity can be only locally naked if a causal geodesic can escape from it (identified by the existence of past incomplete causal geodesic) but later in its path, comes across the trapped surfaces, and falls back to the singularity (identified by a past incomplete causal geodesic that is future complete). The evolution of the apparent horizon is determined by equating the physical radius with the Misner-Sharp mass function as $R = F(r)$, similar to the way it is done in the case of OSD collapse. This, along with Eqs. (10.35) and (10.36) gives us the apparent horizon curve as

$$t_{AH}(r) = \frac{r^{3/2}\mathcal{G}(-rf/F)}{\sqrt{F}} - F\mathcal{G}(-f). \tag{10.37}$$

$F(0) = 0$ implies $t_s(0) = t_{AH}(0)$, thereby creating a possibility for nonspacelike geodesic to have a positive tangent at $r = 0$. Such singularities may be at least locally naked. We will discuss the necessary and sufficient condition for such singularity to be locally naked in the next subsection.

10.3.1 Local Visibility of Singularity in Spherical Collapse

For a singularity to be at least locally naked, there should exist past incomplete causal curves. We have the following result that ensures the existence of such curves.

Theorem *Consider an unhindered gravitational collapse of a spherically symmetric perfect fluid. The singularity formed as an end state of such collapse is at least locally naked if and only if* $\exists\ X_0 \in \mathbb{R}^+$ *as a root of* $V(X)$, *where*

$$V(X) = X - \frac{1}{\alpha}\left(X + \sqrt{\frac{F_0(0)}{X}}\left(\chi_1(0) + 2r\chi_2(0) + 3r^2\chi_3(0)\right)r^{\frac{5-3\alpha}{2}}\right) \times \left(1 - \sqrt{\frac{F_0(0)}{X}}r^{\frac{3-\alpha}{2}}\right). \tag{10.38}$$

Here

$$\alpha \in \left\{\frac{2n}{3} + 1; \quad n \in \mathbb{N}\right\}, \quad \chi_i(v) = \frac{1}{i!}\frac{d^i t(r,v)}{dr^i}\bigg|_{r=0}, \quad F(r,v) = \sum_{i=0}^{\infty} F_i(v) r^{i+3}. \tag{10.39}$$

Proof Refer to [13, 51].

In the case of nonmarginally bound dust, we have $\chi_i(v)$ as follows:

$$\chi_1(v) = -\frac{1}{2}\int_v^1 \frac{\frac{F_1}{v} + f_1}{\left(\frac{M_0}{v} + f_0\right)^{\frac{3}{2}}} dv, \tag{10.40}$$

$$\chi_2(v) = \int_v^1 \left(\frac{3}{8} \frac{\left(\frac{F_1}{v} + f_1\right)^2}{\left(\frac{F_0}{v} + f_0\right)^{\frac{5}{2}}} - \frac{1}{2} \frac{\frac{F_2}{v} + f_2}{\left(\frac{F_0}{v} + f_0\right)^{\frac{3}{2}}} \right) dv, \tag{10.41}$$

and

$$\chi_3 = \int_v^1 \left(\frac{f_1}{\left(\frac{F_0}{v} + f_0\right)^{\frac{3}{2}}} \left(-\frac{5}{16} \left(\frac{f_1}{\frac{F_0}{v} + f_0} \right)^2 + \frac{3}{4} \left(\frac{\frac{F_2}{v} + f_2}{\frac{F_0}{v} + f_0} \right) \right) - \frac{1}{2} \frac{\left(\frac{F_3}{v} + f_3\right)}{\left(\frac{F_0}{v} + f_0\right)^{\frac{3}{2}}} \right) dv. \tag{10.42}$$

Here, we Taylor expand f around $r = 0$ as

$$f(r) = \sum_{i=0}^{\infty} f_i r^{i+2}. \tag{10.43}$$

Let us consider the following functional form of the Misner-Sharp mass function and the velocity function:

$$F(r) = F_0 r^3 + F_3 r^6; \quad F_0 > 0, \quad F_3 < 0 \quad f(r) = f_0 r^2. \tag{10.44}$$

Equation (10.38) for such $F(r)$ and $f(r)$ can be rewritten as $(r, v) \to (0, 0)$:

$$V(X) = 2X^2 + \sqrt{F_0} X^{3/2} - 3\sqrt{F_0}\chi_3(0)\sqrt{X} + 3F_0\chi_3(0), \tag{10.45}$$

where

$$\chi_3(0) = -\frac{1}{2} \int_0^1 \frac{F_3/v}{(F_0/v)^{3/2}} dv. \tag{10.46}$$

We can see numerically for $F_0 = 1$, Eq. (10.45) has a positive real root if and only if $F_3 \lessapprox -25.967$.

10.3.2 Globally Visible Singularity in Spherical Collapse

If there exists past incomplete causal curves, then these curves may or may not be future complete. The former case corresponds to global visibility while the latter case corresponds to only local visibility and not global visibility.

The singularities which are only locally visible may not be of much observational significance. This is because, in such a case, an observer outside the event horizon will not be able to receive any signal escaping from the singularity.

For a singularity to be globally visible, null geodesics originating from the singularity should not only avoid getting trapped by the trapped surfaces but also reach the boundary of the star before the event horizon. It turns out that for globally visible singularity, the latter always implies the former. This is because, at $r = r_c$, the apparent horizon coincides with the event horizon.

We know that the evolution of the event horizon of the collapsing cloud is the same as the evolution of the null geodesic along with the condition that at the boundary of the cloud r_c, the following equality should be satisfied:

$$F(r_c) = R(t, r_c). \tag{10.47}$$

Now, the event horizon cannot start forming after the initiation of the formation of trapped surfaces (or its boundary, i.e., the apparent horizon). This is because any null geodesic, more specifically outgoing null geodesic, forming inside the apparent horizon, will have a negative tangent and fall back into the singularity. The evolution of EH can be thought of as the evolution of the last outgoing radial null geodesic escaping the center without getting trapped and falling back to the singularity. The equation of the null geodesic is given by

$$\frac{dt}{dr} = \frac{R'}{\sqrt{1+f}}. \tag{10.48}$$

In the case of inhomogeneous dust, at $r = 0$, the time of formation of AH is the same as the time of formation of central singularity, as seen from Eqs. (10.36) and (10.37) and the fact that F vanishes. Hence, it can be concluded that the EH starts forming either before or during the formation of the singularity due to the collapse of the central shell i.e.

$$t_{EH}(0) \leq t_s(0). \tag{10.49}$$

Here, $t_{EH}(r)$ is the event horizon curve which is the solution of Eq. (10.48) with the condition given by Eq. (10.47).

If the singularity is locally naked, then it can be globally naked if the past incomplete null geodesic reaches the boundary before the event horizon (Figs. 10.3 and 10.4). This happens if

$$R(t, r_c) > F(r_c). \tag{10.50}$$

If this inequality is satisfied, then the expansion scalar of the outgoing null geodesic congruence Θ_l, at $r = r_c$, is positive. This can be seen as follows:

$$\Theta_l = \frac{2}{R}\left(\sqrt{1+f} - \sqrt{\frac{F}{R} + f}\right). \tag{10.51}$$

Hence, the outgoing null geodesic congruences at $r = r_c$ diverge if Eq. (10.50) is satisfied. We now integrate Eq. (10.35) and get

$$R = \left(\frac{r^{\frac{3}{2}}\mathcal{G}\left(-fr/F\right) - \sqrt{F}t}{\mathcal{G}\left(-fR/F\right)}\right)^{\frac{2}{3}}. \tag{10.52}$$

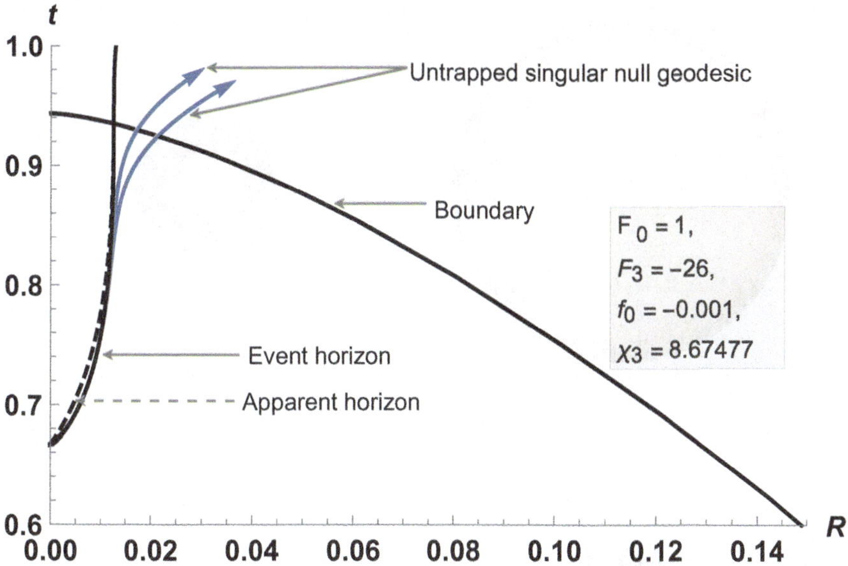

Fig. 10.3 Spacetime plot: Nonmarginally bound dust collapse; R: Physical radius of the collapsing cloud; t: comoving time coordinate. The boundary of the collapsing cloud is where the density ρ vanishes, beyond which the spacetime is governed by the Schwarzschild metric. The Misner-Sharp mass function: $F = F_0 r^3 + F_3 r^6$; the velocity function: $f = f_0 r^2$; solid black curve: event horizon; dashed black curve: evolution of the apparent horizon; solid blue curves: past incomplete null geodesics. $t_{EH}(0) = t_{AH}(0)$, hence the singularity is globally visible

In the case of marginally bound collapse ($f = 0$), this is reduced to

$$R = \left(r^{\frac{3}{2}} - \frac{3}{2}\sqrt{F}t\right)^{\frac{2}{3}}. \tag{10.53}$$

However, in the case of non-marginally bound collapse ($f \neq 0$), the integration of Eq. (10.35) is not so straightforward comparatively. For $0 \leq y < 1$, we Taylor expand $\mathcal{G}(y)$ in Eq. (10.9) around $y = 0$ and obtain

$$\mathcal{G}(y) = \frac{2}{3} + \frac{1}{5}y + \frac{3}{28}y^2 + o(y^3). \tag{10.54}$$

This can then be used in Eq. (10.52) to write R explicitly as

$$R(t,r) = \frac{5F}{2f}\left(1 - \sqrt{1 - o(y_1^3) - \frac{4f}{5F}\left(r^{\frac{3}{2}}\left(1 - \frac{3fr}{10F} + o(y_2^2)\right) - \frac{3}{2}\sqrt{F}t\right)^{\frac{2}{3}}}\right), \tag{10.55}$$

for nonvanishing velocity function, i.e. $f \neq 0$. Here,

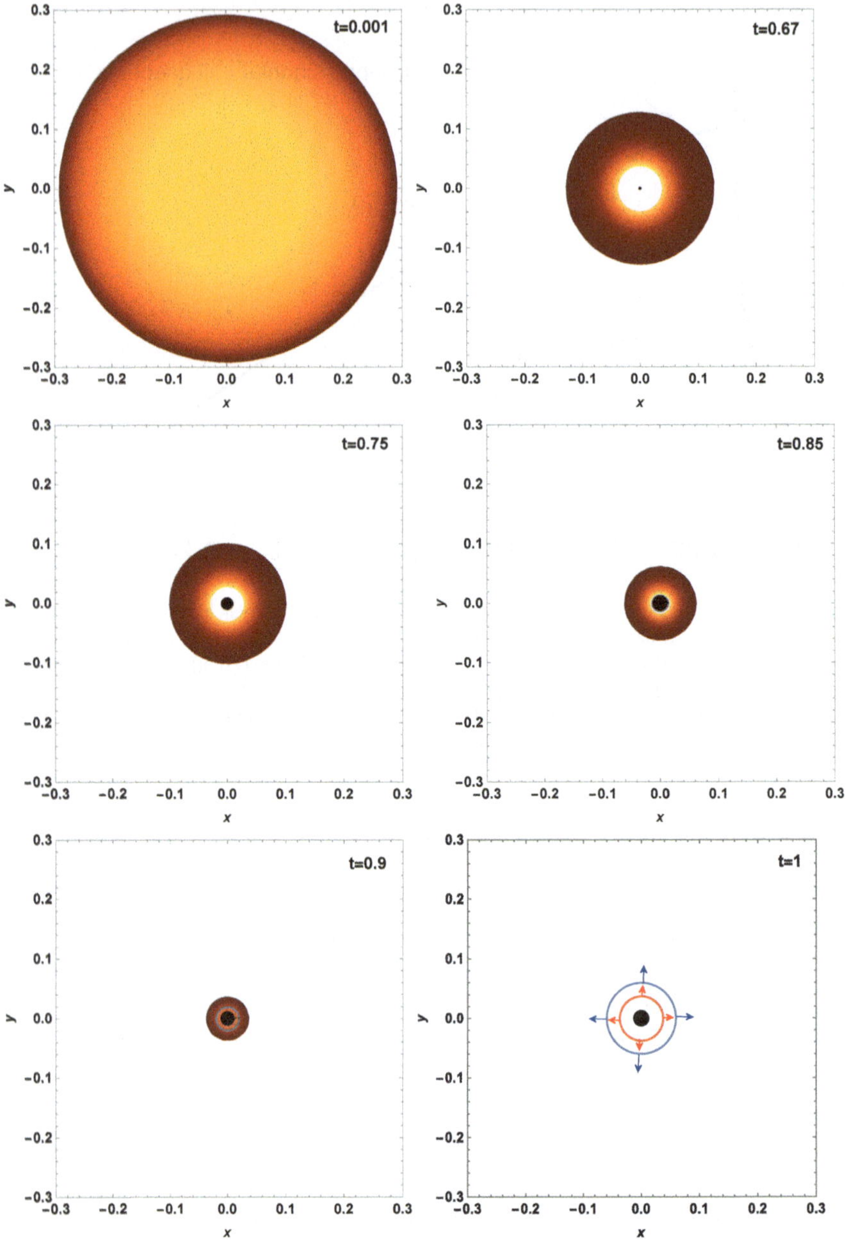

Fig. 10.4 Density plot: Nonmarginally bound dust collapse. The central black disk depicts the event horizon on a time slice. The red and blue circles depict the null geodesic wave fronts that are past incomplete but future complete. $F_0 = 1$, $F_3 = -26$ and $F_i = 0$ for $i \neq 1, 3$. $f_0 = -0.001$ and $f_j = 0 \ \forall \ j \neq 0$. $\chi_1 = \chi_2 = 0$ and $\chi_3 = 8.67447$

$$y_1 = -\frac{fR}{F}, \qquad y_2 = -\frac{fr}{F}. \tag{10.56}$$

Ignoring higher order, i.e. $o(y_1^3)$ and $o(y_2^2)$ in Eq. (10.55) is equivalent to considering the expansion of $\mathscr{G}$ from Eq. (10.54) only up to first order. Hence, large value of the ratio fR/F may not give a good approximation. Therefore, in our investigation, we make sure to keep this ratio small by considering positive velocity function having small deviation from zero.

Deshingkar et al. [52] studied the global causal structure of the end state of marginally bound collapse, wherein three different mass distributions were considered. Here, we analyze the global behavior of the singularity formed by bound collapse. We look at example given by Eq. (10.44). The boundary of the cloud is found such that the density smoothly matches to zero there. Hence, the boundary is given by

$$r_c = \left(-\frac{F_0}{2F_3}\right)^{\frac{1}{3}}. \tag{10.57}$$

We find numerically that for such a choice of free functions

$$t_{EH}(r=0) = t_{AH}(r=0), \tag{10.58}$$

for $F_0 = 1$ and whenever $F_3 < -25.967$. Hence, we have an example of a globally visible singularity.

In the next section, we go byond the assumption of spherical symmetry and consider an exact solution proposed first by Szekeres in 1975 [40].

10.4 Non-spherical Gravitational Collapse of Pressureless Fluid

Studies on spherical collapse help us to understand the basic physics inherent to gravitational collapse since it simplifies the collapsing scenario. However, in nature, perturbations from sphericity are a common phenomenon and that's why studies on non-spherical collapse are very much important. By studying aspherical collapse, one can understand the effect of asphericity on various aspects of gravitational collapse. In this section, we discuss the gravitational collapse of a non-spherical cloud of dust. We show that in the presence of asphericity, not only the position of the spacetime singularity is shifted from the geometric center towards the boundary, but the causal structure singularity is also greatly affected. Therefore, our results indicate that a small perturbation from sphericity can in principle radically alter the visibility of the singularity formed in the gravitational collapse.

Here, we use Szekeres spacetime [40–42] in order to model the aspherical gravitational collapse. The standard form of the line element of the Szekeres metric is the following:

$$ds^2 = -dt^2 + e^{2\nu(t,x,y,z)}dz^2 + e^{2\eta(t,x,y,z)}(dx^2 + dy^2) \,, \tag{10.59}$$

where the relation between real functions $\nu(t, x, y, z)$ and $\eta(t, x, y, z)$ can be determined by the Einstein field equations considering the matter is pressureless. One can define two classes of Szekeres metric where for one $\partial_z \eta = 0$ and for other $\partial_z \eta \neq 0$. Here, we consider $\partial_z \eta \neq 0$. After LTB type parameterization of the above line element, one can get the following modified form of the above metric:

$$ds^2 = -dt^2 + M^2(t, r, \zeta, \bar{\zeta})dr^2 + N^2(t, r, \zeta, \bar{\zeta})d\zeta d\bar{\zeta}, \tag{10.60}$$

where $M = QN'/\sqrt{1 + f(r)}$, $N = S(t, r)/Q(r, \zeta, \bar{\zeta})$, where $f > -1$ is the velocity function, $S(t, r)$ is a positively definite function of (t, r) that serves a role analogous to the physical radius $R(t, r)$ in LTB, and $\zeta = x + iy$ and $\bar{\zeta} = x - iy$ is a pair of conjugate coordinates. As stated previously in LTB collapse, here also f greater than, equal to, and less than zero corresponds to bound, marginally bound, and unbound collapse respectively. Since we consider the case where $\partial_z \eta \neq 0$, for the above LTB type parameterization of Szekeres metric $N' \neq 0$. We also have the following expression of $Q(r, \zeta, \bar{\zeta})$,

$$Q = a(r)\zeta\bar{\zeta} + B(r)\zeta + \bar{B}(r)\bar{\zeta} + c(r), \tag{10.61}$$

where a and c are real, and B is a complex function having the relation $ac - B\bar{B} = \delta/4$, where $\delta = 0, \pm 1$. If we choose $\zeta = \tan(\theta/2)e^{i\phi}$, and $B = 0$, we can write $N^2 d\zeta d\bar{\zeta}$ as:

$$N^2 d\zeta d\bar{\zeta} = R^2(t, r, \theta)\left(d\theta^2 + \sin^2\theta d\phi^2\right), \tag{10.62}$$

where

$$R(t, r, \theta) = \frac{S(t, r)\sec^2(\frac{\theta}{2})}{2(a(r)\tan^2(\frac{\theta}{2}) + c(r))} \,. \tag{10.63}$$

Here, $R(t, r, \theta)$ is the physical radius of the collapsing cloud that shows the physical distance between a point in the matter cloud and the center of mass of the matter cloud. Since we consider $\zeta = \tan(\theta/2)e^{i\phi}$, the physical radius is independent of ϕ. As we know, the energy-momentum tensor of *type I* matter field with vanishing pressure is written as $T^{\mu\nu} = \rho U^\mu U^\nu$, where U^μ are the components of the four-velocity. Using Einstein's field equation, one can obtain the following expressions of the density and pressure of the collapsing cloud modeled by Szekeres metric, respectively:

$$\rho(t, r, \theta) = \frac{QA' - 3AQ'}{S^2(QS' - SQ')}, \tag{10.64}$$

and

$$p(t, r) = -\frac{\dot{A}}{S^2\dot{S}}, \tag{10.65}$$

where

$$A = S(\dot{S}^2 - f). \tag{10.66}$$

Since here we consider only the dust solution, the Misner-Sharp mass-like function A in the above expression becomes a function of r only. Like LTB case, here also we can solve the above differential equation of $S(t, r)$ which is analogous to the physical radius $R(t, r)$ in LTB model. After integrating Eq. (10.66) we obtain

$$t - t_s(r) = -\frac{S^{\frac{3}{2}}\mathcal{G}(-fS/A)}{\sqrt{A}}, \tag{10.67}$$

where $t_s(r)$ is the integration constant and can be identified as the comoving time at which a collapsing shell corresponding to a comoving radius r becomes singular. The functional form of $t_s(r)$ is:

$$t_s(r) = \frac{S_0(r)^{\frac{3}{2}}}{\sqrt{A}}\mathcal{G}\left(-\frac{fS_0(r)}{A}\right), \tag{10.68}$$

where $S_0(r) = S(0, r)$ and in the above two successive equations the function $\mathcal{G}(y)$ is defined as in Eq. (10.9). In this aspherical collapse model, like LTB we define the boundary of the collapsing cloud by the comoving radius r_c at which the density of matter becomes zero. Here, the density depends upon the angular coordinate θ, the comoving radius of the boundary $r_c = r_c(\theta)$, and consequently, from the above expression of $t_s(r)$ we get $t_s(r_c) = t_s(r_c(\theta))$ which implies that the whole boundary of the collapsing matter would not collapse to the singularity simultaneously, but rather falls in the singularity at different times along different angular directions.

Before going into the discussion of local visibility of singularity in aspherical gravitational collapse, we will first discuss the apparent horizon dynamics near $(t = t_s(0), r = 0)$. In the aspherical collapse, on the apparent horizon surface, the following condition should always hold $A = S$ which can be derived from the null value of the expansion scalar of null geodesic congruence. Using this condition along with Eq. (10.67), one can obtain the apparent horizon curve as

$$t_{AH}(r) = t_s(r) - A\mathcal{G}(-f). \tag{10.69}$$

The above equation implies $t_s(0) = t_{AH}(0)$, since $A(0) = 0$, as demanded by the regularity conditions [7]. For the marginally bound case, the above equation becomes:

$$t_{AH}(r) = t_s(r) - \frac{2}{3}A = \frac{2}{3}\left(\frac{r^{\frac{3}{2}}}{\sqrt{A}} - A\right), \tag{10.70}$$

where we consider $S_0(r) = r$. Now, since the matter is pressureless, $A(t, r) = A(r)$, and we consider, $A(r) = A_0 r^3 + A_3 r^6$, where the constant parameters $A_0 > 0$ and $A_3 < 0$. Now, near $r = 0$, we can write down the above equation as,

$$t_{AH}(r) \approx \frac{2}{3}\frac{1}{\sqrt{A_0}} - \left(\frac{A_3}{2A_0^{\frac{3}{2}}} + A_0\right) r^3 \, , \tag{10.71}$$

which implies apparent horizon would expand from the central singularity if $|A_3| > 2A_0^{\frac{5}{2}}$ and would shrink towards singularity otherwise. Now, as discussed in the LTB case, null geodesics can escape from the singularity locally if there exist no trapped surfaces (i.e. the surfaces where $A > S$) around the singularity.

10.4.1 Local Visibility of Singularity in Aspherical Collapse

In non-marginally bound Szekeres spacetime, for a null curve, we can write:

$$\frac{dt}{dr} = \frac{M(t, r, \theta)}{\sqrt{1+f}} = \frac{(QN')}{\sqrt{1+f}} = \frac{\left(S' - S\frac{Q'}{Q}\right)}{\sqrt{1+f}}. \tag{10.72}$$

For LTB collapse, we used the (R, u) frame in order to understand the causal structure of singularity. Here, we will use the (S, u) frame for a similar reason. Using the (S, u) coordinate frame we can write:

$$\begin{aligned} \frac{dS}{du} &= \frac{S'}{\alpha\, r^{\alpha-1}}\left(1 + \frac{\dot{S}}{\sqrt{1+f}}\right) - \frac{S\,\dot{S}}{\alpha\, r^{\alpha}\sqrt{1+f}}\frac{r\,Q'}{Q} \\ &= \frac{H(X,u)}{\alpha}\left(1 - \frac{\sqrt{\frac{A}{S}+f}}{\sqrt{1+f}}\right) + \frac{X}{\alpha}\frac{\sqrt{\frac{A}{S}+f}}{\sqrt{1+f}}\mathcal{N} = U(X,u)\, , \end{aligned} \tag{10.73}$$

where

$$\begin{aligned} X &= \frac{S}{r^{\alpha}}\, , \\ \mathcal{N} &= \frac{r\,Q'}{Q}\, , \\ H &= \frac{S'}{r^{\alpha-1}} = \left(\frac{S}{u} + \frac{\sqrt{w}w'}{\sqrt{\frac{S}{u}}} r^{\frac{5-3\alpha}{2}}\right)\, , \end{aligned} \tag{10.74}$$

where we consider $S(r, t) = r\, w(r, t)$. Like the LTB case, here also the condition for local visibility is:

$$\lim_{(u,S)\to(0,0)} \frac{dS}{du} > 0 \implies \lim_{(u,S)\to(0,0)} \frac{S}{u} > 0 \implies X_0 = U(X_0, 0), \tag{10.75}$$

where $\lim_{(u,S)\to(0,0)} \frac{S}{u} = X_0 > 0$. Therefore, we can define a function $V(X) = U(X,0) - X$, which has a positive root X_0. Now, as discussed in LTB case, if $V(X)$ has a positive root then only the singularity at $(S = 0, u = 0)$ can be locally visible. Here, we consider

$$Q = \exp(r)\tan^2\left(\frac{\theta}{2}\right) + \frac{e^{-r}}{4},$$

therefore,

$$\lim_{(u,S)\to(0,0)} \mathscr{N} = 0.$$

Hence,

$$\begin{aligned} U(X,0) &= \frac{H(X,0)}{\alpha}\left(1 - \frac{\sqrt{\frac{A}{S}+f}}{\sqrt{1+f}}\right) \\ &= \frac{1}{\alpha}\left(X + \sqrt{\frac{A_0(0)}{X}}\left(\chi_1(0) + 2r\chi_2(0) + 3r^2\chi_3(0)\right) r^{\frac{5-3\alpha}{2}}\right)\left(1 - \sqrt{\frac{A_0(0)}{X}} r^{\frac{3-\alpha}{2}}\right), \end{aligned} \tag{10.76}$$

which implies

$$V(X) = X - \frac{1}{\alpha}\left(X + \sqrt{\frac{A_0(0)}{X}}\left(\chi_1(0) + 2r\chi_2(0) + 3r^2\chi_3(0)\right) r^{\frac{5-3\alpha}{2}}\right)\left(1 - \sqrt{\frac{A_0(0)}{X}} r^{\frac{3-\alpha}{2}}\right), \tag{10.77}$$

which is exactly the similar expression of $V(X)$ we obtained previously for LTB, where the χ_1, χ_2 and χ_3 have the following expressions:

$$\begin{aligned} \chi_1(w) &= -\frac{1}{2}\int_w^1 \frac{\frac{A_1}{w} + f_1}{\left(\frac{A_0}{w} + f_0\right)^{\frac{3}{2}}} dw\,, \\ \chi_2(w) &= \int_w^1 \left(\frac{3}{8}\frac{\left(\frac{A_1}{w} + f_1\right)^2}{\left(\frac{A_0}{w} + f_0\right)^{\frac{5}{2}}} - \frac{1}{2}\frac{\frac{A_2}{w} + f_2}{\left(\frac{A_0}{w} + f_0\right)^{\frac{3}{2}}}\right) dw\,, \\ \chi_3(w) &= \int_w^1 \frac{f_1}{\left(\frac{A_0}{w} + f_0\right)^{\frac{3}{2}}}\left(-\frac{5}{16}\left(\frac{f_1}{\frac{A_0}{w} + f_0}\right)^2 + \frac{3}{4}\left(\frac{\frac{A_2}{w} + f_2}{\frac{A_0}{w} + f_0}\right)\right) - \frac{1}{2}\frac{\left(\frac{A_3}{w} + f_3\right)}{\left(\frac{A_0}{w} + f_0\right)^{\frac{3}{2}}} dw\,. \end{aligned} \tag{10.78}$$

Therefore, all the analysis regarding constraining the parameters S_0 and S_3 to get the positive root of $V(X)$ will be the same as in the LTB case, where we constrained the parameter space of F_0 and F_3. Therefore, the above results imply that the asphericity in our model does not have any influence on the local visibility of the singularity.

However, if one considers some other functional form of $Q(r, \zeta, \bar{\zeta})$, then the situation can be different. In that case, asphericity may influence the local visibility of the singularity. Further investigation in this direction is necessary to understand how in general, perturbations on sphericity can alter the causal structure of the spacetime singularity.

10.4.2 Global Visibility of Singularity in Aspherical Collapse

As discussed in the LTB collapse, the global visibility of singularity depends upon the dynamics of the event horizon near the singular point. As we know, the event horizon is a null surface and it follows the following null condition

$$\frac{dt}{dr} = M(t, r, \theta), \tag{10.79}$$

along with the condition,

$$A = S$$

at the boundary $r = r_c$. As stated previously, for a singularity to be visible globally, the formation time of the event horizon and the apparent horizon at the center of mass should be the same (i.e., $t_{AH}(0) = t_{EH}(0)$). The singularity would not be globally visible if the event horizon forms at the center before the formation of the singularity though it may be locally visible. For the scenarios where the singularity is locally visible but globally hidden, the null geodesics escaping from the singularity would ultimately get trapped and fall back inside.

In Fig. 10.5, we depict one example of those scenarios where the singularity is globally visible only at a particular solid angle. The directional visibility of singularity is caused by the variation of the formation time of the event horizon at the center in different angular directions. We consider

$$A = r^3 - 26r^6 , \tag{10.80}$$

$$f = -10^{-3}r^2 , \tag{10.81}$$

$$Q = \exp(r)\tan^2\left(\frac{\theta}{2}\right) + \frac{e^{-r}}{4}. \tag{10.82}$$

Considering these functional forms of A, f, and Q, in Fig. 10.5, we show how the event horizon evolves from the central singularity which is formed near the boundary of the collapsing aspherical matter. It can be seen in that figure that the event horizon during its evolution has an aspherical shape since it forms at the center at different times in different angular directions. The null geodesics should escape from the region sufficiently close to the singularity to contain signatures of the strong gravity regions (mimicking possibly the quantum gravity effects); i.e. the difference in time

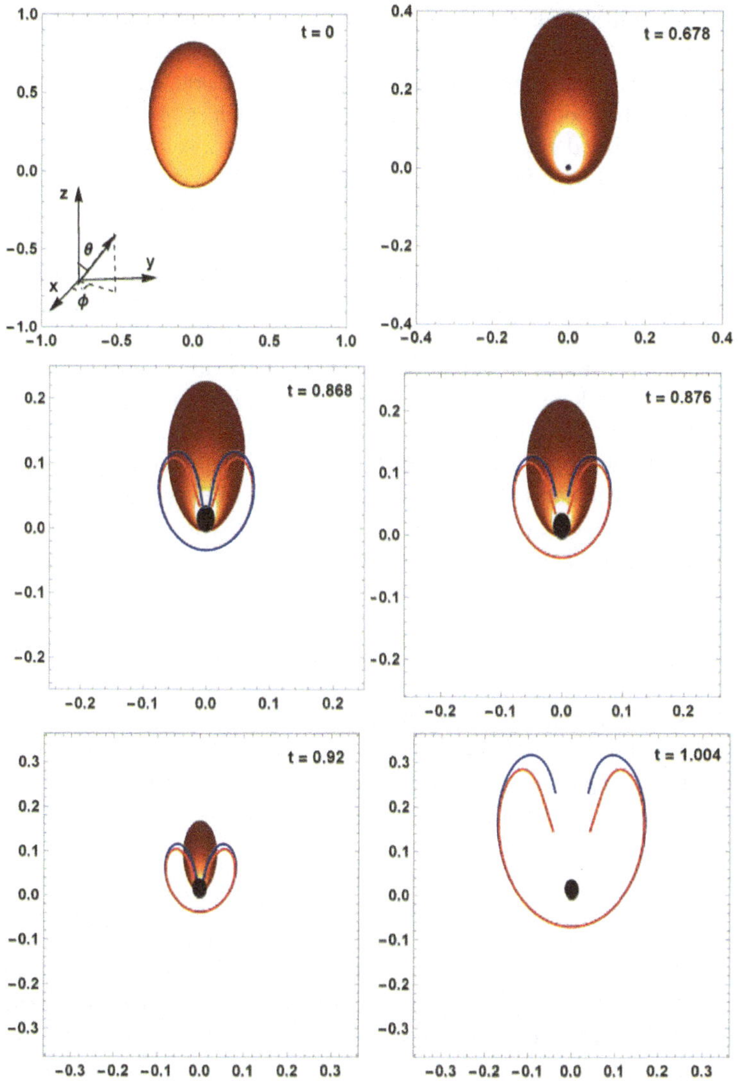

Fig. 10.5 Density plot: Aspherical non-marginally bound dust collapse. The central black region depicts the event horizon on a time slice. The red and blue curves depict the null geodesic wave fronts that are past incomplete but future complete. $A_0 = 1$, $A_3 = -26$ and $A_i = 0$ for $i \neq 1, 3$. $f_0 = -0.001$ and $f_j = 0 \ \forall \ j \neq 0$. $\chi_1 = \chi_2 = 0$ and $\chi_3 = 8.67447$. The figures are shown on $\phi = \frac{\pi}{2}$ plane and are symmetric with change in ϕ. Here, we consider $Q = \exp(r) \tan^2\left(\frac{\theta}{2}\right) + \frac{e^{-r}}{4}$. The evolution of the event horizon depends upon the angular coordinate θ. The aspherical wavefronts of singular null geodesics escape the singularity, thereby making it globally visible. However, part of these outgoing wavefronts which lies in the neighborhood of $\theta = 0$ gets trapped by the trapped surfaces and falls back to the singularity, hence making it only directionally globally naked in the neighborhood around $\theta = \pi$

of escape of the null singularity at $r = 0$ and the time of formation of the singularity at $r = 0$ should be of the order of the Planck time. The outermost aspherical wavefront of the null geodesic satisfying this criterion lies partly inside and partly outside the aspherical event horizon surface, intersecting it such that the locus of the points of intersection is a closed curve. This closed curve creates a solid angle at the center $(0, 0)$ inside which the singular null geodesic wavefront gets trapped and outside which the singular null wavefront escapes. Therefore, the singular region can be visible to an asymptotic observer along a particular solid angle. This phenomenon is depicted in Fig. 10.5, where it can be seen that some parts of the two singular null geodesics' wavefronts (highlighted by blue and red color) can escape out and reach the asymptotic observer. Figure 10.6 shows the same phenomenon but in two dimensional (R, t) frame. It should be noted that here, in the aspherical collapse model we presented here, we don't consider linear perturbations on the sphericity of the LTB metric. The collapsing cloud is highly aspherical and for a non-zero measured set of initial parameters A and f, gives rise to a directional globally visible singularity that is offcentric and close to the boundary of the collapsing cloud.

One can verify that the directionally visible singularity formed in aspherical collapse is Tipler-strong [43] in the sense that at least along one null geodesic with affine parameter λ, the following inequality holds:

$$\lambda^2 R_{ij} K^i K^j > 0,$$

where R_{ij} is the Ricci tensor, and K^α is the null vector (i.e., it follows $K^\alpha K_\alpha = 0$) tangent to the outgoing null geodesics from the singularity. The above condition for Tipler-strong singularity physically implies that the Jacobian volume formed by independent Jacobian fields along the null geodesic vanishes as the geodesic ends at the singularity.

As we have discussed previously, due to the different formation times of the event horizon at the center along different directions, the event horizon has an aspherical shape during its evolution. However, once the matter cloud falls inside the event horizon, the exterior spacetime is aspherical, static, vacuum, and asymptotically flat. In [45, 46], it is shown that a collapsing non-rotating body radiates away small perturbations on sphericity in the form of gravitational waves and takes the shape of minimum curvature (as visualized by Misner [47]), where the authors use perturbation theory. Therefore, the end state is possibly a spherical event horizon. For highly deformed spacetime, the outcome may or may not be the same [47].

It should be noted that the null geodesics escaping from the singularity along $\theta = \pi$ are less scattered since they face less interactions with the collapsing matter, and consequently, they are less distorted. Therefore, the traces of the quantum gravity theory near the singularity would be less contaminated. One might wonder if there is a scenario, a solution of Einstein's field equation, wherein the singularity forms 'at' the boundary (and not close to it) so that the original form of the escaping singular null geodesic is maintained with no distortion at all.

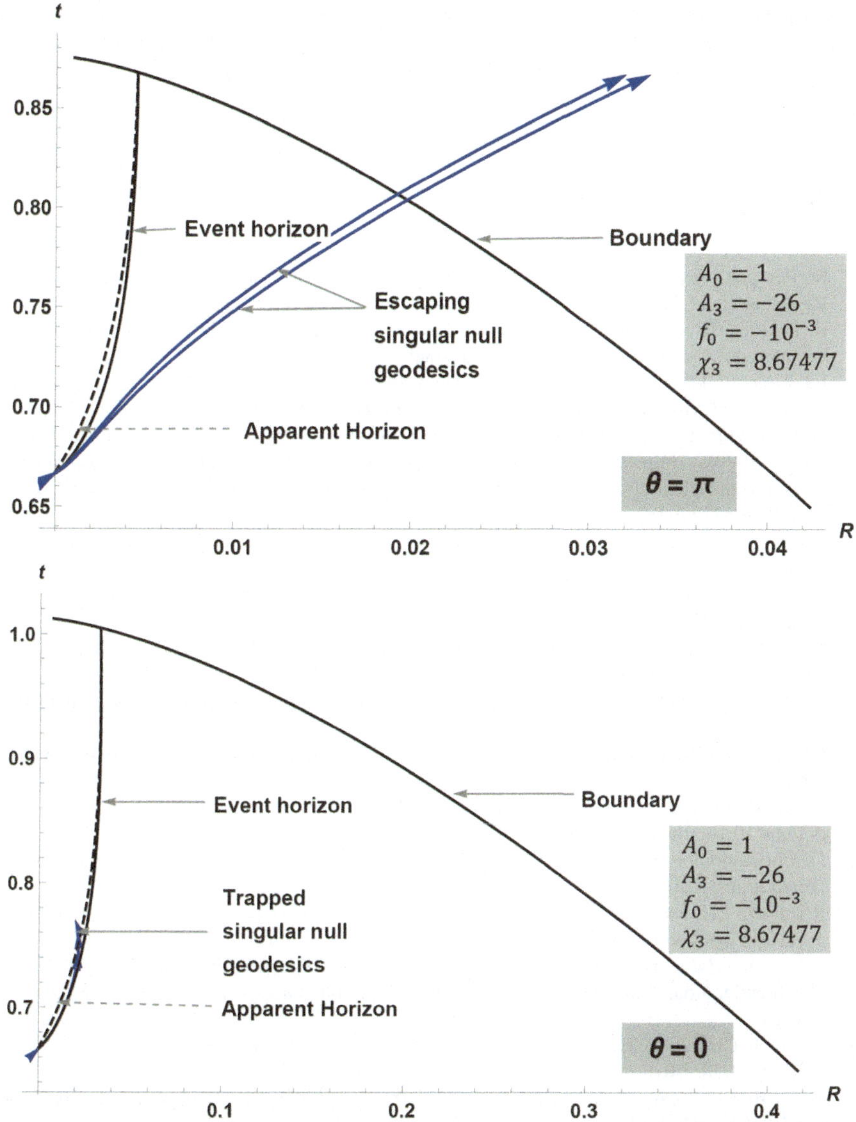

Fig. 10.6 Spacetime plot: Aspherical nonmarginally bound dust collapse; R: Physical radius of the collapsing cloud; t: comoving time coordinate. The boundary of the collapsing cloud is where the density ρ vanishes. The function analogous to Misner-sharp mass: $A = A_0 r^3 + A_3 r^6$; the velocity function: $f = f_0 r^2$; solid black curve: event horizon; dashed black curve: evolution of the apparent horizon; solid blue curves: past incomplete null geodesics. $t_{EH}(0) = t_{AH}(0)$ at $\theta = \pi$ and $t_{EH}(0) < t_{AH}(0)$ at $\theta = 0$, hence the singularity is globally visible along $\theta = \pi$ and only locally visible along $\theta = 0$ angular directions

Acknowledgements DD would like to acknowledge the support of the Atlantic Association for Research in the Mathematical Sciences (AARMS) for funding the work. KM would like to express gratitude to Rituparno Goswami for fruitful discussions.

References

1. R. Penrose, Phys. Rev. Lett. **14**, 57 (1965). https://journals.aps.org/prl/abstract/10.1103/PhysRevLett.14.57
2. R. Penrose, Rivista del Nuovo Cimento, Numero Speziale I **252**, (1969). https://ui.adsabs.harvard.edu/abs/1969NCimR...1..252P/abstract
3. F.J. Tipler, C.J.S. Clarke, G.F.R. Ellis, A. Held (Plenu, New York, 1980)
4. R. Geroch, G. Horowitz (Cambridge University Press, Cambridge, United Kingdom, 1979)
5. S.W. Hawking, W. Israel (eds.) (Cambridge University Press, Cambridge, United Kingdom, 1979). https://www.cambridge.org/us/universitypress/subjects/physics/cosmology-relativity-and-gravitation/general-relativity-einstein-centenary-survey?format=PB&isbn=9780521299282
6. P.S. Joshi, A. Krolak, Class. Quantum Gravity **13**, 3069 (1996). https://iopscience.iop.org/article/10.1088/0264-9381/13/11/020
7. P.S. Joshi, D. Malafarina, Int. J. Mod. Phys. D **20**, 2641 (2011). https://www.worldscientific.com/doi/abs/10.1142/S0218271811020792
8. S.S. Deshingkar, S. Jhingan, P.S. Joshi, Gen. Relativ. Gravit. **30**, 1477 (1998). https://link.springer.com/article/10.1023
9. D. Christodoulou, Commun. Math. Phys. **93**, 171 (1984). https://link.springer.com/article/10.1007
10. D. Christodoulou, Commun. Math. Phys. **109**, 613 (1987). https://link.springer.com/article/10.10070
11. D. Christodoulou, Commun. Pure Appl. Math. **44**, 339 (1991). https://onlinelibrary.wiley.com/doi/abs/10.1002/cpa.3160440305
12. D. Christodoulou, Commun. Pure Appl. Math. **46**, 1131 (1993). https://onlinelibrary.wiley.com/doi/abs/10.1002/cpa.3160460803
13. P.S. Joshi, I.H. Dwivedi, Phys. Rev. D **47**, 5357 (1993). https://journals.aps.org/prd/abstract/10.1103/PhysRevD.47.5357
14. P.S. Joshi, *Global Aspects in Gravitation and Cosmology* (Clendron Press, Oxford, 1993)
15. D. Christodoulou, Ann. Math. (N.Y.) **140**, 607 (1994). https://www.jstor.org/stable/2118619?seq=1
16. G. Magli, Class. Quantum Gravity **14**, 1937 (1997). https://iopscience.iop.org/article/10.1088/0264-9381/14/7/026
17. G. Magli, Class. Quantum Gravity **15**, 3215 (1998). https://iopscience.iop.org/article/10.1088/0264-9381/15/10/022
18. T.P. Singh, J. Astrophys. Astron. **20**, 221 (1999). https://link.springer.com/article/10.1007/BF02702354
19. R. Giambó, F. Giannoni, G. Magli, P. Piccione, Commun. Math. Phys. **235**, 545–563 (2003). https://link.springer.com/article/10.1007/s00220-003-0793-9
20. R. Giambó, J. Math. Phys. **47**, 022501 (2006). https://aip.scitation.org/doi/10.1063/1.2167919
21. P.S. Joshi, *Gravitational Collapse, and Spacetime Singularities* (Cambridge University Press, Cambridge, United Kingdom, 2007)
22. K. Mosani, D. Dey, P.S. Joshi, Phys. Rev. D **102**, 044037 (2020). https://journals.aps.org/prd/abstract/10.1103/PhysRevD.102.044037
23. K. Mosani, K. Mosani, D. Dey, D. Dey, P.S. Joshi, P.S. Joshi, Phys. Rev. D **101**(4), 044052 (2020). https://journals.aps.org/prd/abstract/10.1103/PhysRevD.101.044052

24. K. Mosani, D. Dey, P.S. Joshi, Mon. Not. Roy. Astron. Soc. **504**(4), 4743–4750 (2021). https://academic.oup.com/mnras/article-abstract/504/4/4743/6253197?redirectedFrom=fulltext&login=false
25. K. Mosani, D. Dey, K. Bhattacharya, P.S. Joshi, Phys. Rev. D **105**)(6), 064048(2022). https://journals.aps.org/prd/abstract/10.1103/PhysRevD.105.064048
26. K. Modani, D. Dey, P. Joshi, G.C. Samanta, H. Menon, V. Patel, Class. Quantum Gravity, Accepted (2023). http://iopscience.iop.org/article/10.1088/1361-6382/acd97a
27. D. Dey, P.S. Joshi, K. Mosani, V. Vertogradov, Eur. Phys. J. C **82**(5), 431 (2022). https://link.springer.com/article/10.1140/epjc/s10052-022-10401-1
28. R. Shaikh, P. Kocherlakota, R. Narayan, P.S. Joshi, Mon. Not. Roy. Astron. Soc. **482**, 52 (2019). https://academic.oup.com/mnras/article-abstract/482/1/52/5113467?redirectedFrom=fulltext
29. P. Bambhaniya, A.B. Joshi, D. Dey, P.S. Joshi, Phys. Rev. D **100**, 124020 (2019). https://journals.aps.org/prd/abstract/10.1103/PhysRevD.100.124020
30. D. Dey, P.S. Joshi, A. Joshi, P. Bambhaniya, Int. J. Mod. Phys. D **28**(14), 1930024 (2019). https://www.worldscientific.com/doi/abs/10.1142/S0218271819300246
31. R. Shaikh, P.S. Joshi, J. Cosmol. Astropart. Phys. **10**, 064 (2019). https://iopscience.iop.org/article/10.1088/1475-7516/2019/10/064/meta
32. A.B. Abdikamalov, A.A. Abdujabbarov, D. Ayzenberg, D. Malafarina, C. Bambi, B. Ahmedov, Phys. Rev. D **100**, 024014 (2019). https://journals.aps.org/prd/abstract/10.1103/PhysRevD.100.024014
33. K. Bhattacharya, D. Dey, A. Mazumdar, T. Sarkar, Phys. Rev. D **101**, 043005 (2020). https://journals.aps.org/prd/abstract/10.1103/PhysRevD.101.043005
34. S.W. Hawking G.F.R. Ellis, *The Large Scale Structure of Spacetime* (Cambridge University Press, 1973)
35. J.A. Wheeler, in *Relativity, Groups and Topology*, Les Houches, ed. by C. DeWitt, B. DeWitt (Gordon, and Breach, New York, 1964)
36. S.W. Hawking, in *General Relativity: An Einstein Centenary Survey*, ed. by S.W. Hawking, W. Israel (Cambridge University Press, Cambridge, 1979)
37. P.G. Bergmann, *Some Strangeness in the Proportion*, ed. by H. Woolf (Addison-Wesley Reading, Massachusetts, 1980)
38. C.W. Misner, Phys. Rev. **186**, 1328 (1969). https://journals.aps.org/pr/abstract/10.1103/PhysRev.186.1328
39. L.M. Burko, A. Ori, Introduction to the internal structure of black holes, in *Internal Structure of Black Holes and Spacetime Singularities*, ed. by L.M. Burko, A. Ori (IOP Publishing, Bristol, 1997)
40. P. Szekeres, Phys. Rev. D **12**, 2941 (1975). https://journals.aps.org/prd/abstract/10.1103/PhysRevD.12.2941
41. A.A. Coley, N. Layden, D.D. McNutt, Gen. Rel. Grav. **51**(12), 164 (2019). https://link.springer.com/article/10.1007/s10714-019-2647-6
42. N.T. Layden, A.A. Coley, D.D. McNutt, Gen. Rel. Grav. **54**(8), 75 (2022). https://link.springer.com/article/10.1007/s10714-022-02962-z
43. F.J. Tipler, Phys. Lett. **64A**, 8 (1977). https://www.sciencedirect.com/science/article/abs/pii/0375960177905084
44. C.J.S. Clarke, A. Krolak, J. Geo. Phys. **2**, 127 (1986). https://www.sciencedirect.com/science/article/pii/0393044085900129
45. V. de la Cruz, J.E. Chase, W. Israel, Phys. Rev. Lett. **24**, 423 (1970). https://journals.aps.org/prl/abstract/10.1103/PhysRevLett.24.423
46. R. Price, Ph.D. Thesis, Caltech (1970). https://thesis.library.caltech.edu/5526/1/Price_rh_1971.pdf
47. K.S. Thorne, Comments Astrophys. Space Phys. **2**, 191 (1970). http://adsabs.harvard.edu/full/1970CoASP...2..191T
48. J.R. Oppenheimer, H. Snyder, Phys. Rev. J. Arch. **56**, 455 (1939). https://journals.aps.org/pr/issues/56/5
49. S. Datt, Zs. f. Phys. **108**, 314 (1938)

50. K. Mosani, Ph.D. Thesis. arXiv:2211.06604
51. R. Goswami, P.S. Joshi, Phys. Rev. D **76**, 084026 (2007). https://journals.aps.org/prd/abstract/10.1103/PhysRevD.76.084026
52. S.S. Deshingkar, S. Jhingan, P.S. Joshi, Gen. Relativ. Gravit. **30**, 1477 (1998). https://link.springer.com/article/10.1023/A:1018813108516
53. C. Hellaby, K. Lake, Astrophys. J. **290**, 381 (1985). https://adsabs.harvard.edu/full/1985ApJ...290..381H

Chapter 11
Revisit Cosmic Censorship and Black Holes

Ken-ichi Nakao

In physics we believe very nicely that we have a theory by which if you tell us the initial conditions, we will predict what will happen in the future.

R. Feynman *(in The Quotable Feynman)*

11.1 Introduction

General relativity is the most successful theory of gravity and is one of the paradigms of current physics and astrophysics. However, since it cannot describe gravitational interactions between objects whose motions are described by quantum theory, nobody would disagree with the assertion that a new theory beyond general relativity is needed. With tremendous effort, various attempts in this direction are underway.[1] On the other hand, since, at present, it is, in practice, impossible to see gravitational interactions between quantum objects, we have never encountered any serious problems even without the theory beyond general relativity. There is no serious contradiction between predictions of general relativity and observational and experimental results, and general relativity presents a basis of our understanding about the universe (see, for example, Ref. [1]). Even in such a situation, there are several motivations

[1]There are so many references, but it is not the purpose of this chapter to review their details. You could check how many efforts through, for example, a key word "quantum gravity".

K. Nakao (✉)
Department of Physics, Graduate School of Science, Osaka Metropolitan University, Sugimoto-cho 3-3-138, Osaka, Sumiyoshi-ku 558-8585, Japan
e-mail: knakao@omu.ac.jp

Nambu Yoichiro Institute of Theoretical and Experimental Physics (NITEP), Osaka Metropolitan University, Sugimoto-cho 3-3-138, Osaka, Sumiyoshi-ku 558-8585, Japan

D. Malafarina and P. S. Joshi (eds.), *New Frontiers in Gravitational Collapse and Spacetime Singularities*, Springer Series in Astrophysics and Cosmology,
https://doi.org/10.1007/978-981-97-1172-7_11

for looking for a new theory beyond general relativity, besides the gravitational interactions between quantum objects. The most frequently asserted motivation among them may be that general relativity predicts the existence of spacetime singularities in our universe; many researchers believe that there should be no infinite quantity in physics. However, we should note that if the theory can predict all observable quantities and if there is no serious discrepancy between the predicted values and results of observations or experiments, there is no problem. As mentioned later, from this point of view, the spacetime singularity is not necessarily a serious flaw of general relativity.

By contrast, it is a serious problem for the theory that observable quantities cannot be predicted by the theory. In general relativity, there might be such problematic situations: the predicted spacetimes are not globally hyperbolic, in other words, there is no initial data set which predicts any physical phenomena in whole spacetime by relying on the theories locally consistent to the special relativity. If general relativity predicts that our universe is not globally hyperbolic, the theory beyond general relativity is necessary even in the case that gravity produced by quantum matters is negligible. Famous conjecture proposed by Penrose, which is called the cosmic censorship hypothesis, is deeply related to this issue, and will be reconsidered here.

The black hole is regarded as one of important keys of the door to the theory beyond general relativity through the famous information-loss problem. In this chapter, we reexamine the fact that the existence of a black hole cannot be confirmed by any observations, although this fact is written in orthodox text books of general relativity. The fact that the black hole is unobservable leads to a conclusion that since the quantum theory describes the causal relations between observations and observables, the corresponding concept to the black hole in general relativity cannot exist in the new theory beyond general relativity, if the new theory is established in the quantum framework.

This chapter is organized as follows. In Sect. 11.2, the definition of a black hole in asymptotically flat spacetime is reviewed because of its importance in this chapter. Then, in Sect. 11.3, we review the global hyperbolicity and the cosmic censorship hypothesis. We see a few examples of non-globally hyperbolic spacetimes which are not necessarily counter examples to the cosmic censorship hypothesis in Sect. 11.4. In Sect. 11.5, we reconsider a black hole formed gravitational collapse of a spherically symmetric object and study what observers outside the black hole can observe. What is the black hole shadow will be explained in this section. Section 11.6 is devoted to summary and discussion.

In this chapter, we adopt the unit in which the speed of light and Newton's gravitational constant are one. However if necessary, they are explicitly denoted by c and G, respectively.

11.2 Black Hole

In this section, we briefly review the asymptotic flatness and the definition of the black hole in the asymptotically flat spacetime. This is a necessary concept for the cosmic censorship hypothesis mentioned in the next Sect. 11.3. Then, we will revisit the black hole as a key of the door to the theory beyond general relativity in Sect. 11.5.

11.2.1 Asymptotic Flatness

The spacetime which is a concept introduced in special relativity has evolved in general relativity to have its own physical degrees of freedom corresponding to gravity. The picture of the dynamical evolution of spacetime itself has been very successful in describing physical phenomena involving strong gravity, such as the Big Bang and gravitational collapse of massive objects, which forms black holes and spacetime singularities. In this chapter, The spacetime is assumed to be a four dimensional time orientable manifold $\mathscr{M}$ with a metric tensor of the Lorentzian signature, and we will focus in particular on gravitational collapse in asymptotically flat spacetime explained below.

In order to define the asymptotic flatness, we invoke the conformal compactification of the Minkowski spacetime, which is completed through, for example, the coordinate transformation from the inertial coordinate system with the spherical polar coordinates for the space section,

$$-\infty < t < \infty, \;\; 0 \leq r < \infty, \;\; 0 \leq \theta \leq \pi, \;\; 0 \leq \varphi < 2\pi, \tag{11.1}$$

in which the infinitesimal world interval is represented as

$$ds^2 = -dt^2 + dr^2 + r^2 \left(d\theta^2 + \sin^2\theta d\varphi^2\right), \tag{11.2}$$

to a new coordinate system, (T, R, θ, φ), defined as

$$u := t - r = L \tan \frac{T - R}{2}, \tag{11.3}$$

$$v := t + r = L \tan \frac{T + R}{2}, \tag{11.4}$$

where L is a constant with the dimension of length. The infinitesimal world interval in the new coordinate system is

$$ds^2 = \Omega^{-2} \left[-dT^2 + dR^2 + \sin^2 R \left(d\theta^2 + \sin^2\theta d\varphi^2\right)\right] = \Omega^{-2} d\tilde{s}^2, \tag{11.5}$$

where

$$\Omega = \frac{2}{L}\cos\left(\frac{T-R}{2}\right)\cos\left(\frac{T+R}{2}\right). \tag{11.6}$$

Here, note that $d\tilde{s}^2$ is the infinitesimal world interval of Einstein's static universe whose domain is $-\infty < T < \infty$, $0 \leq R \leq \pi$. The domains of $-\infty < u < \infty$ and $-\infty < v < \infty$ correspond to $-\pi < T - R < \pi$ and $-\pi < T + R < \pi$, respectively. Since $r = \Omega^{-1} \sin R$ is larger than or equal to 0, we have $R \geq 0$. By the coordinate transformation (11.3) and (11.4), the Minkowski spacetime is embedded to the domain

$$-\pi < T - R < \pi, \quad -\pi < T + R < \pi \text{ and } R \geq 0 \tag{11.7}$$

in Einstein's static universe as depicted in Fig. 11.1 which is called the Penrose diagram of the Minkowski spacetime. The conformal factor Ω vanishes on the boundaries $T - R = \pm\pi$ and $T + R = \pm\pi$ which corresponds to infinities $u = \pm\infty$ and $v = \pm\infty$, respectively. Those boundaries are called the conformal infinities. The infinities, $u = -\infty$ and $v = \infty$, are mapped to $T - R = -\pi$ and $T + R = +\pi$ in Einstein's static universe and called the past null infinity denoted by $\mathscr{I}_-$ and the future null infinity $\mathscr{I}_+$, respectively. $T = \pi$ ($T = -\pi$) and $R = 0$ is called the future (past) timelike infinity denoted by i_+ (i_-), whereas $T = 0$ and $R = \pi$ is called the spacelike infinity denoted by i_0. All null geodesics begin at $\mathscr{I}_-$ and end at $\mathscr{I}_+$. By contrast, all timelike geodesics start from i_- and end at i_+, whereas all spacelike geodesics begin at i^0 and end alos at i_0.

Note that null curves in the spacetime with ds^2 are also null curves in the spacetime with $d\tilde{s}^2$. Hence the causal structure of Minkowski spacetime is equivalent to that in the restricted domain (11.7) of Einstein's static universe. In Fig. 11.1, regions shaded by blue are the causal future $J^+(p)$ and the causal past $J^-(p)$ of the event p: the causal future of the event p is a set of events which can be connected with p by a future directed causal curves, where "causal" means "timelike" or "null", whereas the causal past of p is defined in the similar manner to the causal future by interchanging future and past. From an observational perspective, the causal past of the event p corresponds to the view field of an observer at the event p and plays an important role in the later discussion about the observations related to black holes.

The merit to consider the causal structure of Minkowski spacetime by using the Einstein's static universe is that infinities can be regarded as subsets of the "spacetime", although infinities are not included in the Minkowski spacetime. The spacetime in which the spacetime $\mathscr{M}$ is confomally embedded is called the unphysical spacetime denoted by $\tilde{\mathscr{M}}$; in the case of the conformal compactification (11.3) and (11.4) of the Minkowski spacetime, the unphysical spacetime $\tilde{\mathscr{M}}$ is Einstein's static universe. The asymptotically flat spacetime is defined as the one whose conformally compactified manifold has the similar conformal infinity $i_0 \cup \mathscr{I}_+ \cup \mathscr{I}_-$ to that of the Minkowski spacetime (for detail, see, for example, Ref. [2]).

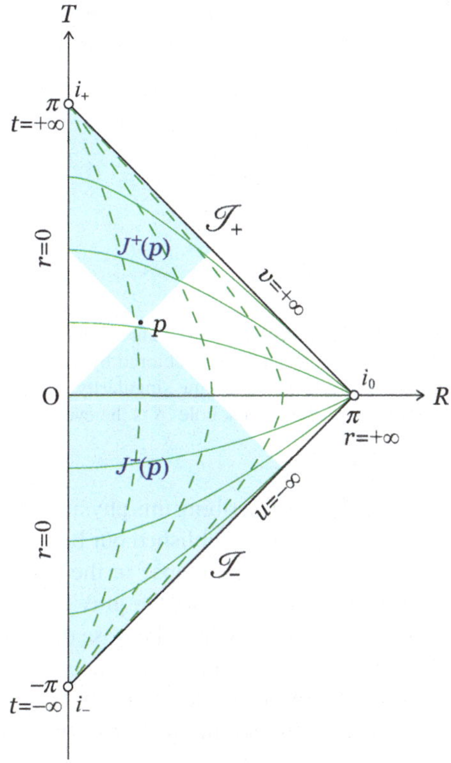

Fig. 11.1 The Penrose diagram of the Minkowski spacetime is depicted. The solid green curves represent spacelike hypersurfaces labeled by t, whereas r is constant on each dashed green curve

11.2.2 Black Hole as a Causal Structure of Spacetime

The black hole $\mathscr{B}$ in the asymptotically flat spacetime is defined as

$$\mathscr{B} = \mathscr{M} - J^-(\mathscr{I}_+). \tag{11.8}$$

This definition of the black hole was presented by Hawking [3] and well represents its nature that is the domains from which nothing comes out *forever*. The boundary of a black hole is called the event horizon $\mathscr{H}$. We have

$$\mathscr{H} = \dot{J}^-(\mathscr{I}_+), \tag{11.9}$$

where $\dot{J}^-(\mathscr{I}_+)$ is a boundary of $J^-(\mathscr{I}_+)$. The definition of the event horizon equivalent to Eq. (11.9) was proposed by Penrose [4]. Since the boundary of the causal future of any set is generated by null geodesic congruence, so is the event horizon.

It is believed that a very massive star loses the stress strong enough to support itself against self-produced gravity in the final stages of its evolution, and hence experiences gravitational collapse to form a black hole $\mathscr{B}$ and spacetime singularities

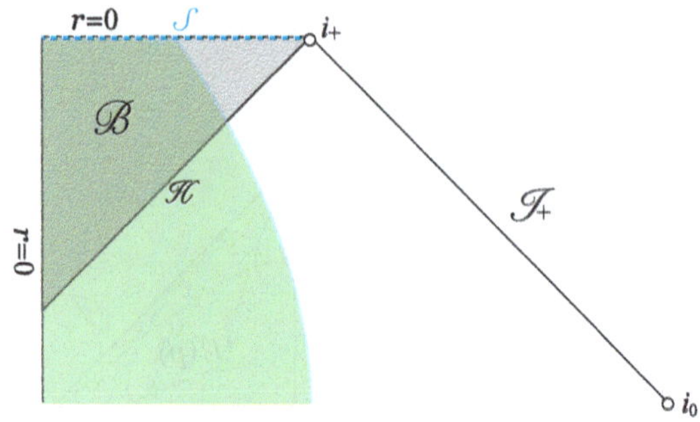

Fig. 11.2 The upper half of the Penrose diagram of the Oppenheimer-Snyder solution is depicted. The dust exists in the domain colored by green, whereas the domain shaded by gray is the black hole $\mathscr{B}$. There is the spacetime singularity $\mathscr{S}$ depicted by the blue dashed line in the black hole. The boundary of the black hole $\mathscr{B}$ is the event horizon $\mathscr{H}$

$\mathscr{S}$. The solution describing this physical process was first given by Oppenheimer and Snyder in 1939 and established our basic understanding of the black hole formation through gravitational collapse in the asymptotically flat spacetime. Oppenheimer-Snyder solution represents the gravitational collapse of a gravitationally bounded uniform dust sphere, whose Penrose diagram is shown in Fig. 11.2: the green area is occupied by the dust, and a spacelike spacetime singularity $\mathscr{S}$ represented by a blue dashed curve forms inside a black hole $\mathscr{B}$ which is an area shaded gray.

By virtue of the definition (11.8), important aspects of black holes were revealed by Hawking. The non-decreasing area of the event horizon is one of the most important features of the black hole, and is known as the area theorem. It plays a crucial role in the black hole energetics, and furthermore, leads to an idea that an entropy of a black hole is proportional to the area of its event horizon [5], which is now regarded as a key to search for the theory beyond general relativity (see for example Ref. [6]).

An asymptotically flat spacetime is an isolated system in the framework of general relativity. Our universe is not exactly asymptotically flat, but there are domains each of which is well approximated by an isolated system. As mentioned in the previous section, the notion of null infinities is introduced so as to define an unambiguous ideal asymptotic flatness, and is used for defining a black hole from which nothing comes out forever. However, there will be no null infinity of the asymptotic flatness in our universe even in approximate sense, although non-asymptotically flat infinities may exist. Recent observational data [7] imply that our present universe is globally well described by the ΛCDM model which describes a spatially flat, homogeneous and isotropic universe occupied mainly by the positive cosmological constant Λ and the cold dark matter. If ΛCDM model globally well describes our universe in the future, the future conformal infinity $\mathscr{J}^+$ similar to the de Sitter spacetime exists, and hence the black hole can be defined as Eq. 11.8. (see, for example, [8] about the black hole in the asymptotically deSitter spacetime.) Thus black holes from which nothing comes out *forever* may exist in our universe. Here, we should note that it is, in principle, impossible to observationally verify the existence of a black hole defined in the manner (11.8) [9, 10], since we cannot get any information about the infinite future $\mathscr{I}_+$ of our universe through any observation by causality. We need to carefully consider what is observable in the black hole spacetime and will revisit this issue later.

11.3 Global Hyperbolicity and Cosmic Censorship Hypothesis

Let S denote a closed achronal set, where a set is said to be achronal if there are no two points connected to each other by any timelike curve. The future domain of dependence of a closed achronal set S is a set of events such that all past inextendible causal caurves through each event intersect S, and is denoted by $D^+(S)$. Within the framework of theories of physics consistent with the relativistic causality, which will, hereafter, be called the relativistic theories, physical phenomena in $D^+(S)$ are predicted as solutions of the Cauchy problem which satisfy initial data on S. The past domain of dependence of S, which is denoted by $D^-(S)$, is defined by the similar manner to that of $D^+(S)$ by interchanging past and future. Physical phenomena in $D^-(S)$ are retrodicted by relativistic theories as solutions of the Cauchy problem which satisfy initial data on S.

The domain of dependence of S denoted by $D(S)$ is defined as

$$D(S) = D^+(S) \cup D^-(S). \tag{11.10}$$

A closed achronal set Σ is called a Cauchy surface if

$$\mathscr{M} = D(\Sigma). \tag{11.11}$$

A spacetime with a Cauchy surface is said to be globally hyperbolic. In the globally hyperbolic spacetime, any physical phenomenon described by relativistic theories can be predicted or retrodicted from initial data on a Cauchy surface. Conversely, if the spacetime is not globally hyperbolic, $\mathscr{M} - D(S)$ is not empty for any closed achronal set S. In this case, the boundary of $D(S)$ is not empty and is called the Cauchy horizon of S, which is denoted by $H(S)$. Relativistic theories cannot predict or retrodict what happens in the domain beyond Cauchy horizon, in non-globally hyperbolic spacetime. Hence, it is a crucial question whether our universe predicted by general relativity is globally hyperbolic. Theories of physics do not need to describe unobservable phenomena but should have ability to predict all of observables. Thus, if the physical phenomena in the domain beyond Cauchy horizon are observable, general relativity should be rejected and a theory beyond general relativity is necessary.

There are several motivations for looking for a new theory beyond general relativity, besides the fact that general relativity cannot describe gravitational interactions between objects whose behaviors are described by the quantum theory. One of them is that general relativity predicts the existence of spacetime singularities in our universe. However, if general relativity predicts that our universe is globally hyperbolic, all observables can be predicted as solutions of Cauchy problems, as long as the gravitational interactions between quantum objects are negligible. Thus the theories are, in principle, falsifiable through experiments or observations, as long as the spacetime is globally hyperbolic. Spacetime singularities are not observable, and hence there is no observable quantity with infinite value, in globally hyperbolic spacetimes. In my personal opinion, it is insufficient motivation for the theory beyond general relativity that general relativity predicts the spacetime singularities, if the predicted spacetime is globally hyperbolic.

Here, one of important assumptions in the proofs of important theorems about black holes (e.g. area theorem mentioned in previous section), called "strongly asymptotically predictable spacetime" should be mentioned. If there is a globally hyperbolic open domain which includes the closure of $\mathscr{M} \cap J^-(\mathscr{I}_+)$ in the unphysical spacetime $\tilde{\mathscr{M}}$, the spacetime is said to be strongly asymptotically predictable. All of physical phenomena which actually occurs outside black holes in the strongly asymptotically predictable spaceitme can be predicted by relativistic theories as solutions of Cauchy problems.

A celebrated conjecture called "cosmic censorship hypothesis" which is related to the global hyperbolicity of the spacetime predicted by general relativity was proposed by Penrose [4, 11]. There are two versions of this hypothesis. One, called the weak version, asserts that the physical asymptotically flat spacetime is strongly asymptotically predictable. Here physical spacetime means that there is a spcelike hypersurface (a closed achronal set without edge) with physically reasonable first and second fundamental forms, and there are initial data of matter fields which satisfy reasonable energy conditions, on the spacelike hypersurface. The other, called the strong version, asserts the physical spacetime is globally hyperbolic. If cosmic censorship hypothesis is true, the formation of spacetime singularities cannot be a motivation for looking for the theory beyond general relativity. Conversely if general relativity predicts that our universe is not globally hyperbolic, a new theory is necessary, since, if so, there are observable phenomena in our universe, which cannot be predicted by relativistic theories.

11.4 Non-globally Hyperbolic Examples

In this section, we see a few examples of non-globally hyperbolic spacetimes which are significant in studying the cosmic censorship hypothesis.

11.4.1 Spherically Symmetric Inhomogeneous Dust

The gravitational collapse of a spherically symmetric inhomogeneous dust is an impressive physical system related to the cosmic censorship hypothesis. It is described by the Lemaître-Tolman-Bondi solution in the framework of general relativity. The Lemaître-Tolman-Bondi solution includes the Oppenheimer-Snyder solution as a special case. If C^∞ is required at the origin, the energy density ρ on a spacelike hypersurface before the singularity fomation is expressed in the form of a Maclaurin series around the symmetry center as

$$\rho(r) = \sum_{n=0}^{\infty} \rho_{(2n)} r^{2n}, \tag{11.12}$$

where r is the radial coordinate of the spherical polar coordinates associated to the spherical symmetry of the system. From the positivity of the energy density, $\rho_{(0)}$ should be positive. Surprisingly, if $\rho_{(2)}$ is negative, the spacetime singularity first formed at the symmetry center, $r = 0$, is a null singularity $\mathscr{S}_\mathrm{N}$ whose causal future $J^+(\mathscr{S}_\mathrm{N})$ is not empty (see Figs. 11.3 and 11.4) [12–15]. The non-empty causal future $J^+(\mathscr{S}_\mathrm{N})$ of $\mathscr{S}_\mathrm{N}$ implies that $\mathscr{S}_\mathrm{N}$ is observable in principle. Such a spacetime singularity is called a naked singularity.

Let us consider a Cauchy problem with the initial condition (11.12) on a spacelike hypersurface, Σ, depicted in Figs. 11.3 and 11.4. Since the domain of dependence $D^+(\Sigma)$ of Σ does not include $J^+(\mathscr{S}_\mathrm{N})$, the solution of this Cauchy problem cannot determine $J^+(\mathscr{S}_\mathrm{N})$. In order to determine what happens in $J^+(\mathscr{S}_\mathrm{N})$, the boundary condition should be imposed on $\mathscr{S}_\mathrm{N}$. Here, we should note that some boundary condition has been implicitly imposed on $\mathscr{S}_\mathrm{N}$ so that $J^+(\mathscr{S}_\mathrm{N})$ is still described by the

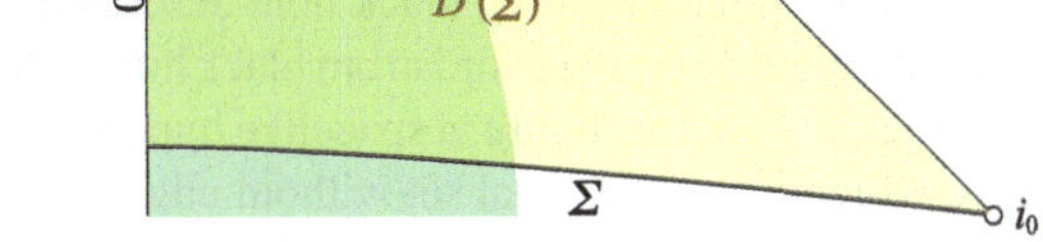

Fig. 11.3 The upper half of the Penrose diagram of the Lemaître-Tolman-Bondi solution is depicted. Both of the blue and the red dashed lines are the spacetime singularity, but the red one $\mathscr{S}_\mathrm{N}$ is the naked singularity whose causal future cannot be predicted by a solution of a Cauchy problem with the initial data on a spacelike hypersurface, Σ, whose future domain of dependence $D^+(\Sigma)$ is shaded by yellow. In this case, the unpredictable domain beyond the Cauchy horizon $H(\Sigma)$ is included in the black hole $\mathscr{B}$

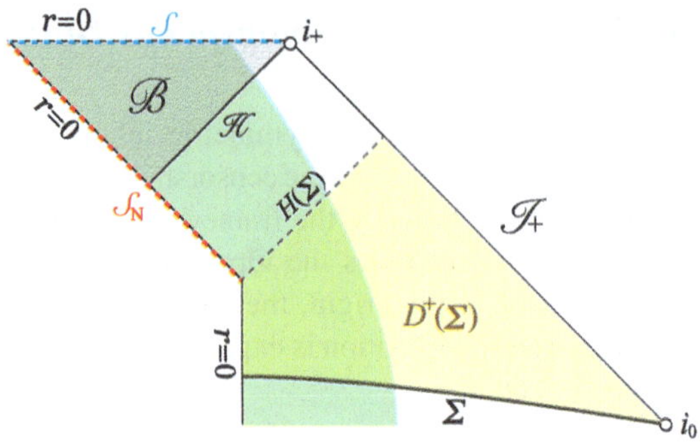

Fig. 11.4 The same as Fig. 11.3, but a part of the naked singularity $\mathscr{S}_N$ depicted by the red dashed line forms before the black hole formation. Hence this is a counter example of both weak and strong versions of the cosmic censorship hypothesis

Lemaître-Tolman-Bondi solution, in order to draw Figs. 11.3 and 11.4. If we assume that radiation emanates from the null singularity $\mathscr{S}_N$, then $J^+(\mathscr{S}_N)$ is different from that described by the Lemaître-Tolman-Bondi solution.

The total mass determines whether the naked singularity is totally contained within a black hole (Fig. 11.3) or forms before the black hole formation (Fig. 11.4). The former is a counterexample of the strong version and is called a local naked singularity, while the latter is a counterexample of both the weak and strong versions and is called a global naked singularity. Is the Lemaître-Tolman-Bondi solution a counter example to the cosmic censorship hypothesis? If we consider that the matter should be fundamental fields, the answer is no. However, this example is still suggestive.

11.4.2 Charged Black Hole and Rotating Black Hole

It has been an issue whether the interior of a charged black hole and a black hole with angular momentum should be regarded as a counter example to the strong version of the cosmic censorship hypothesis. There are timelike naked singularities inside a spherically symmetric charged black hole described by the Reissner-Nordström solution and an axisymmetric stationary rotating black hole described by the Kerr solution. The Penrose diagram of the Reissner-Nordström black hole with non-degenerate horizon is depicted in Fig. 11.5. In this figure, a spacelike hypersurface Σ is depicted as a thick black curve which is an achcronal set without edge. Due to the existence of local naked singularities $\mathscr{S}_N$, a Cauchy horizon $H(\Sigma)$ exists. The causal future of naked singularities is not included in the future domain of dependence of Σ which is denoted by $D^+(\Sigma)$. Thus, if such spacetime structures are realized inside black holes in our universe, then there will exist observable phenomena that cannot be predicted by relativistic theories.

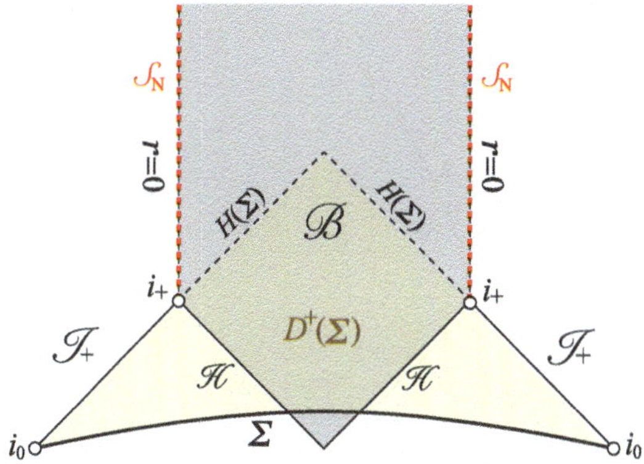

Fig. 11.5 The upper half of the Penrose diagram of the the Reissner-Nordström black hole spacetime with non-degenerate horizons is depicted. Like as in Figs. 11.3 and 11.4, the red dashed lines represent the naked spacetime singularities $\mathscr{S}_N$, whereas the domain shaded by gray is the black hole. The future domain of dependence of a spacelike hypersurface, Σ, which is denoted by $D^+(\Sigma)$, is shaded by yellow. There is a non-empty Cauchy horizon denoted by $H(\Sigma)$ of Σ. This means that there is an unpredictable domain by the solutions of the Cauchy problem with the initial data on Σ

However, this internal structure is unstable against perturbations and the Cauchy horizon $H(\Sigma)$ become a curvature singularity composed of null and spacelike portions. The metric does not exist on the spacelike singularity, but on the null singularity. Hence the C^0-extension is possible beyond the null portion of the singularity [16, 17]. The spacetime with the C^0-extention beyond the null singularity is not globally hyperbolic. However, since the second fundamental form of the null portion of the singularity diverges, any objects with finite size would be destroyed by the infinite tidal force before they go across the null singularities, and, of course, so would the observers and the experimental apparatus. Since no observable phenomenon unpredictable by general relativity exists inside a charged, a rotating or charged rotating black hole, no theory beyond general relativity seems to be necessary.

11.4.3 Regular Black Hole

The spacetime singularity is often regarded as a fatal flaw of general relativity and hence the possibilities of black holes with no spacetime singularity have been studied (see, for example, Ref. [18] and references are therein). The Bardeen black hole is a famous and the first example of a regular black hole whose metric is given as

$$ds^2 = -f(r)dt^2 + \frac{dr^2}{f(r)} + r^2\left(d\theta^2 + \sin^2\theta d\varphi^2\right), \qquad (11.13)$$

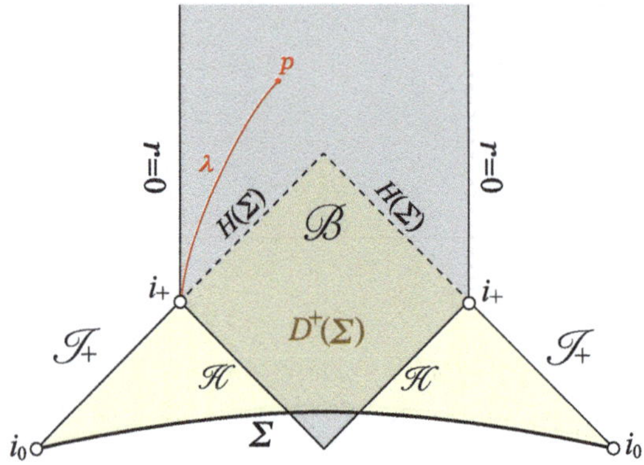

Fig. 11.6 The upper half of the Penrose diagram of the Bardeen black hole spacetime with non-degenerate horizons is depicted. Except for the regularity at $r = 0$, the causal structure of this spacetime is very similar to that of the Reissner-Nordström black hole spacetime shown in Fig. 11.5. There is an event p in the causal future of an infinity i_+, which is depicted as a red point in the figure. By definition, there is a past directed causal curve λ from the event p, which has an end point at an infinity i_+. The curve λ is past inextendible but does not intersect Σ. This fact implies that the event p is not included in $D(\Sigma)$, and hence, as in the case of the Reisner-Nordtröm black hole spacetime, there is a Cauchy horizon $H(\Sigma)$ of a spacelike hypersurface Σ

with

$$f(r) = 1 - \frac{r_g r^2}{\left(r^2 + l^2\right)^{\frac{3}{2}}}, \tag{11.14}$$

where l and r_g are positive constants [19]. In the domain of $r \gg l$, the geometry is nearly equal to that of the Schwarzschild spacetime, i.e., $f(r) \simeq 1 - r_g/r$, whereas, in the domain of $r \ll l$, we have $f(r) \simeq 1 - (r_g/l^3)r^2$ which is nearly equal to that of the de Sitter spacetime, and thus $r = 0$ is not a spacetime singularity.

The case of $l \ll r_g$ is of special interest. In this case, two positive roots of $f(r) = 0$ exist. The larger one corresponds to the event horizon. The upper half of the Penrose diagram of this spacetime is depicted in Fig. 11.6. Although there is no spacetime singularity in the Bardeen black hole, as mentioned below, this spacetime is not globally hyperbolic as in the case of the Reisner-Nordström black hole spacetime whose Penrose diagram is very similar to that of the Bardeen black hole spacetime (see Fig. 11.5).

In Fig. 11.6, we depict a spacelike hypersurface Σ as a thick black curve, which is a closed achronal set without edge. As can be seen from this figure, the causal future of an infinity i_+ which is denoted by $J^+(i_+)$ is not empty and is included in the causal future of Σ. By definition of the causal future, there is a past directed causal curve connected to i_+ from an event p in $J^+(i_+)$, which is depicted as a red curve λ in Fig. 11.6. The curve λ is past inextendible but does not intersect Σ. This fact implies that any event in $J^+(i_+)$ is not included in $D(\Sigma)$, and hence, the Cauchy horizon $H(\Sigma)$ of the spacelike hypersurface Σ exists. The spacelike hypersurface Σ is not a Cauchy surface. We can not find a Cauchy surface in this spacetime, even though there is no naked singularity. However, this spacetime is not a counter example to the strong version of the cosmic censorship, since the reasonable energy conditions are not satisfied (see, for example, Ref. [18]). The Bardeen black hole may as well be regarded as a solution of a theory other than general relativity. However, the non-global hyperbolicity of this spacetime implies the necessity of a theory beyond the theory which predicts the Bardeen black hole.

11.4.4 Evaporation of Black Hole

The information-loss problem accompanied by the evaporation of a black hole was pointed out by Hawking [20] and is still regarded as a key to the new theory beyond general relativity. Figure 11.7 is the Penrose diagram drawn by Hawking to represent the spacetime in which a spherically symmetric black hole evaporates due to quantum radiation [20]. In the massless limit of the Schwarzschild black hole, the curvature on the event horizon diverges, and hence the event at which the black hole completely evaporates is a naked singularity $\mathscr{S}_{\mathrm{N}}$. This spacetime is neither globally hyperbolic nor strongly asymptotically predictable, and hence seems to be a counter example to both weak and strong versions of the cosmic censorship hypothesis.

Let us consider a spacelike hypersurface Σ_{after} after the black hole evaporation (see Fig. 11.8). The causal past of Σ_{after}, which is denoted by $J^-(\Sigma_{\mathrm{after}})$, does not include the black hole, and hence there is no information about the black hole on Σ_{after}. In terminology of quantum theory, this situation is represented by non-unitary evolution. Since, exactly speaking, the back reaction of the quantum radiation cannot be treated in the framework of general relativity, some theory other than general relativity has been implicitly assumed for drawing Figs. 11.7 and 11.8.[2] Hence, it is

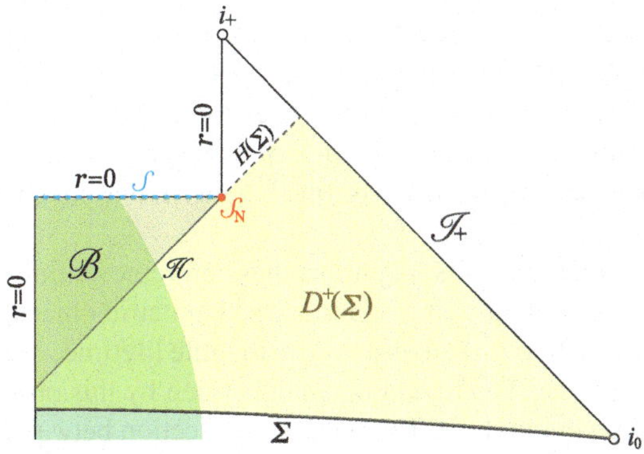

Fig. 11.7 The upper half of the Penrose diagram of the spherically symmetric gravitational collapse to form a black hole which eventually evaporates due to quantum radiation. The future domain of dependence of a spacelike hypersurface, Σ, which is denoted by $D^+(\Sigma)$, is shaded by yellow. This spacetime is not globally hyperbolic

[2] It is inevitable to search for the theory beyond general relativity, since general relativity cannot describe the gravitational interaction between quantum matter.

Fig. 11.8 The same as Fig. 11.7 but a spacelike hypersurface, Σ_{after}, in the domain after the black hole evaporation and its causal past $J^-(\Sigma_{\text{after}})$ are drawn as a black curve and as a domain shaded by blue, respectively. The black hole is not included in $J^-(\Sigma_{\text{after}})$. This means that no information about the black hole exists on Σ_{after}

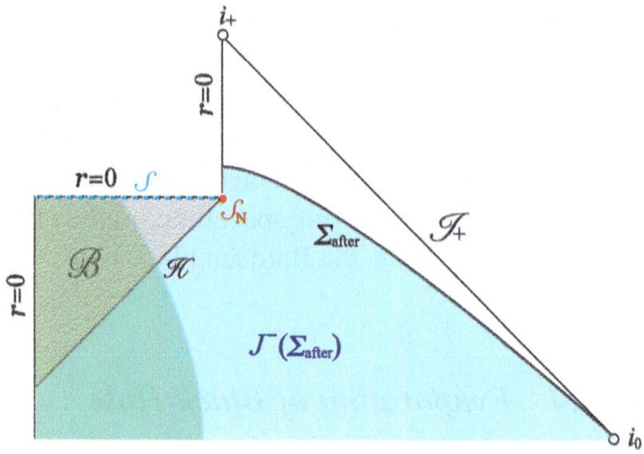

subtle whether we can say that this is a counter example to the cosmic censorship hypothesis. Anyway, as mentioned in the previous subsection about the regular black hole, it is a sufficient condition for rejection of the theory that the theory predicts non-globally hyperbolic "spacetime". The reason for enclosing the word "spacetime" with quotation marks is that the notion of the spacetime may disappear in the theory beyond general relativity. However, the theory of physics is to say the causal relations between observables. If so, a kind of global hyperbolicity is necessary even in the framework of the new theory beyond general relativity.

11.5 Collapsing Object; What Can We See?

Let us consider once again the black hole, assuming the weak version of the cosmic censorship hypothesis. The definition of a black hole in the asymptotically flat spacetime (11.8) is a mathematically rigorous expression of the common understanding about black holes that "nothing can come out from a black hole". As mentioned in Sect. 11.2, although our universe is not asymptotically flat, there will be black holes defined as the same as Eq. (11.8), if, for example, our universe is globally well described by the ΛCDM model, by replacing the future null infinity of the asymptotically flat spacetime by that of the asymptotically de Sitter spacetime. Observations support the existence of black holes in our universe (see for example, Refs. [21–25, 27, 28].

Figure 11.9 is a Penrose diagram representing the formation of a black hole due to the collapse of a spherically symmetric object, as in Fig. 11.2. In this diagram, a world line λ of an observer with finite lifetime, who is located outside the black hole, is drawn. The region that can be seen by this observer is $J^-(\lambda)$ which is shaded by blue in Fig. 11.9. There is no intersection between this observer's observable region and the black hole. It is often misunderstood by researchers unfamiliar with general relativity that the reason why we cannot see a black hole is that it is in our field of view but does not emit anything. However, as Fig. 11.9 shows, the black hole does

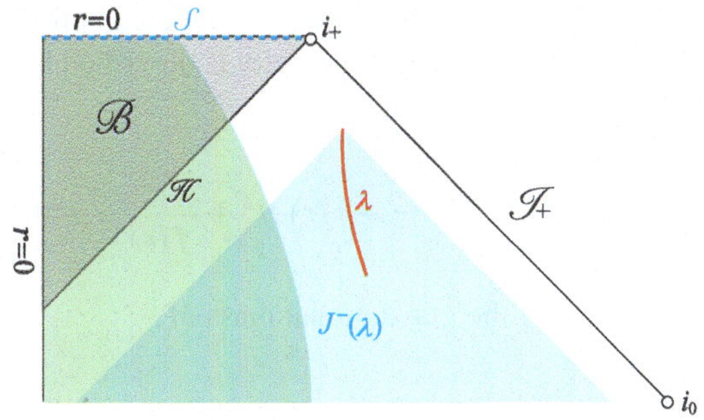

Fig. 11.9 The same as Fig. 11.2, but with a world line λ of an observer whose life time is finite. The world line λ is the red curve, and its causal past $J^-(\lambda)$ is the domain shaded by blue

not exist in the observer's field of view. What is in the observer's field of view is a collapsing object. Even if the observer's lifetime is infinitely long, what the observer outside the black hole can see is an object that continues to collapse forever but not a black hole. Thus, it is, in principle, impossible for any outside observer to confirm the existence of a black hole defined as Eq. (11.8) through any observation. On the other hand, we often see reports that "a black hole has been discovered". What has been found?

11.5.1 Collapse of a Spherically Symmetric Homogeneous Dust

The gravitational collapse of a massive object occurs when the gravity overcomes the force of stress, and hence the pressure free fluid is often a good approximation in the study of black hole formation. In this subsection, we consider the collapse of a spherically symmetric homogeneous dust, i.e., the Oppenheimer-Snyder model mentioned in Sect. 11.2 and study what we can see through the observation of the collapsing object. Here, for simplicity in calculations, we assume the dust ball is marginally bound: in the infinite past, the dust ball has an infinite radius and is stationary.

11.5.1.1 Oppenheimer-Synder Model

Here, as mentioned, we consider the marginally bound dust ball, so that the geometry inside the homogeneous dust ball is described by the Einstein-de Sitter universe whose metric is given as

$$ds^2 = -a^2(\eta)\left[-d\eta^2 + d\chi^2 + \chi^2\left(d\theta^2 + \sin^2\theta d\phi^2\right)\right], \qquad (11.15)$$

where $a(\eta) \propto \eta^2$. The outside of the dust ball is assumed to be vacuum. Then, by Birkhoff's theorem, the geometry is described by the Schwarzschild solution whose metric is given as

$$ds^2 = -f(r)dt^2 + \frac{dr^2}{f(r)} + r^2 \left(d\theta^2 + \sin^2\theta d\phi^2\right), \tag{11.16}$$

where, being the gravitational mass M,

$$f(r) = 1 - \frac{2M}{r}. \tag{11.17}$$

The 1st and 2nd fundamental forms of the surface of the dust ball calculated at both of the inside (11.15) and the outside (11.16) of the dust ball should be common. This condition leads to a fact that the fluid particle on the surface of the dust ball moves along a marginally bound radial timelike geodesic. Inside the dust ball, being τ the proper time, the 4-velocity vector, u^μ, of the surface of the dust ball is given as

$$u^\mu = \left(\frac{d\eta_{\rm sf}}{d\tau}, \frac{d\chi_{\rm sf}}{d\tau}, \frac{d\theta_{\rm sf}}{d\tau}, \frac{d\phi_{\rm sf}}{d\tau}\right) = \left(\frac{1}{a(\eta_{\rm sf})}, 0, 0, 0\right) \tag{11.18}$$

whereas, outside the dust ball, we have

$$u^\mu = \left(\frac{dt_{\rm sf}}{d\tau}, \frac{dr_{\rm sf}}{d\tau}, \frac{d\theta_{\rm sf}}{d\tau}, \frac{d\phi_{\rm sf}}{d\tau}\right) = \left(\frac{1}{f(r_{\rm sf})}, -\sqrt{\frac{r_{\rm g}}{r_{\rm sf}}}, 0, 0\right), \tag{11.19}$$

where

$$r_{\rm g} = 2M \tag{11.20}$$

is the gravitational radius. Equation (11.18) leads to constant $\chi_{\rm sf}$ and then, together with the continuity of the 1st fundamental form, $r_{\rm sf} = a(\eta_{\rm sf})\chi_{\rm sf}$, leads to

$$a(\eta) = \frac{r_{\rm g}}{4\chi_{\rm sf}^3}\eta^2, \tag{11.21}$$

where η is negative and the spacetime singularity forms at $\eta = 0$. The energy density of the dust ball is given as

$$\rho = \frac{3M}{4\pi r_{\rm sf}^3} = \frac{3M}{4\pi \chi_{\rm sf}^3 a^3}. \tag{11.22}$$

11.5.1.2 Propagation of Photon

In the observation of a collapsing object, one of direct observables for us are photons from or through the domain occupied by the collapsing object. A photon moves along a null geodesic. By virtue of the spherical symmetry, without loss of generality, we focus on the null geodesic confined in the equatorial plane $\theta = \pi/2$. Denoting the affine parameter of the null geodesic by λ, the tangent vector, k^μ, of the world line of a photon in the dust ball is given as

$$k^\mu = \left(\frac{d\eta_{\rm ph}}{d\lambda}, \frac{d\chi_{\rm ph}}{d\lambda}, \frac{d\theta_{\rm ph}}{d\lambda}, \frac{d\phi_{\rm ph}}{d\lambda}\right) = \frac{\omega_{\rm c}}{a^2(\eta_{\rm ph})}\left(1, \sigma_\chi\sqrt{1-\frac{b_{\rm c}^2}{\chi_{\rm ph}^2}}, 0, \frac{b_{\rm c}}{\chi_{\rm ph}^2}\right), \tag{11.23}$$

where $\omega_{\rm c}$ and $b_{\rm c}$ are constant. Since the tangent vector is future directed, $\omega_{\rm c}$ should be positive, and thus

$$\sigma_\chi = \mathrm{sgn}[k^\chi]. \tag{11.24}$$

By Eq. (11.23), we have

$$\frac{d\chi_{\rm ph}}{d\eta} = \pm\sqrt{1-\frac{b_{\rm c}^2}{\chi_{\rm ph}^2}}, \tag{11.25}$$

and, by integrating this equation, we have

$$\eta - \eta_{\rm in} = \sqrt{\chi_{\rm sf}^2 - b_{\rm c}^2} - \sqrt{\chi_{\rm ph}^2 - b_{\rm c}^2}, \tag{11.26}$$

$$\eta - \eta_{\rm out} = \sqrt{\chi_{\rm ph}^2 - b_{\rm c}} - \sqrt{\chi_{\rm sf}^2 - b_{\rm c}^2}, \tag{11.27}$$

where $\eta_{\rm in}$ and $\eta_{\rm out}$ are integration constants; $\eta_{\rm in}$ is the conformal time $\eta_{\rm ph}$ at which the photon enters the collapsing domain, whereas $\eta_{\rm out}$ is that at which the photon comes out from the dust ball. These two solutions are connected at $\chi_{\rm ph}^2 = b_{\rm c}^2$ at $\eta_{\rm ph} = \eta_{\rm cn}$, so that

$$\eta_{\rm cn} - \eta_{\rm in} = \sqrt{\chi_{\rm sf}^2 - b_{\rm c}^2} \quad \text{and} \quad \eta_{\rm cn} - \eta_{\rm out} = -\sqrt{\chi_{\rm sf}^2 - b_{\rm c}^2} \tag{11.28}$$

hold. From these equations, we have

$$\eta_{\rm out} - \eta_{\rm in} = 2\sqrt{\chi_{\rm sf}^2 - b_{\rm c}^2}. \tag{11.29}$$

Hereafter, we denote

$$a_{\rm in} = a(\eta_{\rm in}) \quad \text{and} \quad a_{\rm out} = a(\eta_{\rm out}). \tag{11.30}$$

On the other hand, outside the dust ball, the tangent vector, k^μ, of the world line of the photon is given as

$$k^\mu = \left(\frac{dt_{\rm ph}}{d\lambda}, \frac{dr_{\rm ph}}{d\lambda}, \frac{d\theta_{\rm ph}}{d\lambda}, \frac{d\phi_{\rm ph}}{d\lambda}\right) = \omega\left(\frac{1}{f(r)}, \sigma\sqrt{1-\frac{b^2}{r^2}f(r)}, 0, \frac{b}{r^2}\right), \tag{11.31}$$

where ω and b are constant. By the same reason as in the case of ω_c, ω should be positive, and hence

$$\sigma = \mathrm{sgn}[k^r]. \tag{11.32}$$

We regard k^μ is the 4-momentum vector of the photon. Since the coordinate basis vector of ϕ is common to the inside and the outside of the dust ball, $d\phi_{\mathrm{ph}}/d\lambda$ should be continuous and hence we have

$$\omega_c b_c = \omega b. \tag{11.33}$$

11.5.2 Case of a Shining Dust Ball

After sufficiently long time elapses from the beginning of the collapse, any radiation from the collapsing object is, in practice, unobservable due to the redshift effect. Let us evaluate the strength of the redshift for a static observer far from the dust ball. From Eq. (11.19), we have the evolution equation of the radius of the dust ball as

$$\frac{dr_{\mathrm{sf}}}{dt} = \left(\frac{dt_{\mathrm{sf}}}{d\tau}\right)^{-1} \frac{dr_{\mathrm{sf}}}{d\tau} = -f(r_{\mathrm{sf}})\sqrt{\frac{r_g}{r_{\mathrm{sf}}}}. \tag{11.34}$$

The solution of this equation is given as

$$t_{\mathrm{ff}} - t = \frac{2}{3}\sqrt{\frac{r_{\mathrm{sf}}}{r_g}}\left(r_{\mathrm{sf}} + 3r_g\right) + r_g \ln\left|\frac{\sqrt{r_{\mathrm{sf}}} - \sqrt{r_g}}{\sqrt{r_{\mathrm{sf}}} + \sqrt{r_g}}\right|, \tag{11.35}$$

where t_{ff} is a constant of integration.

We assume that the radius of the dust ball at $t = 0$, which is denoted by r_0, is much larger than the gravitational radius r_g. Then, we have

$$t_{\mathrm{ff}} = \frac{2}{3}\sqrt{\frac{r_0}{r_g}}\left(r_0 + 3r_g\right) + \ln\left|\frac{\sqrt{r_0} - \sqrt{r_g}}{\sqrt{r_0} + \sqrt{r_g}}\right| \simeq \frac{2}{3}\sqrt{\frac{r_0^3}{r_g}} = \sqrt{\frac{c^2}{6\pi G\rho_0}}, \tag{11.36}$$

where we have recovered the speed of light c and Newton's gravitational constant G in the last equality, and ρ_0 is the energy density of the dust ball at $t = 0$ (see Eq. (11.22)). The last expression is the so called free-fall time at which the homogeneous dust ball shrinks to a point $r_{\mathrm{sf}} = 0$ in the framework of the Newtonian theory. By contrast, in the framework of general relativity, the radius of the dust ball at $t = t_{\mathrm{ff}}$ does not vanish but is expected to be nearly equal to r_g. In order to roughly evaluate the radius of the dust ball at $t = t_{\mathrm{ff}}$, we introduce a positive parameter ε to represent it as

$$r_{\mathrm{sf}}|_{t=t_{\mathrm{ff}}} = r_g(1+\varepsilon). \tag{11.37}$$

Substituting this expression into Eq. (11.35), we have

$$0 = \frac{2}{3}\sqrt{1+\varepsilon}\,(4+\varepsilon) + \ln\left|\frac{\sqrt{1+\varepsilon}-1}{\sqrt{1+\varepsilon}+1}\right|, \tag{11.38}$$

Keeping the 1st order of ε in this equation, we have

$$\varepsilon \simeq 4e^{-\frac{8}{3}} \simeq 0.2779. \tag{11.39}$$

This result implies that our expectation is not so wrong. After several of the free-fall time from $t = 0$, we may approximate Eq. (11.35) by

$$t - t_{\mathrm{ff}} \simeq -r_{\mathrm{g}} \ln\left|\frac{r_{\mathrm{sf}} - r_{\mathrm{g}}}{4r_{\mathrm{g}}}\right|, \tag{11.40}$$

or equivalently,

$$r_{\mathrm{sf}} - r_{\mathrm{g}} \simeq 4r_{\mathrm{g}} \exp\left(-\frac{c(t-t_{\mathrm{ff}})}{r_{\mathrm{g}}}\right). \tag{11.41}$$

Here note that the time coordinate t is almost equal to the proper time of static observers far from the dust ball.

The free-fall time t_{ff} of $r_0 = N r_{\mathrm{g}}$ with N ($\gg 1$) is written in the form,

$$t_{\mathrm{ff}} \simeq \frac{c}{\sqrt{6\pi G\rho_0}} = \sqrt{\frac{2(Nr_{\mathrm{g}})^3}{9GM}} = 6.7\times 10^{-2}\left(\frac{M}{10M_\odot}\right)\left(\frac{N}{100}\right)^{3/2} \text{ sec}, \tag{11.42}$$

where $M_\odot = 2.0 \times 10^{30}$ kg is the mass of the sun. By using Eq. (11.42), we have

$$f(r_{\mathrm{sf}})|_{t=(n+1)t_{\mathrm{ff}}} \simeq 4e^{-\frac{2}{3}nN^{\frac{3}{2}}} \simeq 4\times 10^{-0.29nN^{\frac{3}{2}}}, \tag{11.43}$$

where n is a number larger than one. The angular frequency ω_{sf} of a photon emitted at the surface of the dust ball is given by

$$\omega_{\mathrm{sf}} = -u^\mu k_\mu = \frac{\omega}{f(r_{\mathrm{sf}})}\left[1 + \sqrt{\frac{r_{\mathrm{g}}}{r_{\mathrm{sf}}}\left(1 - \frac{b^2}{r_{\mathrm{sf}}^2}\right)}\right], \tag{11.44}$$

where we have used Eqs. (11.19) and (11.31) and assumed the outgoing photon, $\sigma = +1$. Since the angular frequency of the photon detected by a very distant static observer is almost equal to ω, the redshift factor is given as

$$1 + z = \frac{\omega_{\mathrm{sf}}}{\omega} = \frac{1}{f(r_{\mathrm{sf}})}\left[1 + \sqrt{\frac{r_{\mathrm{g}}}{r_{\mathrm{sf}}}\left(1 - \frac{b^2}{r_{\mathrm{sf}}^2}\right)}\right]. \tag{11.45}$$

If N is equal to 100, the redshift z suffered by a photon emitted from the surface of the collapsing dust ball and detected by a static observer far from the dust ball at $t = (n + 1)t_{\rm ff}$ is

$$1 + z|_{t=(n+1)t_{\rm ff}} \simeq \frac{2}{f(r_{\rm sf})|_{t=11t_{\rm ff}}} \simeq 10^{290n} > 10^{290}, \tag{11.46}$$

After a several times $t_{\rm ff}$ from $t = 0$, it is an extremely rare event for a distant static observer to detect a photon from the dust ball due to the elongation of the time interval by the redshift effect: one second at the surface of the dust ball corresponds to more than 10^{290} s at the observer, and hence if one photon is emitted from a point at the surface of the dust ball per one second, the observer detect them as one photon per more than 10^{290} s. Even if a photon arrives at the observer, its energy is too low to be detected due to the redshift effect, and hence the observer will recognize the collapsing dust ball as a completely dark object. The situation will be the same even if the collapsing object is not a dust ball. Thus the gravitationally collapsing object whose radius is almost equal to its gravitational radius may be misunderstood as a black hole, but we should note that it is not a black hole defined as Eq. (11.8).

The observer makes an image by detecting photons with various impact parameters. The image of a spherically symmetric dust ball is a disk shape. The impact parameter b of the photon at the center of the image is zero, whereas photons at off-center have non-zero impact parameters. The absolute value of the impact parameter of photons at the rim of the image is the largest. The photons of different impact parameters are emitted from the surface of the dust ball at different time. Hence, the photons with impact parameters different from each other suffer different redshift at the observer. In order to know the dependence of the redshift on the impact parameter in an image, numerical calculations are necessary. In Ref. [29], the time evolution of the image of a gravitationally collapsing dust ball was obtained. It is shown in Fig. 11.10.[3]

11.5.3 Black Hole Shadow

An impressive theoretical prediction related to a black hole is the black hole shadow first obtained in Ref. [26] as far as I know. It was reported that images of black hole shadows at the centers of M87 and Milky Way galaxies were successfully obtained [27, 28]. The black hole shadow is often described as a silhouette of a black hole caused by the bending of the orbit of light by the black hole, but this explanation is completely wrong. By definition of a black hole (11.8), any observable for observers outside the black hole can not relate to the black hole. The black hole shadow is

[3] Although Fig. 11.10 is an image of a bound dust ball obtained in Ref. [29], the image of a marginally bound dust ball considered here is expected to be basically the same.

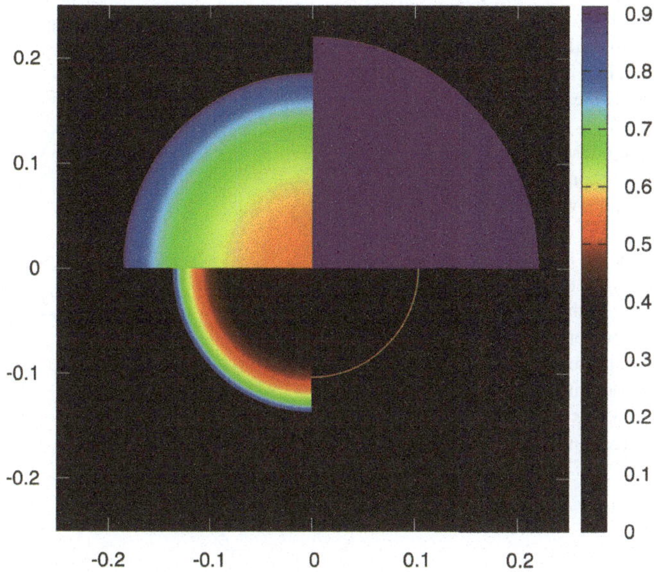

Fig. 11.10 Four images of a dust ball, each of which is taken at $t_o = 0,\ 35M,\ 50M$ and $69M$, are depicted from the first quadrant to forth quadrant on the black background (counterclockwise from top right). The color indicates the inverse of the redshift factor $1 + z$, where the gravitational collapse starts at $t = 0$. The photons near the center of the image are emitted at the moment later than the photons near the rim of the image. Furthermore, the photons near the rim suffer the kinematical blue shift. Hence the reddening near the central part is stronger than that near the rim, and hence the image looks a kind of rainbow (see Ref. [29] for detail)

a phenomenon that a dark image appears on a bright background if there are light sources around a black hole.

It is often explained that the contour of a black hole shadow is determined by the bound orbit of the photon, but as pointed out by R. Takahashi, the shape of the black hole shadow depends on the distribution of light source [30]. We consider the marginally bound Oppenheimer-Snyder model also here and adopt a setting proposed by Grala, Holz, and Wald [31]: the light source is a spherical bright screen surrounding both of the collapsing dust ball and an observer who is located outside the photon sphere which is generated by circular orbits of photons at $r = 3r_g/2$. In this setting, the boundary of the black hole shadow coincides with the photon sphere.[4]

Figure 11.11 shows a Penrose diagram of the situation mentioned above. We assume that the observer detects photons at the event O. The world line of a photon is a null geodesic as long as the photon is not scattered. The world line of the spherical bright screen is the blue curve. Null geodesics intersecting the event O are drawn as green and red curves. In this Penrose diagram, radial null geoodesics are straight

[4] Since we assume that the dust ball is marginally bound in our calculation, both of a spherical bright screen and the observer are in the dust ball in the past. However, we are not interested in such a situation, and focus on the situation in which both of them are outside the dust ball.

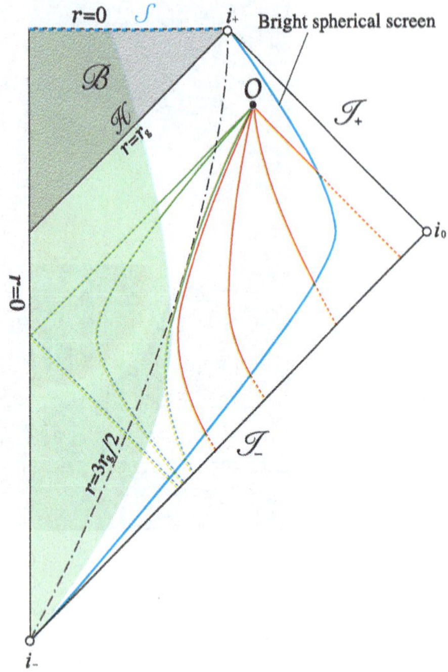

Fig. 11.11 The Penrose diagram of the spacetime in which a spherically symmetric black hole forms by the collapse of a marginally bound dust ball. The black hole is represented by a grayed domain, whereas the dust occupies the domain shaded by green. All null geodesics arriving at the event O intersect with the bright spherical screen. The null geodesics represented by green curves go through the dust ball, whereas those represented by red curves do not

lines with inclination ± 1, while non-radial null geodesics are curves with the absolute value of inclination greater than 1. The green null geodesics pass through the dust ball, whereas the red null geodesics do not. Note that all null geodesics reaching the observation event O intersect the bright screen in the past. If the dust ball is opaque to photons, there will be no photons propagating along the green null geodesics, and the solid angle spanned by those null geodesics will be a dark image. On the other hand, photons propagating along the red null geodesics can reach the observer, so the solid angle spanned by those null geodesics is a bright background. Even if the dust ball is transparent to photons, photons traveling through the dust ball are subjected to the redshift due to the Rees-Sciama effect [32]. We estimate how large it is, below.

The 4-momentum of the photon in the vacuum region is given by Eq. (11.31), but $\omega = \omega_{\rm in}$ and $b = b_{\rm in}$ from the bright screen to the surface of the dust ball, whereas $\omega = \omega_{\rm out}$ and $b = b_{\rm out}$ from the surface of the dust ball to the observer. In order to write down the continuity condition of the 4-momentum of the photon at the surface of the dust ball, we introduce the surface rest frame whose components with respect to the coordinate basis on the vacuum side are

$$e^{(0)}{}_{\mu} = \left(1, \frac{1}{f(r_{\rm sf})}\sqrt{\frac{r_{\rm g}}{r_{\rm sf}}}, 0, 0\right), \tag{11.47}$$

$$e^{(1)}{}_{\mu} = \left(\sqrt{\frac{r_{\rm g}}{r_{\rm sf}}}, \frac{1}{f(r_{\rm sf})}, 0, 0\right), \tag{11.48}$$

$$e^{(2)}{}_{\mu} = (0, 0, r_{\rm sf}, 0), \tag{11.49}$$

$$e^{(3)}{}_{\mu} = (0, 0, 0, r_{\rm sf}\sin\theta), \tag{11.50}$$

whereas those on the dust ball side

$$e^{(0)}{}_{\mu} = \big(a(\eta_{\rm sf}), 0, 0, 0\big), \tag{11.51}$$

$$e^{(1)}{}_{\mu} = \big(0, a(\eta_{\rm sf}), , 0, 0\big), \tag{11.52}$$

$$e^{(2)}{}_{\mu} = \big(0, 0, a(\eta_{\rm sf})\chi_{\rm sf}, 0\big), \tag{11.53}$$

$$e^{(3)}{}_{\mu} = \big(0, 0, 0, a(\eta_{\rm sf})\chi_{\rm sf}\sin\theta\big). \tag{11.54}$$

The radius of the dust ball when the photon enters it is denoted by $r_{\rm in} = a(\eta_{\rm in})\chi_{\rm sf}$, while the radius when the photon comes out from the dust ball is denoted by $r_{\rm out} = a(\eta_{\rm out})\chi_{\rm sf}$, where $\eta_{\rm in}$ and $\eta_{\rm out}$ have been given in Eqs. (11.26) and (11.27) and satisfy the relation (11.29). Then, we have, from the continuity of $k^{\mu}e^{(0)}{}_{\mu}$, $k^{\mu}e^{(1)}{}_{\mu}$ and $k^{\mu}e^{(3)}{}_{\mu}$ at $r_{\rm sf} = r_{\rm out}$,

$$\frac{\omega_{\rm out}}{f_{\rm out}}\left[1 + \sqrt{(1 - f_{\rm out})\left(1 - \frac{b_{\rm out}^2}{r_{\rm out}^2} f_{\rm out}\right)}\right] = \frac{\omega_{\rm c}\chi_{\rm sf}}{r_{\rm out}}, \tag{11.55}$$

$$\frac{\omega_{\rm out}}{f_{\rm out}}\left[\sqrt{1 - f_{\rm out}} + \sqrt{1 - \frac{b_{\rm out}^2}{r_{\rm out}^2} f_{\rm out}}\right] = \frac{\omega_{\rm c}\chi_{\rm sf}}{r_{\rm out}}\sqrt{1 - \frac{b_{\rm c}^2}{\chi_{\rm sf}^2}}, \tag{11.56}$$

$$\omega_{\rm out} b_{\rm out} = \omega_{\rm c} b_{\rm c} \tag{11.57}$$

and, from those at $r_{\rm sf} = r_{\rm in}$,

$$\frac{\omega_{\rm in}}{f_{\rm in}}\left[1 - \sqrt{(1 - f_{\rm in})\left(1 - \frac{b_{\rm in}^2}{r_{\rm in}^2} f_{\rm in}\right)}\right] = \frac{\omega_{\rm c}\chi_{\rm sf}}{r_{\rm in}}, \tag{11.58}$$

$$\frac{\omega_{\rm in}}{f_{\rm in}}\left[\sqrt{1 - f_{\rm in}} - \sqrt{1 - \frac{b_{\rm in}^2}{r_{\rm in}^2} f_{\rm in}}\right] = -\frac{\omega_{\rm c}\chi_{\rm sf}}{r_{\rm in}}\sqrt{1 - \frac{b_{\rm c}^2}{\chi_{\rm sf}^2}}, \tag{11.59}$$

$$\omega_{\rm in} b_{\rm in} = \omega_{\rm c} b_{\rm c}, \tag{11.60}$$

where we have assumed that k^r and k^{χ} are positive at $r = r_{\rm out}$, while those are negative at $r = r_{\rm in}$, and, for notational simplicity, we have introduced

$$f_{\rm out} = f(r_{\rm out}) \quad \text{and} \quad f_{\rm in} = f(r_{\rm in}). \tag{11.61}$$

Our goal is to obtain the radio ω_{in}/ω_{out} as a function of r_{out}, b_{out} and χ_{sf}. From Eqs. (11.55) and (11.58), we have

$$\frac{\omega_{in}}{\omega_{out}} = \frac{r_{out} f_{in}}{r_{in} f_{out}} \frac{1 + \sqrt{(1 - f_{out})\left(1 - \frac{b_{out}^2}{r_{out}^2} f_{out}\right)}}{1 - \sqrt{(1 - f_{in})\left(1 - \frac{b_{in}^2}{r_{in}^2} f_{in}\right)}}. \tag{11.62}$$

In order to achieve our goal, we need to know the dependence of r_{in} and b_{in} on r_{out} and b_{out}. We have

$$\frac{r_{in}}{r_{out}} = \frac{a_{in}}{a_{out}} = \left(\eta_{out} - 2\sqrt{\chi_{sf}^2 - b_c^2}\right)^2 = \left[1 + \sqrt{(1 - f_{out})\left(1 - \frac{b_c^2}{\chi_{sf}^2}\right)}\right]^2. \tag{11.63}$$

where we have used Eq. (11.29) in the 2nd equality and used the relation

$$\eta_{out} = -2\chi_{sf}\sqrt{\frac{r_{out}}{r_g}}, \tag{11.64}$$

which is derived from Eq. (11.21), in the final equality. Since b_c is included in Eq. (11.63), we need to write down b_c as a function of r_{out} and b_{out}. Equation (11.55) leads to

$$\omega_c = \frac{\omega_{out} r_{out}}{\chi_{sf} f_{out}} \left[1 + \sqrt{(1 - f_{out})\left(1 - \frac{b_{out}^2}{r_{out}^2} f_{out}\right)}\right]. \tag{11.65}$$

By substitution of this equation into Eq. (11.57), we have

$$\frac{b_c}{\chi_{sf}} = \frac{b_{out}}{r_{out}} f_{out} \left[1 + \sqrt{(1 - f_{out})\left(1 - \frac{b_{out}^2}{r_{out}^2} f_{out}\right)}\right]^{-1}. \tag{11.66}$$

This equation is just the one we need. Substituting Eq. (11.66) into Eq. (11.63), we have

$$\frac{r_{in}}{r_{out}} = \left(2 - \frac{f_{out}}{1 + \sqrt{(1 - f_{out})\left(1 + \frac{b_{out}^2}{r_{out}^2} f_{out}\right)}}\right)^2. \tag{11.67}$$

A remaining task is to eliminate b_{in} in Eq. (11.62). The summation of Eq. (11.58) multiplied by $\sqrt{1 - b_c^2/\chi_{sf}^2}$ and Eq. (11.59) leads to

$$\sqrt{1-\frac{b_{\rm in}^2}{r_{\rm in}^2}f_{\rm in}}=\frac{\sqrt{1-f_{\rm in}}+\sqrt{1-\dfrac{b_{\rm c}^2}{\chi_{\rm sf}^2}}}{1+\sqrt{(1-f_{\rm in})\left(1-\dfrac{b_{\rm c}^2}{\chi_{\rm sf}^2}\right)}}. \tag{11.68}$$

By using this equation, we can achieve our goal.

We are interested in photons which come out from the dust ball just before the formation of the event horizon, $0 < f_{\rm out} \ll 1$. In this case, Eq. (11.63) leads to $r_{\rm in} \simeq 4r_{\rm g}$ and Eq. (11.66) leads to $|b_{\rm c}/\chi_{\rm sf}| \simeq |b_{\rm out} f_{\rm out}/2r_{\rm g}| \ll 1$. Hence Eq. (11.68) leads to $\sqrt{1-b_{\rm in}^2 f_{\rm in}/r_{\rm in}^2} \simeq 1$. From these results and Eq. (11.62), we have

$$\frac{\omega_{\rm in}}{\omega_{\rm out}} \simeq \frac{3}{4f_{\rm out}} \gg 1. \tag{11.69}$$

The redshift factor $1+z$ of the photon, which is emitted at the bright screen $r = r_{\rm sc}$ and is detected by a static observer at $r = r_{\rm ob}$, is given as

$$1+z=\sqrt{\frac{f(r_{\rm ob})}{f(r_{\rm sc})}\frac{\omega_{\rm in}}{\omega_{\rm out}}}. \tag{11.70}$$

This result implies that photons that come out from the dust ball at the moment after several times the free-fall time from the beginning of the gravitational collapse suffer very large redshift of the same order as Eq. (11.46). Hence, even if the dust ball is transparent to photons, a dark image will appear. Here, it should be noted that, as in the case of the shinning dust ball, numerical study is necessary for drawing the dark image, because the knowledge about the dependence of the redshift on the impact parameter of the photon are required. Anyway, this dark image will be regarded as a black hole shadow.

Here, it should also be noted that there is not redshift but blue shift of a photon detected by a static observer enclosing the bright screen, if the spherical object is not collapsing but static, since $\omega_{\rm out}$ is equal to $\omega_{\rm in}$ in the static case. Anyway, it is an incorrect explanation that a black hole shadow is a silhouette of a black hole. As explained above, the black hole shadow is a silhouette of a collapsing object. Therefore, the black hole shadow will be observed even if the collapse eventually stops by some reason and forms something other than black hole, as long as the object continues to collapse in the observer's field of view.

The eternal black hole is usually assumed in theoretical studies of the black hole shadow. In this case, there is no collapsing object which forms the black hole, and hence the black hole shadow cannot be a silhouette of a collapsing object. The situation is represented by the Penrose diagram in Fig. 11.12 [10, 37]. This figure is basically the same as Fig. 11.11, but there is no collapsing object and the black hole is eternal. By contrast to the case of a black hole formed by a collapsing dust

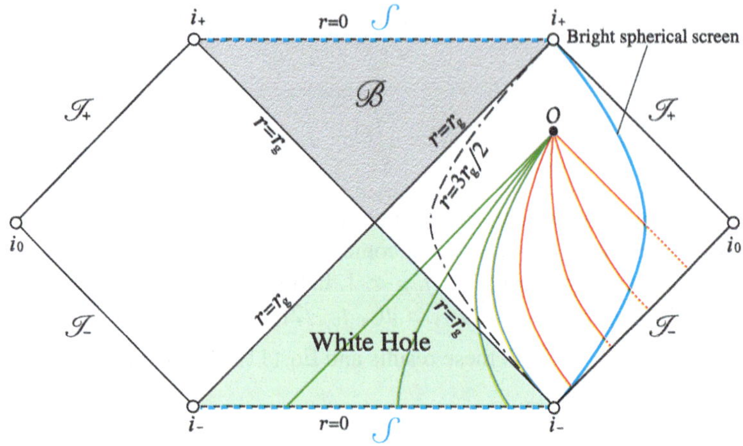

Fig. 11.12 This is the same as Fig. 11.11 but there is no collapsing object and the black hole is eternal. The black hole is represented by a grayed domain, whereas the white hole is the domain shaded by green. Red null geodesics arriving at the event O intersect with the bright spherical screen. By contrast, null geodesics represented by green curves come from the white hole

ball shown in Fig. 11.11, not all null geodesics arriving at the observation event O intersect the spherical bright screen in the past. The green null geodesics come from the white hole domain shaded by green, and there is no photon propagating along these null geodesics if we assume that nothing is emitted from the white hole. Thus, in theoretical studies with the assumption that the white hole emits nothing, the solid angle spanned by these green null geodesics is a dark image, i.e., a black hole shadow, whereas the solid angle spanned by the red null geodesics is a bright background.

11.5.4 Merger of Black Holes

By the astronomy invoking gravitational waves, many mergers of black hole binaries have been found [33]. Usually we think that black holes get mass through not only the accretions of ordinary matter but also mergers with other black holes. But it is worth to note that, exactly speaking, the merger process of black holes is a scenario which depends on the choice of the time coordinate.

In order to describe the merger process of black holes, we need to define a black hole at each moment. We consider the foliation by a family of Cauchy surfaces of the globally hyperbolic domain V introduced in Sect. 11.3, which includes $\mathscr{M} \cap J^-(\mathscr{I}_+)$ as a subset. We often call each Cauchy surface a time slice and call the foliation the time slicing. Then each Cauchy surface defines a moment. Black hole B at a moment equivalent to a time slice Σ is defined as $B := \mathscr{B} \cap \Sigma$. The number of black holes at a moment Σ is that of connected components of B. The merger of black holes is that of connected components of B.

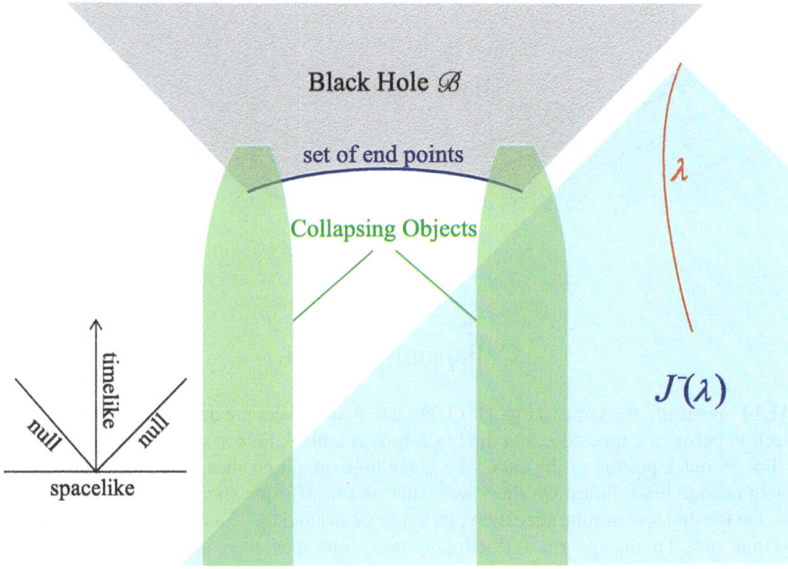

Fig. 11.13 The schematic diagram of a merger of black holes through a head-on collision. The vertical direction is timelike, whereas the horizontal direction is spacelike. The null directions are inclinations ± 1. The future direction is upward. The domain of the black hole is colored by gray. In this case, the black hole forms through the gravitational collapse of two massive objects colored by green. The set of end points of null geodesic generators of the event horizon is acausal and thus is represented by a dark blue spacelike curve. In the view field of the observer whose world line is λ, there is no black hole but two collapsing objects exist

Let us consider the merger of two black holes each of which forms through gravitational collapse of a massive object. A schematic diagram representing a head-on collision of non-rotating black holes is given in Fig. 11.13 in which the future direction is upward. The curves with the inclination equal to ± 1 are null. The surface colored by gray is a black hole. As mentioned, the event horizon, which is the boundary of the black hole, is generated by null geodesics. The null geodesic generators of the event horizon have end points in the past but no end point in the future. It should be noted that the set of the end points of the null geodesic generators is necessarily acausal, where a set is said to be acausal if there are no two points connected to each other by any causal curve [34]. Because of the acausal nature of the set of end points, there exists a time slicing with which two black holes merge with each other and form one black hole as shown in Fig. 11.14, but there exists another time slicing with which one highly deformed black hole forms and becomes eventually spherical (see Fig. 11.15), or with which three black holes first form and then merge with each other to one black hole (see Fig. 11.16). The number of black holes in a merger process is not invariant notion, but the only final number of black holes is invariant. Thus, exactly speaking, "gravitational waves generated by two black holes" is not an invariant statement but a story depending on the choice of the time slicing [35].

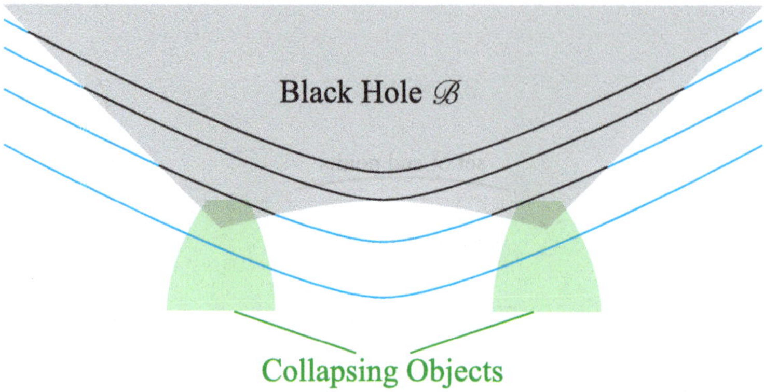

Fig. 11.14 Basically the same as Fig. 11.13, but a few time slices are depicted by solid curves. An intersection between a time slice and the black hole is a black hole at the moment defined by the time slice. A black portion of the curve is a black hole on a time slice, whereas a blue portion is a domain outside black holes. On the lowest (the past most) time slice, there are two collapsing bodies. On the 2nd lowest time slice, two black holes exist and they attach to each other on the 3rd lowest time slice. On the uppermost (the future most) time slice, there is one black hole

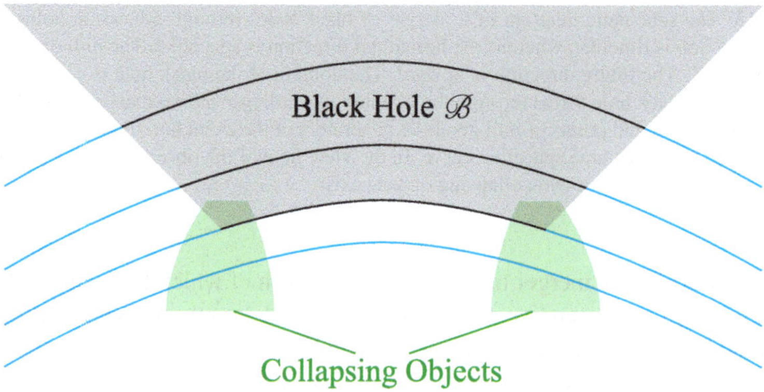

Fig. 11.15 The same black hole as that in Fig. 11.14, but with a different time slicing is depicted. On the lowest time slice, there are two collapsing bodies. On the 2nd lowest time slice, one highly deformed black hole forms and it will become spherical in future

We should also note that a binary of black holes must merge before making a half revolution [36]. This is not a surprising fact. What we observe through gravitational waves is the motion of two bodies each of which is collapsing forever in our view fields; the frequency of gravitational waves emitted in the so called inspiral phase is equal to the two times the orbital frequency of the binary system composed of not black holes but collapsing bodies.

Finally, I would like to point out a fact that black hole shadows of the binary black holes will never merge with each other [37]. This fact is also not so surprising, since the collapsing bodies do not coalesce with each other in the view field of

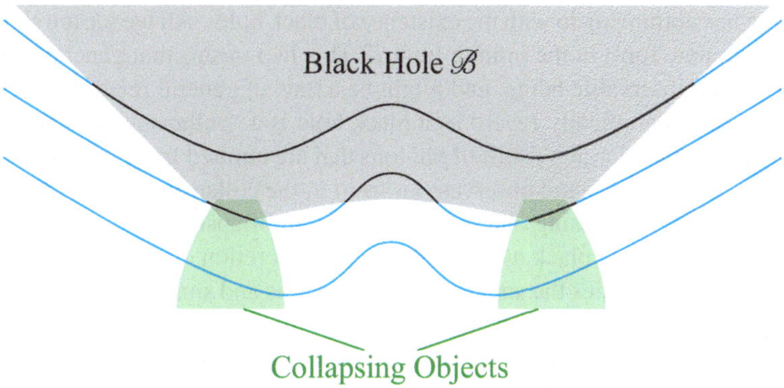

Fig. 11.16 The same black hole as those in Figs. 11.14 and 11.15, but with a different time slicing from them is depicted. On the lowest time slice, there are two collapsing bodies. On the 2nd lowest time slice, three black holes exist, and then they will eventually merge with each other to form a black hole

observer outside the black hole in the case of "the merger of black holes" as shown in Fig. 11.13. A black hole shadow is a silhouette of a collapsing body. However, in the binary case, due to the gravitational lensing effect, each collapsing body causes multiple black hole shadows [38, 39].[5]

11.6 Summary and Discussion

Gravitational collapse accompanied by the formation of a black hole and a spacetime singularity is one of the most interesting physical processes predicted by general relativity and fascinates many researchers for gravity. It is often asserted that the prediction of spacetime singularities is a flaw of general relativity and a motivation of the study about the theory beyond general relativity. From the standpoint of falsifiability [41], however, the formation of a spacetime singularity is not necessarily a serious problem. Even if spacetime singularities exist, all of *observable* phenomena are predicted by relativistic theory in globally hyperbolic spacetime. A fatal flaw is the prediction of non-globally hyperbolic spacetime structure. It is still an issue as a celebrated conjecture called the cosmic censorship hypothesis.

Black holes are the spacetime structure predicted by general relativity, and they are expected to exist in our universe. Results of various observations are often regarded as evidences for the existence of black holes, but they are only suggestions, by its definition. Since black holes never affect outside observers like us, what we are

[5] In Ref. [38, 39], the black hole shadows in the Kator-Traschen multi-black hole solution is studied. The Kastor-Traschen solution describes the merger of eternal black holes which do not form through gravitational collapse of multiple bodies. In this case, a black hole shadow corresponds to an image of a white hole. (see Appendix A in Ref. [37] for details.).

observing has nothing to do with the existence of black holes. All we see are signs that a black hole may form in the infinite future.[6] This fact means that general relativity predicts the unobservable being, and might be a flaw of general relativity [42].

What we astronomically regard as a black hole is a "collapsing domain" which can cause extremely large redshift of photons that are emitted from or going through it and then detected by static observers far from it; the collapsing domain is a region in which there is a collapsing object, or are multiple collapsing objects as in the case of the merger of binary black holes, possibly with accretion of matter and radiation. The large redshift implies the small energy of photons and small number of detected photon per unit time. Thus, as mentioned, after several times the free-fall time from the beginning of gravitational collapse, a distant static observer cannot obtain any information of the collapsing domain like as in the case of a black hole. Here, we should note that if the observer is not static but falling to the collapsing domain with extremely high speed, such moving observer can, in principle, get information about collapsing domain through the detection of photon by virtue of the large blue shift effect due to the kinematic Doppler effect which cancels the gravitational redshift. Thus, the collapsing domain can behave as a black hole for not all observers.The important prediction of general relativity which can be observationally confirmed is that the geometry around the collapsing domain eventually is well described by the Kerr-Newmann solution.

The black hole is regarded as a key in the study about a new theory beyond general relativity. The corresponding concept of a black hole in the framework of the new theory must still be the unobservable being which is the very definition of a black hole in general relativity. The new theory would be established in the quantum formulation. If the quantum theory is to treat only observables, the corresponding notion to the black hole in general relativity does not exist in the new theory.

Acknowledgements I would like to thank Tomohiro Harada, Daisuke Ida, Hideki Ishihara, Hirotaka Yoshino, Kensuke Sueto and colleagues in astrophysics and gravity group at Osaka Metropolitan University for useful discussions. I am deeply grateful would like to Deniele Marafarina for carefully reading the manuscript. This work was partly supported by JSPS KAKENHI Grants No. JP21K03557 and MEXT Promotion of Distinctive Joint Research Center Program JPMXP0723833165.

References

1. C.M. Will, N. Yunes, *Is Einstein Still Right?: Black Holes, Gravitational Waves, and the Quest to Verify Einstein's Greatest Creation* (Oxford University Press, 2020)
2. R.M. Wald, *General Relativity* (University of Chicago Press, 1984)
3. S.W. Hawking, Black holes in general relativity. Commun. Math. Phys. **25**, 152 (1972)

[6] The formation of a black hole is not a necessary condition for the thermal spectrum of the quantum radiation, but a collapsing object which does not necessarily form a black hole causes the quantum radiation with thermal spectrum [40]. Black holes do not affect the observer outside the black holes even through the quantum fields, as long as the quantum fields are described by relativistic theories.

4. R. Penrose, Gravitational collapse: the role of general relativity. Riv. Nuovo Cim. **1**, 252–276 (1969); reprinted in Gen. Relativ. Gravit. **34**, 1141 (2002). https://doi.org/10.1023/A:1016578408204
5. J.D. Bekenstein, Black holes and entropy. Phys. Rev. D **7**, 2333 (1973)
6. T. Padmanabhan, Thermodynamical aspects of gravity: new insights. Rep. Prog. Phys. **73**, 046901 (2010). arXiv: 0911.5004
7. PLANCK collaboration, Planck 2018 results. VI. Cosmological parameters. Astron. Astrophys. **641**, A6 (2018). [Erratum ibid. **652**, C4 (2021)]
8. T. Shiromizu, K. Nakao, H. Kodama, K. Maeda, Can large black holes collide in de Sitter space-time?: an inflationary scenario of an inhomogeneous universe. Phys. Rev. D **47**, R3099 (1993)
9. V. Cardoso, P. Pani, Nat. Astron. **1**, 586 (2017)
10. K. Nakao, C.-M. Yoo, T. Harada, Gravastar formation: what can be the evidence of a black hole? Phys. Rev. D **99**, 044027 (2019)
11. R. Penrose, in *General Relativity, an Einstein Centenary Survey*, ed. by S.W. Hawking, W. Israel (Cambridge University Press, Cambridge, England, 1979), p. 581
12. D.M. Eardley, L. Smarr, Time functions in numerical relativity: marginally bound dust collapse. Phys. Rev. D **19**, 2239 (1979)
13. D. Christodoulou, Violation of cosmic censorship in the gravitational collapse of a dust cloud. Commun. Math. Phys. **93**, 171 (1984)
14. T.P. Singh, P.S. Joshi, The final fate of spherical inhomogeneous dust collapse. Class. Quantum Gravity **13**, 559 (1996)
15. S. Jhingan, P.S. Joshi, The structure of singularity in spherical inhomogeneous dust collapse. Ann. Isr. Phys. Soc. **13**, 357 (1997)
16. M. Dafermos, I. Rodnianski, A proof of Price's law for the collapse of a self-gravitating scalar field. arXiv:gr-qc/0309115
17. M. Dafermos, J. Luk, The interior of dynamical vacuum black holes I: the C^0-stability of the Kerr Cauchy horizon. arXiv:1710.01722
18. H. Maeda, Quest for realistic non-singular black-hole geometries: regular-center type. JHEP **11**, 108 (2022)
19. J.M. Bardeen, Non-singular general-relativistic gravitational collapse, in *Proc. Int. Conf. GR5*, Tbilisi (1968), p. 174
20. S.W. Hawking, Particle creation by black holes. Commun. Math. Phys. **43**, 199 (1975)
21. B.P. Abbott et al., [LIGO Scientific and Virgo], Observation of gravitational waves from a binary black hole merger. Phys. Rev. Lett. **116**, 061102 (2016). https://doi.org/10.1103/PhysRevLett.116.061102. arXiv:1602.03837 [gr-qc]
22. M. Miyoshi, J. Moran, J. Herrnstein, L. Greenhill, N. Nakai, P. Diamond, M. Inoue, Evidence for a black hole from high rotation velocities in a sub-parsec region of NGC4258. Nat. **373**, 127 (1995)
23. R. Bender et al., HST STIS spectroscopy of the triple nucleus of M31: two nested disks in Keplerian rotation around a supermassive black hole. Astrophys. J. **631**, 280–300 (2005)
24. A.M. Ghez, S. Salim, S.D. Hornstein, A. Tanner, J.R. Lu, M. Morris, E.E. Becklin, G. Duchêne, Stellar orbits around the galactic center black hole. Astrophys. J. **620**, 744 (2005)
25. S. Gillessen, F. Eisenhauer, S. Trippe, T. Alexander, R. Genzel, F. Martins, T. Ott, Monitoring stellar orbits around the massive black hole in the galactic center. Astrophys. J. **692**, 1075 (2009)
26. J.P. Luminet, Image of a spherical black hole with thin accretion disk. Astron. Astrophys. **75**, 228 (1979)
27. T.E.H.T. Collaboration, First M87 event horizon telescope results. I. The shadow of the supermassive black hole. Astrophys. J. Lett. **875**, L1 (2019)
28. T.E.H.T. Collaboration, First Sagittarius A* Event Horizon Telescope results. I. The shadow of the supermassive black hole in the center of the Milky Way. **930**, L12 (2022)
29. H. Yoshino, K. Takahashi, K. Nakao, How does a collapsing star look? Phys. Rev. D **100**, 084062 (2019)

30. R. Takahashi, Shapes and positions of black hole shadows in accretion disks and spin parameters of black holes. Astrophys. J. **611**, 996 (2004)
31. S.E. Gralla, D.E. Holz, R.M. Wald, Black hole shadows, photon rings, and lensing rings. Phys. Rev. D **100**, 024018 (2019)
32. M.J. Rees, D.W. Sciama, Nat. **217**, 511 (1968)
33. The LIGO Scientific Collaboration, the Virgo Scientific Collaboration, KAGRE Scientific Collaboration, GWTC-3: compact binary coalescences observed by LIGO and Virgo during the second part of third observing run. arXiv: 2111.03606
34. M. Siino, Topology of event horizons. Phys. Rev. D **58**, 104016 (1995)
35. D. Ida, K. Nakao, S.A. Hayward, M. Siino, Hoop conjecture for colliding black holes. Phys. Rev. D **58**, 121501 (1998)
36. M. Siino, D. Ida, Half-period theorem of binary black holes. Prog. Theor. Phys. **122**, 1015 (2009)
37. K. Okabayashi, N. Asaka, K. Nakao, Do black hole shadows merge? Phys. Rev. D **102**, 044011 (2020)
38. D. Nitta, T. Chiba, N. Sugiyama, Shadows of colliding black holes. Phys. Rev. D **84**, 063008 (2011)
39. A. Yumoto, D. Nitta, T. Chiba, N. Sugiyama, Shadows of multi-black holes: analytic exploration. Phys. Rev. D **86**, 103001 (2012)
40. C. Barceló, S. Liberati, S. Sonego, M. Visser, Hawking-like radiation from evolving black holes and compact horizonless objects. J. High Energy Phys. **1102**, 03 (2011). arXiv:1011.5911 [gr-qc]
41. K.R. Popper, *The Logic of Scientific Discovery* (First published in 1935, English edition in 1959)
42. I. Hacking, *Representing and Intervening* (Cambridge University Press, 1983)

Chapter 12
Naked Singularities and Black Hole Killing Horizons

Daniela Pugliese and Hernando Quevedo

12.1 Introduction

The creation of a naked singularity (**NS**) through accretion onto a (extreme) black hole (**BH**) is forbidden. In particular, the horizon area cannot decrease in any process with suitable conditions on the energy and no repulsive gravity. More precisely, the total horizon area of (classical) **BHs** cannot decrease over time (Hawking's area theorem—[25, 27, 28]). Observational data of GW150914 have been recently considered as an observational confirmation of Hawking's black-hole area theorem—[34]. However, assuming the existence of **NS**s, the reverse process, consisting in the conversion, at some point of the singularity formation or evolution, into a **BH** is still debated in different contexts. A classical method for extracting energy from a spinning **BH** and from a hypothetical **NS** is, for example, the Penrose process, which is based on the frame-dragging effects of the ergoregion of a spinning **BH**—[44]. In this region, a small body may have negative energy, as measured by an observer at infinity and, if fragmented into two parts, and the part with negative-energy will be swallowed by the **BH**, it will decrease its (total) mass, while the other part, with positive energy, can be ejected. In this process, the Kerr **BH** rotational energy decreases and consequently, the Kerr **BH** area (determined by its outer horizon) increases.

In this chapter, we focus on **BH** spacetimes with Killing horizons and on naked singularity solutions of the corresponding field equations. It is possible to study the

D. Pugliese (✉)
Research Centre for Theoretical Physics and Astrophysics, Institute of Physics, Silesian University in Opava, Bezručovo náměstí 13, 74601 Opava, Czech Republic
e-mail: daniela.pugliese@physics.slu.cz

H. Quevedo
Instituto de Ciencias Nucleares, Universidad Nacional Autónoma de México, AP 70543, 04510 Ciudad de México, Mexico
e-mail: quevedo@nucleares.unam.mx

D. Malafarina and P. S. Joshi (eds.), *New Frontiers in Gravitational Collapse and Spacetime Singularities*, Springer Series in Astrophysics and Cosmology,
https://doi.org/10.1007/978-981-97-1172-7_12

singularities using light surfaces related to the background Killing fields, and in particular to the generators of Killing horizons. Special null orbits associated to the Killing vectors of the geometry can generate a useful framework for the study of the more diverse geometries, such as wormholes, accelerating black holes, or binary black hole systems,[1] pointing out the existence of common properties of the light surfaces in different spacetimes.

Here, we will use the properties of the Killing vectors, generators of the **BH** Killing horizons, limiting ourselves to the cases of the Kerr and Kerr–Newman spacetimes.

The Kerr-Newman (**KN**) solution is an electrically charged, electro–vacuum, asymptotically flat, axially symmetric (and stationary) solution of the Einstein–Maxwell equations, used to describe the spacetime around an electrically charged gravitational source, spinning along it symmetry axis. The Kerr solution can be read as a limit of the **KN** metric when the electric charge is null. According to the "no–hair" theorems, the Einstein–Maxwell **BH** solutions are characterized only by mass, electric charge, and angular momentum; therefore the Kerr–Newman spacetime is the most general asymptotically flat stationary **BH** solution of the Einstein–Maxwell equations. On the other hand, there are other stationary axisymmetric solutions of the Einstein–Maxwell equations with additional parameters, such as acceleration, magnetic charge, NUT charge, and cosmological constant, having therefore up to seven parameters. Furthermore, the theorem has been proved to be violated, for example, by the self–gravitating Yang–Mills solitons or dilaton fields with asymptotically–defined global charges. The **KN** spacetime has also as a limiting case the static and electrically charged Reissner–Nordström solution. The Schwarzschild spacetime is a static, neutral and spherically symmetric solution.[2]

---

[1] See [53] for the study of the light surfaces in static and spinning wormhole solutions, black holes immersed in external (perfect fluid) dark matter, spacetimes with (Taub) NUT charge, acceleration, magnetic charge, and cosmological constant, binary Reissner–Nordström black holes, a solution of a (low–energy effective) heterotic string theory, and the $(1+2)$ dimensional BTZ geometry, where there are cosmological and acceleration horizons, or internal solutions matching exact vacuum solutions of Einstein equations. Properties of the **BH** thermodynamics for a Mini-Super-Space, semi-classical polymeric **BH** have been also discussed in [47] in this framework.

[2] In General Relativity, any spherically symmetric solution in vacuum must be static (and asymptotically flat), implying that the vacuum exterior solution of a spherically symmetric gravitating source must be the Schwarzschild metric, which is the content of the well known Birkhoff's theorem [5]. Anticipating what is known today as **BH** uniqueness, the inverse of this formulation was the first formulation (grounded on different conceptual bases) of Israel's (no–hair) theorem [35, 36]. However, Birkhoff's theorem has been generalized in different formulations. In particular, any spherically symmetric and asymptotically flat solution of the Einstein–Maxwell field equations, is static, picturing the case of the Reissner–Nordström electro–vacuum solution. (On the other hand, a not asymptotically flat Einstein–Maxwell spherically symmetric solution, is for example the Bertotti-Robinson universe [56]). A further consequence of the Birkhoff theorem is that a (charged) spherical (uniform) shell of dust has flat spacetime inside the shell. If a star is not static but it is spherically symmetric, the solution outside must be the Schwarzschild solution if electrically neutral. Spherical symmetry implies stationarity. (In other words, the Schwarzschild metric can be derived without imposing the metric stationarity). Furthermore, if the central source has a (radial) time evolution (radially pulsating, spherical collapse, etc.), the exterior metric is the Schwarzschild (static) spacetime. Hence, there are no gravitational waves emitted from a spherically symmetric

The **KN BH** horizons are Killing horizons. A Killing horizon is a hypersurface where a Killing vector of the metric becomes null. Or, more precisely, a Killing horizon is a null hypersurface, whose null tangent vector can be normalized to coincide with a Killing vector field. In other words, a Killing horizon is a null surface, whose normal is a Killing vector field. Null geodesics, whose tangent vectors are normal to a null hypersurface N are called generators of N.[3] In particular, the event horizon of a stationary and asymptotically flat **BH** geometry is a Killing horizon (Hawking rigidity theorem) with respect to a Killing vector, say $\mathscr{L}^a$. We shall consider the null vector $\mathscr{L}^a$ in *all* points of the spacetime, including the case of naked singularities. The light surfaces considered here are null hypersurfaces defined by Killing vectors and characterized by a constant parameter ω, which is the photon orbital frequency.

Within this framework, we can explore properties enfolding along the entire family of geometries, and the notion of Killing horizons are reconsidered from a special perspective, where **NS** solutions are related to **BH** solutions.

The relevance of this new representation, together with the possible conceptual significance, lies also in the fact that the entire family of solutions is considered as a *unique geometric object*. In this new framework, we can find properties of the spacetime geometries where several geometric quantities, such as the **BH** horizons, acquire a completely different significance, when considered for the entire family, where naked singularity solutions, for example, are related to the **BH** horizons.

A key element in the definition of this frame is, therefore, the **BH** Killing horizons. Horizons define **BHs**, fixing their geometrical, thermodynamical and even quantum properties. Bekenstein first suggested the idea that the event horizon defines the **BH** entropy–[2, 3]. The **BH** thermodynamic properties acquire, therefore, a purely geometric meaning. A **BH** macrostate is defined and determined only by the mass M, spin J, and electric charge Q, while the number of **BH** microstates increases with the horizon area, which is a function of the outer horizon. Hence, the **BH** state may correspond to a large number of microstates, leading to a very high **BH** entropy.

In a very multifaceted sense, **BH** horizons define also its quantum properties. The progenitor collapse into a **BH** appears as an irreversible process of information degeneration, where the information becomes inaccessible to the distant observers and the **BH** has no observable hair. However, if **BHs** have an entropy and a temperature, this would imply the possibility of **BH** radiation. The Hawking radiation (tightly connected to the Unruh effect) is explained by effects of quantum fluctuation of matter fields in the region close to the horizon, which possibly could be observed in the thermal profile. The derivation of Hawking radiation is, however, semi-classical and yet does not determine if the **BH** can evaporate completely or some remnants are left (in this case, the emitted radiation may be entangled with the **BH** remnant,

---

system. (A similar symmetry prevents a spherically symmetric distribution of charges or currents to radiate. No monopole electromagnetic radiation is possible, the Coulomb solution is the only spherically symmetric solution of Maxwell's equations in vacuum). On the other hand, it has been also shown that static **BHs** are not necessarily spherically symmetric or axially symmetric and viceversa that non-rotating **BHs** are not necessary static [6, 7, 39, 53, 55].

[3] A portion of a null hypersurface can be also the Killing horizon of two or more independent Killing vectors–[41].

which should have entropy). This fact is at basis of the so called **BH** information paradox. A possibility is that the final radiation of the complete evaporation does not "bring" information of the initial matter state (see, for example, [46] for a review on the **BH** information problem).

However, a **BH** singularity is an "active" gravitational object in the sense that it can interact with its environment constituted by matter and fields. Then, a **BH** transition from a state to another, corresponding to a change of the **BH** characteristic parameters, is regulated by the laws of **BH** thermodynamics. Such transition may follow, for example, by a **BH** energy extraction, which can be detectable by observing jet emissions. Therefore, we will explore in this framework also the concepts of **BH** thermodynamics.

The light surfaces considered here define particular structures known as metric Killing bundles (or more simply metric bundles–**MB**s), enlightening some properties of the spacetime causal structure as spanning in different geometries. Metric bundles collect all the geometries of the same metric family having a particular characteristic in common. In the Kerr spacetime, for example, metric bundles collect all geometries having equal photon circular orbiting frequency. One advantage of the **MB**s is the establishment of a new framework of analysis where the entire family of metric solutions is studied as a unique geometric object, and the single spacetime is a part of the plane (*extended plane*), where **MB**s are defined as curves tangent to the Killing horizons curve, which is the curve in the extended plane representing all the **BH** horizons. In this way, the properties of the spacetime solutions are studied as unfolding across the spacetimes of the plane. Different geometries are, therefore, related through metric bundles. The extended plane establishes also a **BH**–**NS** connection, through the tangency condition, providing a global frame, in particular, for the study of **NS** solutions [47–49, 52, 54]. At the base of this relation, for example, in the Kerr spacetime, is the fundamental property that all the **MB**s photon circular orbits are the frequency of an inner or outer Kerr **BH** horizon. These photon circular orbits are known as horizon *replicas*. A consequence of this fact is that the horizons curves emerge as the envelope surfaces of all the curves associated to the **MB**s in the extended plane.

In details, the structure of this chapter is as follows: in Sect. 12.2, we introduce the Kerr-Newman geometry. In Sect. 12.3, we build the **MB**s framework using the properties of special light surfaces and the **BH** horizons, discussing some of their characteristics and relevance for **BH** and **NS** spacetimes, while in the following sections we detail these definitions for the Kerr geometry. We start in Sect. 12.3.1 examining the Killing horizon definitions in the context of metric bundles. Stationary observers and metric bundles are the focus of Sect. 12.3.2. Light surfaces are investigated in Sect. 12.3.3. **MB**s characteristics in the extended plane of the Kerr geometries are explored in Sect. 12.3.4. Finally, we close this section with the analysis of some limiting cases in Sect. 12.3.5. In Sect. 12.4, we detail the **BH** thermodynamic properties in terms of metric bundles. Sect. 12.4.1 discusses the role of metric bundles in **BH** thermodynamics, focusing on the concept of **BH** surface gravity, the first law of **BH** thermodynamics and concluding with some notes on the **BH** transitions as described by the laws of **BH** thermodynamics. Following this analysis, in

Sect. 12.4.2, we explore **BH** transitions in the extended plane. The **BH** irreducible mass is the subject of Sect. 12.4.3, where we examine the extraction of **BH** rotational energy as constrained with **MB**s. Sect. 12.4.4, **BH** thermodynamics in the extended plane, closes this section. Finally, Sect. 12.5 contains some concluding remarks.

12.2 The Kerr-Newman Metric

Although in this chapter we will focus mainly on the properties of the Kerr metric, it is useful to consider the Kerr geometry as a limiting case of the more general Kerr-Newman solution.

The Kerr-Newman (**KN**) spacetime is a solution of the Einstein–Maxwell equations, for an electro-vacuum, asymptotically flat spacetime describing the geometry surrounding a rotating, electrically charged mass, with mass parameter M, charge parameter Q, and dimensionless spin parameter $a \equiv J/M$ (the specific angular momentum, the rotational parameter associated to the central singularity), while J is the total angular momentum.

The line element in Boyer–Lindquist (**BL**) coordinates (t, r, θ, ϕ) is[4]:

$$ds^2 = \frac{\sin^2\theta\left[d\phi\left(a^2+r^2\right)-adt\right]^2}{\Sigma} - \frac{\Delta\left(ad\phi\sin^2\theta - dt\right)^2}{\Sigma} + \frac{\Sigma dr^2}{\Delta} + \Sigma d\theta^2, \tag{12.1}$$

with $r \in [0, +\infty)$, $t \in [0, +\infty)$, $\theta \in [0, \pi]$ and $\phi \in [0, 2\pi]$, where $G = c = 1$ and

$$\Delta \equiv a^2 + Q^2 + r^2 - 2Mr \quad \text{and} \quad \Sigma \equiv a^2\cos^2\theta + r^2. \tag{12.2}$$

In the following analysis, we shall use also the parameter $\sigma \equiv \sin^2\theta \in [0, 1]$. Let us introduce the total charge:

$$Q_T \equiv \sqrt{a^2 + Q^2}. \tag{12.3}$$

The **KN** metric describes black hole (**BH**) solutions for $Q_T \leq M$, with outer and inner horizons

$$r_\pm \equiv M \pm \sqrt{M^2 - Q_T^2}, \tag{12.4}$$

respectively in terms of the total charge. Extreme **KN BHs** are for $Q_T = M$, where $r_+ = r_- = M$. The metric (12.1) describes **NSs** for $Q_T > M$.

[4] For the seek of simplicity, where convenient, we adopt geometrical units with $c = 1 = G$. The radius r has unit of mass $[M]$, and the angular momentum units of $[M]^2$, the velocities $[u^t] = [u^r] = 1$ and $[u^\phi] = [u^\theta] = [M]^{-1}$ with $[u^\phi/u^t] = [M]^{-1}]$ and $[u_\phi/u_t] = [M]$. Latin and Greek indices run in $\{0, 1, 2, 3\}$.

The **KN** metric reduces to the Kerr solution for $Q = 0$, where $Q_T = a$ (vacuum exact solution of the Einstein equations describing an axisymmetric, stationary, asymptotically flat spacetime).

For $a = 0$ the **KN** metric is the electrically charged, spherically symmetric, asymptotically flat Reissner–Nordström (**RN**) solution, where $Q_T = Q$. The static, electrically neutral and spherically symmetric asymptotically flat Schwarzschild geometry is the limit for $Q = 0$ and $a = 0$ (i.e. $Q_T = 0$) where the **BH** horizon is $r = 2M$.

The ergosurfaces of the **KN** spacetimes are defined by the zeros of the norm of the Killing vector field $\xi^a_{(t)} = \partial_t$, which in **BL** coordinates are

$$r_\epsilon^\pm \equiv M \pm \sqrt{M^2 - Q^2 - a^2 \cos^2\theta}, \tag{12.5}$$

for the outer and inner ergosurfaces, respectively (the Killing vector $\xi^a_{(t)}$ is space-like in the region $r \in]r_\epsilon^-, r_\epsilon^+[$). In the spherically symmetric cases ($a = 0$), we obtain $r_\epsilon^\pm \equiv r_\pm$ and there are no ergosurfaces.

In the following sections, where convenient in complex expressions, we adopt dimensionless quantities where $r \to r/M$ and $a \to a/M$.

12.3 Light-Surfaces and Metric Bundles

In this section, we build the metric bundles (**MB**s) framework using the properties of special light surfaces and the **BH** horizons. We will introduce the concept of metric bundles on the basis of some spacetime light surfaces. We will illustrate some of their characteristics and relevance for **BH** and **NS** spacetimes. In the next sections, we will precise these definitions for the Kerr geometry, discussing details and applications.

First introduced in [49] for the Kerr geometries, framed in the analysis of properties of the Kerr **BHs** and **NSs**, **MB**s can be generally defined in spacetimes with Killing horizons (for generalizations to other spacetimes with Killing fields, see [47, 53]).

The idea is to bundle geometries according to some particular characteristics common to all the geometries of the bundle. This formalism, therefore, allows to explore properties in the bundled metrics from a special, "global", perspective, studying spacetime properties as unfolding along the entire set of solutions. Bundles enlighten properties attributable to different points of spacetime and, on the other hand, connect different geometries, and in particular connect points of **BHs** and **NSs** spacetimes. These properties can be measured through the observation of special light-like orbits.

In this framework, we introduce the concept of extended plane which, in brief, can be defined as a graphic representing the collection of metrics (**MB**s curves) related by a common property. We specify below the details of these definitions, while in the next sections we develop in details the set-up and applications.

For the Kerr geometry, a metric Killing bundle is a collection of Kerr spacetimes characterized by a particular photon circular frequency, ω (bundles characteristic frequency), at which the norm of the stationary observers four-velocity vanishes. We detail this aspect in Sect. 12.3.2. It is straightforward to prove that ω is also the frequency (angular velocity) of a **BH** (inner or outer) horizon of the bundle. A metric bundle is represented by a curve in the *extended plane*, i.e., a plane $\mathscr{P} - r$, where $\mathscr{P}$ is a parameter of the spacetime. In the Kerr geometry $\mathscr{P}$, for example, is the dimensionless spin, where r is the (dimensionless) radial Boyer-Lindquist coordinate. Thus, an extended plane represents all the metrics of the same spacetime family so that, by varying the parameter $\mathscr{P}$, we can extend a particular analysis to include all metrics. In Fig. 12.1–upper panel, we represent the extended plane of the Kerr spacetime on the equatorial plane $\sigma = 1$.

Concentrating on the Kerr spacetime, the metrics of one metric Killing bundle with characteristic orbital frequency $\bar{\omega}$ are all and the only Kerr (**BHs** or **NSs**) spacetimes, where at some point r the light-like (circular) orbital frequency is $\omega = \bar{\omega}$. In the extended plane, all the curves associated to the **MB**s (bundle curves) are tangent to the horizons curve (the curve representing in the extended plane the Killing horizons of all Kerr **BHs**). Then, this tangency condition implies that each bundle characteristic frequency ω coincides with the frequency of a **BH** Killing horizon. Consequently, the horizons in the extended plane emerge as the envelope surface of the collection of all the metric bundles curve–(Fig. 12.1–upper panel).

The metric bundles of the Kerr spacetimes contain either **BHs** or **BHs** *and* **NSs**. Therefore, it is possible to find a **BH-NS** correspondence through the tangency condition of the **MBs** to the horizons curve in the extended plane, providing also an alternative interpretation of **NSs** and **BHs** horizons.

In fact, the exploration of **MBs** singles out some fundamental properties of the **BHs** and **NSs** solutions, which are related, in particular, to the thermodynamic properties of **BH** spacetimes. We shall explore this aspect in Sect. 12.4.

The analysis of light surfaces defining **MB**s opens the possibility to extract information about the **BH** horizons, i.e., to detect properties which are directly attributable to the presence of a **BH** horizon. From Kerr metric bundles, we define the "horizons replicas" special orbits of a Kerr spacetime with photon (circular) orbital frequency equal to the **BH** (inner[5] or outer) horizon frequency, which coincides with the characteristic frequency of (one of the two) bundle(s) crossing the point on the extended plane (r, a) corresponding to the replica orbit r on of the spacetime with spin parameter a (at fixed poloidal angle θ)–Figs. 12.2, 12.3, 12.4.

Therefore, replicas could be detected, for example, from the spectra of electromagnetic emissions coming from **BHs** and, in particular, from locations close to the **BH** rotational axis–[50, 51].

On the other hand, for the Kerr geometry it was pointed out that **NSs** with spin-mass ratio $a = J/M$ close to the value of the extreme **BH**, were related, in the extended plane, to **MBs** curves tangent to a portion of the inner horizon, and faster

[5] Note, the inner horizon has been shown to be unstable and it has been conjectured it could end into a singularity [11, 58].

spinning **NSs**, with $a > 2M$, to the outer horizons curve. This property of the **NS** spacetimes in the extended plane explains a peculiarity emerging in some light surfaces.

The light surfaces defining **MBs**, functions of the spin parameter a and the frequency ω, bound a region in the plane $r - \omega$ for **NS** spacetimes called Killing throats (tunnels). In **NS** geometries, a Killing throat or tunnel is a connected and bounded region in the $r - \omega$ plane, containing all the stationary observers allowed within two limiting frequencies $\omega \in]\omega_-, \omega_+[$. On the other hand, in the case of **BHs**, a Killing throat is either a disconnected region or a region bounded by singular surfaces in the extreme Kerr **BH** spacetime. We detail this aspect in Sect. 12.3.3.

Slowly spinning **NSs** are characterized by "restrictions", called "Killing bottlenecks", of the Killing tunnel, which are related to metric bundles in the extended plane, through the **MB** curve tangency condition to the horizons curve [49]. The **BH** extreme Kerr spacetime, therefore, represents the limiting case of the Killing bottleneck (as defined in the **BL** coordinates), where the tunnel narrowing closes on the **BH** horizon–Fig. 12.1–lower panel. Killing bottlenecks were also connected with the concept of pre-horizon regime introduced in [15, 18]. The pre-horizon regime was analyzed in [18]. It was shown that a gyroscope would conserve a "memory" of the static or stationary initial state [9, 16, 19, 21]. Adopting a similar idea of plasticity, bottlenecks have also been named horizon remnants. Killing throats and bottlenecks were also grouped in [62] in structures named "whale diagrams" of the Kerr and Kerr-Newman spacetimes–see also [42, 65].

Below, we precise the **MBs** definitions for Kerr spacetimes and relate them explicitly to quantities of importance in **BH** thermodynamics. We discuss the relations between **MBs**, stationary observes, and light surfaces, which are used to constrain many processes associated to the physics of jet emission, accretion disks, and energy extraction from **BHs**.

In details, in Sect. 12.3.1, we discuss the Killing horizon definitions in relation to metric bundles. Stationary observers and metric bundles are the focus in Sect. 12.3.2. Light-surfaces and bottlenecks are investigated in Sect. 12.3.3. **MB**s characteristics in the Kerr extended plane are explored in Sect. 12.3.4. Finally, we close this section with the analysis of some limiting cases in Sect. 12.3.5.

12.3.1 Killing Horizons

Let us introduce the Killing vector $\mathscr{L}^a = \xi^a_{(t)} + \omega \xi^a_{(\phi)}$ with $\xi^a_{(t)} = \partial_t$ representing the stationarity of the background and the rotational Killing field $\xi^a_{(\phi)} = \partial_\phi$ of the KN spacetime. The quantity $\mathscr{L}_{\mathscr{N}} \equiv g(\mathscr{L}, \mathscr{L})$ becomes null for photon-like particles with orbital frequencies $\omega = \omega_\pm$.

The limiting frequencies (or relativistic velocities) $\omega_H^\pm \equiv \lim_{r \to r_\pm} \omega_\pm$ are the outer and inner **BH** horizons frequencies (relativistic angular velocities), representing the **BH** rigid rotation.

Thus, the null frequencies $\omega = \omega_{\pm}$ evaluated on the **BH** horizons, are the **BH** inner and outer horizons frequencies, $\omega_H^{\mp}$, respectively. There is $\omega_H^- \geq \omega_H^+$, where $\omega_H^- = \omega_H^+ = 1/2$ for the extreme Kerr **BH**.

The vector fields $\mathcal{L}_H^{\pm} \equiv \mathcal{L}(r_{\pm}) = \xi_{(t)} + \omega_H^{\pm}\xi_{(\phi)}$ define, therefore, the **BH** horizons as Killing horizons.[6]

The Killing horizons are null (light–like) hypersurfaces generated by the flow of a Killing vector, whose null generators coincide with the orbits of a one-parameter group of isometries, i.e., in general, there exists a Killing field $\mathcal{L}$, which is normal to the null surface. The Kerr **BH** horizon r_+ is, therefore, a non-degenerate (bifurcate) Killing horizon generated by the vector field $\mathcal{L}$. In the case $a = 0$ (the Schwarzschild and **RN** spacetimes, where $\omega = 0$), the Killing vector $\xi_{(t)}$, generator of r_+, is hypersurface-orthogonal.[7]

Metric Killing bundles or metric bundles are structures defined as solutions of $\mathcal{L}_{\mathcal{N}} \equiv g(\mathcal{L}, \mathcal{L}) = 0$ with *constant* ω. Therefore, they are the collections of all and only geometries defined as solutions of $g(\mathcal{L}, \mathcal{L}) = 0$, having photon circular orbits with equal (constant) orbital frequency ω, called *bundle characteristic frequency*. It has been proved that (Kerr) metric Killing bundles always contain at least one **BH** geometry. A bundle can contain **BHs** or **BHs** *and* **NSs** solutions.

Notably, many quantities considered in this analysis are conformal invariants of the metric and inherit some of the properties of the Killing vector $\mathcal{L}$ up to a conformal transformation. The simplest case is when one considers a conformal expanded (or contracted) spacetime where $\tilde{\mathcal{L}}_{\mathcal{N}} \equiv \tilde{g}(\mathcal{L}, \mathcal{L}) = \Omega^2 g(\mathcal{L}, \mathcal{L}) = \Omega^2 \mathcal{L}_{\mathcal{N}}$. This holds also for a Killing tensor $\tilde{\mathcal{L}} \equiv \Omega \mathcal{L}$.

Focusing our discussion on the **KN** and Kerr spacetimes, the bundle characteristic frequency coincides (in magnitude) to an (inner or outer) **BH** horizon frequency, ω_H^+ or ω_H^-, of the bundle.

Kerr **MB**s can be found, for example, as solutions for the dimensionless spin $a : \mathcal{L}_{\mathcal{N}} = 0$ for each fixed values of the constant ω. We can consider the co-rotating, $a\omega > 0$, and the counter-rotating, $a\omega < 0$, orbits separately with frequencies that are equal in magnitude to the horizon frequencies. Nevertheless, here we will consider mainly $a \geq 0$ and $\omega \geq 0$.

Below, we explore in details the definitions of **MB**s in relation to stationary observers and horizons replicas, introducing the concept of extended plane allowing the **MB**s representation as curves tangent to the **BH** horizon curves in the plane. We

---

[6] The Killing vector $\mathcal{L}_{\pm}^a \equiv \xi_{(t)}^a + \omega_{\pm}\xi_{(\phi)}^a$, where $\omega_{\pm}$ satisfies the null condition on the norm of $\mathcal{L}^a$, can be interpreted as generator of null curves ($g_{ab}\mathcal{L}_{\pm}^a\mathcal{L}_{\pm}^b = 0$) as the Killing vectors $\mathcal{L}_H^{\pm}$ are also generators of Killing horizons.

[7] The (strong) rigidity theorem connects the event horizon with a Killing horizon–see [22, 31]. The event horizons of a spinning **BH** are Killing horizons with respect to the Killing field $\mathcal{L}_H^{\pm} = \xi_{(t)} + \omega_H^{\pm}\xi_{(\phi)}$, where $\omega_H^{\pm}$ is the angular velocity (frequency) of the horizons, representing the **BH** rigid rotation. Or, the event horizon of a stationary asymptotically flat solution with matter satisfying suitable hyperbolic equations is a Killing horizon. In the limiting case of spherically symmetric, static spacetimes, the event horizons are Killing horizons with respect to the Killing vector ∂_t and the event, apparent, and Killing horizons with respect to the Killing field $\xi_{(t)}$ coincide (we can say that $\omega_H^{\pm} = 0$).

also investigate metric bundles in the Schwarzschild spacetime as limits of bundles of the Kerr and **RN** spacetimes.

12.3.2 Stationary Observers and Metric Bundles

The definition of **MBs** is tightly connected to the definition of stationary observers, whose worldlines have a tangent vector which is a Killing vector. Hence, in the Kerr spacetime, for example, their four-velocity u^a is a linear combination of the two Killing vectors $\xi^a_{(\phi)}$ and $\xi^a_{(t)}$, i.e. $u^a = \Omega \mathscr{L}^a = \Omega(\xi^a_{(t)} + \omega \xi^a_{(\phi)})$, where Ω is a normalization factor and $d\phi/dt = u^\phi/u^t \equiv \omega$ is the stationary observer relativistic angular velocity.[8] The causal structure defined by timelike stationary observers is characterized by a frequency bounded in the range $\omega \in]\omega_-, \omega_+[$ [40]. The limiting frequencies $\omega_\pm$ are then solutions of the condition $\mathscr{L}_\mathcal{N} = 0$, determining the frequencies $\omega_H^\pm$ of the Killing horizons, which coincide with the metric bundles characteristic frequency.

MBs provide, therefore, the limiting orbital frequencies of stationary observers, constraining the timelike (matter circular) motion.

12.3.3 Light Surfaces and Bottlenecks

Light surfaces

For Kerr spacetimes, the solutions $a : \mathscr{L}_\mathcal{N} = 0$ for fixed constant ω define the **MBs** with characteristic frequency ω =constant. For a fixed Kerr spacetime (a/M =constant), the solutions $\omega = \omega_\pm : \mathscr{L}_\mathcal{N} = 0$ define the spacetime stationary observers limiting frequencies $\omega_\pm$ as functions of the orbit r. Also, for a fixed Kerr spacetime (a/M =constant), the solutions $r = r_s^\pm : \mathscr{L}_\mathcal{N} = 0$, define the light surfaces of the Kerr spacetimes having photon-like (circular orbiting) frequencies $\omega_\pm$. Physical (timelike) stationary observers are located between the two light surfaces $]r_s^-, r_s^+[$ with orbital frequencies in the range $\omega \in]\omega_-, \omega_+[$ (where $\omega_- < \omega_+$). We have that $\omega_+ = \omega_-$ on $r = 0$ and $r = r_\pm$. Therefore, at each point of the extended plane, there is a maximum of two **MBs** curves crossing with characteristic frequency ω_+ and ω_-, respectively.

In other words, **MBs** are related to the light surfaces $r_s^\pm$ in a fixed spacetime, defining the photon orbits $r = r_s^\pm$ with orbital frequencies $\omega_\pm$, constraining the stationary observers. **MBs** have characteristic frequencies $\omega_\pm$, therefore, **MBs** relate different geometries through their light surfaces $r_s^\pm$.

[8] Because of the spacetime symmetries, the coordinates r and θ of a stationary observer are constants along its worldline, i.e., a stationary observer does not see the spacetime changing along its trajectory. Static observers are defined by the limiting condition $\omega = 0$ and cannot exist in the ergoregion.

Killing tunnels and Killing bottlenecks
Slowly spinning Kerr **NSs** are characterized by the emergence of special structures in their light surfaces, the Killing bottlenecks, which can be seen through **MB**s.

Let us concentrate on the plane $r - \omega$ for $\sigma = 1$ (equatorial plane), of the **NSs** geometries as represented in Fig. 12.1–lower panel. At fixed spin $a > M$, for each Kerr spacetime, the regions bounded by the light surfaces $r_s^{\pm}$ are called Killing throats or tunnels. Stationary observers orbits of the Kerr **NSs** are defined in the Killing throats.

In the Kerr **BH** spacetimes, there is $r_s^- = r_s^+$ on the singularity $r = 0$ and on the horizons $r_{\pm}$. The surfaces, $r_s^{\pm}$ in Fig. 12.1–lower panel, merge for $a = M$, the extreme Kerr spacetime, where $r_s^{\pm} = r_{\pm} = M$. Increasing the spin $a > M$, the Killing throat is defined by the connected surfaces $r_s^{\pm}$.

The Killing bottleneck is as restriction of the Killing tunnel (and in particular of the orbital range bounded by $r_s^{\pm}$) boundaries for **NSs**, for spins $a =]0, 2M]$, close to the orbits $r = M$. Killing bottlenecks, therefore, are restrictions of the Killing throats appearing in special classes of slowly spinning **NSs** and they are related to the **MB**s features and the ergoregions, where repulsive gravity effects appear. In this region, the stationary observer orbital range restrict to a (non zero) minimum range, which is null in the limit of extreme **BHs**.

Similar structures have been recognized in [15, 17, 18, 20], by using the concept of pre-horizons, introduced in [15]. There is a pre-horizon regime in the spacetime when there are mechanical effects allowing circular orbit observers which can recognize the close presence of an event horizon [16, 19, 21].

The concept of remnants, bundles with close characteristic frequencies values, evokes a sort of spacetime "plasticity" (or memory), which we will deepened in the next section introducing the concept of extended plane. (See also [9], for a discussion on similar surfaces).

12.3.4 MBS Characteristics in the Kerr Extended Plane

Below we list some of the main **MB**s characteristics in the extended plane. The concept of metric bundles and some of their main features are illustrated in Fig. 12.1–upper panel, using the extended plane $a/M - r/M$ of the Kerr geometry on the equatorial plane ($\sigma = \sin^2\theta = 1$).

Metric bundles of the Kerr spacetimes are solutions of $\mathcal{L}_{\mathcal{N}} = 0$, expressed as functions of (ω, r, σ) and can be represented as curves of a plane $a - r$ or $\mathcal{A} - r$ (*extended plane*), where $\mathcal{A} \equiv a\sqrt{\sigma}$ is the origin spin for any plane σ–[47, 49, 52]. Figure 12.1–upper panel represents the extended plane of the Kerr geometry for $\sigma = 1$, where the Kerr bundle curves can be represented through the (dimensionless) spin-function:

$$a(r, \omega) = \frac{2\omega \pm \sqrt{r^2\omega^2\left[1 - r(r+2)\omega^2\right]}}{(r+1)\omega^2}. \tag{12.6}$$

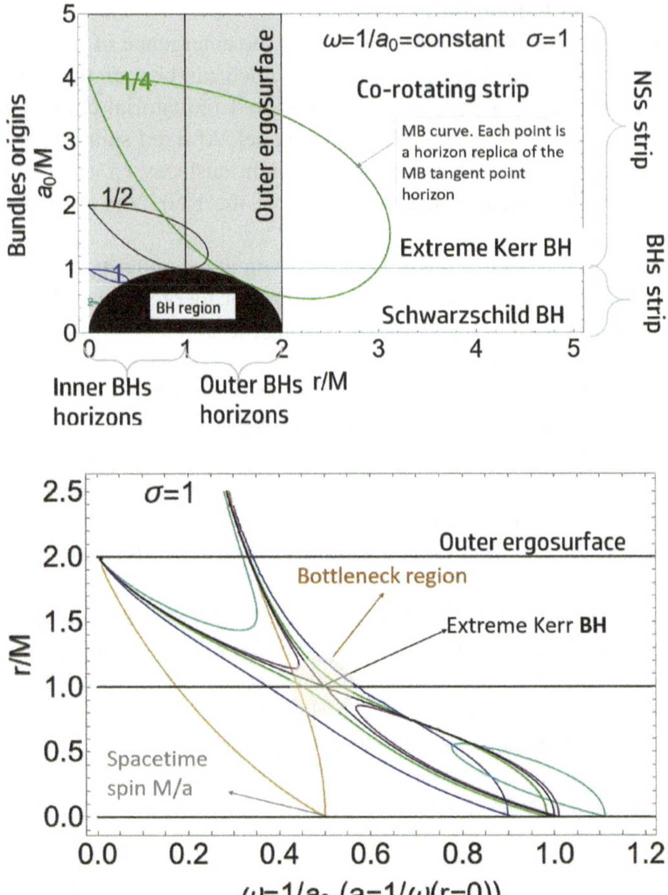

Fig. 12.1 Upper panel: Extended plane $a/M - r/M$ on the equatorial plane ($\sigma = \sin^2\theta = 1$) of the Kerr geometry. Metric bundles (**MBs**) with characteristic frequency $\omega = 1/a_0$ =constant are shown for different values of ω, where a_0 is the bundle origin spin (bundles value at $r = 0$). The black region represents the **BH** region with the outer and inner horizons $a_\pm$, as functions of r/M. A horizontal line on the extended plane at $\sigma = 1$ represents a fixed spacetime a/M =constant. In particular, $a = 0$ corresponds to the Schwarzschild **BH** spacetime and $a = M$ to the extreme Kerr **BH**. Naked singularities correspond to values $a > M$. The extended plane is shown for co-rotating orbits, i.e., for $a_0 \geq 0$ and $\omega \geq 0$. The frequencies of the inner and outer horizons curve are clearly distinguished. The frequency $\omega_H^\pm = 1/2$ corresponds to the extreme Kerr **BH**. On the equatorial plane, the point $r = 2M$ is the outer ergosurface (the Schwarzschild **BH** horizon for $a = 0$). Lower panel: Light surfaces $r_s^\pm(\omega; a)$ on the equatorial plane, solutions of condition $\mathscr{L}_\mathcal{N} = 0$, as functions of the frequency ω, for different spacetimes. Here, $\omega = 1/a_0$ corresponds to $r = 0$. The extreme **BH** spacetime and the bottleneck region are also pointed out

The black region is the **BH** region bounded by the outer ($r = r_+ \in [M, 2M]$) and inner ($r = r_- \in]0, M]$) horizons curves $a_\pm \in [0, M]$ of the extended plane $a - r$. Bundles are not defined in the region $]r_-, r_+[$ of **BH** spacetimes, but they are defined in the **BH** region $r \in [0, r_-]$ and in $r \in [0, 2M]$ of **NS**. The point ($a = 0, r = 0$) is the singularity, ($a = M, r = M$), the maximum of the curve $a_\pm$, is the horizon of the extreme Kerr **BH**, ($a = 0, r = 2M$) is the horizon of the Schwarzschild **BH** . The radius $r = 2M$ in the extended plane coincides also with the outer ergosurface on the equatorial plane of all the Kerr **BHs** and **NSs**. Each line $a_\pm$ =constant in this extended plane is a **BH** geometry. In particular $a = 0$ to the Schwarzchild geometry and the zero **MB**s. Per definition the bundle characteristic frequency ω is constant along the **MB**s curve, in particular, the frequency of the origin $r = 0$ coincides with the bundle frequency ω.

At a point r, in general, there are two different limiting photon frequencies $\omega_\pm$ for the stationary observers; then, it follows that at each point of the extended plane (at fixed σ and with the exception of the horizon curve) there have to be a maximum of two different crossing metric bundles.

Each **MB** curve is tangent to the horizon curves in the extended plane at one point only, defining, therefore, uniquely the **BH** with the **MB** frequency ω, coinciding with the frequency of the horizon at the point where the bundle is tangent to the horizon curve. The fact that metric bundles are tangent to the horizons curve in the extended plane has significant implications.

The tangent point distinguishes the tangent spin $a = a_g \in [0, M]$, on the curve $a_\pm$, but it is defined by the tangent radius $r_g \in [0, M]$ on inner horizons curve or $r_g \in [M, 2M]$ on the outer horizon curve. The horizon curve emerge as the envelope surface of all **MB**s.

Bundles can contain either only **BHs** or **BHs** and **NSs**. In the extended plane, **NSs** are "necessary" for the construction of horizons curve. The outer horizons curve emerges as envelope surface of bundles with origin in **NSs**. Part of the inner horizons curve (on the equatorial plane) emerges from **MB**s with origin in the slowly spinning **NSs**. Bottlenecks in these **NSs** spacetimes coincide in the extended plane with the region where these **MB**s curves are defined. Killing bottlenecks appear in the **NSs** region of the extended plane containing parts of the **MB**s tangent to the inner horizons. (The light surface $r_s(\omega) : \mathscr{L}_\mathcal{N} = 0$ emerge as the collections of all points, crossing of the **MB** curve with a line a =constant on the extended plane.)

MB curves in the extended plane connect points of different (**BH** or **BH** and **NS**) geometries having all the same characteristic null frequency ω, which is the replica of a **BH** horizon. Figure 12.1 shows metric bundles at $\sigma = 1$ for different characteristic frequencies $\omega = 1/a_0$ =constant, where a_0 is the bundle origin on the plane $\sigma = 1$ (corresponding to the solution $a = a_0 : \mathscr{L}_\mathcal{N} = 0$ for $r = 0$ and $\sigma = 1$). In general, $\sigma \in]0, 1]$, the bundle origin spin is $\mathscr{A}_0 \equiv 1/\omega\sqrt{\sigma}$.

Replicas

Replicas are light-like (circular) orbits, of **NSs** as **BH**s spacetimes having the same frequency as a **BH** $\omega_H^\pm$. All the points of a **MB** curve are the *replica* of the **BH** horizon frequency ω_H^+ or ω_H^- of the **BH** individuated by the tangent point of a **MB** curve to the horizon curve (a_g, r_g). The study of replicas is the study of the **MB**s curves in the extended plane.

Per definition all orbits of a bundle are a replica of the horizon the bundle is tangent to. More precisely, replicas connect measurements in different spacetimes characterized by the same value of the property $\mathcal{Q}$ function of ω, by connecting the two null vectors, $\mathcal{L}(r_\star, a_\star, \sigma_\star)$ and $\mathcal{L}(r, a, \sigma)$, to $\mathcal{L}(r_g, a_g, \sigma_g)$ where r_g is the tangent radius of the **MB** to the outer or inner Killing **BH** horizon and (σ_g, a_g), $(\sigma_\star, a_\star)$ are generally different from (σ, a). In some cases, horizons (frequencies) replicas are in the same spacetime ($a = a_\star = a_g$).

More precisely, the orbital frequency ω of a photon on a replica coincides with the **BH** horizon frequency, which is also the bundle characteristic frequency. Although here we restrict our analysis to the co-rotating replicas, i.e. $\omega > 0$ as $a > 0$, the study of counter-rotating photon circular orbits with frequency $\omega < 0$ is possible with **MB**s in the extended plane $a \in \mathrm{R}$ with $\omega > 0$. Notably not all **BH** frequencies are replicated. There is a *confinement*, when a horizon frequency is not replicated i.e. there is no replica connected to that frequency.

On the equatorial plane an exact expression of the Kerr inner horizons replicas in a fixed **BH** spacetime a, $r_-^- : \quad \omega_-(r_-^-) = \omega_-(r_-) = \omega_H^-$ and outer horizons replicas $r_+^+ : \quad \omega_+(r_+^+) = \omega_+(r_+) = \omega_H^+$ are

$$r_\mp^\mp = \frac{1}{2}\left(\sqrt{\frac{32 r_\mp}{a^2} - a^2 \pm 6\sqrt{1-a^2} - 22} - r_\mp\right). \tag{12.7}$$

Since $0 < r_-^- < r_- < r_+ < r_+^+$—see Figs. 12.2, 12.3, 12.4—while the outer horizon replicas can be detected on the equatorial plane, the inner horizon replicas remain confined to the observer.

In the Kerr spacetime, part of the inner horizon frequencies are "confined". However, importantly for the observational point of view, some replicas of the inner horizons are in the proximity of the **BH** poles, while confined on different planes ($\sigma < 1$). In [50, 51], the **BH** poles ($\sigma \approx 0$) have been studied for the observation of the photon orbits with **MB**s characteristic frequency. One of the intriguing applications of **MB**s in **BH** physics is the possibility to explore the regions close to the **BH** rotational axes (poles), where it is possible to measure the Kerr inner horizons replicas–[50, 51]. In Figs. 12.2, 12.3, we show the **BH** horizons replicas in some **NS**s spacetimes of the two bundles of the Kerr extended plane $\mathcal{A}_0 - r$ (at different σ), defined by the characteristic frequencies $\omega_H^\pm$, respectively, for the spacetime with $a = 0.99M$, hence with tangent equal to the tangent spin $a_g = 0.99M$, but different tangent radius $r_g = r_\mp$, corresponding to the inner and outer horizon of the **BH** with spin $a = 0.99M$.

Therefore, replicas in a fixed geometry, allow an observer to detect the horizon frequency on a different orbit. The topology of the **MB**s curves in the extended plane provides information about the local properties of the spacetime replicated in regions more accessible to observes, which can detect the presence of a replica at the point

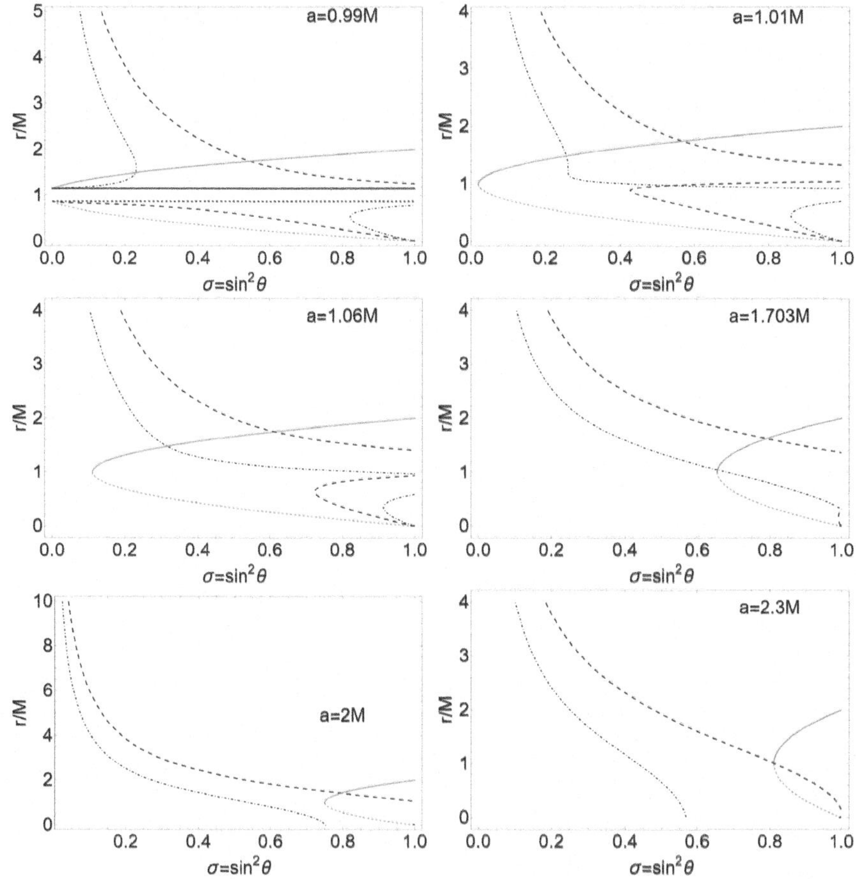

Fig. 12.2 Co-rotating replicas in the Kerr geometry with dimensionless spin a/M. Panels show the replicas radius r on each plane $\sigma \equiv \sin^2\theta \in [0, 1]$ for the **BH** with spin $a = 0.99M$ in the **BH** spacetime (upper left panel) and in different **NSs** spacetimes with spin a/M signed on the panels. Lower right panel is a close-up view of the lower left panel. Gray solid (dashed) curve is the outer (inner) ergosphere $r_\epsilon^+(\sigma)$ $(r_\epsilon^-(\sigma))$. Purple large-dashed (blue dotted-dashed) curves are the outer (inner) horizons co-rotating replicas for the **BH** with spin $a = 0.99M$. See also Figs. 12.3 and 12.4

$(r_\star, \sigma_\star)$ of the **BH** spacetime with spin $a_\star$, by measuring the **BH** horizon frequency $\omega_H^+(a_\star)$ or $\omega_H^-(a_\star)$ (the outer or inner **BH** horizon frequency) at the point $(r_\star, \sigma_\star)$.

The stationary observers orbital frequency is $\omega \in]\omega_-, \omega_+[$ where one of (ω_-, ω_+) is the horizon's frequency ω_H^+, replicated on a pair of orbits (r_+, r_s^*), where r_s^* is a light surface orbit, i.e. $r_s^* = r_s^+$ or $r_s^* = r_s^-$. The second light-like frequency is the frequency of a horizon in a **BH** spacetime.

From Fig. 12.3 is it possible to see that, for slow spinning **NSs**, the inner horizons co-rotating replicas change slowly, increasing the singularity spin-mass ratio. The outermost co-rotating replicas of the outer horizon change slowly, increasing the **NSs**

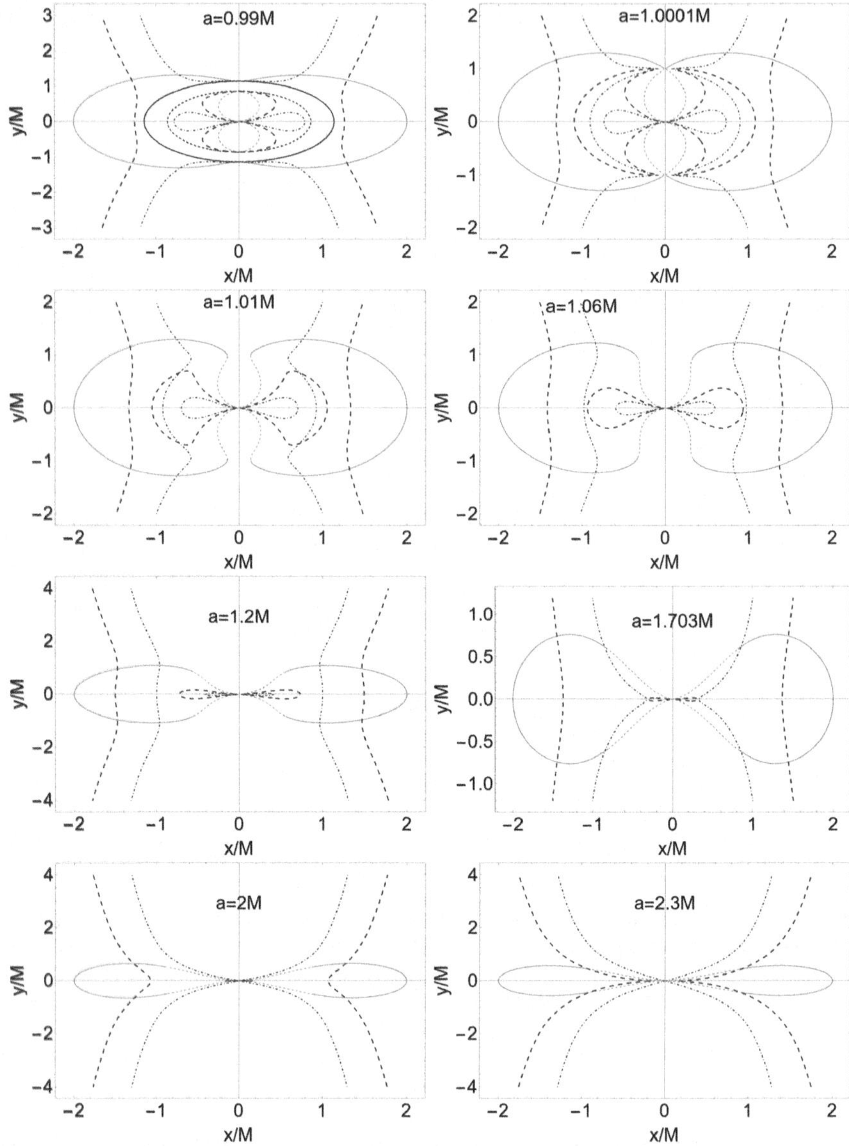

Fig. 12.3 Co-rotating replicas of the **BH** horizons represented in Fig. 12.2 for different values of a/M corresponding to **BH** s with spin $a = 0.99M$ and **NSs** with different spin–mass rations a/M as signed on the panels. Gray solid (dashed) curve is the outer (inner) ergosurface. Black solid (dashed) is the outer (inner) **BH** horizon. Purple large-dashed (blue dotted-dashed) curve is the outer (inner) **BH** horizon co-rotating replica. (Here $x = r \sin\theta$ and $y = r \cos\theta$). See also Figs. 12.2 and 12.4

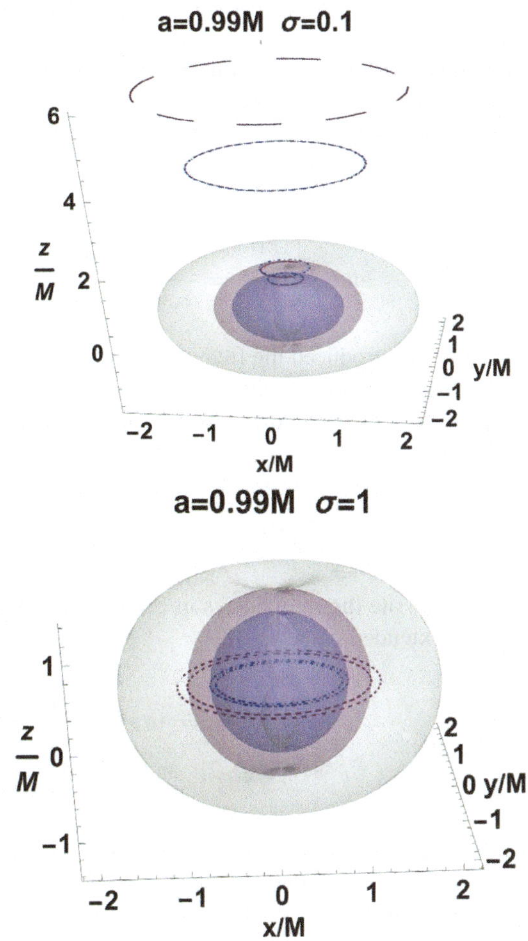

Fig. 12.4 Co-rotating replicas of the **BH** horizons represented in Fig. 12.2 for the **BH** with spin $a/M = 0.99$ on the plane $\sigma = 0.1$ (with $\sigma \equiv \sin^2\theta \in [0, 1]$)–upper panel– and on the equatorial plane $\sigma = 1$–lower panel. Purple large-dashed (blue dotted-dashed) curve is the outer (inner) **BH** horizon co-rotating replicas. Here, $\{z = r\cos\theta,$ $y = r\sin\theta\sin\phi,$ $x = r\sin\theta\cos\phi\}$. The central singularity is the point $(x = 0, y = 0, z = 0)$. Gray surfaces are the outer and inner ergosurfaces. Blue surface is the inner horizon, purple surface is the outer horizon. See also Figs. 12.2 and 12.4

spin with respect to the outermost horizon replicas. The innermost outer horizon replicas change from very slow spinning **NSs** (upper right panel), slow spinning **NSs** and very fast spinning **NSs**. It is clear the change of the **NSs** replicas between in **the** spacetimes of very slow spinning **NSs** (which is similar to the case of **BH** spacetimes), slow spinning **NSs** and faster spinning **NSs**. For very slow spinning **NSs** the replicas are very close to the inner and outer horizons (upper right and left panels), describing the **NSs** bottleneck and horizon remnants.

Horizons curve in the extended plane The horizon curve $a_\pm$, solution of $r_\pm = r$, boundary of the **BH** region in the extended plane, is

$$a_\pm \equiv \sharp\sqrt{r(2-r)}, \quad (\text{where} \quad r \in [0, 2]) \tag{12.8}$$

where $\sharp = \pm$, for the co-rotating ($\omega > 0$) and counter-rotating ($\omega < 0$) orbits, respectively.

Frequencies in the extended plane Explicitly, the limiting frequencies $\omega_\pm$: $\mathcal{L}_\mathcal{N} = g(\mathcal{L}, \mathcal{L}) = 0$ are

$$\omega_\pm \equiv \frac{\pm\sqrt{\sigma\Delta\Sigma^2} - 2ar\sigma}{\sigma\left[a^2\sigma\Delta - \left(a^2+r^2\right)^2\right]}, \quad \text{where} \quad \lim_{r\to\infty}\omega_\pm = 0 \tag{12.9}$$

$$\text{and} \quad \lim_{r\to 0}\omega_\pm = \frac{1}{a\sqrt{\sigma}} = \frac{1}{\mathcal{A}_0} \equiv \omega_0,$$

where we introduced the frequency function ω_0 expressing the bundle characteristic frequency as frequency at the origin $r = 0$. Considering the limiting cases of the static Schwarzschild spacetime and of the faster spinning **NSs** geometry, we obtain

$$\lim_{a\to 0}\omega_\pm = \mp\frac{(r-2)}{\sqrt{(r-2)r^3\sigma}} \quad \text{and} \quad \lim_{a\to\infty}\omega_\pm = 0. \tag{12.10}$$

We can write the frequencies in Eq. (12.9) as frequencies of the horizon curves in the extended plane as

$$\omega_H^\pm(r) \equiv \omega_\pm(a_\pm) = \frac{\sqrt{(2-r)r}}{2r} \quad \text{where,} \tag{12.11}$$

$$r \in [0, 2] \quad \text{and} \quad \lim_{a\to 0}\omega_H^\pm = 0,$$

where $r \in [0, 2]$ is the horizons radius in the extended plane. Note that $\omega_H^\pm = 0$ for $r = 2$. The characteristic frequency ω is related to the bundles origin $a_0 \in [0, +\infty]$ (or $\mathcal{A}_0 \in [0, +\infty]$), and the radius $r \in [0, 2]$ on the horizon curve in the extended plane as follows

$$\omega = \frac{1}{\mathcal{A}_0} = \frac{1}{a_0\sqrt{\sigma}} = \frac{\sqrt{(2-r)r}}{2r} = \omega_H^\pm(r) \quad (\text{where} \quad r \in [0, 2]). \tag{12.12}$$

Then, $\omega_\pm = 1/\sqrt{\sigma}$ for the extreme **BH**, and $\omega_\pm$ has a minimum on the equatorial plane, where it is $\omega_\pm = \omega_0 = M/a$.

Note that the light surfaces $r_s^\pm$ on the Kerr spacetimes equatorial plane can be written as functions of the frequency ω and parametrized in terms of the bundle characteristic frequency of the origin point ω_0 as follows

$$r_s^- \equiv \frac{2\beta_1 \sin\left[\frac{1}{3}\arcsin \beta_0\right]}{\sqrt{3}}, \quad r_s^+ \equiv \frac{2\beta_1 \cos\left[\frac{1}{3}\arccos(-\beta_0)\right]}{\sqrt{3}}, \tag{12.13}$$

$$\text{where} \quad \beta_1 \equiv \sqrt{\frac{1}{\omega^2} - \frac{1}{\omega_0^2}}, \quad \beta_0 \equiv \frac{3\sqrt{3}\beta_1\omega^2}{\left(\frac{\omega}{\omega_0}+1\right)^2}$$

–see Fig. 12.1–lower panel.

MBs curve origin spin Conversely, the origin spin a_0 can be read as

$$a_0 = \frac{2}{\sqrt{\sigma}}\sqrt{\frac{r}{2-r}} = \pm\frac{1}{\omega\sqrt{\sigma}} \quad (\text{where} \quad r \in [0,2]). \tag{12.14}$$

MBs curve tangent radius The **MB** tangent radius $r_g \in [0,2]$ to the horizons curve can be expressed in terms of the frequency ω and the bundle origin spin as

$$r_g(\omega) = \frac{2}{4\omega^2+1} = \frac{2\mathscr{A}_0^2}{\mathscr{A}_0^2+4} \tag{12.15}$$

–Figure 12.5.

MBs curve tangent spin The **MBs** tangent spin reads

$$a_g(a_0) = 4\sqrt{\frac{\omega^2}{\left(4\omega^2+1\right)^2}} = \frac{4\mathscr{A}_0}{\mathscr{A}_0^2+4}, \tag{12.16}$$

–Figure 12.5.

In the next section we shall analyze **MBs** characteristics for the limits of the **KN** solution as seen in the extended plane.

12.3.5 Limiting Cases

It is possible to explain some **MBs** properties and their representation in the extended plane by considering some limiting cases, in particular the zeros of the **MBs** curves in the extended plane. Here, we discuss the limiting cases of the static **RN** solution and electrically neutral and static limit of the Schwarzschild spacetime.

The definition of the extended plane in the Schwarzschild spacetime is complicated due to the identification of the metric parameter defining **MBs**. On the other hand, **MBs** in the Schwarzschild case can be studied as zero limits of the extended planes of the **RN** and Kerr spacetimes (in the Schwarzschild case, the horizon is the axis $r = 2M$ on the Kerr extended plane shown in Fig. 12.1–left panel.). More generally in static geometries, the bundle characteristic frequencies, solutions of $\mathscr{L}_{\mathcal{N}} = 0$, are null and the tangency condition of the metric bundles with respect to the horizons

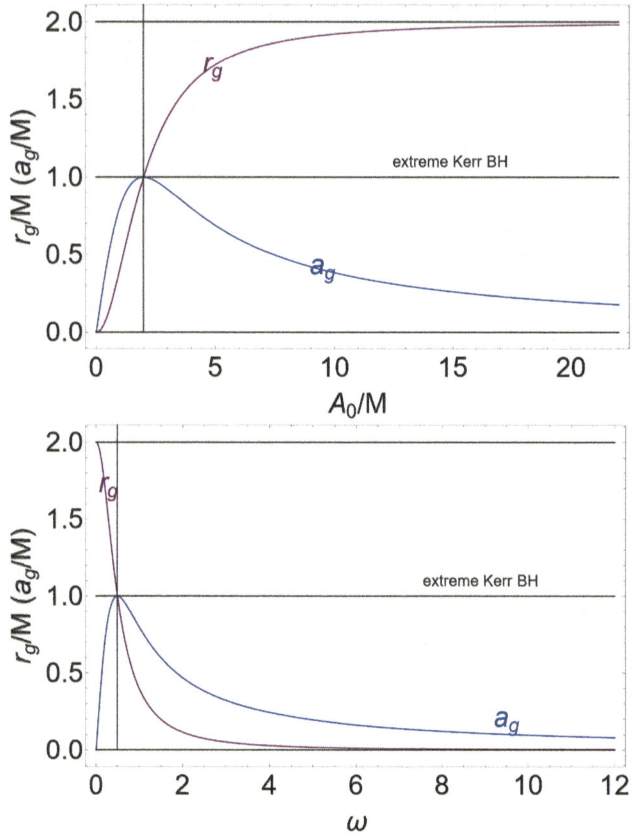

Fig. 12.5 **MB**s tangent radius r_g (purple) and tangent spin a_g (blue) as function of the bundle origin spin $\mathscr{A}_0 \equiv a_0\sqrt{\sigma}$. Here we show $\sigma \equiv \sin^2\theta \in [0, 1]$) (upper panel) and the bundle characteristic frequency ω (lower panel). See discussion of Sect. 12.3.4. The radius r_g is defined in Eq. (12.15) and tangent spin a_g is defined in Eq. (12.16). The extreme Kerr **BH** case, for $a_g = M$, $r_g = M$, $\omega = 1/2$, $\mathscr{A}_0 = 2M$ is also shown

curves is to be understood as an asymptotic condition. In this situation, the horizon replicas are the asymptotic solutions for orbits far from the gravitational source.

Below, first we analyze Schwarzschild **MB**s as the zeros of the extended planes. Then, we shall discuss the Schwarzschild **MB**s as limits of the **RN** **MB**s, concluding this section with the analysis of the Schwarzschild **MB**s as limits of the Kerr **MB**s.

The Schwarzschild MBs as zeros in the extended planes The Schwarzschild metric case corresponds to the zeros of the metric bundles of the Kerr and **RN** spacetimes (solutions with $a = 0$ and $Q = 0$, respectively). The Schwarzschild limiting photon orbital frequencies are

$$\omega = \pm \frac{\sqrt{r-2}}{r^{3/2}\sqrt{\sigma}}, \tag{12.17}$$

where $\omega = 0$ on the **BH** horizon ($r_+ = 2$).

The Schwarzschild MBs as limits of RN MBs Let first discuss the **RN** case, where the horizons are represented as the curve $Q_\pm = \sqrt{r(2-r)}$ in the extended plane $Q - r$. Let us consider the Killing field $\mathscr{L}$ and the quantity $\mathscr{L}_\mathcal{N}$:

$$\mathscr{L}^a \equiv \xi^a_{(t)} + \omega \xi^a_{(\phi)} \quad \mathscr{L}_\mathcal{N} \equiv g(\mathscr{L}, \mathscr{L}) = g_{tt} + g_{\phi\phi}\omega^2. \tag{12.18}$$

From the equation $\mathscr{L}_\mathcal{N} = 0$, we find the limiting (photon orbital) frequencies $\omega_\pm$ for the **RN** geometry:

$$\omega_\pm = \pm \frac{-g_{tt}}{g_{\phi\phi}} = \pm \frac{\sqrt{Q^2 + (r-2)r}}{\sqrt{\sigma} r^2} \tag{12.19}$$

where $\omega_\pm = 0$ on $r_\pm$. Differently from the Kerr case given in Eq. (12.9), the frequencies functions in Eq. (12.19) are even in the electric charge parameter Q. In a static spacetime, we can define relatively counter-rotating photon orbits and following the same procedure in the spherically symmetric background, the expression of the **MBs** in the **RN** spacetime reflects this property. **RN MBs** in the extended plane can be represented as the function

$$Q(r) \equiv \sqrt{r}\sqrt{r^3\sigma\omega^2 - r + 2}. \tag{12.20}$$

The Schwarzschild MBs as limits of the Kerr MBs Metric bundles of the Schwarzschild geometry are the limiting case of the metric bundles of the Kerr geometry for $a = 0$. In this case, $\omega = 0$ on the horizon. From Eqs. (12.15), (12.14) and (12.16), we obtain

$$\lim_{\omega \to 0} r_g(\omega) = 2, \quad \lim_{\omega \to 0} a_0 = \infty \quad \text{and} \quad \lim_{a_0 \to \infty} a_g = 0. \tag{12.21}$$

The first and second limits define the static case for the null horizon frequency $\omega = 0$, described as a limiting situation occurring for the horizon tangent point $r_g = 2$ (the Schwarzschild horizon radius), corresponding to the zero of the (outer) horizon curves in the Kerr extended plane. The frequency $\omega = 0$ corresponds also to the **MBs** origin spin a_0 for extremely fast spinning **NSs**. The **MBs** in the limiting case of very slowly spinning **BHs** have origins corresponding to extreme fast spinning **NSs**, as it is clear from the third limit.

12.4 Black Hole Thermodynamics

Israel's theorem concerns the uniqueness of the Schwarzschild **BH** (the only static and asymptotically flat vacuum spacetime with a regular horizon is the Schwarzschild solution) and establishes that the event horizon must be spherically symmetric, implying that the sphericity of an isolated and static **BH** cannot be broken [26, 31, 35, 36]. This is the first version of the so called "no-hair theorem", later formulated by Wheeler, and which today refers more generally to a set of theorems and conjectures. The first result of Israel established the uniqueness of the Schwarzschild metric, which was then extended to the case of spinning and charged **BH**s [8, 57]. Although some steps have been made towards a full generalization of the Israel theorem, there is still no proof for a general no-hair theorem remaining, therefore, a conjecture.

The "no–hair theorem" states also what is known as the simplicity of the isolated **BH**s in equilibrium, which (in General Relativity) can be fully characterized by the mass M, the angular momentum J, and the electric charge Q. Said differently, all stationary **BH** solutions of the Einstein–Maxwell equations are defined by only three independent (classical and observable) parameters. The term "hairs" refers to any further "information" as, for example, the matter of the **BH** progenitor or other "conserved numbers" of the matter swallowed by the **BH**, which could not be traced back by a far away observer. Hence, the physical **BH** (as final result of a physical process in a progenitor lifetime) after collapse has left with no other defining (independent) "remnant" parameters, when the matter-field information (prior horizon formation /crossing) "disappeared" behind the **BH** horizon, inaccessible to the outside observer and, therefore, lost to the observer's universe.

Hairs, in other words, are fields (different from the electromagnetic ones) and matter, which is associated to and determines stationary **BHs**.

The conjecture also outlines the **BH** information paradox (problem) in its classical and quantum formulation, when considering the possible quantum effects associated to the **BH** and the spacetime close to the horizons. This and the approximation of the general formulation of the conjecture has led to several reformulations, revisions, and modifications to the initial conjecture, as confirmations or exceptions were encountered in relation to specific "hairs–fields" [24, 32, 33].

In the first theorem formulation, the electromagnetic field was singled out (excluding, however, a magnetic charge), and excluding further "degrees of freedom" as the Baryon number, for example, or other quantum numbers (as strangeness, etc.), which, therefore, could not be measured by a (far away) observer, being destroyed during the collapse (or crossing the horizon).

Therefore, the theorem had the further issue to isolate, without physical foundation, the electromagnetic field (gauge fields subjected to Gaussian law) with respect to other fields. Several counter–examples have been found with fields hairs for stationary and stable **BH** solutions. There are stationary **BH**s with hairs as exterior fields constituting exceptions. Stable stationary **BH**s solutions with gauge field hair are also known (the **BH** pierced by cosmic strings are another example of hair as other solutions in non-abelian gauge theories.).

Remarkably, there is no clear, univocal and physically supported indication (or criterion) on the fields to consider as possible **BH** – "hairs", this constituting a deeper and broader issue in the "no-hairs" theorem investigation. It should be noted that mass, electric charge and angular momentum are all conserved quantities holding the Gauss law. An eventual magnetic charge, according to this requirement, would be a "proper" **BH** hair. However, many **BH** solutions with gauge fields (different from the Maxwell fields), subjected to a Gauss law were also found.

Therefore, the formulation of a no-hair theorem itself is under scrutiny, so that more than one theorem is to be proved; it resulted into an idea to be tested on a case-by-case basis which, with negative results, confirms the conjecture for that specific case, or provides a counter-examples by narrowing the field of analysis on a possible geometric justification of the **BH** hair fields, giving raise to a sequence of "no–hair theorems" (there are, for example, non scalar–hair theorem, etc.). It would, therefore, be more correct to say that, in the context of General Relativity, the no-hair theorem refers to a series of statements concerning the "degrees of freedom" defining a **BH** and observable to a distant observer. Another issue of the no-hair theorem is the role of conditions at infinity (and the asymptotic flatness), relevant, for example, in cosmological models, where formulations are known for positive cosmological constant–see [4]. Obviously a no-hair theorem would imply the **BH** "simplicity" and would have consequences on the **BH** uniqueness. A conceptual consequence of the no–hair theorems (**BH** simplicity) is that in this way the **BH** entropy, defined by the **BH** area, would provide the "real" measure of the **BH** degrees of freedom. However, the information on the **BH** "past" (with respect to the state of observation), in the sense of **BH** "hair", is to be considered eliminated or inaccessible to far away observers.

This field of investigation with its problematic aspects is a promising cross-road among **BH** physics, quantum gravity, information theory, and theories of great unification. Nevertheless, this relevant aspect of **BH** physics diverges from the purposes of this chapter and we refer the reader, for further details on this issue, to the extensive literature on the subject.

Here we explore the laws of **BH** thermodynamics in the extended plane, using the Smarr's formula connecting, in the Kerr **BH** spacetime, (M, J, A_{area}), where A_{area} is the (outer) **BH** horizon area [59]. The extended plane and **MBs** can be used to relate different states of a **BH**, interacting with its environment. Replicas can connect **BH** spacetimes related in a transition and governed by the thermodynamic laws in the extended plane. Hence, it is possible to reformulate **BH** thermodynamics in terms of the light surfaces, exploring the thermodynamic properties of the geometries through the metric bundles.

This formalism turns out to be adapted to the description of the **BH** state transitions, constrained by the properties of the null vector $\mathscr{L}$, building up many of the thermodynamic properties of **BHs**, as it enters the definitions of thermodynamic variables and stationary observers.

In this section, we detail the **BH** thermodynamic properties in terms of the null vector $\mathscr{L}$ and, therefore, of the metric bundles. Sect. 12.4.1 discusses the role of metric bundles and the quantity $\mathscr{L}_{\mathcal{N}}$ in **BH** thermodynamics, introducing the concept of **BH** surface gravity, the first law of **BH** thermodynamics, and concluding with some notes on the **BH**s states transitions as described by the laws of **BH** thermodynamics.

In this framework, we explore **BH** states transition in the extended plane in Sect. 12.4.2. The **BH** irreducible mass is the subject of Sect. 12.4.3, where we discuss also the extraction of **BH** rotational energy as constrained with **MB**s. Finally, Sect. 12.4.4, **BH** thermodynamics in the extended plane, closes this section.

12.4.1 Black Hole Thermodynamics, Metric Bundles and the Quantity $\mathscr{L}_{\mathcal{N}}$

In this section, we discuss some concepts of **BH** thermodynamics in terms of metric bundles and the quantity $\mathscr{L}_{\mathcal{N}}$. First, we introduce the concept of **BH** surface gravity, we then discuss the first law of **BH** thermodynamics, concluding the section with some notes on the **BH**s states transitions.

BH surface gravity

We start our analysis of the **BH** thermodynamics by introducing the concept of **BH** surface gravity and derive a surface gravity function in the extended plane, expressed in terms of **MB**s characteristics.

The surface gravity ℓ for a **BH** Killing horizon is defined by the fact that the Killing vector defines a non-affinely parametrized geodesic on the Killing horizon (a global Killing vector field $\mathscr{L}^a$ becomes null on the event horizon). According to the Zeroth Law of **BH** thermodynamic, the surface gravity of a stationary **BH** is constant over the event horizon (somehow fixing a concept of **BH** thermal equilibrium)—see for example [64].

In general terms, given an object and its surface, we can define the surface gravity as the (gravitational) acceleration a test particle in the proximity of the object is subject to. Then, a **BH** surface gravity is the acceleration, as exerted at infinity, of a test particle located in close proximity to the **BH** outer horizon, which is necessary to keep it at the horizon. For a static **BH**, it can be expressed as the force exerted at infinity to hold a test particle at close proximity of the horizon. Notice that the locally exerted force at the horizon is infinite.

A similar definition holds for stationary **BH**s and, in general, for **BH**s with "well defined" Killing horizons. However, surface gravity is a classical geometrical concept, playing a major role in **BH** thermodynamics–[1]. Bekenstein first suggested the analogy leading to the concepts of **BH** entropy and temperature and the formulation of classical **BH** thermodynamics–[2, 3].

Hawking later confirmed Bekenstein's conjecture of a reliability of a **BH** entropy definition, establishing the (correct) constant of proportionality relating the **BH** entropy and **BH** area, following a more complex and complete argument relating **BH** thermodynamics and **BH** energy extraction [28, 29, 60].

The introduction of a temperature leads naturally to inquire about a possible **BH** radiation emission, which would be regulated by the temperature defined by its surface gravity. (However, although the **BH** temperature could be null, the **BH** entropy could not vanish). An interpretation of this suggestion has been realized in the quantum (semi-classical) frame. **BH** thermal radiation has been shown to be the result of quantum mechanical effects governed by the **BH** temperature and, therefore, its surface gravity (Hawking semi-classical radiation). The surface gravity regulates also the probability of a negative energy particle tunneling through the horizon–[28]. Because it radiates, one can expect the **BH** mass to decrease and eventually disappear. Nevertheless, the actual fate of the **BH** and the information carried in the **BH** during and before the evaporation process, the possible existence of remnant at the end of process, and the entropy evolution are matters of an ongoing (vivacious) debate–see for example [30, 46, 61]. The actual fate of a **BH**, following the characteristic thermal emission, is an example of controversial aspects of the surface gravity physics.

Thus, the **BH** (Bekenstein-Hawking) entropy links gravitation, **BH** thermodynamics, geometry, and quantum theory. The so-called (**BH**) information paradox is seen today as the source of one of most promising hints for a quantum theory of gravity.

The relation between area and entropy and the discussion around their constant of proportionality had enormous relevance far beyond the study of **BH**s. The Bekenstein bound was the starting point, for example, of the holographic principle and in general shred light on profound aspects of geometric theories and problems of quantum information theory.

The history of Bekenstein bound confirmations, developments and applications is broad and complex. As it goes beyond the discussion in this chapter, we refer the interested reader to the extensive literature on the subject. Here, we will reformulate the relation (proportionality) between the concept of entropy and **BH** area, in terms of light surfaces through the **MB**s.

Formally, surface gravity may be defined as the rate at which the norm of the Killing vector $\mathcal{L}$ vanishes from outside (i.e. from $r > r_+$). In fact, ℓ is (the constant) defined through the relation $\nabla^a \mathcal{L}_\mathcal{N} = -2\ell\mathcal{L}^a$ (on the outer horizon).

Equivalently, $\mathcal{L}^b \nabla_a \mathcal{L}_b = -\ell \mathcal{L}_a$ and $L_\mathcal{L} \ell = 0$, where $L_\mathcal{L}$ is the Lie derivative,- a non affine geodesic equation, i.e., ℓ =constant on the orbits of $\mathcal{L}$. (The norm $\mathcal{L}_\mathcal{N} \equiv g(\mathcal{L}, \mathcal{L})$ is constant on the horizon.)

For the Kerr spacetime it is $\ell = (r_+ - r_-)/2(r_+^2 + a^2)$. In the extreme Kerr spacetime ($a = M$), where $r_\pm = M$, the surface gravity is zero and the **BH** temperature is also null ($T_H = 0$), but with a *non-vanishing* entropy [63, 64]. This relevant fact establishes a profound topological difference between **BH**s and extreme **BH**s, implying that a **BH** cannot reach the extremal limit in a finite number of steps–(third law of **BH** thermodynamics)—having consequences also regarding the stability against Hawking radiation.

On the other hand, the condition (constancy of) $\nabla^a \mathcal{L} = 0$ when $\ell = 0$ substantially constitutes the definition of the degenerate Killing horizon **BH**. In the case of

Kerr geometries, only the extreme **BH** case is degenerate. Therefore, in the extended plane it corresponds to the point $(a = M, r = M)$. (The **BH** surface gravity ℓ, is also a conformal invariant of the metric. Furthermore the surface gravity re-scales with the conformal Killing vector, i.e. it is not the same on all generators but, because of the symmetries, it is constant along one specific generator–see also [37, 38].)

First law of BH thermodynamics The Kerr **BH** area is given by the function

$$A_{area} = \int_0^{\pi} d\theta \int_0^{2\pi} \sqrt{g_{\phi\phi} g_{\theta\theta}} d\phi = 4\pi \left(a^2 + r^2\right), \tag{12.22}$$

evaluated on the outer horizon $r = r_+$. The **BH** (horizon) area A^+_{area} is related to the outer horizon definition, $A^+_{area} = 8\pi m r_+$, where $M = c^2 m/G = m$ in geometric units. (For simplicity, in some of the expressions we do not consider the factor 8π and in some expressions we write spin, radius, characteristic frequency and origin spin as dimensionless parameters.)

The **BH** horizon area is non-decreasing, a property which is considered as the second law of **BH** thermodynamics (establishing the impossibility to achieve with any physical process a **BH** state with zero surface gravity). Note that the (Hawking) temperature term is related to the surface gravity by $T^+_H = \hbar c \ell^+ / 2\pi k_B$, while the horizon area A^+_{area} defines the **BH** entropy as $S^+ = k_B A^+_{area}/l_P^2$, l_P is the Planck length, $\hbar$ the reduced Planck constant, and c is the speed of light, k_B is the Boltzmann constant. (We use the notation $(+)$ or $\mathcal{Q}^+_H$ for all the quantities $\mathcal{Q}$, evaluated on the outer Killing horizon.)

The first law of **BH** thermodynamics, $\delta M = (1/8\pi)\ell^+_H \delta A^+_{area} + \omega^+_H \delta J$, relates the variation of the mass δM, to the (outer) horizon area δA^+_{area}, and angular momentum δJ with the surface gravity ℓ^+_H and angular velocity ω^+_H on the outer horizon. In this expression, the term $p^+ = -\omega^+_H$, could be seen as "pressure-term", where the internal energy is $U = GM$. The term $\omega^+_H \delta J$ is interpreted as the "work". The volume term is $V = GJ/c^2$ (where $J = amc^3/G$).

We shall consider generalized definitions in the extended plane, where we express some of the concepts of **BH** thermodynamics in terms of the norm $\mathcal{L}_{\mathcal{N}}$, which defines the metric bundles. It will be necessary to consider quantities evaluated on the inner horizon, where we use the notation $(-)$ or $\mathcal{Q}^-_H$ for a quantity $\mathcal{Q}$.

BHs transitions

We can describe the **BH** transition from an initial state (0) to a new state (1) in terms of the characteristic bundle frequency (appearing explicitly in the first law as the work term $\omega^+_H \delta J$). We use the notation (0) and (1) to indicate any quantity related to the initial and final state, respectively; therefore, $\omega^+_H(0)$ is the **MB** frequency tangent to the outer horizon of the **BH** at the initial state (0). The notation $\delta\mathcal{Q} \equiv \mathcal{Q}(1) - \mathcal{Q}(0)$ denotes the change of the quantity $\mathcal{Q}$ from the initial (0) to the final state (1) of the transition. As in the bundles framework we evaluate quantities in the extended plane, we express many quantities considered here as evaluated on the inner horizons. There is a relation between the quantities $(\delta A^-_{area}, \delta J, \delta M)$ and $(\omega^-_H(0), \ell^-(0))$ evaluated on the inner horizon r_-. All the quantities are expressed in terms of bundles at $\omega_H^{\pm}(0)$ =constant and $(\delta A_{area}^{\pm}, \delta J, \delta M)$, describing the transition from the initial to final state.

12.4.2 Masses and BH Thermodynamics

We proceed relating the parameters $(\ell^{\pm}, \omega_H^{\pm})$, regulating the **BH** transition from one state to another, represented as a transition from a point $r(0)$ to a point $r(1)$ along the horizons curve in the extended plane in the following two cases: ***1.*** $(r(0), r(1)) \in [0, 1]$, inner horizon range, and **2.** $(r(0), r(1)) \in [1, 2]$, outer horizon range (note, we included the horizon $r_{\pm} = 1$ for the extreme Kerr spacetime in both the inner and outer horizons ranges). We will connect properties defined on the outer horizons curve of the extended plane with properties defined on the inner horizons curve of the extended plane.

The frequency $\omega_H^{\pm}$ will be considered always positive or null, i.e., $J\omega_H^{\pm} \geq 0$, or $a\omega_H^{\pm} \geq 0$, vanishing only for $J = a = 0$. This implies that we will not consider counter-rotating orbits, nor we shall connect proprieties defined in the positive section of the extended plane to the negative section of the extended plane. However, with $J\omega_H^{\pm} \geq 0$, the surface gravity $\ell_H^{\pm}$ can be negative, when evaluated on a point of the inner horizon $(\ell_H^- \leq 0)$.

We consider the quantity s such that $\mathscr{P}_{\pm}(r(0)) = s\mathscr{P}_{\pm}^{*}(r(1))$, where $\mathscr{P}_{\pm} \in \{\ell_H^{\pm}, \omega_H^{\pm}\}$ and the notation $(*)$ represents an eventual change in sign. Note that $\delta A_{area}^{+} = -\delta A_{area}^{-}$ for $\delta M = 0$ (as, per definition, $\delta M^+ = \delta M^- \equiv \delta M$). For a fixed **BH** spacetime $\delta M^+ = \delta M^-$, but the relation regulating this variation with the other characteristic quantities depends on the point on the horizons curve.

More generally, we can consider $\delta M^{\star}(a(0)) = s\delta M^{*}(a(1))$, where $(a(0), a(1)) \in [0, M]$, s is a constant, and for $a(0) = a(1)$ there is $(\star = * = \pm, s = 1)$. In general, it can be proved that

$$\begin{aligned}\delta M^+ = \ell_H^+\delta A_{area}^+ + \omega_H^+\delta J^+ = \delta M = \\ \ell_H^+\delta A_{area}^+ + \omega_H^+\delta J = \delta M^- \equiv \ell_H^-\delta A_{area}^- + \omega_H^-\delta J^-.\end{aligned} \tag{12.23}$$

Assuming $s \geq 0$ and $s \neq 1$, we consider the relations $\omega_H^{*}(r(0)) = s\omega_H^{\star}(r(1))$ and $\ell_H^{\circ}(r(0)) = -s\ell_H^{\diamond}(r(1))$, where $(\circ, \diamond) = \pm$. Note that s $\ell_H^+(a) \geq 0$ (surface gravity as function of the spin a in the extended plane) and $\ell_H^-(a) \leq 0$, where $\ell_H^+(a) < -\ell_H^-(a)$. Also, $\ell_H^{\circ}(s) < 0$ for $s \in]0, 1[$ and $\ell(r) \gtreqless 0$ for $r \gtreqless 1$ (surface gravity/acceleration as function of the radius r on the extended plane). We obtain a horizon parametrization in terms of the constant s, where $r(0)$ and $r(1)$ correspond to the horizons of the **BH** spacetime with $a(r(0)) = a(r(1))$. However, the limiting case $s = 1$ corresponds to the extreme Kerr **BH**. Also, for $s \in]0, 1[$ we have that $r(1) = r_-(1)$ and $r(0) = r_+(0)$, and for $s > 1$, it holds $r(1) = r_+(1)$ and

$r(0) = r_-(0)$, where

$$\frac{r(1)}{r(0)} = s, \quad r(1) = \frac{2s}{s+1}, \quad \omega(r(1)) = \frac{1}{2\sqrt{s}}, \quad \ell(r(1)) = \frac{s-1}{4s} \tag{12.24}$$
$$\text{where} \quad a(r(1)) = a(r(0)) = 2\sqrt{\frac{s}{(s+1)^2}}$$

(ω is the bundle characteristic frequency). It is noteworthy that the s factor is different according with the spin, i.e., expressed in terms of the spin it is $s = \hat{s}_{\mp} \equiv [(2-a^2) \mp 2\sqrt{(1-a^2)}]/a^2$.

12.4.3 The BH Irreducible Mass

The total **BH** mass M of the Kerr **BH** can be decomposed into the mass M_{irr} and the rotational energy. From the first law of **BH** thermodynamics, we obtain $M^2 = M_{irr}^2 + J^2/4M_{irr}^2$ (note, J has units of mass M^2 and it is measured in the asymptotic flat region). The Kerr **BH** irreducible (or rest) mass $M_{irr} = \sqrt{a^2 + r_+^2}/2$, where $A_{area} = 16\pi M_{irr}^2$ (evaluated on the outer horizon). It follows that

$$\frac{M_{irr}^+}{M} = +\frac{\sqrt{1+\sqrt{1-\frac{J(0)^2}{M(0)^4}}}}{\sqrt{2}}, \tag{12.25}$$

or, more generally the four solutions

$$\frac{M_{irr}}{M} = \flat \frac{\sqrt{1 \natural \sqrt{1-(a_{(\pm)})^2}}}{\sqrt{2}}, \quad \text{where} \quad \flat = \mp; \ \natural = \pm. \tag{12.26}$$

Here, $a_{(\pm)}$ is the dimensionless spin on the horizons curve, where

$$\frac{M_{irr}^+}{M_{irr}^-} = \sqrt{\frac{r_+}{r_-}} = \sqrt{\frac{\omega_H^-}{\omega_H^+}} = \sqrt{\frac{-\ell_H^-}{\ell_H^+}} = \frac{\sqrt{r_+}}{\sqrt{2-r_+}} = \sqrt{\frac{1}{4(\omega_H^+)^2}} = \frac{\mathcal{A}_0}{2}, \tag{12.27}$$

in terms of the bundles origin spin $\mathcal{A}_0 \geq 2$. (The notation $\pm$ refers to the radii $r_\pm$.) (Note that $M_{irr}/M = \{0, 1\}$ for $a = 0$ and $M_{irr}/M = \pm 1/\sqrt{2}$ for $a = M$.).

Equations (12.27) and (12.26) are in terms of quantities defined on the extended plane where, in terms of the irreducible mass M_{irr}, we obtain

$$\mathcal{A}_0(all) = \frac{2M_{irr}}{\sqrt{1-M_{irr}^2}}. \tag{12.28}$$

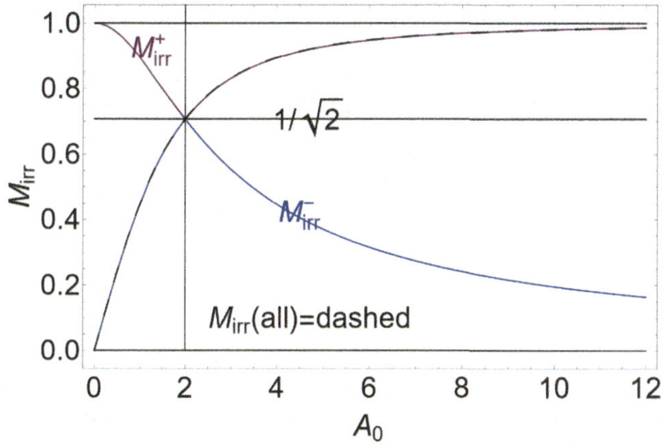

Fig. 12.6 Irreducible mass function M_{irr} versus bundle origin spin $\mathscr{A}_0 \equiv a_0\sqrt{\sigma}$ ($\sigma \equiv \sin^2\theta$). The functions $M^{\pm}_{irr}$ and $M_{irr}(all)$ are defined in Eq. (12.29). The limiting value $M^{\pm}_{irr} = M_{irr}(all) = 1/\sqrt{2}$ for $\mathscr{A}_0 = 2$ (corresponding to the extreme Kerr **BH**) is also shown. For $\mathscr{A}_0 > 2$ ($\mathscr{A}_0 < 2$) metric bundles are tangent to the outer (inner) horizons curves in the extended plane. See the analysis of Sect. 12.4.3. All quantities are dimensionless

Here and in the following analysis, for $\mathscr{Q}(all)$ it is intended a generic quantity $\mathscr{Q}$, considering the entire horizon curve in the positive section of the extended plane.

Therefore, the **BH** irreducible mass can be written as follows:

$$M_{irr}(all) = \sqrt{\frac{\mathscr{A}_0^2}{\mathscr{A}_0^2+4}} = \sqrt{\frac{r}{2}} = \sqrt{\frac{1}{4\omega^2+1}}, \tag{12.29}$$

$$\text{and} \quad M^{\mp}_{irr} = \sqrt{\frac{1}{2} \mp \frac{1}{2}\sqrt{\frac{\left(\mathscr{A}_0^2-4\right)^2}{\left(\mathscr{A}_0^2+4\right)^2}}} = \sqrt{\frac{r_{\mp}}{2}},$$

–Figure 12.6– (r is a point on the horizon curve in the extended plane).

The dependence on σ, implicit in the bundle origin definition $\mathscr{A}_0 \equiv a\sqrt{\sigma}$, does not contradict the **BH** horizon rigidity, as it has to be considered as a representation of the light surfaces frequencies on different poloidal angle σs. In Eq. (12.29), we note the special value $\mathscr{A}_0 = 2$, corresponding to a bundle characteristic frequency $\omega = 1/2$, tangent to the horizon curve point, corresponding to the extreme Kerr **BH** ($r = r_{\pm} = 1$), where $M_{irr} = 1/\sqrt{2}$.

In terms of the bundle origin spin, we obtain the limiting values,

$$\lim_{\mathscr{A}_0\to\mathbf{X}} M_{irr}(all) = \left\{1, 0, \frac{1}{\sqrt{5}}, \frac{1}{\sqrt{2}}\right\}, \quad \text{and} \tag{12.30}$$
$$\lim_{\mathscr{A}_0\to\mathbf{X}} M_{irr}(r_+) = \left\{1, 1, \frac{2}{\sqrt{5}}, \frac{1}{\sqrt{2}}\right\}$$

for $\mathbf{X} = \{+\infty, 0, 1, 2\}$, respectively.

The rotational energy ξ

The upper limit of the the amount of energy that can be extracted from a Kerr **BH** is the *total* rotational energy, occurring when the Kerr **BH** final state is a Schwarzschild **BH** [10]. The total mass M at the end of a (stationary) process of the energy extraction has to be with M_{irr}, which is the maximum rotational energy that can be extracted from the black hole is $\xi \propto (M - M_{irr})$. Hence, the maximum rotational energy which can be extracted is the limit of $\xi = \xi_\ell \equiv \left(2-\sqrt{2}\right)/2$, where at the state (0) (prior to the extraction) there is an extreme Kerr spacetime (with spin $a = M$). In general, $\xi \in [0, \xi_\ell]$.

The extracted rotational energy ξ is

$$\frac{\xi}{M(0)} = 1 - \frac{\sqrt{1+\sqrt{1-\frac{J(0)^2}{M(0)^4}}}}{\sqrt{2}}.$$

In the extended plane, we can express the tangent (dimensionless) spin as function of the rotational energy parameter ξ:

$$a_\xi(\xi) \equiv 2\sqrt{(2-\xi)(\xi-1)^2\xi}, \tag{12.31}$$

linking the former state spin $a(0)$ to the rotational extraction in the subsequent phase, where the **BH** is settled in a Schwarzschild spacetime. More generally, solving $a_\xi(\xi) \equiv a_{(\pm)}$ for ξ, we obtain the four solutions

$$\xi_s^{(\pm)} = 1\flat\frac{\sqrt{1\natural\sqrt{1-a_{(\pm)}^2}}}{\sqrt{2}}, \quad \text{where} \quad \flat = \mp; \quad \natural = \pm, \tag{12.32}$$

where $a_{(\pm)}$ is the dimensionless spin on the horizons curve.

The dimensionless spin $a/M \equiv J(0)/M(0)^2$ refers to an initial state before the transition, defined by the mass $M(0)$ and spin $J(0)$, function of the extracted rotational energy $\xi/M(0) = 1 - M_{irr}(0)/M(0)$, and, therefore, of the ratio $M_{irr}(0)/M(0)$. Measuring ξ, therefore, will provide and indication of the **BH** spin [12–14, 23, 54], relating the dimensionless **BH** spin a/M to the dimensionless ratio ξ (as total released rotational energy versus **BH** mass, measured by an observer at infinity, and assuming a process ending with the total extraction of the rotational energy of the central Kerr **BH**).

We have that $a_\xi(\xi) \in [0, M]$ (dimensionless) with $\xi \in [0, \xi_\ell]$, limiting, therefore, the (rotational) energy extracted (through a classical process) to a maximum of $\approx$ 29% of the mass M. More in details: $\xi = 1$ for $a_\xi(\xi) = 0$ (the static limit corresponds also to the values $\xi = \{0, 2\}$). For the extreme case, where $a = M$ and the frequency is $\omega = 1/2$, it is $\xi = 1 \pm 1/\sqrt{2}$. There is a maximum $\xi_\ell \equiv \left(2 - \sqrt{2}\right)/2$ (and $\xi_m \equiv \left(2 + \sqrt{2}\right)/2$), where $a_\xi(\xi) = M$ and $r_+ = M$ (the extreme Kerr **BH**).

We can relate the rotational energy parameter ξ to the horizons, in the extension of the plane for extended values of ξ, where the following two cases are possible

$$\begin{aligned}
&(\mathbf{1})\quad \xi_\pm^\mp = 1 \pm \sqrt{\frac{r_\mp}{2}}, \quad \text{and} \quad \xi_\nu^\mp \equiv 1 \mp \sqrt{\frac{r}{2}}, \quad \text{then} \quad \xi_\tau^\mp \equiv 1 \mp \frac{1}{\sqrt{4\omega^2+1}},\\
&(\mathbf{2})\quad \xi_\mu^\mp = 1 \mp \sqrt{1 - \frac{r}{2}}, \quad \text{then} \quad \xi_{\tau\tau}^\mp \equiv 1 \mp \frac{2\omega}{\sqrt{4\omega^2+1}}
\end{aligned} \tag{12.33}$$

–Figure 12.7. Here, r is a point of the horizon curve in the extended plane. However, as $\xi \propto M - M_{irr}$ the solution ξ_-^+ has to be considered.

Within this parametrization, the frequency $\omega_H^+ = 1/2\sqrt{2}$ for the outer horizon $r/M = 4/3$ (**BH** with dimensionless spin $a/M = 2\sqrt{2}/3$) is also the saddle point of $\xi_\tau^\pm$ as function of ω. The maximum extractable rotational energy decreases/increases faster with the horizon (and **MB**s) frequencies constrained by the limiting values $\omega_H^+ = 1/2\sqrt{2}$.

The characteristic bundle (horizons) frequencies, can be expressed in terms of the dimensionless parameter ξ as

$$\omega_\pm \equiv \frac{1 \pm \sqrt{4(\xi-2)\xi(\xi-1)^2+1}}{4\sqrt{(2-\xi)(\xi-1)^2\xi}}. \tag{12.34}$$

With the irreducible mass definitions and Eq. (12.33) we obtain, respectively

$$\begin{aligned}
&(\mathbf{1})\quad \frac{M_{irr}}{M} = 1 - \xi_\pm^\mp = \mp\sqrt{\frac{r_\mp}{2}}, \quad \text{where} \quad \frac{M_{irr}(\xi_\nu^\mp)}{M} \equiv 1 - \xi_\nu^\mp = \pm\sqrt{\frac{r}{2}}\\
&\text{and} \quad \frac{M_{irr}(\xi_\tau^\mp)}{M} \equiv 1 - \xi_\tau^\mp = \pm\frac{1}{\sqrt{4\omega^2+1}}\\
&(\mathbf{2})\quad \frac{M_{irr}(\xi_\mu^\mp)}{M} \equiv 1 - \xi_\mu^\mp \equiv \pm\sqrt{1 - \frac{r}{2}},\\
&\text{and} \quad \frac{M_{irr}(\xi_{\tau\tau}^\mp)}{M} \equiv 1 - \xi_{\tau\tau}^\mp = \pm\frac{2\omega}{\sqrt{4\omega^2+1}}
\end{aligned} \tag{12.35}$$

–Figure 12.7. For $r = 0$, according to Eq. (12.33), it is $M_{irr}/M = \{1, 0\}$, with $M_{irr}/M = 1/\sqrt{2}$ for $r = M$, and $M_{irr}/M = 0$ for $r = 2M$.

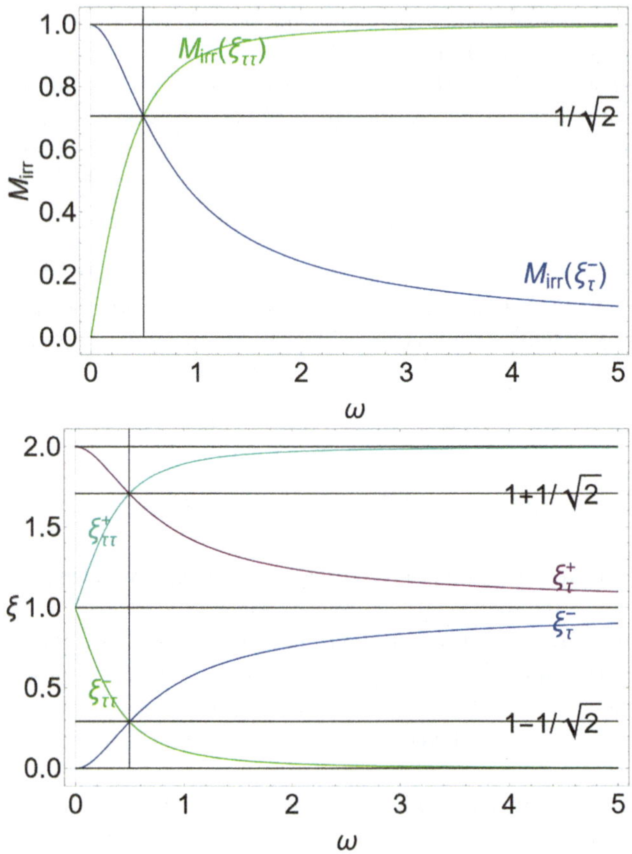

Fig. 12.7 Black hole irreducible mass functions M_{irr} (upper panel) and maximum extractable rotational energy ξ (lower panel) versus bundle characteristic frequency ω. See the analysis of Sect. 12.4.3. The functions $\xi_\tau^\pm$ and $\xi_{\tau\tau}^\pm$ are in Eq. (12.33), with limiting values $\xi_\tau^+ = \xi_{\tau\tau}^+ = 1 + 1/\sqrt{2}$ and $\xi_\tau^- = \xi_{\tau\tau}^- = 1 - 1/\sqrt{2}$ for $\omega = 1/2$ (corresponding to the extreme Kerr **BH**) are also shown. The functions $M_{irr}(\xi_\tau^-)$ and $M_{irr}(\xi_{\tau\tau}^-)$ are in Eq. (12.35). The limiting value $M_{irr}(\xi_\tau^-) = M_{irr}(\xi_{\tau\tau}^-) = 1/\sqrt{2}$ for $\omega = 1/2$ (corresponding to the extreme **BH**) is shown. All quantities are dimensionless

12.4.4 BH Thermodynamics in the Extended Plane

Let us start from the relation

$$\frac{\delta M_{irr}}{M_{irr}} = \frac{\delta M - \delta J \omega_H^+(0)}{\sqrt{M(0)^2 - \frac{J(0)^2}{M(0)^2}}},$$

where $\delta M_{irr} \geq 0$, thus $(\delta M - \delta J \omega_H^+(0)) \geq 0$ and all quantities are evaluated on the outer Killing horizon for the initial **BH**.

Note, the upper bound $\delta M_{irr} \geq 0$ sets the limit

$$\frac{\delta J}{\delta M} \leq \frac{1}{\omega_H^+(0)} = \mathscr{A}_0(0) = \frac{2r(0)}{\sqrt{(2-r(0))r(0)}}$$
$$(\text{where} \quad \mathscr{A}_0(0) \geq 2, \quad r(0) \in [1,2]) \tag{12.36}$$

to any spin variation of the **BH**, where we have used Eq. (12.12) in terms of the dimensionless origin spin of the bundle and its dimensionless tangent radius to the outer horizon curve. It is a well known fact that limits of **BH** energy extraction are imposed by the **BH** horizon. Indeed, massive particles or photons with momentum p^a that cross the outer horizon r_+ of a Kerr **BH** should satisfy the inequality $-p_a(\mathscr{L}_H^+)^a \geq 0$, This implies that $\mathscr{E} - \omega_H^+ \mathscr{L} \geq 0$, for $\mathscr{E} \equiv -p_a \xi_{(t)}^a$ (energy of the particle as measured at infinity), where $\mathscr{L} = p_a \xi_{(\phi)}^a$ is the ϕ component of the particle (photon) angular momentum. Thus, the specific angular momentum $\mathscr{L}/\mathscr{E} \leq 1/\omega_H^+ = \mathscr{A}_0$. If the energy $\mathscr{E}$ is negative, $\mathscr{L}$ is negative and the **BH** spin is reduced (**BH** spin–down). (The bound regulates also the super-irradiance as the analogue of the Penrose process for radiation scattering by a Kerr **BH**. For a wave–mode of angular-frequency ω, ω is amplified if $\omega \in]0, z/\mathscr{A}_0[$, where $z \in Z - \{0\}$ is the wave angular momentum number).

More generally, the **BH** area is a function of the horizon radius as $A_{area}(a_\pm) = 8\pi r$.[9] Considering the variation of the horizon area (and the irreducible masses) for the ADM mass M, the area A_{area} and the momentum J, we obtain

$$\delta M^\mp = \frac{\ell^\mp}{8\pi} \delta A_{area}^\mp + \omega_H^\mp \delta J^\mp, \tag{12.37}$$

where the following relation holds

$$\frac{da_\pm}{dr} = \mp \left. \frac{\ell}{\omega_\pm} \right|_{a=a_\pm}, \tag{12.38}$$

in terms of frequency and surface gravity function evaluated on the horizon curve $a_\pm$.

Explicitly, using the expressions for the bundle frequencies and the surface gravity function,

$$\omega = \frac{\sqrt{2-r}}{2\sqrt{r}} = \frac{1}{2}\sqrt{1-4\ell}, \quad \text{and} \quad \ell(r, a_\pm) = \frac{r-1}{2r},$$

[9] i.e. $A_{area}^\pm(r_\pm) = \pm 8\pi r_\pm$. With $\delta A_{area}^- = -\delta A_{area}^+$ or $\partial_a A_{area}^+ \delta a = -\partial_a A_{area}^- \delta a$. From the relation $\delta r_+ = -\delta r_-$, (since $r_+ r_- = a^2$ and $r_+ + r_- = 2M$) and, therefore, it holds if $\delta M = 0$.

which are related by $\ell = \omega_{\mathbf{EBH}}^2 - \omega^2$, where $\omega_{\mathbf{EBH}} = 1/2$ is the extreme **BH** horizon frequency.

Finally, in the extended plane, we obtain

$$\delta M = \frac{\delta A_{area}(r-1)}{2r} + \frac{\delta J \sqrt{2-r}}{2\sqrt{r}}, \quad \text{and} \tag{12.39}$$

$$\delta M = \delta A_{area}\left(\frac{1}{4} - \omega^2\right) + \delta J \omega, \quad \text{or} \quad \delta M = \left(\frac{1}{4} - \frac{1}{\mathscr{A}_0^2}\right)\delta A_{area} + \frac{\delta J}{\mathscr{A}_0}$$

in terms of the **MBs** tangent radius r, characteristic frequency ω, and origin spin $\mathscr{A}_0$, where the limiting case $\mathscr{A}_0 = 2, \omega = 1/2$ and $r = 1$ corresponds here to the extreme Kerr **BH**, distinguishing the relations defined on the inner and outer horizons curves in the extended plane, where the terms relative to the surface gravity change sign.

From $A_{area}^- = -A_{area}^+ + 4M^2$, we can write $\delta A_{area}^- = -\delta A_{area}^+ + 8M\delta M$, (with $\delta J = 0$ then $\delta M = \delta A_{area}^+ \ell_H^+/2M$).

In the special case of $\delta M = 0$ (invariant **BH** mass), $\delta A_{area}^+ = -\delta A_{area}^-$ and

$$\frac{\delta J}{\delta A_{area}^+} = \frac{\ell^- + \ell^+}{\omega_H^- - \omega_H^+}. \tag{12.40}$$

Then, using quantities as defined in the extended plane, we obtain

$$\frac{\delta J}{\delta A_{area}^{\mp}} = (\pm)\frac{\sqrt{1-a^2}}{a} = \frac{4-\mathscr{A}_0^2}{4\mathscr{A}_0} = \frac{4\omega^2 - 1}{4\omega} = \frac{1-r}{\sqrt{r(2-r)}}, \tag{12.41}$$

which represents the variation of the angular momentum versus area in terms of the initial bundle tangent point of the horizon (a, r), or, alternately, the characteristic frequency ω or its origin spin $\mathscr{A}_0$ in dependence from the poloidal angle σ. Note that the limiting case $\mathscr{A}_0 = 2$ (a **NS**) is equivalent to $\omega = 1/2$ for the extreme Kerr spacetime i.e. $(a = 1, r = 1)$.

12.5 Final Remarks

Metric bundles relate all the geometries and all the orbits with equal photon circular orbital frequency. These geometries always include **BHs** and in some cases also **NSs**. In the Kerr background, for example, the bundles characteristic frequencies coincide with the frequencies of the Kerr Killing horizons. In this way, **NS** solutions are related to **BH** solutions (in the bundle). So, the **BH** horizon distinguishing the characteristic bundle frequency and some properties typical of the **NS** light surfaces can be explained in terms of **MBs**, and their relation to the Killing horizons are expressed in the extended plane.

MBs have a natural application in **BH** astrophysics since they are constructed by photon (circular) orbits which can be measured by observers.[10] Some aspects of the background causal structure are determined by the crossings of metric bundles in the extended plane, which has been proved to be essentially determined by the horizon curves. Replicas connect different spacetime regions allowing to explore, for example, regions close to the **BH** horizon.

Horizon replicas are bundle orbits, other than that of **BH** horizons, but characterized by photons with orbital frequency equal to the (inner or outer) horizon frequency of the **BH** spacetime. From an observational viewpoint, it is worth to note that **MBs** can be used also to characterize the geometry and causal structure in the regions close to the **BH** poles and rotational axis–[50, 51]. In fact, this constitutes an aspect of information extraction of the **BH** properties into the region accessible to far away observers. Horizon replicas depend on the angle ($\sigma \in [0, 1]$) with respect to the rotational axis, and the exploration of this region may have important implications for the knowledge of spacetimes structures closed to the singularity.

Replicas can also be defined for counter-rotating photons, i.e., negative frequencies with respect to the positive frequencies in the positive section of the extended plane. Extended planes and metric bundles allow us to connect different points of one geometry, but also different geometries. Through the notion of replica we highlight those properties of a **BH** horizon, which can be replicated in other points of the same or different spacetimes, providing a new and global frame for the interpretation of these metrics and, in particular, of **NS** solutions which result connected, through **MBs** curves tangency properties, to the **BH** horizons in the extended plane. This property establishes a **BHs–NSs** connection, highlighting important properties of the Kerr geometries. The extended plane is, therefore, equivalent to a function relating the characteristic bundle (**BHs** horizon frequencies) to the (bundles origin) spin leading to an alternative definition of the Killing horizons.[11] The bottlenecks and remnants (or pre-horizons regime) can also be interpreted in terms of metric bundles, describing some properties of the Killing horizons in axially symmetric spacetimes and event horizons in the spherically symmetric case.

We have shown that it is possible to write aspects of classic black hole thermodynamics reformulated in terms of light surfaces and horizons, relating the initial and final states of a **BH** transition, expressed by the laws of **BH** thermodynamics in terms of the light surfaces.

The extended plane can represent a significant global frame also for the analysis of **BHs** transitions, where the **MBs** utility lies in enlightening spacetime proper-

[10] Furthermore, the frequencies of stationary observers, bounded by the light surfaces defining the bundles, determine many aspects of **BH** accretion configurations and jets launching and collimation, Blandford-Znajek process (a further mechanism for the extraction of rotational energy from a spinning **BH**, where there is a magnetized accretion disk and a base engine powering jet launching around super-massive Kerr **BHs**), accretion disks or the Grad-Shafranov equation for the force free magnetosphere around **BHs**.

[11] Replica are studied with the analysis of self-intersections of the bundles curves in the extended plane, in the same geometry (horizon confinement) or intersection of bundles curves in different geometries.

ties emerging in the extended plane, related to the local causal structure and **BH** thermodynamics. A (stationary) **BH** transition from a state to another is defined by a transition of its characteristic parameters, and it is regulated by the laws of **BH** thermodynamics, governed by the values **BH** state horizon frequency and surface gravity, before the transition. We can express the **BH** surface gravity using the light surfaces of the corresponding geometry for a **BH** in equilibrium, i.e., stationary.[12] Initial and final states of **BH** transitions can be related using **MB**s, or rephrased alternatively, the **BH** states are given in terms of the bundle characteristic frequency. The analysis through bundles enlighten the possible existence of privileged or not allowed state transitions.

References

1. J.M. Bardeen, B. Carter, S.W. Hawking, Commun. Math. Phys. **31**, 2 (1973)
2. J.D. Bekenstein, Phys. Rev. D **7**, 2333 (1973)
3. J.D. Bekenstein, Phys. Rev. D **12**, 3077 (1975)
4. S. Bhattacharya, A. Lahiri, Phys. Rev. Lett. **99**(20), 201101 (2007)
5. G.D. Birkhoff, *Relativity and Modern Physics* (Harvard University Press, Cambridge, Massachusetts, 1923). LCCN 23008297
6. O. Brodbeck, M. Heusler, N. Straumann, Phys. Rev. D **53**, 754–761 (1996)
7. O. Brodbeck, M. Heusler, N. Straumann, M. Volkov, Phys. Rev. Lett. **79**, 4310–4313 (1997)
8. B. Carter, Phys. Rev. Lett. **26**(6), 331–333 (1971)
9. C. Chakraborty, M. Patil et al., Phys. Rev. D **95**(8), 084024 (2017)
10. D.M. Christodoulou, R. Ruffini, Phys. Rev. D **4**, 3552 (1971)
11. M. Dafermos, J. Luk, arXiv:1710.01722 [gr-qc]
12. R.A. Daly, Astrophys. J. **691**(L72–L76), 1 (2009)
13. R.A. Daly, Mont. Notice R. Astron. Soc. **414**, 1253–1262 (2011)
14. R.A. Daly, T.B. Sprinkle, Mont. Notice R. Astron. Soc. **438**, 3233–3242 (2014)
15. F. de Felice, Mont. Notice R. Astron. Soc. **252**, 197–202 (1991)
16. F. de Felice, L. Sigalotti, Astrophys. J. **389**, 386–391 (1992)
17. F. de Felice, Class. Quantum Gravity **11**, 1283–1292 (1994)
18. F. de Felice, S. Usseglio-Tomasset, Class. Quantum Gravity **8**, 1871–1880 (1991)
19. F. de Felice, S. Usseglio-Tomasset, Gen. Relativ. Gravit. **24**, 10 (1992)
20. F. de Felice, S. Usseglio-Tomasset, Gen. Relativ. Gravit. **28**, 2 (1996)
21. F. de Felice, Y. Yunqiang, Class. Quantum Gravity **10**, 353–364 (1993)
22. H. Friedrich, I. Racz, R. Wald, Commun. Math. Phys. **204**, 691–707 (1999)
23. D. Garofalo, D.A. Evans, R.M. Sambruna, Mont. Notice R. Astron. Soc. **406**, 975–986 (2010)
24. S. Haco, S.W. Hawking, M.J. Perry, A. Strominger, JHEP **(12)**, 98 (2018)
25. S.W. Hawking, Phys. Rev. Lett. **26**, 1344 (1971)
26. S.W. Hawking, Commun. Math. Phys. **25**, 152 (1972)
27. S.W. Hawking, Nature **248**, 30–31 (1974)
28. S.W. Hawking, Comm. Math. Phys. **43**, 199 (1975); Erratum - ibidem **46**, 206 (1976)
29. S.W. Hawking, Sci. Am. **236**(1), 34–40 (1977)
30. S.W. Hawking, Phys. Rev. D. **72**(8), 4 (2005)

[12] In fact, while surface gravity for stationary **BH**s is well defined (as there is a well defined Killing horizon, where the Killing vector is normalized to unit at spatial infinity), this is not the case for dynamical **BH**s [43, 45].

31. S.W. Hawking, G.F.R. Ellis, *The Large Scale Structure of Space-Time* (Cambridge University Press, Cambridge, 1973)
32. S.W. Hawking, M.J. Perry, A. Strominger, Phys. Rev. Lett. **116**(23), 231301 (2016)
33. G.T. Horowitz, Physics **9** (2016)
34. M. Isi, W.M. Farr, M. Giesler et al., Phys. Rev. Lett. **127**(1), 011103 (2021)
35. W. Israel, Phys. Rev. **164**(5), 1776–1779 (1967)
36. W. Israel, Commun. Math. Phys. **8**(3), 245–260 (1968)
37. T. Jacobson, T.P. Sotiriou, Phys. Rev. Lett. **103**, 141101 (2009)
38. T. Jacobson, T.P. Sotiriou, J. Phys. Conf. Ser. **222**, 012041 (2010)
39. B. Kleihaus, J. Kunz, Phys. Rev. Lett. **79**, 1595–1598 (1997)
40. D.B. Malament, J. Math. Phys. **18**, 1399 (1977)
41. M. Mars et al., Class. Quantum Gravity **35**, 155015 (2018)
42. S. Mukherjee, R.K. Nayak, Astrophys. Space Sci. **363**(8), 163 (2018)
43. A.Y. Nielsen, Class. Quantum Gravity **25**(8), 085010 (2008)
44. R. Penrose, Riv. Nuovo Cim. **1**, 252–276 (1969)
45. M. Pielahn, G. Kunstatter, A.B. Nielsen, Phys. Rev. D. **84**(10, 11), 104008 (2011)
46. J. Polchinski, Contribution to: TASI, pp. 353–397 (2015)
47. D. Pugliese, G. Montani, Entropy **22**(4), 402 (2020)
48. D. Pugliese, H. Quevedo, Eur. Phys. J. C **78**(1), 69 (2018)
49. D. Pugliese, H. Quevedo, Eur. Phys. J. C **79**(3), 209 (2019)
50. D. Pugliese, H. Quevedo, Nucl. Phys. B **972**, 115544 (2021)
51. D. Pugliese, H. Quevedo, Gen. Relativ. Gravit. **53**(10), 89 (2021)
52. D. Pugliese, H. Quevedo, Eur. Phys. J. C **81**(3), 258 (2021)
53. D. Pugliese, H. Quevedo, Eur. Phys. J. C **82**, 1090 (2022)
54. D. Pugliese, Z. Stuchlík, Class. Quantum Gravity **38**(14), 14 (2021)
55. S.A. Ridgway, E.J. Weinberg, Phys. Rev. D **52**, 3440–3456 (1995)
56. R.J. Riegert, Phys. Rev. Lett. **53**, 315 (1984)
57. D.C. Robinson, Gen. Relativ. Gravit. **8**, 695–698 (1977)
58. M. Simpson, R. Penrose, Int. J. Theor. Phys. **7** (1973)
59. L. Smarr, Phys. Rev. Lett. **30**, 71 (1973); Erratum: [Phys. Rev. Lett. **30**, 521 (1973)]
60. A. Strominger, C. Vafa, Phys. Lett. B. **379**(1–4), 99–104 (1996)
61. L. Susskind, *The Black Hole War: My Battle with Stephen Hawking to Make the World Safe for Quantum Mechanics* (Little, Brown, 2008)
62. I.V. Tanatarov, O.B. Zaslavskii, Gen. Relativ. Gravit. **49**(9), 119 (2017). [Wald:1999xu]
63. R.M. Wald, Class. Quantum Gravity **16**, A177 (1999)
64. R.M. Wald, Living Rev. Relativ. **4**(1), 6 (2001)
65. O.B. Zaslavskii, Phys. Rev. D **98**(10), 104030 (2018)

The manufacturer's authorised representative in the EU is Springer Nature Customer Service Centre GmbH, Europaplatz 3, 69115 Heidelberg, Germany. If you have any concerns regarding our products, please contact ProductSafety@springernature.com

Printed and bound by CPI Group (UK) Ltd, Croydon, CR0 4YY
30/09/2026
02225318-0016